Intelligent Sensors for Human Motion Analysis

Intelligent Sensors for Human Motion Analysis

Editors

Tomasz Krzeszowski
Adam Switonski
Michal Kepski
Carlos Tavares Calafate

MDPI • Basel • Beijing • Wuhan • Barcelona • Belgrade • Manchester • Tokyo • Cluj • Tianjin

Editors
Tomasz Krzeszowski
Rzeszow University of Technology
Poland

Adam Switonski
Silesian University of Technology
Poland

Michal Kepski
University of Rzeszow
Poland

Carlos Tavares Calafate
Universitat Politècnica de València (UPV)
Spain

Editorial Office
MDPI
St. Alban-Anlage 66
4052 Basel, Switzerland

This is a reprint of articles from the Special Issue published online in the open access journal *Sensors* (ISSN 1424-8220) (available at: https://www.mdpi.com/journal/sensors/special_issues/motion_anal).

For citation purposes, cite each article independently as indicated on the article page online and as indicated below:

LastName, A.A.; LastName, B.B.; LastName, C.C. Article Title. *Journal Name* **Year**, *Volume Number*, Page Range.

ISBN 978-3-0365-5073-2 (Hbk)
ISBN 978-3-0365-5074-9 (PDF)

Contents

About the Editors

Tomasz Krzeszowski

Tomasz Krzeszowski is an associate professor in the Faculty of Electrical and Computer Engineering at the Rzeszow University of Technology in Poland. He received his M.Sc. (Eng.) degree in Computer Science from the Rzeszow University of Technology in 2009. In 2013, he received his Ph.D. with honors at the Silesian University of Technology. His areas of interest lie in computer vision, human motion tracking, machine learning, and particle swarm optimization algorithms. To date, he has published more than 45 articles, several of which in journals including *Neural Computing and Applications* (Springer), *Soft Computing* (Springer), *International Journal of Fuzzy Systems* (Springer) *Multimedia Tools and Applications* (Springer), *IET Biometrics*, and *Computer Methods in Biomechanics and Biomedical Engineering* (Taylor & Francis). He is an associate editor in the Journal of Real-Time Image Processing (Springer) and has been a member of organizing committees at seven international conferences.

Adam Switonski

Adam Switonski is an associate professor in the Department of Graphics, Computer Vision and Digital Systems at the Silesian University of Technology in Gliwice, Poland. He received a Ph.D. in Computer Science from the Silesian University of Technology and a D.Sc. from the Polish-Japanese Academy of Information Technology, Warsaw, Poland. His scientific activity is primarily related to the acquisition, analysis and classification of multimodal sequences of motion data using machine learning and feature extraction. He also has experience in the fields of processing and recognition of multispectral and hyperspectral images and computer vision. His latest research concerns the challenges of human gait identification, assessment of gait abnormalities in selected disorders, discriminative human movements analysis, 3D drone detection based on multicamera registration, as well as the application of multispectral imaging in retinal and photodynamic diagnoses.

Michal Kepski

Michal Kepski is an assistant professor at the Institute of Computer Science, University of Rzeszow, Poland. He received his Ph.D. in 2016 with honors from the AGH University of Science and Technology in Krakow. His research interests include computer vision (in particular, action and fall detection), machine learning, and computer graphics. To date, he has published more than 20 scientific articles and has been the head of grants funded by regional and national institutions (including The National Centre for Research and Development).

Carlos Tavares Calafate

Carlos Tavares Calafate is a full professor in the Department of Computer Engineering at the Technical University of Valencia (UPV) in Spain. He graduated with honors in Electrical and Computer Engineering at the University of Oporto (Portugal) in 2001. He received his Cum Laude Ph.D. degree in Informatics from the Technical University of Valencia in 2006, where he has worked since 2002. His research interests include ad hoc and vehicular networks, UAVs, Smart Cities & IoT, QoS, network protocols, video streaming, and network security. To date, he has published more than 450 articles, several of which in journals including *IEEE Transactions on Vehicular Technology, IEEE Transactions on Mobile Computing, IEEE/ACM Transactions on Networking, Elsevier Ad hoc Networks* and *IEEE Communications Magazine*. He is associate editor for several international journals from editorials including Elsevier, Hindawi, MDPI, IET, and SAGE, and has participated in the TPC of more than 250 international conferences. He is ranked among the World's Top 2% Scientists, and also among the top 100 Spanish researchers in the Computer Science & Electronics field. He is a founding member of the IEEE SIG on Big Data with Computational Intelligence and the IEEE SIG on Green Internet of Vehicles.

sensors

MDPI

Editorial

Intelligent Sensors for Human Motion Analysis

Tomasz Krzeszowski [1,*], Adam Switonski [2], Michal Kepski [3] and Carlos T. Calafate [4]

1 Faculty of Electrical and Computer Engineering, Rzeszow University of Technology, 35-959 Rzeszow, Poland
2 Department of Graphics, Computer Vision and Digital Systems, Silesian University of Technology, 44-100 Gliwice, Poland; adam.switonski@polsl.pl
3 Institute of Computer Science, University of Rzeszów, 35-310 Rzeszow, Poland; mkepski@ur.edu.pl
4 Computer Engineering Department, Universitat Politècnica de València (UPV), 46022 Valencia, Spain; calafate@disca.upv.es
* Correspondence: tkrzeszo@prz.edu.pl

Currently, the analysis of human motion is one of the most interesting and active research topics in computer science, especially in computer vision. The great interest in this area is due to the wide range of promising applications in many fields, such as medicine, surveillance systems, sports performance analysis, virtual reality, human–computer interaction, etc. Human motion analysis concerns the detection, tracking, and recognition of people and their activities based on data recorded by various types of sensors. In these studies, RGB and depth cameras are often used. Additionally, research aimed at developing gait and action recognition methods often uses motion capture systems based on active or passive markers and IMU sensors. These systems are challenging to develop, but also offer possibilities to solve advanced research problems, especially when only visual data are used. Other types of sensors that are used in motion analysis are pressure platforms and EMG sensors.

The Special Issue (SI) entitled "Intelligent Sensors for Human Motion Analysis" focuses on many aspects of human motion analysis. The Issue raised concerns of, among others, pose estimation, action and gait recognition, fall detection, as well as EMG signal processing, pressure platform construction, and issues related to improving motion capture acquisition.

As mentioned, the analysis of human movement is an important and extensive research problem, with many potential applications. This is the subject of the paper [1], which focuses on a review of the applications of pose estimation in human health and performance throughout life. The authors provided many examples of the usage of this type of system, but focused specifically on applications in the areas of human development, performance optimization, injury prevention, and motor evaluation of people with neurologic damage or disease. An extensive review of 125 scientific papers includes an overview of available tools, their use in improving human health and performance, and a discussion of the limitations and implementation problems associated with pose estimation. Moreover, the authors anticipate that, despite the existence of many limitations, the applications of pose estimation in human health and performance will continue to expand in the coming years, and that these technologies will provide powerful tools to capture significant aspects of human movement that have been difficult to register using conventional techniques.

The issues related to the estimation of the pose were also discussed in papers [2,3]. In [2], Rapczyński et al. investigated the commonly used datasets, discussed their biases and used them in cross-database experiments. They also proposed a method to harmonize the definitions of skeleton joints specific to the dataset and a scale normalization method that significantly improves generalization across cameras, subjects, and databases by up to 50%. The experiments carried out showed the negative effect of the biases of the dataset on generalization, as well as the positive impact of the proposed method of scale normalization. The authors also investigated the effect of using more or fewer cameras (also virtual cameras), training with multiple datasets, and using the OpenPose library.

Citation: Krzeszowski, T.; Switonski, A.; Kepski, M.; Calafate, C.T. Intelligent Sensors for Human Motion Analysis. *Sensors* **2022**, *22*, 4952. https://doi.org/10.3390/s22134952

Received: 24 June 2022
Accepted: 27 June 2022
Published: 30 June 2022

The more difficult challenge is to estimate the pose using a monocular camera. In [3], a real-time framework is proposed for the estimation of 3D absolute poses of multiple people using a monocular camera. The developed system, called Root-GAST-Net, combines a human detector, a 2D pose estimator, a 3D root-relative pose reconstructor, and a root depth estimator in a top-down approach. The framework is based on modified versions of the RootNet and GAST-Net networks. In the experiments, the proposed method outperformed the state-of-the-art method. Furthermore, real-time processing was achieved using the Nvidia GeForce GTX 1080.

Another research area that has been widely studied is the problem of gait recognition. In [4], hybrid methods are proposed that combine regularized discriminant analysis (RDA) and swarm intelligence techniques for gait recognition. These techniques are utilized to tune the observation weights and hyperparameters of the RDA method to minimize the objective function. In the investigation, three well-known optimization algorithms were used: particle swarm optimization (PSO), grey wolf optimization (GWO), and whale optimization algorithm (WOA). The experiments carried out confirmed the usefulness of the developed methods.

In turn, Moro et al. [5] presented an approach for markerless gait analysis based on RGB video data and deep learning algorithms. To detect 2D feature points in the image, the AdaFuse algorithm was used. Then, the acquired 2D points were used to determine 3D points and generate the human biomechanical skeleton models. The results obtained by the proposed method were compared with the data registered by the marker motion capture system.

Data augmentation is an important technique in machine learning, focusing on the enhancement of the size and quality of training datasets. In [6], a new method for action recognition time series augmentation is introduced. The method determines constraints on the generated data using statistics for a class and its representatives. The method has been compared with other approaches on eight datasets from the action recognition domain.

Recognizing and monitoring activities of daily living are an important part of understanding human behavior. Several approaches emerged to distinguish between activities of daily living and falls, focusing mainly on camera-based and inertial measurements. Some techniques analyze not only a person's movement, but also their static pose, the correct recognition of which can carry important cues for fall detection. In [7], the recognition of the lying pose from a depth map is approached with a new hybrid FRSystem. Due to the application of the LEM2 algorithm, it was possible to reduce the number of rules almost twofold, making the inference system more interpretable by a human expert.

Detection is not the only research topic related to human falls another important issue is the assessment of physical function and the risk of falls. An automated system proposed in [8] predicts the patients' score on the well-known Berg Balance Scale (BBS) using motion data captured by a multiple camera system. Furthermore, machine learning methods were used to develop fall risk predictors that reduce the number of tasks required to assess fall risk, without compromising the accuracy of the classic BBS assessment.

Human motion analysis may be applied not only to individuals, but also to groups and gatherings. In [9], different depth sensors (Kinect v2, Azure Kinect, and Zed 2) were evaluated in terms of accuracy to assess body orientation angles to detect spaces occupied by social groups using the F-Formations model. In addition, the advantages and disadvantages per device in determining the body orientation were discussed, and an experimental setup for such tasks was presented.

In [10], a deep learning approach is proposed for the human action recognition problem, utilizing existing architectures and transfer learning. The solution consists of multiple steps, including feature mapping, feature fusion, and feature selection. Deep features are fused using the Serial-based Extended (SbE) approach, and the best features are selected using kurtosis-controlled weighted KNN.

In [11], the authors proposed a non-contact monitoring and classification system for breathing patterns using the XGBoost classifier and Mel-frequency cepstral coefficient

(MFCC) feature extraction. Breathing patterns are observed using FMCW radar technology that can be used to develop non-contact medical devices. The authors discuss data analysis, as well as the detailed implementation of hardware-based signal processing. The results of the respiratory pattern classification were presented on a dataset consisting of 4000 samples imitating five breathing patterns, where an 87.375% accuracy was achieved.

The most precise measurements of human movements are provided by optical motion capture systems. The acquisition is based on the calibrated multicamera setup that tracks the 3D coordinates of markers attached to the human body. Although the registration accuracy is high, motion capture systems are not error-free. In fact, occlusions can cause markers to become undetectable. The time instants of a motion sequence with missing markers are called gaps. They require some kind of post-processing to reconstruct missing data, a process that can be performed manually by humans. However, it is a time-consuming operation, and it can be completed only by the experienced and skilled staff of a motion capture laboratory. Thus, automatic methods of gap reconstruction are highly demanded. In [12], feed-forward neural networks, three variants of recurrent networks (gated recurrent unit, long-short-term memory, and bidirectional LSTM), and interpolation techniques (linear, spline, modified Akima, piecewise cubic Hermite, and polynomial), as well as low-rank matrix completion techniques, are employed to predict trajectories of the lost markers.

The applied reconstruction techniques for mocap data and acquisition noise can result in another issue—momentary systematic errors called artefact distortion. They introduce trajectory modifications of different types and scales. In [13], four existing types of artefacts are detected, classified, and removed. The proposed algorithm is based on the derivative, low-pass filtering, mathematical morphology, loose predictor, and applicability analysis. In the validation, multiple simulations using synthetically distorted sequences are used. The outcomes are compared to human performance in the detection and removal of artefact distortion.

The optical marker-based motion capture acquisition has serious limitations as regards its practical applications. The multicamera system has to be mounted in the laboratory and calibrated prior to being used. Moreover, markers are attached to the human body before registration. Thus, the research on the effective markerless acquisition of motion data is of great importance. In [14], the challenge of three-dimensional human mesh reconstruction from a single video is faced. The human pose refinement network based on a non-local attention mechanism is applied to refine the noisy sequence of 3D human poses. It consistently improves the performance of existing state-of-the-art methods.

Another widely studied research area is the problem of recognition of facial emotions, which are expressed by human mimicry. In [15], grammatical facial expressions especially important for sign languages are recognized. The proposed approach extracts time sequences containing selected action units and facial landmarks using the OpenFace library, and classifies them by the chosen deep neural networks. Another contribution of the paper is related to the collected LSE_GFE dataset. It contains isolated signs, expressive sentences, interviews, and annotations for some grammatical facial expressions.

Human motion can also be described by EMG data. They describe the electrical activities of muscles in successive time instants. In [16], the human–machine interface based on the EMG registration is designed and successfully applied to control the robotic manipulator. The interface utilizes a multilayer neural network that identifies four different classes of muscle contraction, and a state machine for the transition change of the manipulator.

In another variant, motion is represented by the ground reaction forces. They describe the reaction of the ground to the body in contact. In [17], a low-cost wearable insole unit is developed that measures plantar pressure. It is based on the principle of photoelectric sensing and performs measurements for six selected key points of the human foot.

The SI entitled "Intelligent Sensors for Human Motion Analysis" comprises 17 articles on numerous aspects related to human motion analysis, which were briefly overviewed above. New techniques and methods for pose estimation, gait recognition, and fall detection have been proposed and verified. Some of them will trigger further research, and some may become the backbone of commercial systems.

It can be noticed that human motion analysis and related matters are challenging and important hot topics. There are still a lot of issues to be addressed, and so an exciting future is expected for this research area.

Author Contributions: T.K., A.S. and M.K. writing—original draft preparation, T.K., A.S., M.K. and C.T.C. writing—review and editing. All authors have read and agreed to the published version of the manuscript.

Funding: This research received no external funding.

Acknowledgments: The guest editors of this SI would like to thank all authors who have submitted their manuscripts for consideration and the reviewers for their hard work during the review process.

Conflicts of Interest: The authors declare no conflict of interest.

References

1. Stenum, J.; Cherry-Allen, K.M.; Pyles, C.O.; Reetzke, R.D.; Vignos, M.F.; Roemmich, R.T. Applications of Pose Estimation in Human Health and Performance across the Lifespan. *Sensors* **2021**, *21*, 7315. [CrossRef] [PubMed]
2. Rapczyński, M.; Werner, P.; Handrich, S.; Al-Hamadi, A. A Baseline for Cross-Database 3D Human Pose Estimation. *Sensors* **2021**, *21*, 3769. [CrossRef] [PubMed]
3. El Kaid, A.; Brazey, D.; Barra, V.; Baïna, K. Top-Down System for Multi-Person 3D Absolute Pose Estimation from Monocular Videos. *Sensors* **2022**, *22*, 4109. [CrossRef] [PubMed]
4. Krzeszowski, T.; Wiktorowicz, K. Combined Regularized Discriminant Analysis and Swarm Intelligence Techniques for Gait Recognition. *Sensors* **2020**, *20*, 6794. [CrossRef] [PubMed]
5. Moro, M.; Marchesi, G.; Hesse, F.; Odone, F.; Casadio, M. Markerless vs. Marker-Based Gait Analysis: A Proof of Concept Study. *Sensors* **2022**, *22*, 2011. [CrossRef] [PubMed]
6. Warchoł, D.; Oszust, M. Augmentation of Human Action Datasets with Suboptimal Warping and Representative Data Samples. *Sensors* **2022**, *22*, 2947. [CrossRef] [PubMed]
7. Pekala, B.; Mroczek, T.; Gil, D.; Kepski, M. Application of Fuzzy and Rough Logic to Posture Recognition in Fall Detection System. *Sensors* **2022**, *22*, 1602. [CrossRef] [PubMed]
8. Eichler, N.; Raz, S.; Toledano-Shubi, A.; Livne, D.; Shimshoni, I.; Hel-Or, H. Automatic and Efficient Fall Risk Assessment Based on Machine Learning. *Sensors* **2022**, *22*, 1557. [CrossRef] [PubMed]
9. Sosa-León, V.A.L.; Schwering, A. Evaluating Automatic Body Orientation Detection for Indoor Location from Skeleton Tracking Data to Detect Socially Occupied Spaces Using the Kinect v2, Azure Kinect and Zed 2i. *Sensors* **2022**, *22*, 3798. [CrossRef] [PubMed]
10. Khan, S.; Khan, M.A.; Alhaisoni, M.; Tariq, U.; Yong, H.S.; Armghan, A.; Alenezi, F. Human Action Recognition: A Paradigm of Best Deep Learning Features Selection and Serial Based Extended Fusion. *Sensors* **2021**, *21*, 7941. [CrossRef] [PubMed]
11. Purnomo, A.T.; Lin, D.B.; Adiprabowo, T.; Hendria, W.F. Non-Contact Monitoring and Classification of Breathing Pattern for the Supervision of People Infected by COVID-19. *Sensors* **2021**, *21*, 3172. [CrossRef] [PubMed]
12. Skurowski, P.; Pawlyta, M. Gap Reconstruction in Optical Motion Capture Sequences Using Neural Networks. *Sensors* **2021**, *21*, 6115. [CrossRef] [PubMed]
13. Skurowski, P.; Pawlyta, M. Detection and Classification of Artifact Distortions in Optical Motion Capture Sequences. *Sensors* **2022**, *22*, 4076. [CrossRef] [PubMed]
14. Kim, D.Y.; Chang, J.Y. Attention-Based 3D Human Pose Sequence Refinement Network. *Sensors* **2021**, *21*, 4572. [CrossRef] [PubMed]
15. Porta-Lorenzo, M.; Vázquez-Enríquez, M.; Pérez-Pérez, A.; Alba-Castro, J.L.; Docío-Fernández, L. Facial Motion Analysis beyond Emotional Expressions. *Sensors* **2022**, *22*, 3839. [CrossRef] [PubMed]
16. Pérez-Reynoso, F.; Farrera-Vazquez, N.; Capetillo, C.; Méndez-Lozano, N.; González-Gutiérrez, C.; López-Neri, E. Pattern Recognition of EMG Signals by Machine Learning for the Control of a Manipulator Robot. *Sensors* **2022**, *22*, 3424. [CrossRef] [PubMed]
17. Ren, B.; Liu, J. Design of a Plantar Pressure Insole Measuring System Based on Modular Photoelectric Pressure Sensor Unit. *Sensors* **2021**, *21*, 3780. [CrossRef] [PubMed]

sensors

MDPI

Review

Applications of Pose Estimation in Human Health and Performance across the Lifespan

Jan Stenum [1,2], Kendra M. Cherry-Allen [2], Connor O. Pyles [3], Rachel D. Reetzke [4,5], Michael F. Vignos [3] and Ryan T. Roemmich [1,2,*]

1 Center for Movement Studies, Kennedy Krieger Institute, Baltimore, MD 21205, USA; jstenum1@jhmi.edu
2 Department of Physical Medicine and Rehabilitation, Johns Hopkins University School of Medicine, Baltimore, MD 21205, USA; kcherry6@jhu.edu
3 Johns Hopkins Applied Physics Laboratory, Laurel, MD 20723, USA; Connor.Pyles@jhuapl.edu (C.O.P.); Mike.Vignos@jhuapl.edu (M.F.V.)
4 Center for Autism and Related Disorders, Kennedy Krieger Institute, Baltimore, MD 21211, USA; Reetzke@kennedykrieger.org
5 Department of Psychiatry and Behavioral Sciences, Johns Hopkins University School of Medicine, Baltimore, MD 21205, USA
* Correspondence: rroemmi1@jhmi.edu

Citation: Stenum, J.; Cherry-Allen, K.M.; Pyles, C.O.; Reetzke, R.D.; Vignos, M.F.; Roemmich, R.T. Applications of Pose Estimation in Human Health and Performance across the Lifespan. *Sensors* **2021**, *21*, 7315. https://doi.org/10.3390/s21217315

Academic Editor: Angelo Maria Sabatini

Received: 24 September 2021
Accepted: 31 October 2021
Published: 3 November 2021

Abstract: The emergence of pose estimation algorithms represents a potential paradigm shift in the study and assessment of human movement. Human pose estimation algorithms leverage advances in computer vision to track human movement automatically from simple videos recorded using common household devices with relatively low-cost cameras (e.g., smartphones, tablets, laptop computers). In our view, these technologies offer clear and exciting potential to make measurement of human movement substantially more accessible; for example, a clinician could perform a quantitative motor assessment directly in a patient's home, a researcher without access to expensive motion capture equipment could analyze movement kinematics using a smartphone video, and a coach could evaluate player performance with video recordings directly from the field. In this review, we combine expertise and perspectives from physical therapy, speech-language pathology, movement science, and engineering to provide insight into applications of pose estimation in human health and performance. We focus specifically on applications in areas of human development, performance optimization, injury prevention, and motor assessment of persons with neurologic damage or disease. We review relevant literature, share interdisciplinary viewpoints on future applications of these technologies to improve human health and performance, and discuss perceived limitations.

Keywords: pose estimation; movement tracking; computer vision; artificial intelligence; markerless motion capture; assessment; kinematics; development; machine learning

1. Introduction

Humans have long been interested in quantitative measurement of our movements [1,2]. This is evident in many aspects of life: an Olympic judge scrutinizes and scores a figure skater's performance; a physical therapist measures a patient's walking speed to assess mobility; a running coach inspects and adjusts a distance runner's foot-strike pattern to prevent injury. We also interpret the movements of others to communicate (e.g., sign language) or make inferences about emotional state (i.e., "reading body language"; [3–5]).

In this review, we focus on applications of human pose estimation, an emerging technology for quantitative measurement of human movement kinematics [6–13]. Pose estimation algorithms use computer vision to identify key landmarks on the body (e.g., fingertip, elbow, knee) from simple digital videos that can be recorded using common household devices (example workflow and applications are shown in Figure 1A,B, respectively). This simplicity offers exciting potential for measuring whole-body kinematics in nearly any setting, with minimal costs of money, time, and effort. We also see significant opportunities for the ongoing maturation and validation of these approaches to offer robust supplements or alternatives to subjective visual motor assessments and to improve accessibility to measurement of movement kinematics by removing long-standing barriers. The ability to capture quantitative, whole-body kinematics using a household device could substantially reduce reliance on traditional methods that are inaccessible or data-limited, such as expensive research-grade motion capture systems or wearable devices.

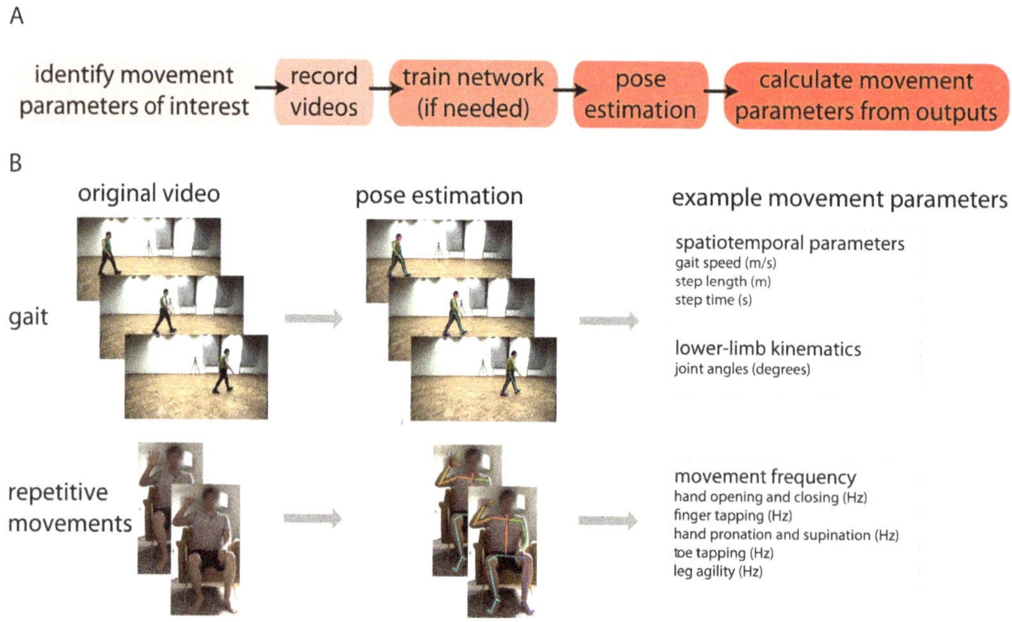

Figure 1. (**A**) Basic workflow for using pose estimation to measure movement kinematics from video; (**B**) Example applications of using pose estimation to quantify spatiotemporal and kinematic gait parameters (top) and frequencies of repetitive upper and lower extremity movements (bottom). These applications are described in greater detail in [14,15]. The gait images shown in (**B**) are taken from the GPJATK dataset [16].

We focus specifically on applications of human pose estimation for improving human health and performance. We note that pose estimation algorithms are used for many other applications (e.g., intelligent video surveillance [17], activity recognition [18], sign language translation [19]), and prior reviews have discussed technical aspects of various algorithms and their perceived advantages and disadvantages [20–22]. Here, we focus less on the technical aspects of pose estimation and instead discuss applications of these algorithms, both in terms of current applications and those that we perceive may be possible in the future. We cover areas of application across the human lifespan, including human development, human performance optimization, musculoskeletal injury prevention, and motor assessment of persons with neurologic damage or disease.

We also integrate the clinical perspective on pose estimation applications. Much prior work on human pose estimation (including our own) has suggested promise for clinical application. However, in our view, the clinician's (i.e., end user) viewpoint on these potential applications has not received adequate consideration or representation, and applications of pose estimation have not been contextualized within current models of clinical care. We aim to address these issues by providing an interdisciplinary perspective that integrates views from physical therapy, speech-language pathology, movement science, and engineering.

2. What Is Pose Estimation?

Markerless human pose estimation relies on recent advances in computer vision to automatically track anatomical landmarks—so-called keypoints—of the human body from digital videos. Examples of possible tracked keypoints include the ankle, knee, hip, wrist, elbow, shoulder, foot (e.g., heel, big toe, and small toe), hand (e.g., tip and three joints of every finger), and face (e.g., ears, eyes, nose, and mouth). Current state-of-the-art algorithms used to track human poses have been trained on large datasets of digital images and/or videos of human movement in which keypoints have been manually annotated [23,24]. The trained algorithms can then track new, unlabeled videos of humans. This enables automated, video-based human movement tracking, with the greatest accuracy achieved for movements similar to those in the training dataset.

The primary output from pose estimation is a series of two-dimensional pixel coordinates of the tracked keypoints, as they appear projected onto the image sensor of the camera. From the two-dimensional pixel coordinates, different approaches of analyzing and processing data have been reported, and fall into three broad categories. First, some studies use the output to represent planar two-dimensional kinematics of human movement, from which specific metrics of interest can be calculated [15,25–28]. An example of an instance in which this approach may be appropriate is capturing a video of the sagittal view of human locomotion and subsequently calculating sagittal gait kinematics (e.g., lower limb joint angles). Second, it is possible to reconstruct three-dimensional kinematics of human movement if capturing videos from multiple viewpoints using at least two cameras [29–31]. This approach offers significant advantages over a single camera view, in part because occlusions occur and out-of-plane motions are not well-captured by a single camera; however, this approach also has potential drawbacks associated with setup and computational complexity. Last, it is also possible to use the pose estimation output as an input for further processing by neural networks designed to predict specific metrics of interest [32–34]. Subsequent processing by neural networks may be appropriate when predicting a scalar value such as peak knee flexion during walking or clinical ratings, but this approach may be less accurate when predicting frame-by-frame time-series data. This inaccuracy is commonly due to the fact that most algorithms do not aim to minimize frame-to-frame variation when performing pose estimation with video data.

These diverse approaches to data analysis of pose estimation of human movement make it possible to obtain many parameters associated with movement. For example, pose estimation has been used to study human locomotion [15,34,35] and provide kinematic measures such as lower limb joint angles; spatiotemporal measures such as gait speed, step length, and step time; and clinical ratings such as the Gait Deviation Index in patients with cerebral palsy or MDS–UPDRS gait scores for persons with Parkinson's disease. Other studies have used pose estimation to assess neuromotor risk and development in human infants [36,37]. These areas of application are introduced briefly here, but will be covered in greater detail in later sections of this manuscript.

3. What Tools Are Available?

Several different algorithms for pose estimation have been published over the past decade (e.g., OpenPose [13], DeepLabCut [12], DeepPose [10], DeeperCut [8], AlphaPose [38], ArtTrack [7]). Using these algorithms, it is possible to take advantage of pretrained networks that are freely available, or train new networks customized for various research or clinical needs. For example, a commonly used pretrained network is the human pretrained demo of OpenPose that includes keypoints of the body, feet, hands, and face [13,39] and has been used in several recent studies for quantitative analysis of human movement [15,26,29,31,34,40].

The computations needed for training a new network and tracking new videos often require intensive computing capabilities. Therefore, the computing power of a graphics processing unit (GPU) may be necessary in order for processing times to reach acceptable limits (many algorithms provide documentation with hardware recommendations, as in [11]). If a user does not have their own GPU, some computing environments (e.g., Google Colaboratory) provide GPU access for faster processing; however, these may not be suitable for applications involving protected health information because the processing occurs externally. Processing without a GPU is slower but may be sufficient depending on the user's time constraints and processing needs (e.g., length of videos, number of people tracked, number of keypoints tracked). Furthermore, it is also possible to use pose estimation for real-time movement tracking (as is available with OpenPose, for example [39]). This capability may be particularly useful to some users, as it could be implemented to provide real-time biofeedback for various applications. Beyond these increasingly popular deep learning approaches, other approaches also use optimization [41–43] and filtering [44,45] techniques to perform pose estimation.

4. How Can These Tools Be Used to Improve Human Health and Performance?

In the following subsections, we will focus on three specific areas of application across the human lifespan: (1) human development, (2) performance optimization and injury prevention, and (3) motor assessment of persons with neurologic damage or disease (Figure 2). Certainly, many additional areas of application exist beyond the scope of this review. We focus on these applications due to the emerging nature of the relevant literature and the expertise of the authors. We expect that many of the principles discussed below are likely to generalize to other applications and/or populations of interest.

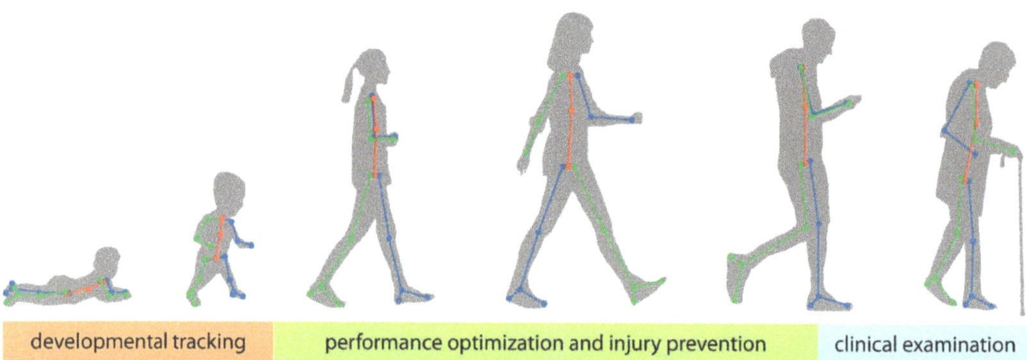

Figure 2. In this manuscript, we focus on three general areas of applications of pose estimation in human health and performance across the lifespan: tracking of motor and non-motor development in young children (orange), performance optimization and injury prevention in athletes and other populations that are primarily young or middle-aged adults (green), and clinical examinations of persons with neurologic damage or disease who are primarily older adults (blue).

4.1. *Tracking General Motor Development*

Developmental scientists study the emergence of specific behaviors from infancy to adolescence in many different settings, including the laboratory, home environment, clinic, and classroom. Accordingly, video recordings are an integral component of most, if not all, developmental research programs. Video-based approaches have been used to study multiple domains of development, including gross and fine motor development as well as social, language, and play development [46–49]. One major limitation of current video-based approaches is the time-intensive but necessary process of manually coding child behaviors of interest by clinicians and researchers. Pose estimation technologies offer a much-needed opportunity to accelerate video coding to capture specific behaviors of interest in such developmental investigations. Due to the extensive manual video coding that has been done in the field over decades, there are large existing video databases that have already undergone human coding/reliability checks and can provide a valuable source of ground truth data for training and validation of machine learning models of development (e.g., [50]). Such approaches could further help decrease reliance on assessment tools that require the expertise and time of trained clinicians for interpretation and, in turn, offer cost-effective and scalable alternatives to more subjective measures of typical and atypical development.

Although in the early stages of application, pose estimation approaches are beginning to be applied to the study of general motor development [36,51] (Figure 3A). For example, pose estimation has been used to detect normal writhing movements (i.e., typical spontaneous movements produced by newborns) vs. abnormal movements from video recordings of newborns in their first days of life [51]. Preliminary findings are promising and suggest that normal vs. abnormal writhing movements can be automatically classified with 80% accuracy, a percentage comparable to expert human classification.

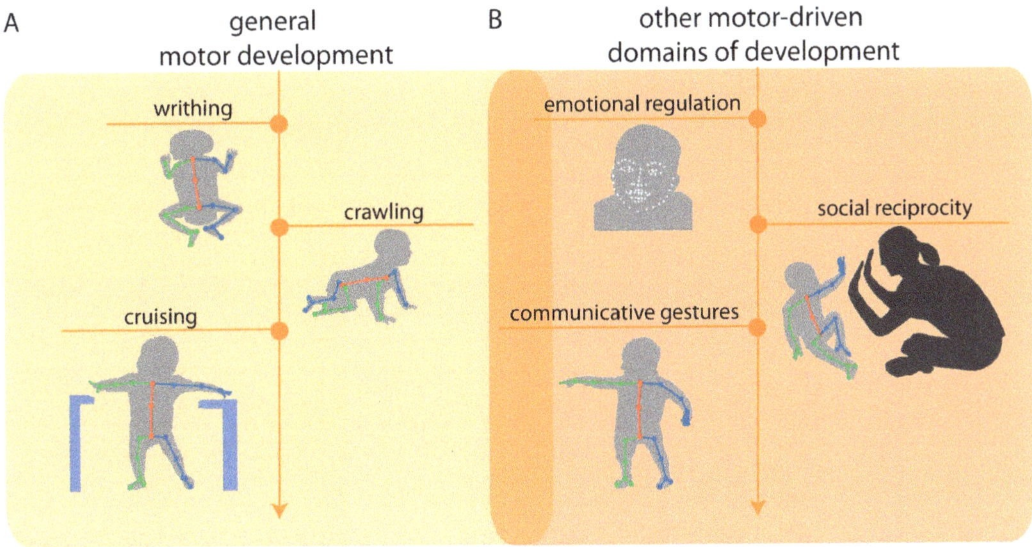

Figure 3. (**A**) Example applications of pose estimation to quantify early motor developmental milestones (left), including writhing movements (e.g., [51]), crawling, and cruising (e.g., [36]); and (**B**) other motor-driven domains of development, including emotional regulation, social reciprocity, and communicative gestures. Overlap between (**A**,**B**) denotes that these areas of development are intimately linked with one another. Arrow indicates that application of pose estimation is not restricted to these examples and can be applied to quantify later motor and motor-driven developmental milestones.

As infants progress in their gross motor development, the onsets of crawling and walking—gross motor advances that allow infants to explore and learn from their environment—have been found to be intimately linked with growth in other developmental domains [52,53]. Indeed, findings from developmental science literature suggest that delays in the onset of walking may result in limited opportunities for exploration and input from caregivers and family members, leading to subsequent delays in language and social communication development [48,54,55]. As a result, it is critical to improve the early detection of delays in locomotor development in order to intervene prior to any cascading effects on other domains of development.

Researchers have begun to implement pose estimation as a useful tool for quantitative tracking of infant locomotor development. For example, Ossmy and Adolph [36] used a combination of pose estimation, machine learning, and time-series analyses to examine the role of experience in infant acquisition of interlimb coordination based on video recordings of the infants "cruising" (i.e., side-stepping with support of the upper extremities)—which is the transitional behavior between crawling and walking—at 11 months of age. More specifically, the authors used pose estimation to track frame-by-frame body movements and subsequently calculated the distance between the limbs (i.e., the distance between the hands and the distance between the feet) for each tracked video frame to extract the coordination pattern for cruising. The results of this study provided insight into the mechanisms by which infants learn to optimally cruise and, as a result, may hold implications for future work aiming to investigate early detection and intervention for delays in locomotor development.

4.2. Clinical Use in Pediatric Populations

Early detection of atypical development is critical for the diagnosis of congenital movement-based disorders (e.g., cerebral palsy) and neurodevelopmental disorders (e.g., autism spectrum disorder) to ensure timely access to early intervention services to improve motor outcomes (e.g., coordination, postural support) and other domains of development (e.g., social, language). Advances in pose estimation approaches and the emergence of novel machine learning-based models offer exciting potential for the assessment of movement-based predictors of clinical disorders. For example, pose estimation is beginning to be applied, not only to measure predictors of later motor-based disorders, but also predictors of other motor-driven domains of development (social communication; Figure 3B). In this subsection, we provide examples of these advances.

Cerebral palsy (CP) is the most common movement disorder in childhood, caused by abnormal neural development or injury that impairs the ability to control movement and posture [56]. Diagnosis of CP using conventional assessments typically occurs between age 12 and 24 months; however, using a combination of standardized assessments and neonatal magnetic resonance imaging (MRI), CP can be accurately predicted before 6 months corrected age [57]. Yet, there remain significant drawbacks to this approach: standardized assessments are based on subjective human observation that requires substantial training and clinical expertise, and neonatal MRI is expensive and often inaccessible in low-resource areas [58].

Recent research efforts have attempted to address these shortcomings by aiming to use video recordings to implement low-cost, automatic, objective alternatives for the detection of CP risk. Such investigations have succeeded in predicting CP based on automatic movement assessment from infant video recordings with performance comparable to standardized CP risk measures [59–61]. For example, in a multi-site cohort investigation, an automated, objective, movement assessment of infant video recordings was compared to standard risk assessment measures (i.e., the General Movement Assessment and neonatal neuroimaging) at 9–15 weeks corrected age to predict CP status and motor function at approximately 3.7 years of age. The results of this investigation found that the automated, video-based approach exhibited sensitivity and specificity comparable to standard measures used to predict CP [61].

There are also clear applications for pose estimation to potentially improve the early identification of neurodevelopmental disorders, such as the early detection of autism spectrum disorder (ASD). Although parents often report first concerns about ASD when their child is between 12 to 14 months of age [62,63] and reliable ASD diagnosis is possible by age 2, the majority of children with ASD remain undiagnosed until 4 years of age [64]. Shortages of ASD expert clinicians and limited capacities at autism tertiary diagnostic centers contribute to the long wait times for families [65]. Families living in rural and low-resource communities are often required to travel long distances to receive diagnostic services, placing them at an even greater disadvantage in accessing services. Indeed, a recent report indicates that approximately 84% (2635/3142) of U.S. counties do not have the necessary ASD diagnostic resources [66]. Given these barriers to a timely diagnosis, a significant portion of children with ASD are missing a critical window for early intervention services, as evidence shows that intervention before the age of 2 significantly improves behavioral and developmental outcomes for children with ASD [67–69]. The detrimental impact of diagnostic delays has resulted in federal prioritization of early identification of ASD and an urgency to develop accessible and accurate early screening methods [64].

Leveraging advances in machine learning, efforts have been made to develop scalable, video-based ASD screeners to improve access to diagnostic and early intervention services. For example, Crippa et al. developed an algorithm to examine the predictive value of motor behavioral biomarker measures in ASD to discriminate preschool children with ASD from children with typical development using a simple upper-limb reach-to-drop task [70]. The resulting model showed an accuracy rate of 96.7%, suggesting that video-based approaches combined with machine learning can be a useful method of classification and discrimination in the diagnostic process [70].

The emerging evidence supporting the application of automated, video-based assessments to monitor general gross motor development and promote early detection of both motor-based and neurodevelopmental disorders is promising. In order to establish the clinical utility of pose estimation, future work is needed to examine the feasibility and acceptability of clinician use of such techniques.

4.3. *Human Performance Optimization, Injury Prevention, and Safety*

Numerous applications of pose estimation exist within optimization of human performance and safety, with these applications spanning injury risk assessment, rehabilitation, and enhancing human performance. This application space commonly consists of some type of instructor, such as a coach, trainer, or clinician, attempting to assess an individual's movement patterns to determine whether the individual is at an increased risk for injury, is moving differently from a healthy, uninjured individual, or is moving with some level of inefficiency that can be modified to improve performance. Within injury assessment, common applications of pose estimation have been to evaluate an individual's risk for specific musculoskeletal injuries and to perform a post-hoc analysis following the occurrence of an injury. For example, two-dimensional pose estimation techniques have been applied to develop proof-of-concept screening technologies that detect abnormal gait patterns during walking and running [71–75], fall detection [76–78], abnormal movements that are indicative of injury risk in manual labor work environments [79–81], and risk of sports-related injury, such as anterior cruciate ligament rupture [82–84]. Post-hoc analysis following an injury has primarily been targeted towards sports performance applications and focused on understanding mechanisms of injury, with the ultimate goal of developing techniques to mitigate injury risk [85,86].

Applications of pose estimation to rehabilitation following injury or surgery typically focus on using these techniques to monitor an individual's return to normal movement patterns and to guide the motion of rehabilitation technology that is designed to interface with a patient. Pose estimation techniques have been used to measure a patient's range of motion and movement during functional exercises and assess their progression towards a healthy range of motion [87–89]. In particular, there has been an emphasis on the use of pose estimation to monitor rehabilitation progress outside of the clinic, such as in home or on an athletic field [90–93]. Additionally, many technologies have been designed to actively interface with an individual to either support their movement during rehabilitation or to help provide a mechanical stimulus to enhance rehabilitation. These technologies are commonly referred to as rehabilitation robotics, and techniques have been developed that leverage pose estimation to inform the movement of these systems [94–97].

The use of pose estimation for enhancing human performance remains a challenging application, given the large range of joint articulation, out of plane motion, and fast movements that can be difficult to capture with the relatively slow sampling rates of common video recording devices and risk of occlusion that occurs in these applications [98,99]. However, a number of proof-of-concept systems have been developed to inform pose of an athlete during training, particularly for sports in which success for the athlete is directly linked to pose (e.g., gymnastics and skiing) [100–102]. Development of new pose estimation techniques for human performance applications have focused on achieving high accuracy with 'in the wild' pose estimations, given the importance of performing these measurements outside of the lab in these applications [11,103,104]. While this previous research has demonstrated applications that may be made possible with pose estimation, very few of these proof-of-concept technologies have made the transition to regular use in a clinical, athletic, or other relevant environments. This likely derives from the fact that many unique requirements arise when attempting to apply these techniques to human performance applications outside of the laboratory.

For pose estimation to influence the broader human performance community, including non-clinical populations, research must drive towards robust 'in the wild' pose estimation encompassing a range of environments and populations. To this end, we will define desirable components of an ideal dataset for pose estimation algorithm development, training, and validation. Future studies should focus on capturing and making available these datasets to expand the application space of pose estimation or define functional limitations of the current hardware or software technology.

Many injury and performance evaluations are based on highly dynamic motion analysis [85,86,105], requiring that any pose estimation validation datasets should include accurate ground truth measurements of human joint kinematics for as many degrees of freedom as feasible. Ideally, this will include kinematics of complex joints, such as the ankle, wrists, intervertebral joints, and scapular motion—all of which play a key role in many injuries and are not estimated in most existing pose estimation techniques. Linear kinematics of the various body components should also be reported on, especially in relation to conditions that result from impact injuries (e.g., traumatic brain injury, chronic traumatic encephalopathy) [106]. Optical motion tracking is currently the gold standard for such ground truth measurements, but further accuracy (and cost) improvements are desirable due to artifacts arising from relative marker motion with respect to the underlying bony anatomy [107]. Therefore, researchers should aim to account for these artifacts within the pose estimation process.

Validation datasets should be captured outside of laboratory environments and include complexities such as partial occlusion (self-occlusion, inter-subject occlusion, environmental occlusion), various illuminations, loose-fitting clothing, and multiple camera standoffs or viewing angles. Recent examples of pose estimation outside of the lab are primarily based on monocular RGB images [108–111]. However, these techniques are generally less accurate—especially in three dimensions—when compared to laboratory pose estimation. The fusion of other pose estimation modalities, including inertial measurement units and infrared imaging, with single or multi-view RGB images is a promising direction for improved pose estimation [112], and should be included in validation datasets, such as those provided by Malleson et al. [113].

As new pose estimation algorithms are developed for human performance applications, special consideration should be given to the evaluation metrics reported. Motion type classification is of limited usefulness for in-depth biomechanical analysis and, instead, joint kinematic errors should be reported for each degree of freedom. Furthermore, estimation accuracies should be reported under varying conditions, including differences between lab-based and outdoor estimations. Finally, the computational cost per frame of pose estimation should be reported to understand applicability to real-time, highly dynamic application spaces [113].

4.4. *Clinical Motor Assessment in Adult Neurologic Conditions*

Clinical assessments and the resulting outcome measures are critical to motor rehabilitation in adults with neurologic conditions. These clinical assessments are typically administered to capture either a patient's status at a specific point in time or to track their motor function longitudinally. When administered at a single time point, assessments are used to classify the severity of an individual's deficits. When administered longitudinally, assessments are commonly used to track disease progression/regression, measure recovery, or evaluate the effectiveness of an intervention.

The International Classification of Functioning, Disability and Health (ICF) is a common, widely accepted framework developed by the World Health Organization for describing health and disability at individual and population levels [114]. It provides standard language and has a wide range of uses across different sectors by identifying three primary levels of human functioning:

1. *Body structures and functions* are anatomical parts of the body and physiological functions of the body systems, respectively. The term impairment refers to problems in *body structure or function*.
2. *Activity* is the execution of a task or action by an individual. The term activity limitation describes difficulties with completion of an *activity*.
3. *Participation* is involvement in a life situation. Participation restrictions are problems that an individual encounters during *participation* in real-world situations.

To provide a concrete example of how this framework is used, consider a person who has experienced a stroke. This person might experience changes in all three levels of human functioning: the impairment of left-sided hemiparesis (*body structures and functions* level), the activity limitation of difficulty walking (*activity* level), and the participation restriction of inability to attend their desired religious activities (*participation* level). One can quickly observe that, while the three levels may be related to one another, there are independent needs for quantitative measurement within each level. In other words, there are needs for quantitative measurement of the hemiparesis, daily walking activity, and the inability to attend religious activities in this particular example.

Clinical outcome measures for each level of the ICF are administered as a part of routine clinical practice. Current measures of impairment involve a skilled clinician observing a patient as they perform a series of movements designed to expose deficits in *body structure and function*. For instance, one item on the Fugl–Meyer Assessment—a widely used quantitative measure of motor impairment after stroke—involves asking the patient to move their hand from the contralateral knee to ipsilateral ear while individual elements (e.g., shoulder retraction, shoulder elevation, elbow flexion, forearm supination) of this movement are scored subjectively from 0 to 2 [115]. Measures of activity limitations involve the patient performing one or more tasks that simulate *activities* encountered in daily life. An example of an ecologically valid task is the water pouring item of the Action Research Arm Test—an extensively used activity level measure for people with stroke [116]—where the person pours water from one glass to another. Lastly are measures of participation restrictions, which are often self-reported measures of the person's perceptions of their movement abilities and resulting impact on their quality of life (e.g., the Stroke Impact Scale [117], a self-report questionnaire that evaluates disability and health-related quality of life after stroke) and daily *participation*. The data gathered from existing outcome measures are valuable for their use in diagnosing movement disorders, establish rehabilitation goals, and track changes in patient status.

Pose estimation tools have the potential to address two important challenges that exist within current clinical assessments spanning all three levels of the ICF (Figure 4). First, they can increase the accuracy, precision, and frequency with which movement kinematics are measured and assessed. Presently, *body structure/function* and *activity* level assessments primarily rely on visual observation of movement or task performance, and many are scored on ordinal scales that require a clinic visit or other similarly time-consuming interaction for both patients and their providers. Pose estimation offers the potential to provide precise, quantitative, and continuous data about single joint or whole-body movements through short video recordings that could be recorded in virtually any setting with much higher frequency. This opportunity to obtain frequent, quantitative motor assessments could significantly enhance the abilities of clinicians to detect and track impairments and activity limitations in their patients longitudinally. Second, current assessments of participation restrictions are almost exclusively self-reported. The self-report format has been necessary due to the difficulty of measuring movement kinematics in the home, but many self-report measures lack reliability and often do not correlate with clinically-administered motor assessments. There is clear potential for the propagation of telerehabilitation and pose estimation tools to make a significant impact in this area by providing significantly improved accessibility for clinicians and researchers to obtain quantitative data about how people move and participate in their home and community environments.

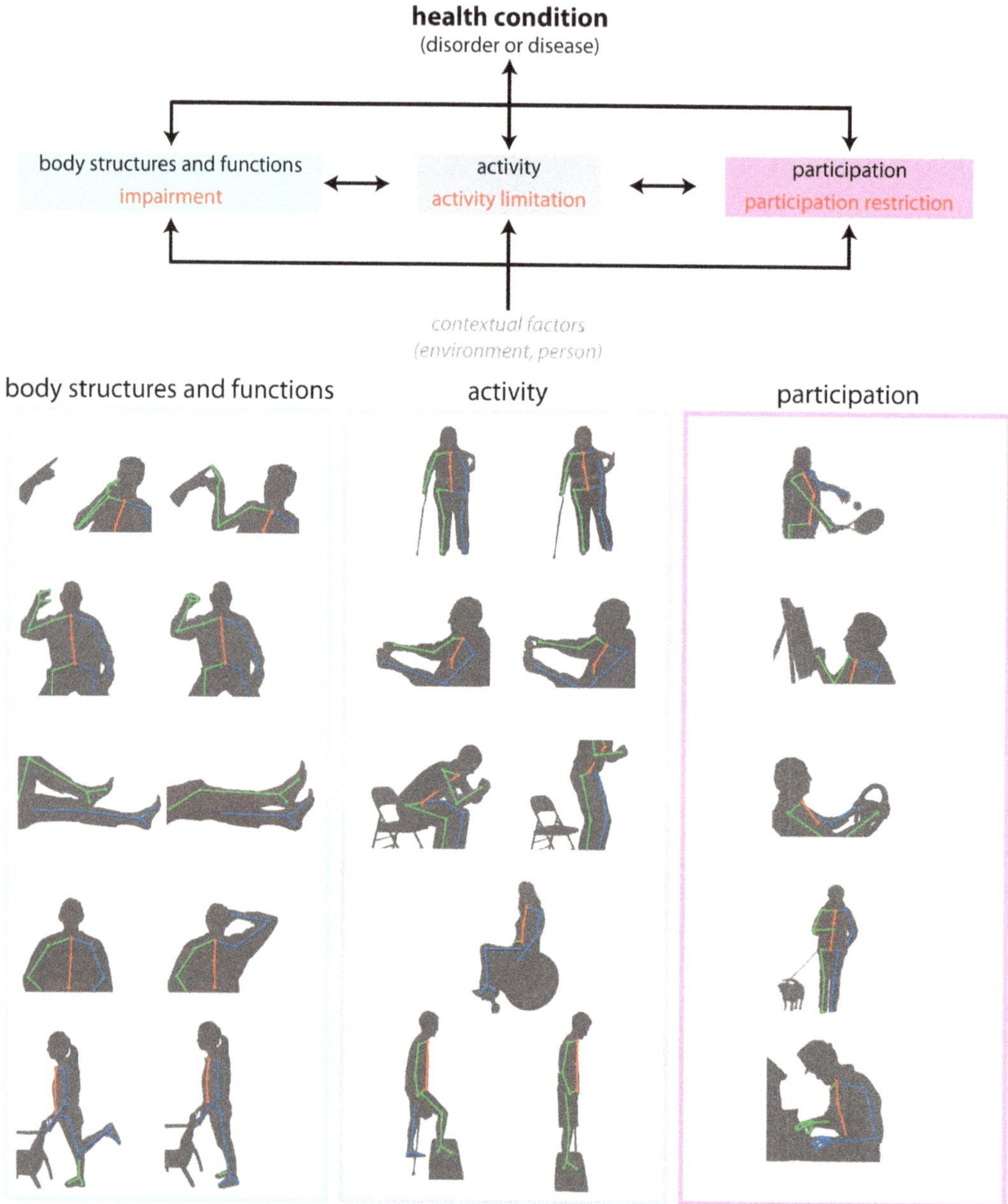

Figure 4. Depiction of potential applications of pose estimation for movement tracking during clinical assessments across the domains of the International Classification of Functioning, Disability and Health (ICF) model. For instance, finger–nose coordination testing the body structures and functions domain (**left**), walking assessment in the activity domain (**middle**), and playing tennis in the participation domain (**right**).

The uses of pose estimation in clinical populations are expanding, but ultimately remain in the beginning stages. At the *body structure/function* level, early work has involved detecting hallmark motor signs in persons with Parkinson's disease (PD). For instance, dyskinesia is an involuntary movement of the head, arm, leg, or entire body. Dyskinesia is commonly seen in persons with PD, often as a side effect of long-term levodopa treatment. A number of recent studies have used pose estimation to assess dyskinesias in persons with PD and found similar or superior performance with standard clinical assessments [118–120]. Bradykinesia, or slowness of movement, is another cardinal motor sign of PD. Liu et al. report that their computer vision-based method was 89.7% accurate in quantifying bradykinesia severity in people with PD as they performed repetitive movements including finger tapping, hand clasping, and alternating hand pronation/supination movements [121].

There are also a number of studies that have begun to use pose estimation to measure *activity*-level behaviors. Gait assessment, in particular, has been an early clinical target for these evolving tools. Video-based tools have been used to successfully capture gait parameters such as step lengths, step width, step time, stride length, gait velocity, and cadence in people with stroke [122], PD [25,123] or dementia [124]. Beyond gait, the timed up and go is a widely accepted assessment of functional mobility in patients with a range of neurological disorders or disease. Li et al. recently validated and used a video-based activity classification to automatize timed-up-and-go sub-task segmentation (sit-to-stand, walk, turn, walk-back, sit-back) in people with PD [125].

Future work should focus on further validation of pose estimation with gold standard kinematic tools and interpretability alongside standard clinical assessments. Additional patient populations with a wide range of different movement patterns should be included in these investigations in order to develop algorithms that are broadly applicable. The potential of video-based analysis and pose estimation to quantitatively measure participation-level data in the home and the community should also be a top priority. Precise data captured in the real world not only will provide clinicians with important data from which they can make clinical decisions, but this may also facilitate early diagnosis of movement disorders and the ability to track movement patterns throughout a disease course.

We summarize many of the applications discussed in Section 4 in Table 1 below.

Table 1. Summary of example applications of pose estimation in human health and performance across the lifespan.

Domain	Behavior/Movement Pattern Tracked	References
Motor and non-motor development	Infant cruising (early locomotion)	[36]
	Infant play/general movement	[37]
	Infant writhing	[51]
Human performance optimization, injury prevention, and safety	Healthy repetitive movements	[14]
	Healthy gait	[15,26,29–31,35,40]
	Sign language	[19]
	Healthy running	[27,35]
	Bilateral squat	[28]
	Healthy gait/jumping/throwing	[29]
	Lifting	[79,84]
	Various unsafe working behaviors	[80,81]
	ACL injury risk	[82,85,86]
	Handcart pushing and pulling	[83]
	Ergonomic postural assessment	[87]
	Remotely-delivered rehabilitation	[88,91–93]
	Healthy finger movements	[90]
	Rehabilitation robotics	[94–97]
	Athletic training	[100,101]
	Swimming	[102]
Clinical motor assessment	Gait in Parkinson's disease	[25,33,123]
	Knee kinetics in osteoarthritis	[32]
	Gait in cerebral palsy	[34]
	Simulated abnormal gait	[72,74]
	Gait in older adults	[73]
	Fall detection	[76–78]
	Dyskinesias in Parkinson's disease	[118–120]
	Gait in older adults with dementia	[124]
	Timed up-and-go in Parkinson's disease	[125]

5. What Are the Limitations of Pose Estimation?

While many of our perspectives on the limitations of human pose estimation algorithms with regard to applications in human health and performance are embedded within the sections above, we considered that it may be helpful to include a condensed summary section here. As mentioned previously, technical limitations have been discussed extensively in prior reviews [20,21]. Here, we list perceived limitations in two general areas: *application limitations* and *barriers to implementation*. We consider application limitations to be those associated with obtaining high quality, usable data from video recordings via pose estimation (some are also discussed in [21]) and barriers to implementation to be limitations associated with the uptake and implementation of pose estimation approaches for common use among clinicians and researchers (with an emphasis on implementation in clinical settings).

5.1. Application Limitations

- Occlusions: these occur when one or more of the anatomical locations desired to be tracked are not visible. This may be due to occlusion by other body segments, by other people in the frame, or by inanimate objects (e.g., assistive devices—canes, walkers, crutches, orthoses, robotics; clinical objects—beds, hospital gowns, medical devices; sporting equipment—helmets, balls, bats, sticks).
- Limited training data: networks that are trained on sets of images that lack diversity (e.g., clothing, poses, illuminations, viewpoints, unusual postures associated with clinical conditions) may not perform well in applications where the videos are quite different from those included in the training set. Applications of current techniques that require a training dataset may require creation of a new training dataset if movements/images of a patient population are substantially different from those included in the existing training dataset (e.g., abnormal hand postures after stroke). This is particularly important given that most training datasets are biased toward healthy movement patterns.
- Capture errors: pose estimation algorithms may identify and track unwanted human or human-like figures in the field of view (e.g., people in the background, images on posters or artwork).
- Positional errors: tracking may be difficult when conditions introduce uncertainty into the positions of anatomical locations within the image (e.g., wearing a dress, hospital gown, athletic uniform or padding). This may also occur when attempting to track a movement from a suboptimal viewpoint (e.g., measuring knee flexion from a frontal view).
- Limitations of recording devices: use of devices with low sampling rates (e.g., the sampling rate of common video recording devices is often approximately 30 Hz) may be unable to capture accurate movement kinematics of movements that occur at high speeds or high frequencies. The aperture and shutter speed of recording devices can also impact image quality and introduce blurring, which can impact the quality of the tracking achieved through pose estimation.

Examples of application limitations are depicted in Figure 5.

5.2. Barriers to Implementation

- User-friendliness: we currently lack plug-and-play options for pose estimation. While we certainly understand and acknowledge the many reasons for this, pose estimation is unlikely to be used widely in clinical settings in particular until user-friendliness improves. We outline several relevant components to user-friendliness below:
    - Set up time: in our experience, many users want point-and-click capability. They want to be able to carry a recording device in their pocket, use it to record a quick video of their patient or research participant when needed, and ultimately obtain meaningful information about movement kinematics. Alternatively, they want a reserved space where a recording device could be permanently mounted and

easily started and stopped (e.g., a tablet mounted to a wall). Any configuration that requires multi-camera calibration or prolonged set up time is unlikely to be adopted for widespread clinical use.
    - Delayed results: many users want results in near real-time. There is a need for fast, automated approaches that immediately process the pose estimation outputs, calculate relevant movement parameters, and return interpretable data.
    - Programming and training requirements: some existing pose estimation options are very easy to download, install, and use for users with basic technical expertise. However, even these can remain prohibitively daunting for clinicians and researchers without technical backgrounds. Technologies that require any amount of programming or significant training are unlikely to reach widespread use in clinical settings.
- Outcome measure challenges: in some cases, users want to use movement data to improve clinical or performance-related decision-making, but it is not immediately clear what parameters of the movement will lead to improved outcomes (e.g., a user may express interest in measuring "walking" but is not sure which specific gait parameters are most relevant to their research study or clinical intervention). Therefore, there is a desire to collect kinematic data, but how these data should be used is not well-defined. Similarly, in the case of clinical assessments, there needs to be a clear link to relevant clinical and translational outcomes—the users should have input as to what output metrics are important.
- Limited hardware infrastructure: as described above, some applications of pose estimation for human movement tracking require significant computational power. Some clinical and research settings are unlikely to have access to the hardware (e.g., GPUs) needed to execute their desired applications in a timely manner.
- Technology challenges: many technologies that promise potential for clinical or human performance impact are made available before they are fully developed. This can lead to buggy software and frequent updating, which harms trust and credibility among users. This can, in turn, exacerbate the hesitancy in adopting new technologies present in some clinical and research communities, especially in artificial intelligence technologies (such as pose estimation) that are purported to supplement or even replace expert human assessment.
- Lack of validation and feasibility data: there is a need for large-scale studies to validate pose estimation outputs against ground truth measures in a wide range of different populations. This may be accomplished in a variety of ways, including (but not limited to) comparisons with three-dimensional motion capture, wearable devices with proven accuracy, expert clinical ratings and/or assessments, or even possibly other pose estimation algorithms. The error (relative to the ground truth measurement) that is deemed acceptable is likely to depend on the use case and the metrics being used. In our experience, users who study very specific movements of joints or other anatomical landmarks (e.g., biomechanics or motor control researchers) are likely to seek greater accuracy than, for example, a clinician who may wish to incorporate a video-based assessment of walking speed as part of a larger clinical examination. It may be desirable to begin to develop field-specific accuracy standards for some applications.

There is also a need for testing of sensitivity, specificity, feasibility, and reliability. When a new clinical outcome measure is developed, a first step should be to establish criterion-validity or construct validity between the pose estimated measures and age-concurrent, clinician-coded, gold-standard clinical measures. Next, using receiver operating characteristic (ROC) analysis, sensitivity and specificity should be compared to assess the ability of the new pose estimated measure in predicting dichotomous outcomes (e.g., motor impaired vs. motor unimpaired). Area under the curve (AUC) should further be computed as a measure of the ability to distinguish between groups. Finally, it is important to evaluate the feasibility and acceptability of the new pose estimation protocol. One way to assess feasibility is to assess the number of completed and submitted usable videos by patients (i.e., the total number of videos submitted divided by the number expected, multiplied by 100). One way to assess acceptability is through satisfaction questionnaires/surveys. For example, after video submission, patients, families of patients (if patients are children), and clinicians can complete a brief satisfaction questionnaire/survey regarding their experience using the pose estimation protocol.

These potential pitfalls along the path to implementation are shown in Figure 6.

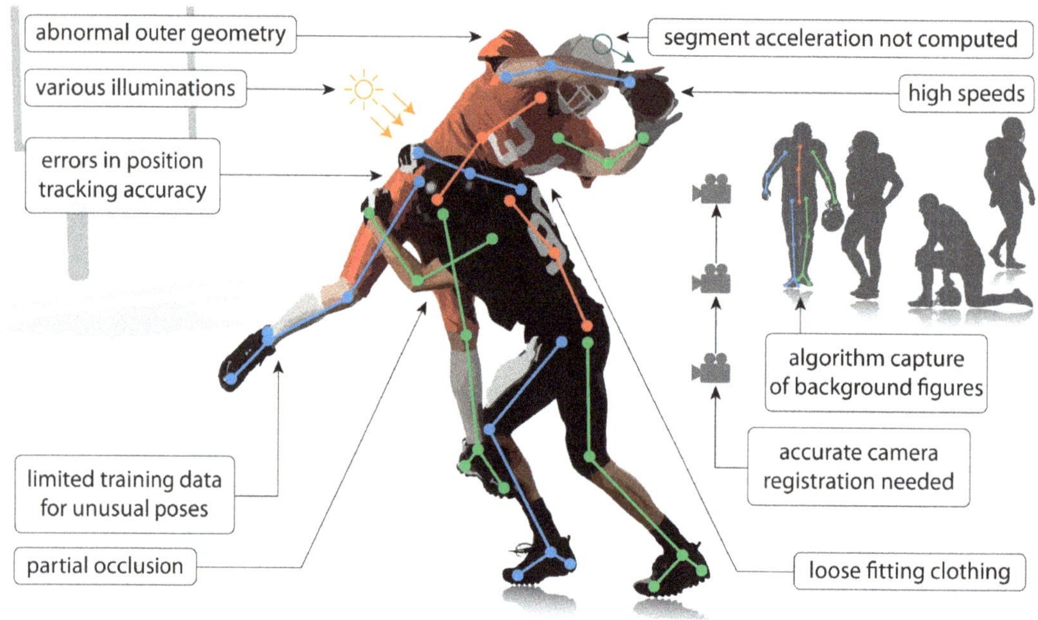

Figure 5. Common application limitations with current pose estimation algorithms and challenges with using these algorithms outside of the laboratory. These applications commonly require three-dimensional kinematics of multiple people moving at relatively high speeds to be tracked in environments with background figures (e.g., irrelevant people and objects shaped similarly to people). This leads to challenges with segment occlusion, unintentional capture of background figures, and registration of multiple cameras. Additionally, using current algorithms for scenarios different than the training dataset (e.g., different movements, different types of clothing or equipment being worn, different lighting) may lead to reduced accuracy in the predicted kinematics or, potentially, failure of the algorithm. Finally, most algorithms do not predict kinematic metrics that are required for some applications (e.g., head acceleration to assess concussion risk), and limitations with using current algorithms on time-series data make it challenging to accurately derive these metrics.

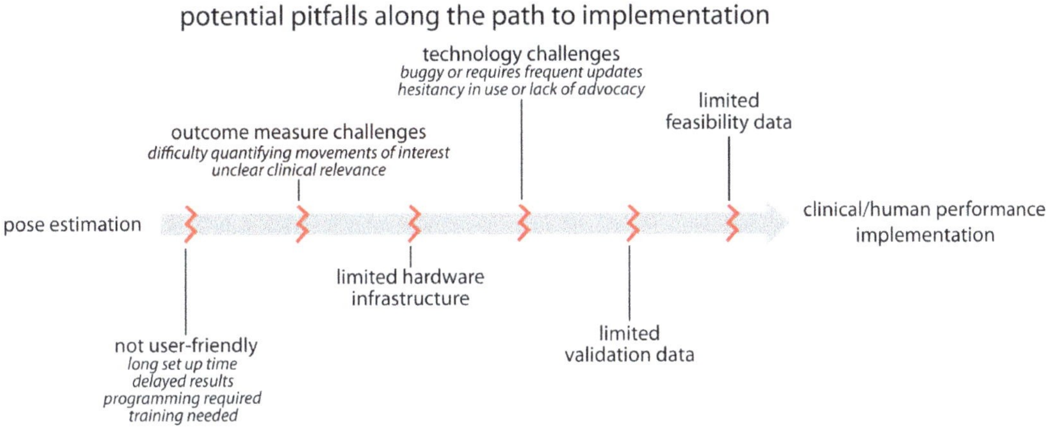

Figure 6. Common pitfalls that must be avoided on the path to widespread implementation of pose estimation applications for human health and performance. These are covered in greater detail in the "**What are limitations of pose estimation?**" section of the manuscript.

6. Conclusions

The emergence and continued development of human pose estimation approaches offer exciting potential for making quantitative assessments of human movement kinematics significantly more accessible. Pose estimation algorithms directly address an important and widespread need for low cost, easy to use, accessible technologies that enable human movement tracking in virtually any environment, including the home, clinic, classroom, playing field, and other 'in the wild' settings. Applications in health and human performance have begun to emerge in the literature, but we perceive that these technologies are still in their relative infancy with regard to the potential for research and clinical implementation. Many limitations persist, and it is important that users are aware of these and adjust expectations accordingly. However, we anticipate that applications of pose estimation in human health and performance will continue to expand in coming years, and these technologies will provide powerful tools for capturing meaningful aspects of human movement that have been difficult to capture with conventional techniques.

Author Contributions: Conceptualization: J.S., K.M.C.-A., C.O.P., R.D.R., M.F.V. and R.T.R.; writing—original draft preparation: J.S., K.M.C.-A., C.O.P., R.D.R., M.F.V. and R.T.R.; writing—review and editing: J.S., K.M.C.-A., C.O.P., R.D.R., M.F.V. and R.T.R.; funding acquisition: C.O.P., R.D.R., M.F.V. and R.T.R. All authors have read and agreed to the published version of the manuscript.

Funding: This research was funded by a Kennedy Krieger Institute Goldstein Innovation Grant to RDR, NIH grant R21 AG059184 to RTR.

Institutional Review Board Statement: Not applicable.

Informed Consent Statement: Not applicable.

Data Availability Statement: Not applicable.

Acknowledgments: We would like to thank the Janney Program within the Johns Hopkins University Applied Physics Laboratory for providing partial funding for this work, which nurtures a culture of discovery, embraces risk, and welcomes being at the center of a vibrant innovation ecosystem.

Conflicts of Interest: The authors declare no conflict of interest. The funders had no role in the design of the study; in the collection, analyses, or interpretation of data; in the writing of the manuscript, or in the decision to publish the results. All authors have read and agreed to the published version of the manuscript.

References

1. Mündermann, L.; Corazza, S.; Andriacchi, T.P. The Evolution of Methods for the Capture of Human Movement Leading to Markerless Motion Capture for Biomechanical Applications. *J. NeuroEng. Rehabil.* **2006**, *3*, 6. [CrossRef]
2. Baker, R. The History of Gait Analysis before the Advent of Modern Computers. *Gait Posture* **2007**, *26*, 23–28. [CrossRef] [PubMed]
3. Roether, C.L.; Omlor, L.; Christensen, A.; Giese, M.A. Critical Features for the Perception of Emotion from Gait. *J. Vis.* **2009**, *9*, 1–32. [CrossRef] [PubMed]
4. Michalak, J.; Troje, N.F.; Fischer, J.; Vollmar, P.; Heidenreich, T.; Schulte, D. Embodiment of Sadness and Depression-Gait Patterns Associated with Dysphoric Mood. *Psychosom. Med.* **2009**, *71*, 580–587. [CrossRef] [PubMed]
5. Kendon, A. Movement Coordination in Social Interaction: Some Examples Described. *Acta Psychol.* **1970**, *32*, 101–125. [CrossRef]
6. Martinez, G.H.; Raaj, Y.; Idrees, H.; Xiang, D.; Joo, H.; Simon, T.; Sheikh, Y. Single-Network Whole-Body Pose Estimation. In Proceedings of the IEEE International Conference on Computer Vision, Seoul, Korea, 27 October–2 November 2019; pp. 6982–6991. [CrossRef]
7. Insafutdinov, E.; Andriluka, M.; Pishchulin, L.; Tang, S.; Levinkov, E.; Andres, B.; Schiele, B. ArtTrack: Articulated Multi-Person Tracking in the Wild. In Proceedings of the 30th IEEE Conference on Computer Vision and Pattern Recognition, Honolulu, HI, USA, 21–26 July 2017; pp. 6457–6465. [CrossRef]
8. Insafutdinov, E.; Pishchulin, L.; Andres, B.; Andriluka, M.; Schiele, B. Deepercut: A Deeper, Stronger, and Faster Multi-Person Pose Estimation Model. In *Computer Vision—ECCV 2016;* Lecture Notes in Computer Science (Including Subseries Lecture Notes in Artificial Intelligence and Lecture Notes in Bioinformatics); Springer International Publishing: Cham, Switzerland, 2016; pp. 34–50. [CrossRef]
9. Pishchulin, L.; Insafutdinov, E.; Tang, S.; Andres, B.; Andriluka, M.; Gehler, P.; Schiele, B. DeepCut: Joint Subset Partition and Labeling for Multi Person Pose Estimation. In Proceedings of the IEEE Computer Society Conference on Computer Vision and Pattern Recognition, Las Vegas, NV, USA, 27–30 June 2016; pp. 4929–4937. [CrossRef]
10. Toshev, A.; Szegedy, C. DeepPose: Human Pose Estimation via Deep Neural Networks. In Proceedings of the IEEE Computer Society Conference on Computer Vision and Pattern Recognition, Columbus, OH, USA, 23–28 June 2014; pp. 1653–1660. [CrossRef]
11. Nath, T.; Mathis, A.; Chen, A.C.; Patel, A.; Bethge, M.; Mathis, M.W. Using DeepLabCut for 3D Markerless Pose Estimation across Species and Behaviors. *Nat. Protoc.* **2019**, *14*, 2152–2176. [CrossRef] [PubMed]
12. Mathis, A.; Mamidanna, P.; Cury, K.M.; Abe, T.; Murthy, V.N.; Mathis, M.W.; Bethge, M. DeepLabCut: Markerless Pose Estimation of User-Defined Body Parts with Deep Learning. *Nat. Neurosci.* **2018**, *21*, 1281–1289. [CrossRef] [PubMed]
13. Cao, Z.; Simon, T.; Wei, S.E.; Sheikh, Y. Realtime Multi-Person 2D Pose Estimation Using Part Affinity Fields. In Proceedings of the 30th IEEE Conference on Computer Vision and Pattern Recognition, Honolulu, HI, USA, 21–26 July 2017; pp. 1302–1310. [CrossRef]
14. Cornman, H.L.; Stenum, J.; Roemmich, R.T. Video-Based Quantification of Human Movement Frequency Using Pose Estimation. *bioRxiv* **2021**. [CrossRef]
15. Stenum, J.; Rossi, C.; Roemmich, R.T. Two-Dimensional Video-Based Analysis of Human Gait Using Pose Estimation. *PLoS Comput. Biol.* **2021**, *17*, e1008935. [CrossRef]
16. Kwolek, B.; Michalczuk, A.; Krzeszowski, T.; Switonski, A.; Josinski, H.; Wojciechowski, K. Calibrated and Synchronized Multi-View Video and Motion Capture Dataset for Evaluation of Gait Recognition. *Multimed. Tools Appl.* **2019**, *78*, 32437–32465. [CrossRef]
17. Wang, L.; Tan, T.; Ning, H.; Hu, W. Silhouette Analysis-Based Gait Recognition for Human Identification. *IEEE Trans. Pattern Anal. Mach. Intell.* **2003**, *25*, 1505–1518. [CrossRef]
18. Holte, M.B.; Cuong, T.; Trivedi, M.M.; Moeslund, T.B. Human Pose Estimation and Activity Recognition from Multi-View Videos: Comparative Explorations of Recent Developments. *IEEE J. Sel. Top. Signal Process.* **2012**, *6*, 538–552. [CrossRef]
19. Isaacs, J.; Foo, S. Hand Pose Estimation for American Sign Language Recognition. In Proceedings of the Thirty-Sixth Southeastern Symposium on System Theory, Atlanta, GA, USA, 16 March 2004; pp. 132–136. [CrossRef]
20. Cronin, N.J. Using Deep Neural Networks for Kinematic Analysis: Challenges and Opportunities. *J. Biomech.* **2021**, *123*, 110460. [CrossRef] [PubMed]
21. Seethapathi, N.; Wang, S.; Saluja, R.; Blohm, G.; Kording, K.P. Movement Science Needs Different Pose Tracking Algorithms. *arXiv* **2019**, arXiv:1907.10226.
22. Arac, A. Machine Learning for 3D Kinematic Analysis of Movements in Neurorehabilitation. *Curr. Neurol. Neurosci. Rep.* **2020**, *20*, 29. [CrossRef]
23. Andriluka, M.; Pishchulin, L.; Gehler, P.; Schiele, B. 2D Human Pose Estimation: New Benchmark and State of the Art Analysis. In Proceedings of the IEEE Computer Society Conference on Computer Vision and Pattern Recognition, Columbus, OH, USA, 23–28 June 2014; pp. 3683–3693. [CrossRef]
24. Lin, T.Y.; Maire, M.; Belongie, S.; Hays, J.; Perona, P.; Ramanan, D.; Dollár, P.; Zitnick, C.L. Microsoft COCO: Common Objects in Context. In *Computer Vision—ECCV 2014;* Lecture Notes in Computer Science (Including Subseries Lecture Notes in Artificial Intelligence and Lecture Notes in Bioinformatics); Springer International Publishing: Cham, Switzerland, 2014; pp. 740–755. [CrossRef]

25. Sato, K.; Nagashima, Y.; Mano, T.; Iwata, A.; Toda, T. Quantifying Normal and Parkinsonian Gait Features from Home Movies: Practical Application of a Deep Learning–Based 2D Pose Estimator. *PLoS ONE* **2019**, *14*, e0223549. [CrossRef]
26. Chambers, C.; Kong, G.; Wei, K.; Kording, K. Pose Estimates from Online Videos Show That Side-by-Side Walkers Synchronize Movement under Naturalistic Conditions. *PLoS ONE* **2019**, *14*, e0217861. [CrossRef]
27. Cronin, N.J.; Rantalainen, T.; Ahtiainen, J.P.; Hynynen, E.; Waller, B. Markerless 2D Kinematic Analysis of Underwater Running: A Deep Learning Approach. *J. Biomech.* **2019**, *87*, 75–82. [CrossRef]
28. Ota, M.; Tateuchi, H.; Hashiguchi, T.; Kato, T.; Ogino, Y.; Yamagata, M.; Ichihashi, N. Verification of Reliability and Validity of Motion Analysis Systems during Bilateral Squat Using Human Pose Tracking Algorithm. *Gait Posture* **2020**, *80*, 62–67. [CrossRef]
29. Nakano, N.; Sakura, T.; Ueda, K.; Omura, L.; Kimura, A.; Iino, Y.; Fukashiro, S.; Yoshioka, S. Evaluation of 3D Markerless Motion Capture Accuracy Using OpenPose With Multiple Video Cameras. *Front. Sports Act. Living* **2020**, *2*, 50. [CrossRef]
30. Zago, M.; Luzzago, M.; Marangoni, T.; De Cecco, M.; Tarabini, M.; Galli, M. 3D Tracking of Human Motion Using Visual Skeletonization and Stereoscopic Vision. *Front. Bioeng. Biotechnol.* **2020**, *8*, 181. [CrossRef] [PubMed]
31. D'Antonio, E.; Taborri, J.; Palermo, E.; Rossi, S.; Patanè, F. A Markerless System for Gait Analysis Based on OpenPose Library. In Proceedings of the 2020 IEEE International Instrumentation and Measurement Technology Conference (I2MTC), Dubrovnik, Croatia, 25–28 May 2020; pp. 1–6. [CrossRef]
32. Boswell, M.A.; Uhlrich, S.D.; Kidziński, Ł.; Thomas, K.; Kolesar, J.A.; Gold, G.E.; Beaupre, G.S.; Delp, S.L. A Neural Network to Predict the Knee Adduction Moment in Patients with Osteoarthritis Using Anatomical Landmarks Obtainable from 2D Video Analysis. *Osteoarthr. Cartil.* **2021**, *29*, 346–356. [CrossRef] [PubMed]
33. Lu, M.; Poston, K.; Pfefferbaum, A.; Sullivan, E.V.; Fei-Fei, L.; Pohl, K.M.; Niebles, J.C.; Adeli, E. Vision-Based Estimation of MDS-UPDRS Gait Scores for Assessing Parkinson's Disease Motor Severity. In *Medical Image Computing and Computer Assisted Intervention—MICCAI 2020*; Springer International Publishing: Cham, Switzerland, 2020; Volume 12263, pp. 637–647. [CrossRef]
34. Kidziński, Ł.; Yang, B.; Hicks, J.L.; Rajagopal, A.; Delp, S.L.; Schwartz, M.H. Deep Neural Networks Enable Quantitative Movement Analysis Using Single-Camera Videos. *Nat. Commun.* **2020**, *11*, 4054. [CrossRef] [PubMed]
35. Ota, M.; Tateuchi, H.; Hashiguchi, T.; Ichihashi, N. Verification of Validity of Gait Analysis Systems during Treadmill Walking and Running Using Human Pose Tracking Algorithm. *Gait Posture* **2021**, *85*, 290–297. [CrossRef]
36. Ossmy, O.; Adolph, K.E. Real-Time Assembly of Coordination Patterns in Human Infants. *Curr. Biol.* **2020**, *30*, 4553–4562. [CrossRef] [PubMed]
37. Chambers, C.; Seethapathi, N.; Saluja, R.; Loeb, H.; Pierce, S.R.; Bogen, D.K.; Prosser, L.; Johnson, M.J.; Kording, K.P. Computer Vision to Automatically Assess Infant Neuromotor Risk. *IEEE Trans. Neural Syst. Rehabil. Eng.* **2020**, *28*, 2431–2442. [CrossRef]
38. Fang, H.; Xie, S.; Lu, C. RMPE: Regional Multi-Person Pose Estimation. *arXiv* **2018**, arXiv:1612.001737.
39. Cao, Z.; Hidalgo Martinez, G.; Simon, T.; Wei, S.-E.; Sheikh, Y.A. OpenPose: Realtime Multi-Person 2D Pose Estimation Using Part Affinity Fields. *IEEE Trans. Pattern Anal. Mach. Intell.* **2021**, *43*, 172–186. [CrossRef]
40. Viswakumar, A.; Rajagopalan, V.; Ray, T.; Parimi, C. Human Gait Analysis Using OpenPose. In Proceedings of the IEEE International Conference Image Information Processing, Shimla, India, 15–17 November 2019; pp. 310–314. [CrossRef]
41. Ye, Q.; Yuan, S.; Kim, T.-K. Spatial Attention Deep Net with Partial PSO for Hierarchical Hybrid Hand Pose Estimation. *arXiv* **2016**, arXiv:1604.03334.
42. Ivekovic, S.; Trucco, E. Human Body Pose Estimation with PSO. In Proceedings of the 2006 IEEE International Conference on Evolutionary Computation, Vancouver, BC, Canada, 16–21 July 2006; pp. 1256–1263. [CrossRef]
43. Lee, K.-Z.; Liu, T.-W.; Ho, S.-Y. Model-Based Pose Estimation of Human Motion Using Orthogonal Simulated Annealing. In *Intelligent Data Engineering and Automated Learning*; Springer: Berlin/Heidelberg, Germany, 2003. [CrossRef]
44. Halvorsen, K.; Söderström, T.; Stokes, V.; Lanshammar, H. Using an Extended Kalman Filter for Rigid Body Pose Estimation. *J. Biomech. Eng.* **2004**, *127*, 475–483. [CrossRef] [PubMed]
45. Janabi-Sharifi, F.; Marey, M. A Kalman-Filter-Based Method for Pose Estimation in Visual Servoing. *IEEE Trans. Robot.* **2010**, *26*, 939–947. [CrossRef]
46. Fanning, P.A.J.; Sparaci, L.; Dissanayake, C.; Hocking, D.R.; Vivanti, G. Functional Play in Young Children with Autism and Williams Syndrome: A Cross-Syndrome Comparison. *Child Neuropsychol. J. Norm. Abnorm. Dev. Child. Adolesc.* **2021**, *27*, 125–149. [CrossRef]
47. Kretch, K.S.; Franchak, J.M.; Adolph, K.E. Crawling and Walking Infants See the World Differently. *Child Dev.* **2014**, *85*, 1503–1518. [CrossRef] [PubMed]
48. LeBarton, E.S.; Iverson, J.M. Fine Motor Skill Predicts Expressive Language in Infant Siblings of Children with Autism. *Dev. Sci.* **2013**, *16*, 815–827. [CrossRef]
49. Masek, L.R.; Paterson, S.J.; Golinkoff, R.M.; Bakeman, R.; Adamson, L.B.; Owen, M.T.; Pace, A.; Hirsh-Pasek, K. Beyond Talk: Contributions of Quantity and Quality of Communication to Language Success across Socioeconomic Strata. *Infancy Off. J. Int. Soc. Infant Stud.* **2021**, *26*, 123–147. [CrossRef]
50. Le, H.; Hoch, J.E.; Ossmy, O.; Adolph, K.E.; Fern, X.; Fern, A. Modeling Infant Free Play Using Hidden Markov Models. In Proceedings of the 2021 IEEE International Conference on Development and Learning (ICDL), Beijing, China, 23–26 August 2021; pp. 1–6. [CrossRef]

51. Doroniewicz, I.; Ledwoń, D.J.; Affanasowicz, A.; Kieszczyńska, K.; Latos, D.; Matyja, M.; Mitas, A.W.; Myśliwiec, A. Writhing Movement Detection in Newborns on the Second and Third Day of Life Using Pose-Based Feature Machine Learning Classification. *Sensors* **2020**, *20*, 5986. [CrossRef] [PubMed]
52. Iverson, J.M.; Shic, F.; Wall, C.A.; Chawarska, K.; Curtin, S.; Estes, A.; Gardner, J.M.; Hutman, T.; Landa, R.J.; Levin, A.R. Early Motor Abilities in Infants at Heightened versus Low Risk for ASD: A Baby Siblings Research Consortium (BSRC) Study. *J. Abnorm. Psychol.* **2019**, *128*, 69. [CrossRef]
53. Iverson, J.M. Developing Language in a Developing Body: The Relationship between Motor Development and Language Development. *J. Child Lang.* **2010**, *37*, 229–261. [CrossRef]
54. Alcock, K.J.; Krawczyk, K. Individual Differences in Language Development: Relationship with Motor Skill at 21 Months. *Dev. Sci.* **2010**, *13*, 677–691. [CrossRef]
55. Adolph, K.E.; Hoch, J.E. Motor Development: Embodied, Embedded, Enculturated, and Enabling. *Annu. Rev. Psychol.* **2019**, *70*, 141–164. [CrossRef] [PubMed]
56. Rosenbaum, P.; Paneth, N.; Leviton, A.; Goldstein, M.; Bax, M.; Damiano, D.; Dan, B.; Jacobsson, B. A Report: The Definition and Classification of Cerebral Palsy April 2006. *Dev. Med. Child Neurol. Suppl.* **2007**, *109*, 8–14. [PubMed]
57. Novak, I.; Morgan, C.; Adde, L.; Blackman, J.; Boyd, R.N.; Brunstrom-Hernandez, J.; Cioni, G.; Damiano, D.; Darrah, J.; Eliasson, A.-C. Early, Accurate Diagnosis and Early Intervention in Cerebral Palsy: Advances in Diagnosis and Treatment. *JAMA Pediatr.* **2017**, *171*, 897–907. [CrossRef] [PubMed]
58. Geethanath, S.; Vaughan, J.T.J. Accessible Magnetic Resonance Imaging: A Review. *J. Magn. Reson. Imaging JMRI* **2019**, *49*, e65–e77. [CrossRef] [PubMed]
59. Adde, L.; Helbostad, J.L.; Jensenius, A.R.; Taraldsen, G.; Grunewaldt, K.H.; Støen, R. Early Prediction of Cerebral Palsy by Computer-based Video Analysis of General Movements: A Feasibility Study. *Dev. Med. Child Neurol.* **2010**, *52*, 773–778. [CrossRef]
60. Rahmati, H.; Aamo, O.M.; Stavdahl, Ø.; Dragon, R.; Adde, L. Video-Based Early Cerebral Palsy Prediction Using Motion Segmentation. *Annu. Int. Conf. IEEE Eng. Med. Biol. Soc.* **2014**, 3779–3783. [CrossRef]
61. Ihlen, E.A.; Støen, R.; Boswell, L.; de Regnier, R.-A.; Fjørtoft, T.; Gaebler-Spira, D.; Labori, C.; Loennecken, M.C.; Msall, M.E.; Möinichen, U.I. Machine Learning of Infant Spontaneous Movements for the Early Prediction of Cerebral Palsy: A Multi-Site Cohort Study. *J. Clin. Med.* **2020**, *9*, 5. [CrossRef] [PubMed]
62. Chawarska, K.; Klin, A.; Paul, R.; Volkmar, F. Autism Spectrum Disorder in the Second Year: Stability and Change in Syndrome Expression. *J. Child Psychol. Psychiatry* **2007**, *48*, 128–138. [CrossRef] [PubMed]
63. Landa, R.; Garrett-Mayer, E. Development in Infants with Autism Spectrum Disorders: A Prospective Study. *J. Child Psychol. Psychiatry* **2006**, *47*, 629–638. [CrossRef]
64. Maenner, M.J.; Shaw, K.A.; Baio, J. Prevalence of Autism Spectrum Disorder among Children Aged 8 Years—Autism and Developmental Disabilities Monitoring Network, 11 Sites, United States, 2016. *MMWR Surveill. Summ.* **2020**, *69*, 1–12. [CrossRef]
65. Gordon-Lipkin, E.; Foster, J.; Peacock, G. Whittling down the Wait Time: Exploring Models to Minimize the Delay from Initial Concern to Diagnosis and Treatment of Autism Spectrum Disorder. *Pediatr. Clin.* **2016**, *63*, 851–859. [CrossRef]
66. Ning, M.; Daniels, J.; Schwartz, J.; Dunlap, K.; Washington, P.; Kalantarian, H.; Du, M.; Wall, D.P. Identification and Quantification of Gaps in Access to Autism Resources in the United States: An Infodemiological Study. *J. Med. Internet Res.* **2019**, *21*, e13094. [CrossRef] [PubMed]
67. Brian, J.A.; Smith, I.M.; Zwaigenbaum, L.; Bryson, S.E. Cross-site Randomized Control Trial of the Social ABCs Caregiver-mediated Intervention for Toddlers with Autism Spectrum Disorder. *Autism Res.* **2017**, *10*, 1700–1711. [CrossRef]
68. Dawson, G.; Rogers, S.; Munson, J.; Smith, M.; Winter, J.; Greenson, J.; Donaldson, A.; Varley, J. Randomized, Controlled Trial of an Intervention for Toddlers with Autism: The Early Start Denver Model. *Pediatrics* **2010**, *125*, e17–e23. [CrossRef] [PubMed]
69. Landa, R.J.; Holman, K.C.; O'Neill, A.H.; Stuart, E.A. Intervention Targeting Development of Socially Synchronous Engagement in Toddlers with Autism Spectrum Disorder: A Randomized Controlled Trial. *J. Child Psychol. Psychiatry* **2011**, *52*, 13–21. [CrossRef]
70. Crippa, A.; Salvatore, C.; Perego, P.; Forti, S.; Nobile, M.; Molteni, M.; Castiglioni, I. Use of Machine Learning to Identify Children with Autism and Their Motor Abnormalities. *J. Autism Dev. Disord.* **2015**, *45*, 2146–2156. [CrossRef]
71. Karatsidis, A.; Richards, R.E.; Konrath, J.M.; van den Noort, J.C.; Schepers, H.M.; Bellusci, G.; Harlaar, J.; Veltink, P.H. Validation of Wearable Visual Feedback for Retraining Foot Progression Angle Using Inertial Sensors and an Augmented Reality Headset. *J. NeuroEng. Rehabil.* **2018**, *15*, 78. [CrossRef] [PubMed]
72. Guo, Y.; Deligianni, F.; Gu, X.; Yang, G.-Z. 3-D Canonical Pose Estimation and Abnormal Gait Recognition with a Single RGB-D Camera. *IEEE Robot. Autom. Lett.* **2019**, *4*, 3617–3624. [CrossRef]
73. Kondragunta, J.; Hirtz, G. Gait Parameter Estimation of Elderly People Using 3D Human Pose Estimation in Early Detection of Dementia. *Annu. Int. Conf. IEEE Eng. Med. Biol. Soc.* **2020**, 5798–5801. [CrossRef]
74. Chaaraoui, A.A.; Padilla-López, J.R.; Flórez-Revuelta, F. Abnormal Gait Detection with RGB-D Devices Using Joint Motion History Features. In Proceedings of the 2015 11th IEEE International Conference and Workshops on Automatic Face and Gesture Recognition (FG), Ljubljana, Slovenia, 4–8 May 2015; pp. 1–6. [CrossRef]
75. Li, G.; Liu, T.; Yi, J. Wearable Sensor System for Detecting Gait Parameters of Abnormal Gaits: A Feasibility Study. *IEEE Sens. J.* **2018**, *18*, 4234–4241. [CrossRef]

76. Chen, Y.; Du, R.; Luo, K.; Xiao, Y. Fall Detection System Based on Real-Time Pose Estimation and SVM. In Proceedings of the 2021 IEEE 2nd International Conference on Big Data, Artificial Intelligence and Internet of Things Engineering (ICBAIE), Nanchang, China, 26–28 March 2021; pp. 990–993. [CrossRef]
77. Bian, Z.; Hou, J.; Chau, L.; Magnenat-Thalmann, N. Fall Detection Based on Body Part Tracking Using a Depth Camera. *IEEE J. Biomed. Health Inform.* **2015**, *19*, 430–439. [CrossRef]
78. Huang, Z.; Liu, Y.; Fang, Y.; Horn, B.K.P. Video-Based Fall Detection for Seniors with Human Pose Estimation. In Proceedings of the 2018 4th International Conference on Universal Village (UV), Boston, MA, USA, 21–24 October 2018; pp. 1–4. [CrossRef]
79. Mehrizi, R.; Peng, X.; Tang, Z.; Xu, X.; Metaxas, D.; Li, K. Toward Marker-Free 3D Pose Estimation in Lifting: A Deep Multi-View Solution. *arXiv* **2018**, arXiv:1802.01741.
80. Han, S.; Lee, S. A Vision-Based Motion Capture and Recognition Framework for Behavior-Based Safety Management. *Autom. Constr.* **2013**, *35*, 131–141. [CrossRef]
81. Han, S.; Achar, M.; Lee, S.; Peña-Mora, F. Empirical Assessment of a RGB-D Sensor on Motion Capture and Action Recognition for Construction Worker Monitoring. *Vis. Eng.* **2013**, *1*, 6. [CrossRef]
82. Blanchard, N.; Skinner, K.; Kemp, A.; Scheirer, W.; Flynn, P. "Keep Me In, Coach!": A Computer Vision Perspective on Assessing ACL Injury Risk in Female Athletes. In Proceedings of the 2019 IEEE Winter Conference on Applications of Computer Vision (WACV), Waikoloa, HI, USA, 7–11 January 2019; pp. 1366–1374. [CrossRef]
83. Vukicevic, A.M.; Macuzic, I.; Mijailovic, N.; Peulic, A.; Radovic, M. Assessment of the Handcart Pushing and Pulling Safety by Using Deep Learning 3D Pose Estimation and IoT Force Sensors. *Expert Syst. Appl.* **2021**, *183*, 115371. [CrossRef]
84. Mehrizi, R.; Peng, X.; Metaxas, D.N.; Xu, X.; Zhang, S.; Li, K. Predicting 3-D Lower Back Joint Load in Lifting: A Deep Pose Estimation Approach. *IEEE Trans. Hum. Mach. Syst.* **2019**, *49*, 85–94. [CrossRef]
85. Krosshaug, T.; Nakamae, A.; Boden, B.P.; Engebretsen, L.; Smith, G.; Slauterbeck, J.R.; Hewett, T.E.; Bahr, R. Mechanisms of Anterior Cruciate Ligament Injury in Basketball: Video Analysis of 39 Cases. *Am. J. Sports Med.* **2007**, *35*, 359–367. [CrossRef]
86. Olsen, O.-E.; Myklebust, G.; Engebretsen, L.; Bahr, R. Injury Mechanisms for Anterior Cruciate Ligament Injuries in Team Handball: A Systematic Video Analysis. *Am. J. Sports Med.* **2004**, *32*, 1002–1012. [CrossRef] [PubMed]
87. Kim, W.; Sung, J.; Saakes, D.; Huang, C.; Xiong, S. Ergonomic Postural Assessment Using a New Open-Source Human Pose Estimation Technology (OpenPose). *Int. J. Ind. Ergon.* **2021**, *84*, 103164. [CrossRef]
88. Li, Y.; Wang, C.; Cao, Y.; Liu, B.; Tan, J.; Luo, Y. Human Pose Estimation Based In-Home Lower Body Rehabilitation System. In Proceedings of the 2020 International Joint Conference on Neural Networks (IJCNN), Glasgow, UK, 19–24 July 2020; pp. 1–8. [CrossRef]
89. Cordella, F.; Di Corato, F.; Zollo, L.; Siciliano, B. A Robust Hand Pose Estimation Algorithm for Hand Rehabilitation. In *New Trends in Image Analysis and Processing—ICIAP 2013*; Springer: Berlin/Heidelberg, Germany, 2013; Volume 8158, pp. 1–10. [CrossRef]
90. Zhu, Y.; Lu, W.; Gan, W.; Hou, W. A Contactless Method to Measure Real-Time Finger Motion Using Depth-Based Pose Estimation. *Comput. Biol. Med.* **2021**, *131*, 104282. [CrossRef]
91. Milosevic, B.; Leardini, A.; Farella, E. Kinect and Wearable Inertial Sensors for Motor Rehabilitation Programs at Home: State of the Art and an Experimental Comparison. *Biomed. Eng. Online* **2020**, *19*, 25. [CrossRef]
92. Tao, Y.; Hu, H.; Zhou, H. Integration of Vision and Inertial Sensors for 3D Arm Motion Tracking in Home-Based Rehabilitation. *Int. J. Robot. Res.* **2007**, *26*, 607–624. [CrossRef]
93. Ranasinghe, I.; Dantu, R.; Albert, M.V.; Watts, S.; Ocana, R. Cyber-Physiotherapy: Rehabilitation to Training. In Proceedings of the 2021 IFIP/IEEE International Symposium on Integrated Network Management (IM), Bordeaux, France, 17–21 May 2021; pp. 1054–1057.
94. Tao, T.; Yang, X.; Xu, J.; Wang, W.; Zhang, S.; Li, M.; Xu, G. Trajectory Planning of Upper Limb Rehabilitation Robot Based on Human Pose Estimation. In Proceedings of the 2020 17th International Conference on Ubiquitous Robots (UR), Kyoto, Japan, 22–26 June 2020; pp. 333–338. [CrossRef]
95. Palermo, M.; Moccia, S.; Migliorelli, L.; Frontoni, E.; Santos, C.P. Real-Time Human Pose Estimation on a Smart Walker Using Convolutional Neural Networks. *Expert Syst. Appl.* **2021**, *184*, 115498. [CrossRef]
96. Airò Farulla, G.; Pianu, D.; Cempini, M.; Cortese, M.; Russo, L.O.; Indaco, M.; Nerino, R.; Chimienti, A.; Oddo, C.M.; Vitiello, N. Vision-Based Pose Estimation for Robot-Mediated Hand Telerehabilitation. *Sensors* **2016**, *16*, 208. [CrossRef]
97. Sarsfield, J.; Brown, D.; Sherkat, N.; Langensiepen, C.; Lewis, J.; Taheri, M.; McCollin, C.; Barnett, C.; Selwood, L.; Standen, P.; et al. Clinical Assessment of Depth Sensor Based Pose Estimation Algorithms for Technology Supervised Rehabilitation Applications. *Int. J. Med. Inf.* **2019**, *121*, 30–38. [CrossRef] [PubMed]
98. Xu, W.; Chatterjee, A.; Zollhoefer, M.; Rhodin, H.; Mehta, D.; Seidel, H.-P.; Theobalt, C. MonoPerfCap: Human Performance Capture from Monocular Video. *arXiv* **2018**, arXiv:1708.02136.
99. Habermann, M.; Xu, W.; Zollhoefer, M.; Pons-Moll, G.; Theobalt, C. LiveCap: Real-Time Human Performance Capture from Monocular Video. *arXiv* **2019**, arXiv:1810.02648.
100. Wang, J.; Qiu, K.; Peng, H.; Fu, J.; Zhu, J. AI Coach: Deep Human Pose Estimation and Analysis for Personalized Athletic Training Assistance. In Proceedings of the 27th ACM International Conference on Multimedia, Nice, France, 21–25 October 2019; pp. 374–382. [CrossRef]

101. Einfalt, M.; Dampeyrou, C.; Zecha, D.; Lienhart, R. Frame-Level Event Detection in Athletics Videos with Pose-Based Convolutional Sequence Networks. In Proceedings of the 27th ACM International Conference on Multimedia, Nice, France, 21–25 October 2019; pp. 42–50. [CrossRef]
102. Einfalt, M.; Zecha, D.; Lienhart, R. Activity-Conditioned Continuous Human Pose Estimation for Performance Analysis of Athletes Using the Example of Swimming. In Proceedings of the 2018 IEEE Winter Conference on Applications of Computer Vision (WACV), Lake Tahoe, NV, USA, 12–15 March 2018; pp. 446–455. [CrossRef]
103. Güler, R.A.; Neverova, N.; Kokkinos, I. DensePose: Dense Human Pose Estimation in the Wild. *arXiv* **2018**, arXiv:1802.00434.
104. Patacchiola, M.; Cangelosi, A. Head Pose Estimation in the Wild Using Convolutional Neural Networks and Adaptive Gradient Methods. *Pattern Recognit.* **2017**, *71*, 132–143. [CrossRef]
105. Fong, C.-M.; Blackburn, J.T.; Norcross, M.F.; McGrath, M.; Padua, D.A. Ankle-Dorsiflexion Range of Motion and Landing Biomechanics. *J. Athl. Train.* **2011**, *46*, 5–10. [CrossRef]
106. Caccese, J.B.; Buckley, T.A.; Tierney, R.T.; Rose, W.C.; Glutting, J.J.; Kaminski, T.W. Sex and Age Differences in Head Acceleration during Purposeful Soccer Heading. *Res. Sports Med.* **2018**, *26*, 64–74. [CrossRef] [PubMed]
107. Cerveri, P.; Pedotti, A.; Ferrigno, G. Kinematical Models to Reduce the Effect of Skin Artifacts on Marker-Based Human Motion Estimation. *J. Biomech.* **2005**, *38*, 2228–2236. [CrossRef]
108. Joo, H.; Neverova, N.; Vedaldi, A. Exemplar Fine-Tuning for 3D Human Model Fitting Towards In-the-Wild 3D Human Pose Estimation. *arXiv* **2020**, arXiv:200403686.
109. Habibie, I.; Xu, W.; Mehta, D.; Pons-Moll, G.; Theobalt, C. In the Wild Human Pose Estimation Using Explicit 2D Features and Intermediate 3D Representations. In Proceedings of the 2019 IEEE/CVF Conference on Computer Vision and Pattern Recognition (CVPR), Long Beach, CA, USA, 15–20 June 2019; pp. 10897–10906. [CrossRef]
110. Mehta, D.; Rhodin, H.; Casas, D.; Fua, P.; Sotnychenko, O.; Xu, W.; Theobalt, C. Monocular 3D Human Pose Estimation in The Wild Using Improved CNN Supervision. *arXiv* **2017**, arXiv:1611.09813.
111. Mehta, D.; Sridhar, S.; Sotnychenko, O.; Rhodin, H.; Shafiei, M.; Seidel, H.-P.; Xu, W.; Casas, D.; Theobalt, C. VNect: Real-Time 3D Human Pose Estimation with a Single RGB Camera. *ACM Trans. Graph.* **2017**, *36*, 1–14. [CrossRef]
112. Gilbert, A.; Trumble, M.; Malleson, C.; Hilton, A.; Collomosse, J. Fusing Visual and Inertial Sensors with Semantics for 3D Human Pose Estimation. *Int. J. Comput. Vis.* **2019**, *127*, 381–397. [CrossRef]
113. Malleson, C.; Gilbert, A.; Trumble, M.; Collomosse, J.; Hilton, A.; Volino, M. Real-Time Full-Body Motion Capture from Video and IMUs. In Proceedings of the 2017 International Conference on 3D Vision (3DV), Qingdao, China, 10–12 October 2017; pp. 449–457. [CrossRef]
114. WHO. *International Classification of Functioning, Disability and Health: ICF 2001, Title of Beta 2, Full Version: International Classification of Functioning and Disability: ICIDH-2 (WHO Document no. WHO/HSC/ACE/99.2)*; WHO: Geneva, Switzerland, 2001.
115. Fugl-Meyer, A.R.; Jääskö, L.; Leyman, I.; Olsson, S.; Steglind, S. The Post-Stroke Hemiplegic Patient. 1. a Method for Evaluation of Physical Performance. *Scand. J. Rehabil. Med.* **1975**, *7*, 13–31. [PubMed]
116. Yozbatiran, N.; Der-Yeghiaian, L.; Cramer, S.C. A Standardized Approach to Performing the Action Research Arm Test. *Neurorehabil. Neural Repair* **2008**, *22*, 78–90. [CrossRef] [PubMed]
117. Duncan, P.W.; Wallace, D.; Lai, S.M.; Johnson, D.; Embretson, S.; Laster, L.J. The Stroke Impact Scale Version 2.0. *Stroke* **1999**, *30*, 2131–2140. [CrossRef]
118. Li, M.H.; Mestre, T.A.; Fox, S.H.; Taati, B. Automated Assessment of Levodopa-Induced Dyskinesia: Evaluating the Responsiveness of Video-Based Features. *Parkinsonism Relat. Disord.* **2018**, *53*, 42–45. [CrossRef]
119. Li, M.H.; Mestre, T.A.; Fox, S.H.; Taati, B. Automated Vision-Based Analysis of Levodopa-Induced Dyskinesia with Deep Learning. In Proceedings of the 39th Annual International Conference of the IEEE Engineering in Medicine and Biology Society (EMBC), Jeju, Korea, 11–15 July 2017. [CrossRef] [PubMed]
120. Li, M.H.; Mestre, T.A.; Fox, S.H.; Taati, B. Vision-Based Assessment of Parkinsonism and Levodopa-Induced Dyskinesia with Pose Estimation. *J. Neuroeng. Rehabil.* **2018**, *15*, 97. [CrossRef] [PubMed]
121. Liu, Y.; Chen, J.; Hu, C.; Ma, Y.; Ge, D.; Miao, S.; Xue, Y.; Li, L. Vision-Based Method for Automatic Quantification of Parkinsonian Bradykinesia. *IEEE Trans. Neural Syst. Rehabil. Eng.* **2019**, *27*, 1952–1961. [CrossRef] [PubMed]
122. Aung, N.; Bovonsunthonchai, S.; Hiengkaew, V.; Tretriluxana, J.; Rojasavastera, R.; Pheung-Phrarattanatrai, A. Concurrent Validity and Intratester Reliability of the Video-Based System for Measuring Gait Poststroke. *Physiother. Res. Int. J. Res. Clin. Phys. Ther.* **2020**, *25*, e1803. [CrossRef]
123. Shin, J.H.; Yu, R.; Ong, J.N.; Lee, C.Y.; Jeon, S.H.; Park, H.; Kim, H.-J.; Lee, J.; Jeon, B. Quantitative Gait Analysis Using a Pose-Estimation Algorithm with a Single 2D-Video of Parkinson's Disease Patients. *J. Park. Dis.* **2021**, *11*, 1271–1283. [CrossRef]
124. Ng, K.-D.; Mehdizadeh, S.; Iaboni, A.; Mansfield, A.; Flint, A.; Taati, B. Measuring Gait Variables Using Computer Vision to Assess Mobility and Fall Risk in Older Adults With Dementia. *IEEE J. Transl. Eng. Health Med.* **2020**, *8*, 2100609. [CrossRef]
125. Li, T.; Chen, J.; Hu, C.; Ma, Y.; Wu, Z.; Wan, W.; Huang, Y.; Jia, F.; Gong, C.; Wan, S.; et al. Automatic Timed Up-and-Go Sub-Task Segmentation for Parkinson's Disease Patients Using Video-Based Activity Classification. *IEEE Trans. Neural Syst. Rehabil. Eng.* **2018**, *26*, 2189–2199. [CrossRef]

sensors

MDPI

Article

Combined Regularized Discriminant Analysis and Swarm Intelligence Techniques for Gait Recognition

Tomasz Krzeszowski * and Krzysztof Wiktorowicz

Faculty of Electrical and Computer Engineering, Rzeszow University of Technology, al. Powstancow Warszawy 12, 35-959 Rzeszow, Poland; kwiktor@prz.edu.pl

* Correspondence: tkrzeszo@prz.edu.pl

Received: 19 October 2020; Accepted: 26 November 2020; Published: 27 November 2020

Abstract: In the gait recognition problem, most studies are devoted to developing gait descriptors rather than introducing new classification methods. This paper proposes hybrid methods that combine regularized discriminant analysis (RDA) and swarm intelligence techniques for gait recognition. The purpose of this study is to develop strategies that will achieve better gait recognition results than those achieved by classical classification methods. In our approach, particle swarm optimization (PSO), grey wolf optimization (GWO), and whale optimization algorithm (WOA) are used. These techniques tune the observation weights and hyperparameters of the RDA method to minimize the objective function. The experiments conducted on the GPJATK dataset proved the validity of the proposed concept.

Keywords: gait recognition; biometrics; regularized discriminant analysis; particle swarm optimization; grey wolf optimization; whale optimization algorithm

1. Introduction

Biometric authentication (also known as biometrics) refers to identifying or verifying individuals based on their biological or behavioral traits [1]. There are many different biometric traits among which can be distinguished the face, iris, fingerprint, palm print, voice, signature, or gait. Typically, gait is a manifestation of an individual's walking style; hence, its recognition means identifying a person by his/her way of walking. The major advantages of gait are: noninvasive, can be captured at a distance, hard to conceal, and non-cooperative. These advantages make it an ideal trait for visual surveillance systems [2]. However, the recognition performance of existing methods is limited by the influence of a large number of covariate factors affecting both appearance and dynamics of the gait, e.g., variations in footwear and clothing, viewpoint variations, changes in the characteristics of the surface on which movement occurs, various carrying conditions, injuries affecting movement, and so on. These are the reasons why gait recognition has been extensively studied in recent years.

Gait recognition methods [3] can be categorized as model-free (appearance-based) [2,4–12] and model-based [13–22]. Most gait recognition studies are based on model-free approaches that employ the whole motion pattern of the human body. Several techniques were proposed to characterize this motion pattern, such as the gait energy image (GEI) [2,5,7,10], which is a spatio-temporal gait representation, GEI region bounded by legs (RBL) [8], human body contours [9], and dense optical flow field [4]. These methods are strongly based on silhouette extraction and therefore are not resistant to changing clothes or carrying luggage. It is also worth noting that most of them can only achieve correct results from a specific point of view, usually side view [4,5,7–9,16]. The recent research on model-free methods has focused on eliminating these drawbacks [6,10–12]. Model-based methods infer gait signature directly by modeling the underlying kinematics of human motion. The methods of this approach initially focused on using only static body parameters for recognition, such as stride length, which were updated over time [13,14]. Yam et al. [14] have extended this concept by analyzing the movement of the legs and the angles between them. In Ref. [15], the authors proposed a method that uses a motion-based model and elliptic Fourier descriptors to extract the key features of gait. Deng et al. [18] proposed a method that combines spatio-temporal and kinematic gait features. The fusion of two different features gives a comprehensive characterization of gait dynamics, which is less sensitive to walking conditions. In [21], the authors presented a gait recognition method that uses a 3D model of the human body and particle swarm optimization to obtain gait features. The number of obtained features was reduced using the multilinear principal component analysis (MPCA).

In the recognition process, various classification techniques are used; most often, they are classical methods such as k-nearest neighbors (kNN) [2,4,10,14–17,19,23], multilayer perceptron (MLP) [21], support vector machine [9,16,24] linear discriminant analysis (LDA) [16], and radial basis function neural networks [18]. The main focus in these papers is on developing descriptors that better describe gait features, rather than introducing new classification methods with better recognition ability. This is a traditional approach that is based on a clear separation between the descriptors and classifier model. On the other hand, in the papers of recent years, the introduction of new classification methods such as deep learning [11,20,22] or hybrid methods [12], in which the description and classification steps cannot be easily distinguished, is increasingly visible. In [20], the authors utilized a 3D convolutional neural network (CNN) and long short-term memory neural networks for training the classification models. They then used a grey wolf optimizer to tune the fusion parameters of each modality to boost the recognition performance of the system. Chao et al. [11] proposed a deep learning model called GaitSet. In this method, the CNN is used to extract frame-level features from each silhouette independently. Next, an operation called set pooling is used to aggregate frame-level features into a single set-level feature. In the end, a structure called horizontal pyramid mapping is used to map the set-level feature into a more discriminative space to obtain the final representation. The proposed method can extract spatial and temporal information more effectively than other methods regarding gait as a template or sequence. In turn, in the paper [12], the authors used a hybrid approach and combine the improved local coupled extreme learning machine and PSO for the classification process. A Gabor filter was used to extract gait features from the GEI and linear discriminant analysis was used to dimensionality reduction.

From the literature review, it is seen that most studies considering the traditional approach are devoted to developing gait descriptors, rather than introducing new classification methods. In this paper, we propose hybrid methods that combine the RDA and swarm intelligence techniques for gait recognition. In our approach, the PSO, GWO, and WOA are used to tune the observation weights and hyperparameters of the RDA model. To the best of our knowledge, the GWO and WOA algorithms have not been used before for this purpose. In the learning process, the confusion value is used as an objective function. The proposed methods are tested on a database of 414 gait cycles belonging to 32 different persons [21]. Summarizing, the main contributions of this paper can be stated as:

- proposing a combination of regularized discriminant analysis and particle swarm optimization for gait recognition,
- proposing a combination of regularized discriminant analysis and grey wolf optimization,
- proposing a combination of regularized discriminant analysis and whale optimization algorithm,
- comparing and improving the results obtained in the paper [21].

The structure of this article is as follows: Section 2 contains the description of the dataset and methods used in the recognition process. In particular, the structure of the gait recognition system, building classification models, and swarm intelligence methods are described. The experimental results are presented in Section 3. The results obtained by eight methods are presented, of which three were proposed by the authors. Section 4 contains the discussion of the achieved results. Finally, the conclusions are given in Section 5.

2. Material and Methods

2.1. Gait Dataset

The publicly available gait dataset (GPJATK) was used in the experiments [21]. The dataset consists of 166 data sequences (414 gait cycles) representing 32 people (10 women and 22 men). The sequences are divided into three subsets: 128 sequences (325 gait cycles) in which each of 32 individuals was dressed in his/her clothes; 24 sequences (58 gait cycles) in which 6 of 32 individuals (person #26–#31) changed clothes; and 14 sequences (31 gait cycles) in which 7 of the individuals (person #26–#32) had a backpack on his/her back. Each sequence contains video data (960x540@25fps) recorded using four calibrated and synchronized cameras and data from a markerless and marker-based motion capture systems. The synchronization between videos and motion capture data has been realized using Vicon MX Giganet. Our research is based on data obtained by a markerless motion capture system [25], which uses the annealed PSO algorithm in the motion capture process and data from four synchronized and calibrated cameras.

2.2. Gait Recognition System

A typical model-based system for gait recognition is presented in Figure 1. Such a system consists of gait capture, a feature extraction module, and a classifier. The objective of the system is to determine the identity of a gait sample using a database consisting of gait patterns from a set of known subjects. In the first step, one or more video-cameras are used to register the user's image in a scene. In a preprocessing phase, image processing, i.e., background subtraction, body silhouette extraction, and edges extraction, is performed. The kinematic model of human motion is used in the next step to extract gait features that will define a gait signature. In the used gait dataset [21], each gait cycle is treated as a data sample represented by a third-order tensor with the dimension $32 \times 11 \times 3$. The first dimension, equal to 32, is the average time of the gait cycle. The motion data was filtered using a moving average of length nine samples to the original data. The second dimension of the tensor is equal to the number of bones (excluding pelvis), i.e., 10 plus one element for storing a person height and distance between ankles. The third dimension relates to three angles, except the 11th vector that contains a person's height, distance between ankles, and value of zero to maintain alignment with the rest of the vectors. Such a gait signature is then reduced using the MPCA algorithm [26]. The last element of the system is the classifier block that we focus in this article.

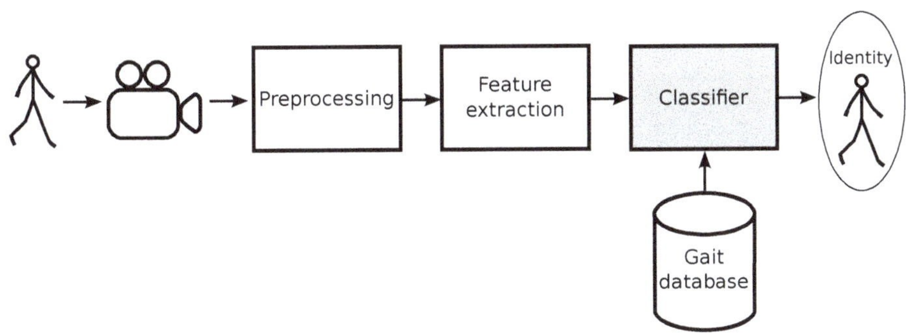

Figure 1. Structure of a gait recognition system.

2.3. Building Classification Model for Gait Recognition

Building a gait classification model involves two main stages, which include training the model and testing it. For this purpose, the gait sequence database is divided into three sets: the training set, validation set, and test set (Figure 2). These three datasets are commonly used in different stages of the model building. Separating the dataset into these three subsets is used to avoid overfitting of the model. Initially, the model is fit on the training set, which is a set of observations used to fit the parameters of the RDA model. It should be noted that, in [21], only two sets were defined: training and testing. However, in the proposed approach, due to the optimization of the classifier parameters, an additional validation set is separated from the training data. The validation dataset is used for an evaluation of the fitted model while training the model's hyperparameters. After building the model, the test set is used for testing, that is, for predicting the classifier's output for data that has never been used in the training phase. Model training is carried out using one of the hybrid methods, in which swarm intelligence techniques optimize the observation weights and hyperparameters of the RDA. This problem is well suited for swarm optimization techniques because it creates a large search space to be explored. This search space is determined by the number of observation weights, the number of hyperparameters, and by the fact that all these variables are real values in the specified intervals. During the optimization, the following objective function, expressed as the confusion value, is used:

$$\text{objective function (confusion)} = \frac{\text{number of samples misclassified}}{\text{number of all samples}} \quad (1)$$

where the samples are taken from the validation set. This objective function is minimized using one of the swarm intelligence methods, i.e., particle swarm optimization, grey wolf optimization, or whale optimization algorithm.

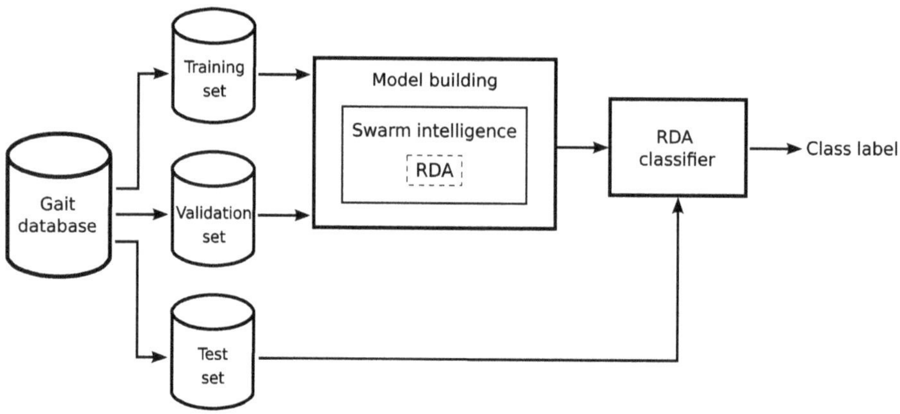

Figure 2. The idea of building the RDA classification model.

After building the RDA classifier, it is used to determine the correct classified ratio (CCR). The CCR is a ratio of correctly classified samples to the total number of samples in the test subset.

2.4. Regularized Discriminant Analysis

Linear discriminant analysis was developed by Sir Ronald Fisher in 1936 [27]. The original method proposed by Fisher was described for a 2-class problem, and it was in 1948 generalized as multi-class problems by Rao [28]. The LDA is a transformation technique used in statistics and machine learning to find linear combinations of features that separate classes of objects. The combinations obtained by this method may be used as:

- dimensionality reduction and feature extraction before classification,
- a linear classifier (considered in this paper).

The LDA consists of statistical properties of data calculated for each class. For a single variable, these are the mean and the variance of the variable. For multiple variables, these are the means and the covariance matrix. These statistical properties are estimated from the data and used to formulate an equation for making predictions. It should be emphasized that the use of the LDA is not associated with problems when the number of observations is greater than the dimension of each observation. Problems arise when the opposite is true, which makes the covariance matrix singular and cannot be inverted. To resolve this problem, instead of using the covariance matrix directly, a regularization of this matrix is used. This approach is applied to the regularized discriminant analysis method, in which the regularized covariance matrix $\widehat{\mathbf{\Sigma}}_\gamma$ is given by [29]:

$$\widehat{\mathbf{\Sigma}}_\gamma = (1-\gamma)\widehat{\mathbf{\Sigma}} + \gamma \mathbf{I} \tag{2}$$

where $\widehat{\mathbf{\Sigma}}$ is the covariance matrix, $\mathbf{I}$ is the identity matrix, and $\gamma \in [0,1]$ is the amount of regularization. The RDA introduces regularization into the covariance matrix estimate, enabling a solution to be obtained and allowing different influences of variables on the classification model. In addition to the parameter γ, the RDA model uses the parameter δ that acts as a threshold: if a model coefficient has the magnitude smaller than δ, the RDA sets this coefficient to zero, and the corresponding predictor can be eliminated from the model.

The output of the RDA classifier $\widehat{y}$ is calculated so as to minimize the classification cost [30]:

$$\widehat{y} = \underset{y=1,\ldots,K}{\arg\min} \sum_{k=1}^{K} \widehat{P}(x|k)C(y|k) \tag{3}$$

where K is the number of classes, $\widehat{P}(x|k)$ is the posterior probability of class k for observation x, $C(y|k)$ is the cost of classifying an observation as y when its true class is k. The RDA used in this paper constructs weighted classifiers using the following scheme. Suppose $\mathbf{M}$ is an N-by-K class membership matrix such that $M_{nk} = 1$ if observation n is from class k, $M_{nk} = 0$, otherwise. The estimate of the class mean for weighted data with positive weights w_n is [30]

$$\widehat{\mu}_k = \frac{\sum_{n=1}^{N} M_{nk} w_n x_n}{\sum_{n=1}^{N} M_{nk} w_n} \tag{4}$$

The estimate of the covariance matrix is

$$\widehat{\Sigma} = \frac{\sum_{n=1}^{N} \sum_{k=1}^{K} M_{nk} w_n (x_n - \widehat{\mu}_k)(x_n - \widehat{\mu}_k)^T}{1 - \sum_{k=1}^{K} \frac{W_k^{(2)}}{W_k}} \tag{5}$$

where $W_k = \sum_{n=1}^{N} M_{nk} w_n$ is the sum of the weights for class k, and $W_k^{(2)} = \sum_{n=1}^{N} M_{nk} w_n^2$ is the sum of squared weights per class.

2.5. Particle Swarm Optimization

A particle swarm optimization algorithm was developed by Kennedy and Eberhart [31]. This algorithm is based on the social behavior of organisms living in large groups. In the PSO, a group of agents called particles forms a swarm, where each particle represents a point in a multidimensional space. The particles explore this space in order to find the optimal solution. Each particle in the swarm is attracted both to its best position and the best position found by other particles. The best solution is obtained by minimizing the objective function.

Each particle has its position (**x**) and velocity (**v**). The velocity $\mathbf{v}_k$ of the kth particle is determined using the following equation [31]:

$$\mathbf{v}_k(t+1) = \omega\mathbf{v}_k(t) + c_1\mathbf{r}_1(\mathbf{pbest}_k(t) - \mathbf{x}_k(t)) + c_2\mathbf{r}_2(\mathbf{gbest}(t) - \mathbf{x}_k(t)) \tag{6}$$

where t is the current iteration number, ω is the inertia weight, $\mathbf{r}_1$, $\mathbf{r}_2$ are vectors of random numbers in the range [0,1], c_1 is the cognitive coefficient, and c_2 is the social coefficient. It is seen that the update of the velocity is a weighted sum of the previous velocity $\mathbf{v}_k(t)$, the difference between the current position and the personal best position (**pbest**), and the difference between the current position and the global best position (**gbest**). The position $\mathbf{x}_k$ of the kth particle is updated according to the equation

$$\mathbf{x}_k(t+1) = \mathbf{x}_k(t) + \mathbf{v}_k(t) \tag{7}$$

After updating the velocity and the position, the objective function is calculated to determine the personal and global positions.

2.6. Grey Wolf Optimization

A grey wolf optimizer is another swarm intelligence technique used to solve optimization problems [32]. The GWO algorithm is inspired by the behavior and hierarchy of grey wolves in nature, searching for the optimal way to attack their prey. In the hierarchy of grey wolves, the most dominating is alpha (α), which leads the entire group. The other wolves are beta (β) and delta (δ), which help to control the rest of the wolves considered as omega (ω). The omega wolves have the lowest ranking in the hierarchy. The main phases of grey wolf hunting are: (a) tracking, chasing, and approaching; (b) chasing, encircling, and harassing; (c) attacking.

2.6.1. Encircling Prey

Grey wolves encircle the prey during the hunt, which can be mathematically modeled by the following equation [32]:

$$\mathbf{X}(t+1) = \mathbf{X}_p(t) - \mathbf{A} \cdot \mathbf{D} \tag{8}$$

where

$$\mathbf{D} = |\mathbf{C} \cdot \mathbf{X}_p(t) - \mathbf{X}(t)| \tag{9}$$

and $\mathbf{X}(t)$ is the current position of a grey wolf at iteration t, $\mathbf{X}_p(t)$ is the position of the prey, and "$\cdot$" is an element-by-element multiplication. The coefficient vectors **A**, **C** are determined as follows:

$$\mathbf{A} = 2\mathbf{a} \cdot \mathbf{r}_1 - \mathbf{a} \tag{10}$$

$$\mathbf{C} = 2\mathbf{r}_2 \tag{11}$$

where components of **a** are linearly decreased from 2 to 0 through iterations and $\mathbf{r}_1$, $\mathbf{r}_2$ are random vectors in $[0, 1]$.

2.6.2. Hunting

In the GWO, we assume that the α, β, and δ are the best solutions for the entire population. Therefore, the other wolves should update their position according to the positions of the three agents. The following formula is used to calculate the positions of search agents [32]:

$$\mathbf{X}(t+1) = \frac{1}{3}(\mathbf{X}_1(t) + \mathbf{X}_2(t) + \mathbf{X}_3(t)) \tag{12}$$

where

$$\mathbf{X}_1 = \mathbf{X}_\alpha(t) - \mathbf{A}_1 \cdot \mathbf{D}_\alpha \tag{13}$$
$$\mathbf{X}_2 = \mathbf{X}_\beta(t) - \mathbf{A}_2 \cdot \mathbf{D}_\beta \tag{14}$$
$$\mathbf{X}_3 = \mathbf{X}_\delta(t) - \mathbf{A}_3 \cdot \mathbf{D}_\delta \tag{15}$$

and

$$\mathbf{D}_\alpha = |\mathbf{C}_1 \cdot \mathbf{X}_\alpha - \mathbf{X}| \tag{16}$$
$$\mathbf{D}_\beta = |\mathbf{C}_2 \cdot \mathbf{X}_\beta - \mathbf{X}| \tag{17}$$
$$\mathbf{D}_\delta = |\mathbf{C}_3 \cdot \mathbf{X}_\delta - \mathbf{X}| \tag{18}$$

The vectors $\mathbf{A}_1$, $\mathbf{A}_2$, $\mathbf{A}_3$ are obtained using Equation (10), while $\mathbf{C}_1$, $\mathbf{C}_2$, $\mathbf{C}_3$ are obtained using Equation (11).

2.6.3. Attacking Prey (Exploitation) and Search for Prey (Exploration)

The grey wolves start the attack, once the prey stops moving. To model the process of approaching the prey, the GWO linearly decrease all the values of **a** from 2 to 0 according to the equation [32]

$$\mathbf{a} = 2 - \frac{2t}{T_{max}} \tag{19}$$

where T_{max} is the total number of iterations of the algorithm. The change of **a** affects the coefficient vector **A**, which controls the behavior of search agents. If $|\mathbf{A}| < 1$, the wolf will move towards the prey; on the other hand, if $|\mathbf{A}| > 1$, the wolf will diverge from the prey in order to find new better prey. In addition, the vector **C** which contains a random value in the range [0,2] is employed to help the algorithm to avoid being trapped in the local optima.

2.7. Whale Optimization Algorithm

The whale optimization algorithm is a nature-inspired metaheuristic technique for solving optimization problems [33]. This algorithm mimics the social behavior of humpback whales realized in the bubble-net hunting strategy. The WOA is based on three operators to simulate the search for prey, encircling prey, and bubble-net foraging.

2.7.1. Encircling Prey

Humpback whales encircle pray after recognizing its position. In this phase, the search agents attempt to change their locations towards the best search agents. This behavior is expressed by the following formula [33]:

$$\mathbf{X}(t+1) = \mathbf{X}^*(t) - \mathbf{A} \cdot \mathbf{D} \tag{20}$$

where

$$\mathbf{D} = |\mathbf{C} \cdot \mathbf{X}^*(t) - \mathbf{X}(t)| \tag{21}$$

and $\mathbf{X}(t)$ is the current position vector at iteration t, $\mathbf{X}^*(t)$ is the best position obtained so far, "$\cdot$" is an element-by-element multiplication. The coefficient $\mathbf{A}$, $\mathbf{C}$ are determined from the formulas:

$$\mathbf{A} = 2\mathbf{a} \cdot \mathbf{r}_1 - \mathbf{a} \tag{22}$$

$$\mathbf{C} = 2\mathbf{r}_2 \tag{23}$$

where components of $\mathbf{a}$ are linearly decreased from 2 to 0 through iterations and $\mathbf{r}_1$, $\mathbf{r}_2$ are random vectors in $[0, 1]$.

2.7.2. Bubble-Net Attacking (Exploitation Phase)

In this phase, humpback whales swim around the prey within a helix-shaped path. To model this behavior, it is assumed that there is a 50% chance to choose between the shrinking encircling or spiral movements [33]:

$$\mathbf{X}(t+1) = \begin{cases} \mathbf{X}^*(t) - \mathbf{A} \cdot \mathbf{D} & \text{if } p < 0.5 \\ \mathbf{D}^* \cdot \exp(bk) \cdot \cos(2\pi k) + \mathbf{X}^*(t) & \text{if } p \geq 0.5 \end{cases} \tag{24}$$

where $\mathbf{D}^* = |\mathbf{X}^*(t) - \mathbf{X}(t)|$ is the distance of the ith whale to the prey, b defines the shape of the spiral, k is a random number in $[-1, 1]$, and p is a random number in $[0, 1]$.

2.7.3. Search for Prey (Exploration Phase)

In this phase, humpback whales search the pray according to the position of each other. The location of a search agent is calculated according to randomly selected search agent rather than the best search agent as in the exploitation phase. The mathematical model is determined as follows [33]:

$$\mathbf{X}(t+1) = \mathbf{X}_{rand}(t) - \mathbf{A} \cdot \mathbf{D} \tag{25}$$

where

$$\mathbf{D} = |\mathbf{C} \cdot \mathbf{X}_{rand}(t) - \mathbf{X}(t)| \tag{26}$$

and $\mathbf{X}_{rand}$ is the random position vector chosen from the current population.

2.8. Integration of Swarm Intelligence Techniques with Regularized Discriminant Analysis

The idea of integrating swarm intelligence methods with the RDA classifier is shown in Figure 3. This figure presents in the form of a block diagram main stages of optimization of the RDA model using swarm algorithms and the method of determining the objective function. The task of particle swarm optimization, grey wolf optimization, or whale optimization algorithm is to select the RDA parameters, which are [30]:

- $w_1, w_2, \ldots, w_n$ — the observation weights,
- δ — the linear coefficient threshold,
- γ — the parameter for regularizing the covariance matrix of the predictors,

where n is the number of observations in the training set. For this purpose, the observation weights and two hyperparameters δ, γ are placed as elements of the agent (candidate) vector, which has the form

$$\overline{|\ w_1 \ |\ w_2\ |\ \dots\ |\ w_n\ |\ \delta\ |\ \gamma\ |} \tag{27}$$

At the beginning of the algorithm, the agents are initialized. In the next step, the value of the objective function for all agents is calculated. Then, the stop condition is checked, if it is not reached, the agents are updated and the value of the objective function is recalculated. The swarm optimization algorithm generates many hypothetical solutions (which are represented by agents) and the best solution is selected in the optimization process. These operations are repeated until the stop condition is reached. In the objective function block, the RDA model is determined for the parameters proposed by the agent and for the training data. The output of this model is then calculated on the validation data and the value of the objective function is determined based on formula (1). The weights $w_1, w_2, \dots, w_n$, hyperparameters δ and γ of the RDA classifier are limited during optimization in given ranges (see Section 3). The algorithm returns the optimal parameters of the RDA classifier in the best agent vector. This result is included in the block 'Return the global best solution' in Figure 3. On this basis, the objective function value for the test data are calculated.

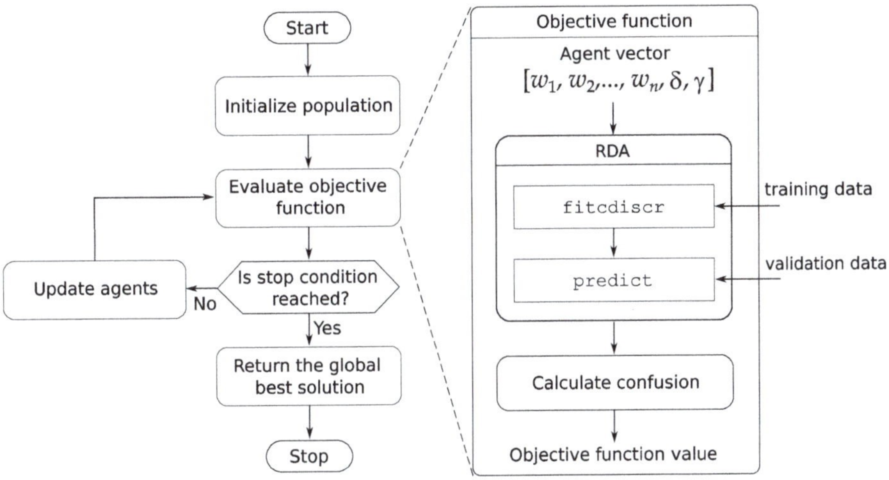

Figure 3. The idea of integration of swarm intelligence techniques with the RDA.

All the proposed hybrid methods have been implemented in Matlab equipped with additional toolboxes. Using the function `fitcdiscr` from the Statistics and Machine Learning Toolbox [30], the RDA classifier model is created, while the function `predict` from the same package is used to determine class predictions (Figure 3). The PSO method has been implemented using the function `particleswarm` from the Global Optimization Toolbox [34] and the GWO and WOA methods using software developed by Mirjalili [32,33].

3. Results

The division of the gait dataset into training, validation, and test sets in four experiments is presented in Table 1. In the first experiment (Set #1) and the fourth experiment (Set #4), all persons in the collection were wearing clothes number 1. In the second experiment (Set #2), persons in the training and validation sets were in clothing number 1, while in the test set in clothing number 2. In the third experiment (Set #3), persons in the training and validation sets were in clothing number 1, while in the test set they wore a backpack. The number of identities in training/validation/test subsets was: in Set #1—32/32/32, Set #2—32/32/6, Set #3—32/32/7, and Set #4—32/32/32. The samples were not repeated in the subsets. The procedure of separating the samples from the training set to the validation set was as follows: for Sets #1 and #4, one sample was taken for each class (the last sample was always taken), for Sets #2 and #3, two samples were taken for each class because there were more training data (the last two samples for each class were taken). Experiments 1–3 were performed in such a way that the classification model and its testing error were determined 10 times, and then the average of the results was calculated. In the fourth experiment (Set #4), the 10-fold cross-validation was performed to obtain an average score. In this method, the original dataset is partitioned into 10 equal size subsets. A single subset is retained as the test data, and the remaining nine subsets are used as training data. The cross-validation process is repeated 10 times (the folds) for the test data, and the 10 results are averaged. In the RDA-PSO, RDA-GWO, and RDA-WOA methods, the number of agents was equal to 30, and the number of iterations was equal to 25. These parameters of optimization techniques were selected experimentally. The weights of the observations were limited to the range $[10^{-8}, 1]$, while the hyperparameters δ and γ were limited to the range $[0, 1]$.

Table 1. Division of the gait dataset into training, validation, and test sets.

Experiment	Subset	Classical Methods	Hybrid Methods
1: Set #1	train	169	137
	validation	–	32
	test	156	156
2: Set #2	train	325	261
	validation	–	64
	test	58	58
3: Set #3	train	325	261
	validation	–	64
	test	31	31
4: Set #4	train	90% (≈293)	80% (≈261)
	validation	–	10% (≈32)
	test	10% (≈32)	10% (≈32)

Table 2 contains the correct classified ratio of the gait recognition for the considered methods. These are four classical methods (kNN [35], NB [35], support vector machines with sequential minimal optimization (SMO) [36], and MLP [35]) taken from [21], linear discriminant analysis (non-regularized) [27], and three proposed hybrid methods. Figure 4 presents the confusion matrices for the best models in the considered experiments. These matrices show the percentage of correct class recognition by the arrangement of the colored elements. The closer the color is to dark red on the diagonal, the more accurate the class recognition. The colors changing from dark red to white mean worse and worse recognition.

Table 2. Correct classified ratio [%].

Experiment	kNN [21]	NB [21]	SMO [21]	MLP [21]	LDA	RDA-PSO	RDA-GWO	RDA-WOA
1: Set #1	47.44	55.77	67.95	80.13	45.51	**87.05**	86.28	86.92
2: Set #2	37.93	56.90	63.79	75.86	79.31	**85.34**	84.48	84.48
3: Set #3	38.71	70.97	67.74	77.42	**93.55**	88.39	**93.55**	**93.55**
4: Set #4	56.92	79.69	84.31	89.85	91.99	95.07	**95.39**	95.09

The best result is marked in bold font.

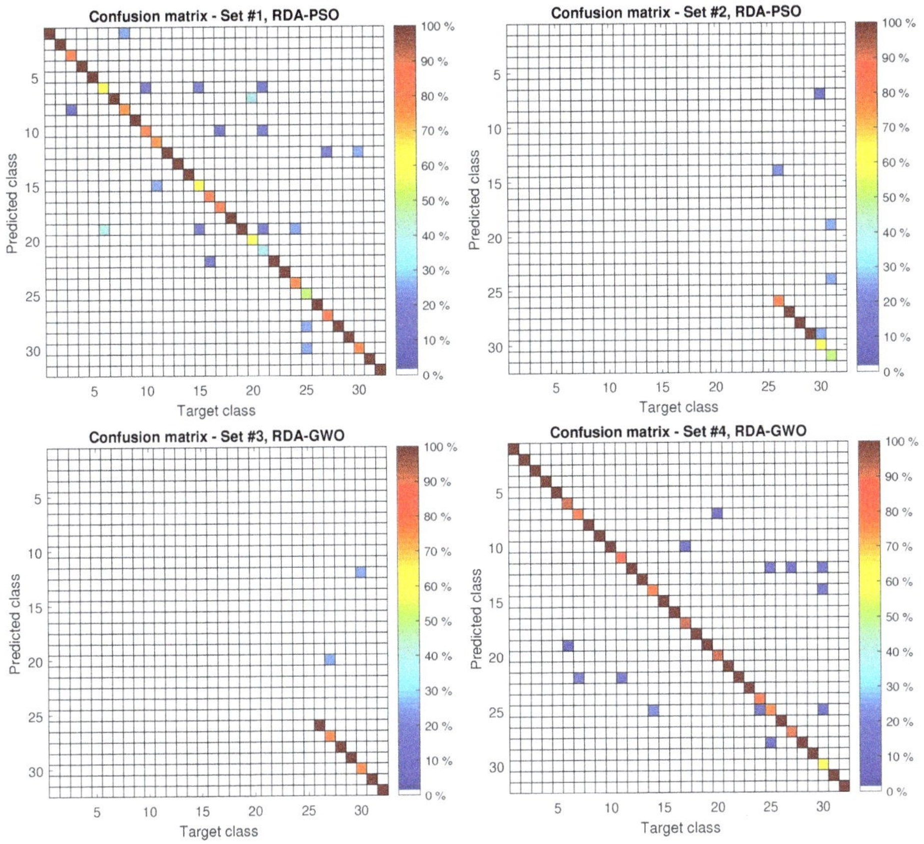

Figure 4. Confusion matrices for best models in each experiment.

4. Discussion

To compare the results, the classical methods taken from the paper [21] and the LDA were considered. This comparison with the methods proposed by the authors (RDA-PSO, RDA-GWO, RDA-WOA) is provided in Table 2. It can be seen that, in Experiment 1 and Experiment 2, the RDA-PSO method proposed by the authors has the highest CCR index. In Experiment 3, which considered clothing with a backpack, three of the analyzed methods (LDA, RDA-GWO, RDA-WOA) achieved the same CCR at the level of 94%. In the final fourth experiment, the method proposed by the authors was again the best, in this case combining RDA with GWO. It should also be noted that the LDA obtained the worst result of all methods for Experiment 1; this is most likely due to insufficient training data. However, the proposed methods obtained very good results for this experiment, at the level of about 86–87%. The use of the regularized discriminant analysis combined with swarm intelligence techniques improved the quality of gait recognition. The developed methods outperformed the classical methods and improved the recognition results achieved by the best of them (MLP) by 6 to 16%, depending on the experiment. When comparing the proposed hybrid methods between each other, it should be noted that the results are inconclusive and it seems that these methods are equivalent in this application.

When analyzing the confusion matrix shown in Figure 4, it can be seen that most people are recognized with high accuracy, but there are classes with which the methods have difficulties. For example, in Experiment 1 (RDA-PSO method) for classes 6, 15, 20, 21, and 25, the recognition efficiency drops to 40–60%. It is most likely caused by too little training data (about five gait cycles for each class). When 10-fold cross-validation is used (about nine gait cycles for each class), the recognition efficiency increases significantly. It should also be noted that the data used in experiments were recorded with a markerless motion capture system, which is not perfect and generates noise [21]. It certainly has an impact on the achieved results. For Experiment 2, for which learning and testing were performed on the sequences for which the clothes were changed, a significant deterioration in the results for persons #30 and #31 can be observed. When analyzing these video sequences, it can be noticed that the heavy shoes changed to sandals. This could be the cause of the deterioration in the results in this experiment. On the other hand, for person #29, there was a change of footwear from sports shoes to shoes with a heel, and still 100% detection rate was achieved.

5. Conclusions

The hybrid methods that combine regularized discriminant analysis and swarm intelligence techniques for gait recognition have been proposed. In the presented approach, particle swarm optimization, grey wolf optimization, and whale optimization algorithm are used. These techniques optimize the observation weights and hyperparameters of the regularized discriminant analysis. The proposed methods were compared with five methods found in the literature. In the learning process, the confusion value was used as an objective function. The conducted experiments on the GPJATK dataset proved the validity of the proposed concept. Future work will focus on improving the proposed concept by replacing the MPCA method with another method of dimensionality reduction. Moreover, some work will be carried out to add new features to gait signatures.

Author Contributions: Conceptualization, T.K. and K.W.; methodology, T.K. and K.W.; software, T.K. and K.W.; validation, K.W.; formal analysis, T.K. and K.W.; investigation, T.K. and K.W.; resources, T.K.; writing—original draft preparation, T.K. and K.W.; writing—review and editing, T.K. and K.W.; visualization, T.K. and K.W. All authors have read and agreed to the published version of the manuscript.

Funding: This research received no external funding.

Conflicts of Interest: The authors declare no conflict of interest.

References

1. Jain, A.K.; Flynn, P.; Ross, A.A. *Handbook of Biometrics*; Springer US: Boston, MA, USA, 2008; [CrossRef]
2. Matovski, D.S.; Nixon, M.S.; Mahmoodi, S.; Carter, J.N. The Effect of Time on Gait Recognition Performance. *IEEE Trans. Inf. Forensics Secur.* **2012**, *7*, 543–552, [CrossRef]
3. Wan, C.; Wang, L.; Phoha, V.V. A survey on gait recognition. *ACM Comput. Surv.* **2018**, *51*, [CrossRef]
4. Little, J.J.; Boyd, J.E. Recognizing People by Their Gait : The Shape of Motion. *J. Comput. Vis. Res.* **1998**, *1*, 1–33.
5. Han, J.; Bhanu, B. Individual recognition using gait energy image. *IEEE Trans. Pattern Anal. Mach. Intell.* **2006**, *28*, 316–322. [CrossRef]
6. Kusakunniran, W.; Wu, Q.; Zhang, J.; Li, H. Support vector regression for multi-view gait recognition based on local motion feature selection. In Proceedings of the IEEE Computer Society Conference on Computer Vision and Pattern Recognition, San Francisco, CA, USA, 13–18 June 2010; pp. 974–981, [CrossRef]
7. Li, X.; Chen, Y. Gait recognition based on structural Gait energy image. *J. Comput. Inf. Syst.* **2013**, *9*, 121–126.
8. Mohan Kumar, H.P.; Nagendraswamy, H.S. LBP for gait recognition: A symbolic approach based on GEI plus RBL of GEI. In Proceedings of the 2014 International Conference on Electronics and Communication Systems (ICECS), Prague, Czech Republic, 2–4 April 2014; pp. 1–5.
9. Wang, H.; Fan, Y.; Fang, B.; Dai, S. Generalized linear discriminant analysis based on euclidean norm for gait recognition. *Int. J. Mach. Learn. Cybern.* **2018**, [CrossRef]
10. Lishani, A.O.; Boubchir, L.; Khalifa, E.; Bouridane, A. Human gait recognition using GEI-based local multi-scale feature descriptors. *Multimed. Tools Appl.* **2019**, *78*, 5715–5730, [CrossRef]

11. Chao, H.; He, Y.; Zhang, J.; Feng, J. GaitSet: Regarding gait as a set for cross-view gait recognition. In Proceedings of the 33rd AAAI Conference on Artificial Intelligence, Hilton Hawaiian Village, Honolulu, HI, USA, 27 January–1 February 2019; pp. 8126–8133, [CrossRef]
12. Guo, H.; Li, B.; Zhang, Y.; Zhang, Y.; Li, W.; Qiao, F.; Rong, X.; Zhou, S. Gait Recognition Based on the Feature Extraction of Gabor Filter and Linear Discriminant Analysis and Improved Local Coupled Extreme Learning Machine. *Math. Probl. Eng.* **2020**, *2020*, [CrossRef]
13. BenAbdelkader, C.; Cutler, R.; Davis, L. Stride and cadence as a biometric in automatic person identification and verification. In Proceedings of the 5th IEEE International Conference on Automatic Face Gesture Recognition, FGR 2002, Washington, DC, USA, 21–21 May 2002; pp. 372–377, [CrossRef]
14. Yam, C.Y.Y.; Nixon, M.S.; Carter, J.N. Automated person recognition by walking and running via model-based approaches. *Pattern Recognit.* **2004**, *37*, 1057–1072, [CrossRef]
15. Bouchrika, I.; Nixon, M.S. Model-based feature extraction for gait analysis and recognition. *Lect. Notes Comput. Sci.* **2007**, *4418 LNCS*, 150–160, [CrossRef]
16. Ng, H.; Ton, H.L.; Tan, W.H.; Yap, T.T.V.; Chong, P.F.; Abdullah, J. Human Identification Based on Extracted Gait Features. *Int. J. New Comput. Archit. Their Appl.* **2011**, *1*, 358–370.
17. Ariyanto, G.; Nixon, M.S. Model-based 3D gait biometrics. In Proceedings of the International Joint Conference on Biometrics (IJCB), Washington, DC, USA, 11–13 October 2011; pp. 1–7, [CrossRef]
18. Deng, M.; Wang, C.; Cheng, F.; Zeng, W. Fusion of spatial-temporal and kinematic features for gait recognition with deterministic learning. *Pattern Recognit.* **2017**, *67*, 186–200, [CrossRef]
19. Switonski, A.; Krzeszowski, T.; Josinski, H.; Kwolek, B.; Wojciechowski, K. Gait recognition on the basis of markerless motion tracking and DTW transform. *IET Biom.* **2018**, *7*, 415–422, [CrossRef]
20. Kumar, P.; Mukherjee, S.; Saini, R.; Kaushik, P.; Roy, P.P.; Dogra, D.P. Multimodal Gait Recognition with Inertial Sensor Data and Video Using Evolutionary Algorithm. *IEEE Trans. Fuzzy Syst.* **2019**, *27*, 956–965, [CrossRef]
21. Kwolek, B.; Michalczuk, A.; Krzeszowski, T.; Switonski, A.; Josinski, H.; Wojciechowski, K. Calibrated and synchronized multi-view video and motion capture dataset for evaluation of gait recognition. *Multimed. Tools Appl.* **2019**, *78*, 32437–32465, [CrossRef]
22. Liao, R.; Yu, S.; An, W.; Huang, Y. A model-based gait recognition method with body pose and human prior knowledge. *Pattern Recognit.* **2020**, *98*, 107069, [CrossRef]
23. Balazia, M.; Sojka, P. Gait Recognition from Motion Capture Data. *ACM Trans. Multimed. Comput. Commun. Appl.* **2018**, *14*, [CrossRef]
24. Castro, F.; Marín-Jiménez, M.; Mata, N.; Muñoz-Salinas, R. Fisher Motion Descriptor for Multiview Gait Recognition. *Int. J. Pattern Recognit. Artif. Intell.* **2017**, *31*, 1756002, [CrossRef]
25. Kwolek, B.; Krzeszowski, T.; Gagalowicz, A.; Wojciechowski, K.; Josinski, H. Real-Time Multi-view Human Motion Tracking Using Particle Swarm Optimization with Resampling. In *Articulated Motion and Deformable Objects*; Perales, F.J., Fisher, R.B., Moeslund, T.B., Eds.; Springer: Berlin/Heidelberg, Germany, 2012; pp. 92–101.
26. Lu, H.; Plataniotis, K.N.; Venetsanopoulos, A.N. MPCA: Multilinear Principal Component Analysis of Tensor Objects. *IEEE Trans. Neural Netw.* **2008**, *19*, 18–39.
27. Fisher, R.A. The use of multiple measurements in taxonomic problems. *Ann. Eugen.* **1936**, *7*, 179–188, [CrossRef]
28. Rao, C.R. The Utilization of Multiple Measurements in Problems of Biological Classification. *J. R. Stat. Soc. Ser.* **1948**, *10*, 159–203. [CrossRef]
29. Guo, Y.; Hastie, T.; Tibshirani, R. Regularized linear discriminant analysis and its application in microarrays. *Biostatistics* **2006**, *8*, 86–100, [CrossRef] [PubMed]
30. MathWorks. *Statistics and Machine Learning Toolbox: User's Guide*; The MathWorks, Inc.: Natick, MA, USA, 2020.
31. Kennedy, J.; Eberhart, R. Particle swarm optimization. In Proceedings of the IEEE International Conference on Neural Networks, Perth, Australia, 27 November–1 December 1995; Volume 4, pp. 1942–1948.
32. Mirjalili, S.; Mirjalili, S.M.; Lewis, A. Grey Wolf Optimizer. *Adv. Eng. Softw.* **2014**, *69*, 46–61, [CrossRef]
33. Mirjalili, S.; Lewis, A. The Whale Optimization Algorithm. *Adv. Eng. Softw.* **2016**, *95*, 51–67, [CrossRef]
34. MathWorks. *Global Optimization Toolbox: User's Guide*; The MathWorks, Inc.: Natick, MA, USA, 2020.

35. Bishop, C.M. *Pattern Recognition and Machine Learning*; Information Science and Statistics; Springer: Secaucus, NJ, USA, 2006.
36. Platt, J. *Sequential Minimal Optimization: A Fast Algorithm for Training Support Vector Machines*; Technical Report MSR-TR-98-14; Microsoft: Albuquerque, NM, USA, 1998.

sensors

MDPI

Article

Non-Contact Monitoring and Classification of Breathing Pattern for the Supervision of People Infected by COVID-19

Ariana Tulus Purnomo [1,*], Ding-Bing Lin [1,*], Tjahjo Adiprabowo [1] and Willy Fitra Hendria [2]

[1] Department of Electronic and Computer Engineering, National Taiwan University of Science and Technology, Taipei 10607, Taiwan; d10602804@mail.ntust.edu.tw

[2] Department of Intelligent Mechatronics Engineering, Sejong University, Seoul 05006, Korea; willyfitrahendria@sju.ac.kr

* Correspondence: d10602808@mail.ntust.edu.tw (A.T.P.); dblin@mail.ntust.edu.tw (D.-B.L.)

Abstract: During the pandemic of coronavirus disease-2019 (COVID-19), medical practitioners need non-contact devices to reduce the risk of spreading the virus. People with COVID-19 usually experience fever and have difficulty breathing. Unsupervised care to patients with respiratory problems will be the main reason for the rising death rate. Periodic linearly increasing frequency chirp, known as frequency-modulated continuous wave (FMCW), is one of the radar technologies with a low-power operation and high-resolution detection which can detect any tiny movement. In this study, we use FMCW to develop a non-contact medical device that monitors and classifies the breathing pattern in real time. Patients with a breathing disorder have an unusual breathing characteristic that cannot be represented using the breathing rate. Thus, we created an Xtreme Gradient Boosting (XGBoost) classification model and adopted Mel-frequency cepstral coefficient (MFCC) feature extraction to classify the breathing pattern behavior. XGBoost is an ensemble machine-learning technique with a fast execution time and good scalability for predictions. In this study, MFCC feature extraction assists machine learning in extracting the features of the breathing signal. Based on the results, the system obtained an acceptable accuracy. Thus, our proposed system could potentially be used to detect and monitor the presence of respiratory problems in patients with COVID-19, asthma, etc.

Keywords: FMCW; vital sign; XGBoost; MFCC; COVID-19

Citation: Purnomo, A.T.; Lin, D.-B.; Adiprabowo, T.; Hendria, W.F. Non-Contact Monitoring and Classification of Breathing Pattern for the Supervision of People Infected by COVID-19. *Sensors* **2021**, *21*, 3172. https://doi.org/10.3390/s21093172

Academic Editor: Tomasz Krzeszowski

Received: 1 April 2021
Accepted: 28 April 2021
Published: 3 May 2021

1. Introduction

On 30 January 2020, the World Health Organization (WHO) officially confirmed that the spread of COVID-19 had caused a global pandemic for countries around the world [1,2]. This pandemic was caused by the SARS-CoV-2 virus [3], which is highly contagious and causes rapid spread through droplets [4,5]. The droplets can spread through the eyes, mouth, or nose within a radius of one or two meters from a person with COVID-19 [6]. The biggest challenge for this pandemic is to control the spread of the virus, and the best strategy to reduce the virus is by preventing direct contact and ensuring social distancing [7,8].

People with COVID-19 usually experience fever and have difficulty in breathing that causes coughing with rapid and short breath (tachypnoea) [9–13]. Therefore, one of the critical conditions that needs to be monitored is the respiration pattern [14–18]. Since pandemic issues, hospitals are always busy and full of patients. Limited medical personnel cause unsupervised care in a hospital [18], whereas some patients suffering from a respiration problem need special or supervised care. Hence, a non-contact respiration monitoring device that can be accessed from a central room in real time is necessary. Thus, radar technology, which provides non-contact detection, has a great opportunity to be developed in the medical field.

Radar sensor has attractive advantages over camera-based systems in terms of light and privacy [18–27]. Periodic linearly increasing frequency chirp, known as FMCW, is one of the radar technologies that uses a wide frequency bandwidth without requiring wideband processing. FMCW has a simple transceiver architecture, low sampling-rate requirements, low power operation, easier proximity detection, high resolution, and the ability to detect small movements [19,20,28–31]. Therefore, FMCW radar is capable of detecting the vibration of chest displacement [19,20], which is the result of the lungs' and heart's mechanical activity [22].

Several studies have been conducted to obtain an accurate respiration rate [32,33] from chest displacement information. However, patients with a respiration disorder or COVID-19 have an unusual respiration characteristic pattern [13] that cannot be represented by using the respiration rate. Therefore, machine-learning assistance in classifying the breathing pattern plays an important role in detecting respiratory disorder. The addition of machine learning will significantly contribute to the automation of a more sophisticated and more intelligent system. Thus, we tried to incorporate radar technology with machine learning to build a system that can detect and classify the breathing pattern disorder.

Based on the background mentioned earlier, we propose a non-contact breathing pattern detection using FMCW radar with XGBoost classifier and MFCC feature extraction in an indoor environment. Some signal processing steps are implemented to extract the breathing information from chest displacement information. XGBoost classifier and MFCC feature extraction are used to classify the breathing class. XGBoost is often used in machine-learning problems because it combines boosting and gradient boosting so that it can process data quickly [34,35]. Moreover, MFCC feature extraction [36,37] helps the XGBoost to identify, minimize and capture important parts of the signal.

The proposed system will not be a perfect substitute for a professional doctor. However, it is hoped that our research can help to screen and monitor patients infected by COVID-19.

The classification model was evaluated and obtained a reasonable accuracy—87.375%. The implementation of the proposed system was tested for a real-time operation and successfully detected five different classes of breathing waveform.

The rest of the chapter is summarized as follows: Section 2 describes the related work, Section 3 explains the proposed method, Section 4 demonstrates the experimental result, and Section 5 concludes the work.

2. Related Work

The listening technique to listen to the breath sounds using a stethoscope is known as the auscultation technique. The auscultation technique is the basic technique used by doctors to evaluate breath sounds. This technique is quite simple and inexpensive but has a weakness; the analysis results are subjective [38]. Due to these factors, misdiagnosis may occur if the auscultation procedure is not performed properly.

Several studies have been conducted to detect and monitor human body conditions without physical touch, such as using CT scan, X-ray, camera, thermal camera, photoplethysmography technology [39], ultrasound technology, Wi-Fi [40–42], radar [43–49], thermography, etc. [50,51].

CT scan [52] and X-ray technology [53] have a high image precision and resolution, but it is quite expensive. CT scanners and X-ray machines are quite large and not portable. It takes a professional to analyze the images. Furthermore, the negative impact is that the patients are exposed to radiation, which is bad for their bodies.

Depth camera technology can be used to observe the chest displacements by recording video footage of the chest movements [13,14,54–56]. However, the camera has limitations in terms of light and privacy [18,21–26].

In thermography, infrared radiation is commonly used to measure the human body temperature [57]. An infection will usually cause the body temperature to be abnormal. [58]. Additionally, in general, COVID-19 patients have a body temperature above 37 degrees Celsius [59,60].

In [39], non-contact photoplethysmography technology is used to monitor oxygen saturation in the blood (SpO_2). In estimating SpO_2, real-time face video monitoring of the patient is carried out with a camera. An abnormal SpO_2 value is a sign of potential COVID-19 infection.

Another study developed ultrasonic waves for monitoring the movements of organs [61]. The disadvantage of this technology is that patients are not allowed to eat for several hours before the monitoring process is carried out [62].

The breathing rate measurement using Wi-Fi was successfully conducted by using peak detection and with CSI amplitude [41], CSI phase [42], and RSS [40]. Unfortunately, RSS and the amplitude of CSI are not sensitive to the chest motion [40,41]. Furthermore, the measurement accuracy decreases dramatically if the patient location is outside of the specified distance [40,42].

Radar sensor has attractive advantages in monitoring the breathing pattern [63] over camera-based systems in terms of light and privacy [18,21–26]. In [63], non-contact vital-sign detection using radar has been developed, and Lee et al. [64] used radar to observe the different breathing patterns. They [63,64] used Doppler radar to capture various breathing patterns, but did not classify them. Ultra-wideband radar (UWB) [65–68], continuous wave (CW) [21,68–71], and FMCW are the radar technologies that can be used to develop non-contact medical devices [72,73]. UWB radar has a high resolution and low level of radiation [74]. However, high power is required to transmit the signal during a short pulse period. Meanwhile, CW is unable to detect vibration, making it difficult to detect a small movement. In [21], Doppler radar-based continuous-wave (CW) was used for the automatic breathing pattern classification system using the SVM classifier. The CW radar can measure the relative velocity accurately at a very low transmit power and tiny equipment size. When operating at low transmit power, the range is limited. CW has a weakness in measuring tiny position changes because the signal is not modulated. Besides, other moving objects in front of and behind the target will interfere with the CW signal, making it difficult to distinguish the target from the disturbing object [75].

As mentioned earlier, FMCW has a low-power operation and easier proximity detection [30]. It has a high-resolution speed and the ability to detect tiny movements [76]. One of the advantages of using FMCW is that it has the ability to filter interrupting objects in the range domain. All targets ahead of and behind the selected range can be eliminated through the monitoring process in the frequency domain. The FMCW radar can measure small movements as the signal is modulated. The respiration rate detector with FMCW performs the measurement based on the variation of the phases due to the chest displacement [47–49]. Initially, frequency analysis was applied to estimate the distance between the subject and the FMCW radar. Furthermore, feature detection and frequency analysis of phase variance at estimated distances are implemented. In a frequency analysis-based method, the breathing rate is estimated by detecting the peaks due to respiration over a spectrum [46]. The studies on estimating the breathing rate using radar have been investigated extensively with Doppler radar [21,43,45,46] and FMCW radar [47–49]. Previous studies on Doppler and FMCW radar provide an accurate estimation of respiration rate [43–49].

The current state-of-art literature shows that CT scan and X-ray have a good precision but are expensive and cannot be used in real time; cameras, thermal cameras, and photoplethysmography can be used in real time but are not good in terms of privacy and require good lighting; ultrasound technology and Wi-Fi technology are less sensitive and not easy to use; UWB and CW radars are sensitive but require a lot of power. Thus, the aforementioned solutions are less applicable for real-time monitoring of the condition of COVID-19 patients in quarantines or hospitals.

On the other hand, FMCW radar technology allows real-time and non-contact measurement, maintains privacy, is not affected by light, has a simple transceiver architecture, has a wide frequency with low power consumption, has a low sampling-rate requirement, has easier proximity detection, can filter interrupt objects, and has a high resolution, which is very important for detecting vibration.

For this reason, the most suitable sensor to overcome all of the aforementioned problems is to use FMCW radar technology. FMCW is a good choice for implementing non-contact respiration detection for COVID-19 patients.

3. Proposed System

This section explains how the non-contact monitoring and classification of breathing patterns using the XGBoost classifier and MFCC feature extraction using FMCW works. Before we begin, we start by formally defining five classes of breathing patterns as follows:

- Class 1—normal breathing: normal breathing has a constant breathing waveform and similar pattern during the time, shown in Figure 1a.
- Class 2—deep and quick breathing: deep and quick breathing has a large amplitude with a high frequency (high respiration rate), shown in Figure 1d.
- Class 3—deep breathing: deep breathing has a large amplitude with a normal respiration rate, shown in Figure 1c.
- Class 4—quick breathing: quick breathing has a small amplitude (short breath) with high frequency (high respiration rate), shown in Figure 1d.
- Class 5—holding the breath: the breathing waveform is almost disappeared, and the amplitudes are close to zero, shown in Figure 1e.

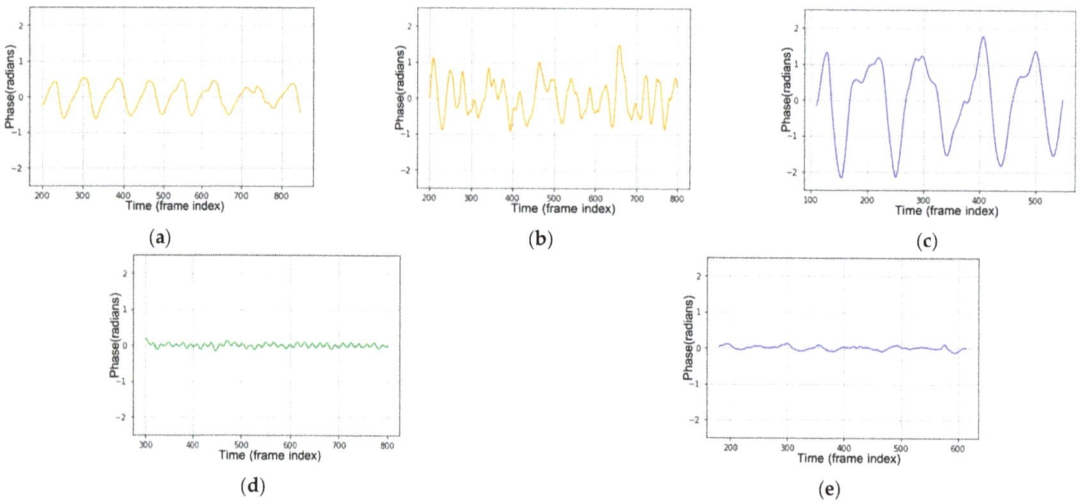

Figure 1. Breathing waveform in the time domain for (**a**) normal breathing; (**b**) deep and quick breathing; (**c**) deep breathing; (**d**) quick breathing; (**e**) holding the breath, recorded by TI-IWR 1443.

Class 1 shows us the normal breathing of an adult. In general, 12 to 20 breaths per minute is the average respiration rate for a relaxed adult. For class 2 to 5, we chose those four breathing patterns because each class has similarities with the symptoms of several diseases.

Breathing pattern disorders are abnormal breathing patterns associated with excessive breathing. They range from simple upper-chest breathing to the most extreme scale, hyperventilation (HVS) [77]. Usually, hyperventilation sufferers experience deep and rapid breathing such as class 2, deep and quick breathing. In general, sufferers of this respiration pattern disorder experience chronic or recurring changes in their breathing patterns that cannot be attributed to a specific medical diagnosis. When ventilation exceeds metabolic requirements, it results in chemical and hemodynamic changes that lead to a breathing pattern disorder. Class 2 (deep and quick breathing) can be found in Kussmaul and Biot patients. The Kussmaul and Biot breathing occur in patients who experience deep and rapid breathing. This indicates that the organs are becoming too acidic. It is caused by kidney failure, metabolic acidosis, and diabetic ketoacidosis. The body breathes quickly and deeply to release carbon dioxide, which is an acidic compound in the blood [78].

In the medical field, class 3 (deep breathing) is known as hyperpnea. Hyperpnea is an increasing depth of breath at normal frequencies.

Asthma starts with a cough or a wheeze. Usually, the chest feels tight, the breathing speeds up, and it becomes shallower. It will cause the person to feel short of breath. These are common symptoms of an asthma attack, which is related to class 4—quick breathing. COVID-19 and tachypnoea patients sometimes have unexpectedly short breathing at unexpected times related to class 4, quick breathing, or short breathing. This kind of patient needs supervised care because short breathing may occur suddenly. This critical condition is very risky for their life.

Bradypnea is a decreased frequency of breath or slowed breathing related to class 5—holding the breath. This situation is found in respiratory center depression. Bradypnea is usually found in patients who use alcohol or narcotics and in patients with tumors. Besides, patients who have difficulties in breathing and are about to die also have a breathing waveform such as class 5.

Now, we will explain how our proposed system works. Figure 2 illustrates the system model that detects and classifies the breathing pattern based on FMCW radar. In general, we have three modules. The first module is the FMCW module that generates and receives the FMCW signal. The first module is explained in the first sub-section. The second module, which is presented in the second sub-section, is the signal processing module that processes and extracts the signal into a breathing waveform. The third sub-section explains the last module, the machine learning module. The machine learning module trains and tests the data and generates the machine-learning model for classification.

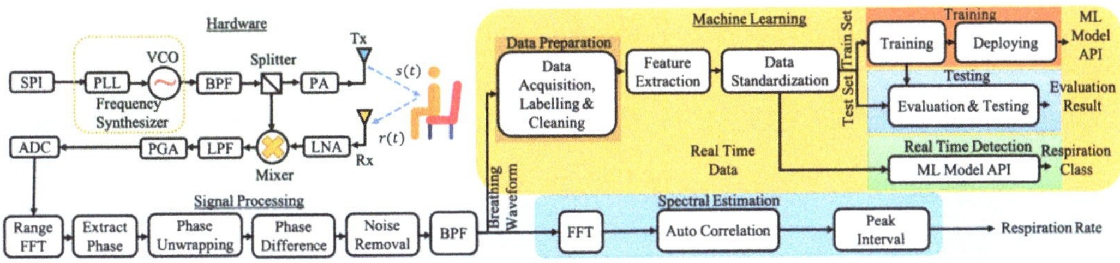

Figure 2. Block diagram of the proposed system.

3.1. FMCW Module

3.1.1. Signal Processing in Hardware

In this part, we explain the signal processing step for generating the FMCW signal and obtaining the reflected signal. The steps are described in the following sub-paragraph.

- The process begins when the user instructs the microcontroller unit (MCU).
- The instruction is transmitted through a serial peripheral interface (SPI), serial communication for short-distance communication.
- FMCW uses a continuous signal that has modulated frequency. Thus, we need a frequency synthesizer that generates the modulated frequency signal.
- A phase-locked loop (PLL) is a feedback control system that compares the phase of two input signals in a frequency synthesizer. It produces an error signal proportional to the difference between their steps.
- The error signal is then passed through the low-pass filter (LPF) and is used to drive the voltage control oscillator (VCO).
- The VCO produces the output frequency. VCO increases the frequency by increasing the voltage.

- Bandpass filter (BPF) is then used to filter the signal. The signal is passed through a BPF so that only the main frequency is used and the harmonic frequency is ignored.
- The splitter is used to split the signal for the mixer and the transmitter.
- A power amplifier (PA) amplifies the signal before being transmitted by the transmitter antenna (Tx).
- Tx emits a modulated signal $s(t)$ towards the object. The object will reflect the signal, and the receiver will receive the reflection.
- The signal $r(t)$ received by the receiver will have a difference in frequency compared to the signal emitted by the transmitter. This difference describes the time for the signal to travel from the transmitter to the object. The object distance is obtained from the traveling time.
- As the received signal is very weak, we use a low noise amplifier (LNA) that amplifies the received signal $r(t)$.
- The mixer will mix the transmitted signal $s(t)$ and received signal $r(t)$.
- We only need the signal with low frequency; we pass the signal through LPF to obtain the low-frequency signal and remove the high-frequency signal.
- PGA is a programmable gain amplifier that can control the gain.
- Finally, the data is transmitted to the MCU.
- The analog-to-digital converter (ADC) will convert the analog signal to the digital signal.

This study uses a TI-IWR 1443 mm-Wave sensor from Texas Instruments [79] to measure the chest displacements. This study was carried out using FMCW radar with a starting frequency of 77 GHz and a chirp frequency of 4 GHz.

As mentioned in the previous section, FMCW has the ability to detect the presence of very small displacements. Usually, the chest displacement has an amplitude below 10 mm with a low frequency of less than 4 Hz. Therefore, there is no large phase change during the time (fast time). Phase changes can be seen from successive chirps (slow time). In Equation (12) of paper [44], if an object is at a distance R, then:

$$\phi_b = \phi_c + \frac{4\pi R}{\lambda}, \tag{1}$$

where ϕ_b is the phase shift at the receiver; ϕ_c is the phase, which is constant for a fixed object; and λ is the wavelength. From Equation (1), it is shown that a smaller λ will result in a larger phase change. This explains why 77 GHz, the smaller wavelength millimeter-wave radar ($\approx$3.9 mm), can measure ten-micron vibrations caused by the lungs and heart. For an object with static angles placed at a fixed distance, they [44] experimentally determined the phase sensitivity by measuring the phase variation across the object-bin range as a function of time. Their study showed that at SNR >40 dB, phase sensitivity <7 milli-radians corresponds to a displacement sensitivity of $\approx$2 microns. Thus, we know that 77 GHz wave radar has greater sensitivity in measuring small displacements. This gives us confidence that using the same device, the system is capable of measuring 10-micron vibrations for breathing measurements. In order to detect the small scale of displacement, the sensor measures the change in phase of the FMCW signal. The sensor detects the chest displacement when it is located nearby the person sitting around the sensor.

3.1.2. FMCW Signal Model

In theory, the FMCW signal model has been explored in several previous studies [80]. This part will briefly explain the basic FMCW signal model that we use in the system. FMCW signal transmits a signal with periodic frequency modulation. The frequency increases linearly over the length of the sweep time T, as shown in Figure 3.

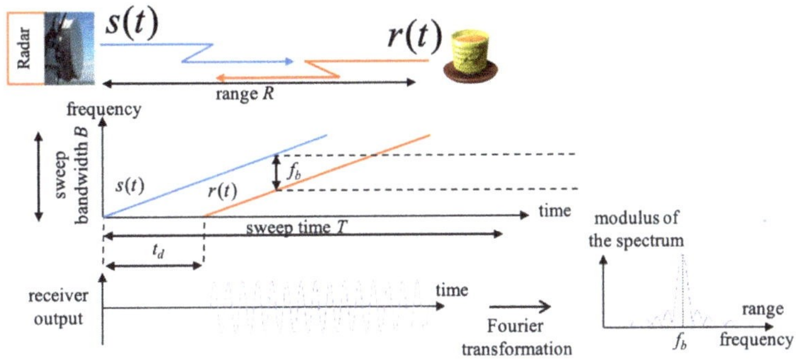

Figure 3. FMCW basic concept.

Based on similar triangles in Figure 3, we have the received time $t_d = \frac{2R}{c}$, so $\frac{t_d}{T} = \frac{f_b}{B}$, where R is the distance, f_b is the beat frequency, c is the light speed and B is the sweep bandwidth. The distance $R = \frac{cTf_b}{2B}$ can be obtained from $f_b = \frac{1}{T}$. The distance resolution is $dR = \frac{c}{2B}$. Note that frequency $\frac{1}{2\pi}\frac{d}{dt}\left(2\pi f_c t + \pi\frac{B}{T}t^2\right) = f_c + \frac{B}{T}t$. The transmitted FMCW signal is expressed as follows:

$$s(t) = A_T \cos\left(2\pi f_c t + 2\pi\frac{B}{T}\int_{-\infty}^{t}\tau\, d\tau + \phi(t)\right) = A_T \cos\left(2\pi f_c t + \pi\frac{B}{T}t^2 + \phi(t)\right), \tag{2}$$

where A_T is the transmitted signal power, f_c is the starting frequency of the chirp, and $\phi(t)$ is the phase. The received signal is the delay time of the transmitted signal defined as:

$$r(t) = \alpha A_T \cos\left(2\pi f_c(t - t_d) + \pi\frac{B}{T}(t - t_d)^2 + \phi(t - t_d)\right), \tag{3}$$

with α as the resized scale. The mixer output is:

$$\begin{aligned} s(t)r(t) &= \alpha(A_T)^2 \cos\left(2\pi f_c t + \pi\tfrac{B}{T}t^2 + \phi(t)\right)\cos\left(2\pi f_c(t - t_d) + \pi\tfrac{B}{T}(t - t_d)^2 + \phi(t - t_d)\right) \\ &= \tfrac{\alpha(A_T)^2}{2}\left(\cos\left(4\pi f_c t - 2\pi\tfrac{Bt_d}{T}t + 2\pi\tfrac{B}{T}t^2 + \pi\tfrac{B}{T}(t_d)^2 - 2\pi f_c t_d + \phi(t) + \phi(t - t_d)\right)\right. \\ &\quad \left. + \cos\left(2\pi\tfrac{Bt_d}{T}t + 2\pi f_c t_d - \pi\tfrac{B}{T}(t_d)^2 + \phi(t) - \phi(t - t_{td})\right)\right) \end{aligned} \tag{4}$$

The LPF output is:

$$b(t) = LPF(s(t)r(t)) = \frac{\alpha(A_T)^2}{2}\cos\left(2\pi\frac{Bt_d}{T}t + 2\pi f_c t_d - \pi\frac{B}{T}(t_d)^2 + \Delta\phi\right), \tag{5}$$

where $\Delta\phi$ is the residual phase noise. Suppose that the target is stationary, let $t_d = \frac{2R}{c}$, $f_c = \frac{c}{\lambda}$ into $b(t)$

$$\begin{aligned} &= \tfrac{\alpha(A_T)^2}{2}\cos\left(2\pi\tfrac{B}{T}\tfrac{2R}{c}t + 2\pi f_c\tfrac{2R}{c} - \pi\tfrac{B}{T}\left(\tfrac{2R}{c}\right)^2 + \Delta\phi\right) \approx \tfrac{\alpha(A_T)^2}{2}\cos\left(2\pi\tfrac{B}{T}\tfrac{2R}{c}t + \tfrac{4\pi f_c R}{c} + \Delta\phi\right) \\ &= \tfrac{\alpha(A_T)^2}{2}\cos\left(2\pi\tfrac{B}{T}\tfrac{2R}{c}t + \tfrac{4\pi R}{\lambda} + \Delta\phi\right) = \tfrac{\alpha(A_T)^2}{2}\cos(2\pi f_b t + \Delta\phi). \end{aligned} \tag{6}$$

Note that frequency: $f_b = \frac{1}{2\pi}\frac{d}{dt}\left(2\pi\frac{B}{T}\frac{2R}{c}t + \frac{4\pi R}{\lambda}\right) = \frac{B}{T}\frac{2R}{c}$. The beat signal means the receiving signal, which is the result of mixer and LPF filter. Thus, we have:

$$b(t) \approx \frac{\alpha(A_T)^2}{2}\cos\left(2\pi\frac{B}{T}\frac{2R}{c}t + \frac{4\pi R}{\lambda} + \Delta\phi\right) = \frac{\alpha(A_T)^2}{2}\cos(2\pi f_b t + \phi_b + \Delta\phi), \tag{7}$$

where ϕ_b is the phase of the beat signal. The beat signal $b(t)$ contains information about the frequency difference, determining the distance R between the radar and the target. The maximum detection range of FMCW is $R_{max} = \frac{cTf_b}{2B}$ and the minimum detection range of FMCW is $R_{min} = \frac{c}{2B}$.

3.2. *Signal Processing Module*

3.2.1. Range FFT

After passing through the low-pass filter mentioned above, the beat signal is sampled in the fast time-frequency. Then, the fast Fourier transformation (FFT) range is implemented to obtain the spectrum. The peak value of the signal spectrum defines the target distance. Peak detection is performed to determine the difference in frequency and distance between the radar and the target. Chest movements caused by heart and lung activity can be observed when the body is in a constant state. The phase change $\Delta\phi_b$ of the beat signal can represent the small-scale vibration ΔR because it has a positive linear relationship.

FFT ranges are referred to as complex span profiles. These FFT ranges are aggregated in a slow-span time matrix for each time T. The variation in the distance from the radar to the chest surface is proportional to the change in phase received by the receiver. The slow time span matrix is then sent to the processor on the PC, and signal extraction is performed, as shown in Figure 2.

The chest surface displacement due to vital organ vibrations has a small amplitude ranging from <12 mm with a low frequency of <4 Hz. This indicates no drastic change in phase during the span of the chirp time (fast time axis) so that the chest movement can be observed by measuring the phase change between successive chirps (slow time axis).

This paragraph describes the slow-time axis sampling rate considerations. Following the Nyquist criteria [44] of the theory, the sampling rate must be twice the sampling rate of maximum frequency to prevent noise aliasing. As the observed range of vibrations is between 0.1 and 4 Hz, the used sampling rate is 20 Hz. On the other hand, the sampling rate must be large to cover up the phase redundancy. In theory, for an object vibrating with $A\ sin\ (2\pi f_m t)$, the selected slow-axis sample must satisfy $Fs > \frac{8\pi f_m A}{\lambda}$. A is the amplitude and f_m is the vibration frequency. For the chirp duration, we chose 50 µs for one chirp range. Based on the theory, SNR and displacement sensitivity are better achieved when the chirp duration is higher.

3.2.2. Extraction and Unwrapping

To obtain information on the value of the displacement distance, arctangent and unwrapping operation on the phase value are calculated as $\varphi\ (m) = unwrap\left[\tan^{-1}\frac{Q}{I}\right]$. I and Q are measured signals of I channel and Q channel, respectively.

The obtained phase is in radian. This phase information can be any real value wrapped into the interval 2π with domain $]-\pi,\ \pi]$ by the *arctan* operator. This information is limited between $-\pi$ and π. This condition causes a phase ambiguity for calculating the phase cycle. To solve this problem, an unwrapping process, a process to eliminate the phase ambiguity, should be carried out so that an absolute phase is obtained. Phase unwrapping reconstructs a continuous signal by removing some 2π ambiguity.

To measure tiny vibrations, the change of the signal within time is measured. From Equation (7), if an object changes position along ΔR, then the phase change between successive measurements is given by:

$$\Delta\phi_b = \frac{4\pi\Delta R}{\lambda} \tag{8}$$

where $\Delta\phi_b$ is the phase change of the beat signal, ΔR is the change of the distance, and λ. is the wavelength. The phase can be measured by taking the FFT of signal $b(t)$ and calculating the phase over the object range. The distance can be calculated by the equation

$R = \lambda\ (\phi + k)$ where k is the phase ambiguity that must be obtained through the phase unwrapping process in order to obtain the absolute phase.

"Itoh's condition" theory [81], adopted by most phase unwrapped strategies [82], is that the absolute value of the phase difference between adjacent neighbors in a continuous phase signal is less than π for unambiguous phase wrapping. When Itoh's condition is not violated, it is possible to obtain absolute and constant values easily. Let us define the wrapper operator $W\ (\cdot)$ that wraps any phase ϕ. into $]-\pi,\ \pi]$ by

$$\begin{aligned} W:\ \mathbb{R} &\longmapsto\ ]\pi,\ \pi] \\ \phi &\hookrightarrow \phi - 2\pi k, \end{aligned} \tag{9}$$

where $k \in \mathbb{Z}$, such that it follows the following rule:

$$\phi(t-1,t) = \begin{cases} \Delta\phi_t, & if\ |\Delta\phi_t| < \pi \\ \Delta\phi_t + 2\pi & if\ \Delta\phi_t \leq -\pi \\ \Delta\phi_t - 2\pi & if\ \Delta\phi_t \geq -\pi \end{cases} \tag{10}$$

$$\Delta\phi_t = \phi(t) - \phi(t-1), \tag{11}$$

where $\phi(t)$ is the current phase and $\phi(t-1)$ is the previous phase. Thus, Itoh's condition [81] can be represented by:

$$|\Delta\phi_t| \leq \pi. \tag{12}$$

Then, we have:

$$\sum_{t=1}^{m} \Delta\phi_t = \phi(m) - \phi(0). \tag{13}$$

From Equation (9), we have $W(\phi(t)) = \phi(t) - 2\pi k_t$, with $k_t \in \mathbb{Z}$, so:

$$\Delta W(\phi(t)) = \phi(t) - \phi(t-1) - 2\pi(k_t - k_{t-1}), \tag{14}$$

where $k_t - k_{t-1} \in \mathbb{Z}$. Then, we can write Equation (11) as:

$$\underbrace{W[\Delta W(\phi(t))]}_{p} = \Delta\phi_t \underbrace{-2\pi(k_t - k_{t-1}) - 2\pi k,}_{q} \tag{15}$$

where $k_t, k_{t-1}, k \in \mathbb{Z}$, and $2\pi k$ is the proper 2π multiple to bring p into the principal interval. From Equation (12) and $|p| \leq \pi$, we have $q = 0$, so that we can write:

$$W[\Delta W(\phi(t))] = \Delta\phi_t \tag{16}$$

Finally, from Equations (13) and (16), we obtain:

$$\phi(m) = \sum_{t=1}^{m} W[\Delta W(\phi(t))] + \phi(0). \tag{17}$$

From Equation (17), we can obtain the unwrapped phase at any time t from the wrapped phase value, with its absolute phase value $\phi(0)$. Thus, we can calculate the absolute phase value for each time when Itoh's condition is met. Lastly, the phase difference between successive unwrapped phases is calculated.

3.2.3. Noise Removal

Noise-induced phase wrapping error might corrupt the un-wrapped differential phase $a\ (m)$, especially in phases around $-\pi$ or π. By calculating the phase difference backwards $a\ (m) - a\ (m-1)$ and forwards $a\ (m) - a\ (m+1)$, impulse-like noise can be eliminated. If the phase exceeds a certain limit, the $a\ (m)$ value is replaced with an interpolation value.

3.2.4. IIR BPF Using Cascaded Bi-Quad

The chest displacement due to cardiac and breathing activity is represented by two overlapping sinusoidal signals, where one represents the heart waveform and the other represents the respiratory waveform. Generally, the adult chest moves due to the process of respiration activity with an amplitude of 4 to 12 mm at a frequency between 0.1 and 0.5 Hz, and cardiac activity with a frequency between 0.8 and 4 Hz with an amplitude of 0.2 to 0.5 mm [83]. Chest surface fluctuations caused by pulmonary and cardiac motion are modeled as a signal [22], as follows:

$$x(t) = \sum_{i=1}^{J} a_{ri} \cos(2\pi i f_r t + \theta_{ri}) + \sum_{i=1}^{K} a_{hi} \cos(2\pi i f_h t + \theta_{hi}). \tag{18}$$

The amplitude of respiration and heart waveform for the i-th harmonic component is denoted as a_{ri} and a_{hi}, respectively. f_r is the base frequency of the respiratory waveform and f_h is the base frequency of the heart waveform. The harmonic phase sequences of the respiratory and heart signal are θ_{ri} and θ_{hi}, respectively. Finally, J and K are the total numbers of components.

As mentioned earlier, the respiration and heart waveform have different frequency bands so that suitable frequency filters can separate them. In this study, a fourth-order IIR cascade Bi-quad BPF was used to obtain a respiratory signal in the frequency range between 0.1 and 0.5 Hz.

The BPF is a frequency filter that passes signals within a certain frequency range. The signal is passed between the lower limit frequency to the upper limit frequency. In other words, this BPF will reject or attenuate frequency signals that are outside the specified range.

Increasing the Butterworth filter order allows for faster roll-off around the cutoff frequency while maintaining flatness in the stopband and passband. However, direct application of a high-order recursive filter will cause different coefficients in many order quantities. Besides, this makes the practical application difficult [84]. Thus, a cascaded bi-quad is used to avoid the use of a high-order filter.

In this section, we explain how the Bi-quad BPF works. The pole-zero form of the BPF response [85] is described as follows:

$$H(z) = K\frac{(z+1)^N(z-1)^N}{(z-p_1)(z-p_2)\ldots(z-p_{2N})}. \tag{19}$$

N is the order of the BPF. Next, $H(z)$ is converted into cascaded sections (bi-quads). Thus, $H(z)$ can be written as the product of N sections with complex-conjugate poles as follows:

$$H(z) = K_1\frac{(z+1)(z-1)}{(z-p_1)(z-p_1^*)}.K_2\frac{(z+1)(z-1)}{(z-p_2)(z-p_2^*)}. \;\;\ldots\;\; .K_N\frac{(z+1)(z-1)}{(z-p_N)(z-p_N^*)}. \tag{20}$$

p_k^* is the complex conjugate of p_k. At each bi-quad, a zero is assigned at $z = +1$ and $z = -1$. We label each term in the equation as biquadratic because it has a quadratic numerator and denominator. Furthermore, we can extend the numerator and denominator of the k-th bi-quad section as follows:

$$H_k(z) = K_k\frac{z^2-1}{z^2-(p_k+p_k^*)z+p_kp_k^*} = K_k\frac{z^2-1}{z^2+a_1z+a_2}. \tag{21}$$

$a_1 = -2 * real(pk)$ and $a_2 = |pk|^2$. After dividing the numerator and denominator by z^2, we form the following equation:

$$H_k(z) = K_k \frac{1 - z^{-2}}{1 + a_1 z^{-1} + a_2 z^{-2}}. \quad (22)$$

Since the same zero is assigned for each bi-quad, the feed-forward (numerator coefficient) $b = [1\ 0\ -1]$ will be the same for all N bi-quads. So, we get:

$$a = \left[1\ \ -2 * real(pk)\ \ |pk|^2\right],\ b = [1\,0\ -1]. \quad (23)$$

A pair of complex conjugate poles is not sufficient to define a second-order polynomial. For a BPF, after bilinear transformation, the output has to be scaled to achieve unity gain in the passband [85]. Each bi-quad is allowed to have a gain of 1 at the filter geometric mean frequency f_0. for finding the gain K_k. Then, $H_k(z)$ is evaluated at f_0 and K_k [85], set as follows:

$$K_k = \frac{1}{|H(f_0)|}. \quad (24)$$

To find f_0, we define $f_1 = f_{center} - \frac{bw}{2}$ and $f_2 = f_{center} + \frac{bw}{2}$. Thus, $f_0 = \sqrt{f_1 * f_2}$. For a narrowband filter, f_0 is close to f_{center}. In theory, we can arrange the sequence of the bi-quad freely. However, to reduce and minimize the possibility of clipping, a bi-quad with the peaking response should be put at the end.

As our system uses a fourth-order IIR cascaded Bi-quad BPF, we need to cascade two IIR bi-quad BPF, as shown in Figure 4. Based on Equations (23) and (24), we have the denominator coefficient a, nominator coefficient b and the gain K_k, respectively, as follows:

$$a = \begin{bmatrix} 1 & a_{1,1} & a_{1,2} \\ 1 & a_{2,1} & a_{2,2} \end{bmatrix} = \begin{bmatrix} 1 & -1.963 & 0.964 \\ 1 & -1.85 & 0.868 \end{bmatrix},\ b = [b_0\ b_1\ b_2] = [1\ \ 0\ -1] \quad (25)$$

$$K = [\ 0.116\quad 0.031\ ] \quad (26)$$

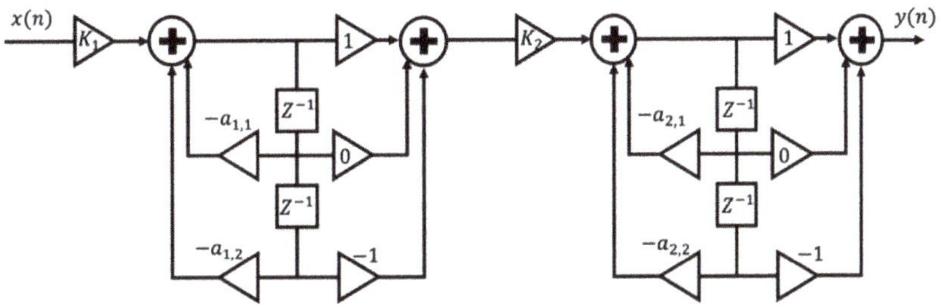

Figure 4. Fourth order of IIR BPF using cascaded bi-quad.

Figure 5a presents the pole-zero plot, and Figure 5b illustrates the frequency response of the fourth-order IIR cascaded Bi-quad BPF for frequency 0.1 to 0.5 Hz.

One of the measurement samples shows the unwrapped phase after the phase differences operation, and noise removal is represented in Figure 6a as the chest displacement. Then, the signal is passed through the fourth-order of IIR BPF using a cascaded bi-quad. Note that the breathing waveform becomes more obvious, as shown in Figure 6b.

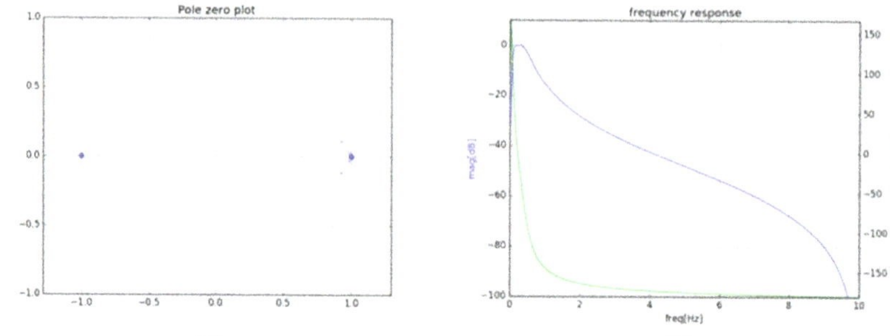

(**a**) (**b**)

Figure 5. (**a**) Pole-zero plot, and (**b**) frequency response for fourth-order IIR cascaded Bi-quad BPF.

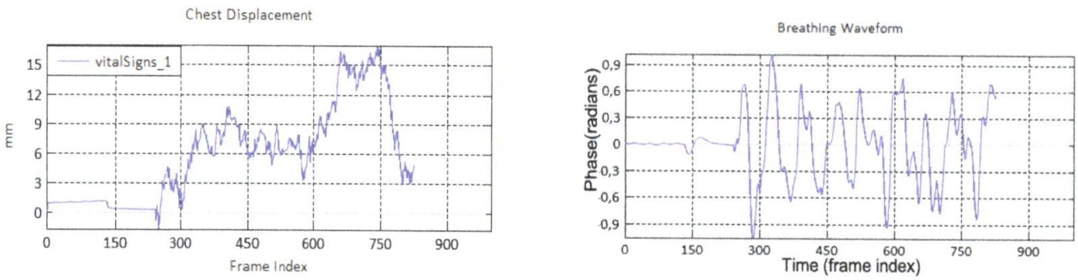

(**a**) (**b**)

Figure 6. (**a**) Unwrapped phase after the phase difference and noise removal, labelled as the chest displacement; (**b**) the output of IIR BPF is the breathing waveform.

3.2.5. Respiration Rate

In order to verify that the breathing waveform is correct, we also calculate the breathing rate. The breathing waveform was passed through a spectrum estimation, autocorrelation, and interpeak distance block to estimate the breathing rate. The BPF is employed to eliminate the noise. Peak detection is performed to determine the difference in frequency and the distance between the radar and the target. The respiration rate value is obtained by calculating the distance between the respiratory wave signal peaks in the time domain.

3.3. Machine Learning (Classification Method) Module

The proposed method uses the XGBoost model as the classifier and MFCC as the feature extraction. We explain the machine-learning module as several parts as follows.

3.3.1. Pre-Processing

When recording the respiratory data, some pieces of data have 0 values or missing values. Besides, some data do not represent the desired class. For example, when the system started to record, the subject had not started imitating the suitable breath class. Thus, data that does not represent the suitable class has 0 value or missing values and is discarded from the data set.

Data sets contain some features that differ in unit and range. Before the data processed by a machine-learning algorithm, data sets must be converted into a proper format. If standardization is not implemented, large numbers and a wide range of features will reach more weight than features with a small number and small range. It means that features with a large number and range will obtain more priority. Therefore, to suppress all these effects, it is necessary to scale the feature with a standardization process. Standardization facilitates faster convergence of loss functions for some algorithms.

$$z = \frac{x - \mu}{\sigma} \quad (27)$$

For each piece of data, we limit each window to 5 s and segment it with 85 step size.

3.3.2. MFCC Feature Extraction

The Mel-frequency cepstral coefficient (MFCC) is a feature extraction introduced by Davis and Mermelstein around 1980 [36,37]. In order to improve the classification accuracy, MFCC feature extraction converts signal waves into cepstral coefficients. It converts the signal into several vectors to generate vector features [86]. The MFCC of a signal is a small set of features with a value between 10 and 20, representing a spectral envelope of the overall shape. The advantage of MFCC is that it can minimize and capture the important parts of the signal. MFCC works based on the differences in frequency [87,88]. MFCC is widely used in audio/speech recognition. We adopt MFCC because the breathing waveform is similar to the audio signal, which has a three-dimensional signal in time, amplitude, and frequency, as shown in Figure 7. Most of the audio recognition studies use MFCC because it has the best performance in extracting the signal. The study in [89] shows good training and test results in speech recognition using MFCC [89]. Thus, in our study, we employ MFCC to assist machine learning in extracting the breathing waveform.

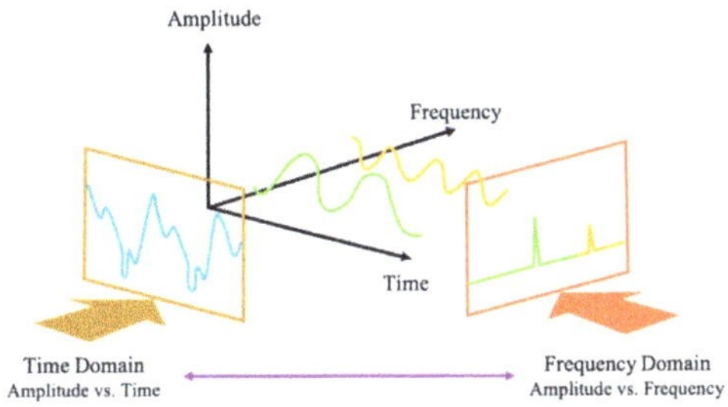

Figure 7. Component of breathing waveform.

MFCC stages, shown in Figure 8, start from frame blocking, windowing, FFT, Mel-frequency wrapping (MFW), discrete cosine transform (DCT), and cepstral liftering.

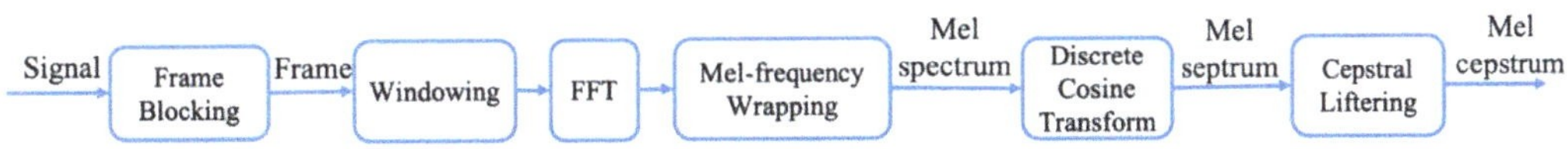

Figure 8. MFCC feature extraction technique.

1. Frame Blocking

Frame blocking divides the signal into several frames then makes the frames overlap each other. The signal is divided into U samples and shifted by V samples so that $U = 2V$ with $V < U$. The width of the frames is denoted by U, while the width of each frame is shifted as V. The overlap width is calculated as the difference of $U - V$.

2. Windowing

Windowing is necessary because the effect of frame blocking on signals causes discontinuity. One way to avoid a discontinuity at the end of the window is to tap the signal to zero or near zero, thereby reducing errors.

3. Fast Fourier Transform (FFT)

After passing through frame blocking and windowing, FFT is applied to the signal. FFT converts the signal from the time domain to the frequency domain as the spectrum.

4. Mel-frequency Wrapping (MFW)

Mel-frequency wrapping is processed based on a filter bank and produces a mel spectrum. A filter bank is a filter to determine the amount of energy from a certain frequency band. The mel frequency scale is a linear frequency scale at frequencies below 1000 Hz and is a logarithmic scale at frequencies above 1000 Hz. This block wraps the resulting spectrum from FFT so that it becomes a mel scale. The inner frequency range is very wide, and the signal does not follow a linear scale, so the computed spectrum of data is mapped in mel scale using overlapping triangular filters. MFW calculation [36] follows:

$$Y[i] = \sum_{j=1}^{G} T[j]H_i[j]. \quad (28)$$

$Y[i]$ is the calculation result of the mel frequency wrapping at i-th, where $1 \leq i \leq E$; E is the number of filter bank channels. G is the total magnitude spectrum; $T[j]$ is the result of FFT; $Hi[j]$ is the filter bank coefficient at frequency j. In this case, mel uses a frequency with the mel scale [90] that follows:

$$mel(f) = 2595 \log_{10}\left(1 + \frac{f}{700}\right), \quad (29)$$

with f as the frequency.

5. Discrete Cosine Transform (DCT)

The DCT produces septrum mel. DCT is assumed to replace the inverse Fourier transform in the MFCC feature extraction process. DCT has the aim of creating septrum mel to improve the quality of recognition. DCT [36] uses the following equation:

$$C_m = \sum_{k=1}^{K} (\log_{10} Y[k] \cos(m\left(k - \frac{1}{2}\right)\frac{\pi}{K})), \quad (30)$$

where $m = 1, 2, \ldots, K$. C_m is the coefficient, $Y[k]$ is the output of the filter bank process on the index k, m is the number of coefficients, and K is the number of expected coefficients.

6. Cepstral Liftering

Cepstral liftering is the last MFCC process that converts the frequency domain signal into the time domain. The cepstral coefficient uses the following equation:

$$w(b) = 1 + \frac{C}{2} \sin\frac{b\pi}{C}, \quad (31)$$

where $b = 1, 2, \ldots, C.w(b)$ is the window function to the cepstral features, C is the cepstral coefficients, and b is the cepstral coefficients index. The cepstral liftering is obtained in the form of frames and cepstral coefficients.

3.3.3. Classification Using XGBoost Classifier

One technique that can be used to improve the performance and the confidence level of learning outcomes is using more than one learning algorithm. In ensemble learning, similar learning algorithms generate several hypotheses, and the results are combined to make the predictions. This combination method can minimize learning errors caused by noise, bias, and variations. Usually, these errors occur in learning processes that use unstable classifiers, such as decision trees [91]. XGBoost, which stands for extreme gradient boosting, is an ensemble machine learning technique that uses a gradient enhancement framework for machine learning predictions [34]. XGBoost has a fast execution time and good scalability. XGBoost is a special implementation of gradient boosting. It is called gradient boosting because gradient descent is used. It minimizes errors when forming a new model. By adding the boosting method, it is expected that the classifier performance will increase. Improving the boosting technique at the training stage helps to optimize the weight gain process in machine learning [91].

To understand how XGBoost works, first, we need to understand how the adaptive boosting (AdaBoost) and gradient boosting machine (GBM) algorithms work, which are the basis of XGBoost. AdaBoost works by constructing a weak learner model, namely a tree, and giving each observation the same weight [91]. The obtained tree is then evaluated to see its predictive ability. There will be some incorrect observations for the prediction tree. The weight of incorrect observation will be increased in the next iteration. Thus, we hope that it will be able to predict in the next iteration model accurately. The procedure is repeated so that 10 to hundreds of weak learners are obtained. The final model is decided by combining various trees obtained by a certain weighting mechanism. This AdaBoost approach is classified as a sequential learning process because it sequentially changes the weak learner model. It does not process the parallel tree, such as the random forest algorithm [91]. The GBM algorithm also performs the iterative and sequential method as well as AdaBoost. The prediction of one iteration is obtained by combining the models from the previous iterations.

Furthermore, in each iteration, the model attempts to correct the previous error. The residue of the previous prediction model is used as the response variable for the next model. At each iteration, a loss function is minimized according to the needs of the user for obtaining a classification model. For modeling the regression, the loss function can be estimated by calculating the error sum of squares, whereas, in the general classification, the logarithmic loss function is used. The final prediction is determined by combining all model predictions from all iterations. XGBoost is an extension of the GBM algorithm with several additional features that are useful in speeding up the computation process and preventing overfitting. XGboost can optimize memory and cache usage on a computer so that it can work efficiently, even dealing with large data sizes [34,35]. This feature allows XGboost to run faster than other advanced models such as deep learning and random forest. Meanwhile, the prevention of overfitting is carried out by providing a penalty component to the loss function. In this way, the algorithm will avoid too complex models but poorly perform in predicting events with new data.

In this part, we explain the choice of our machine learning algorithm. In [35], six different classification algorithms were compared for emotion recognition from Electroencephalography (EEG) signals. The EEG signal used was a one-dimensional signal that changed with the time, as well as our breathing waveform. In their paper, they explained that the algorithm they needed was a fast and accurate algorithm for a real-time prediction. From the Naive Bayes, KNN, C4.5, Decision Tree, Random Forest, and XGBoost algorithms, XGboost achieves the best accuracy for classifying four classes compared to five other classification algorithms [35]. Additionally, in [92], the performance between XGBoost and Light GBM was tested, showing that XGboost has shown much better accuracy and outperforms existing boosting algorithms [92]. XGBoost combines several algorithm techniques that can minimize the learning error. As mentioned in the previous paragraph, XGboost uses the concept of AdaBoost and GBM. It does not process parallel trees such as random forest [91,93]. It uses a sequential learning process that sequentially changes the weak learner model, and the final prediction is determined by combining all model predictions from all iterations. Tianqi Chen claims that XGBoost has better performance because it has an overfitting control feature [34]. As time goes by, XGBoost has often become a champion in various data science competitions. Based on the explanation above, our system requires an algorithm that is able to classify accurately and quickly in real time. The suitable algorithm that meets our system requirement is XGBoost. Thus, we employ the XGBoost algorithm for classifying the breathing waveform in a real-time system.

4. Experimental and Analysis Results

The first part of this section provides selected parameters on the FMCW sensor. The second part describes the data collection and labeling. The last part is the experimental result and data analysis.

4.1. Experimental Setup

This study was carried out using an FMCW IWR 1443 mmWave radar platform from Texas Instruments (TI) [79] with a starting frequency of 77 GHz and a chirp frequency of 4 GHz. The chirp duration is 50 μs with the chirp rate 2 MHz and 250 samples per chirp. Each frame is configured to have two chirps. The details are shown in Table 1.

Table 1. Radar parameter setting.

Starting Frequency	Bandwidth	Chirp Rate	Samples Per-Chirp	Chirps Per-Frame	Chirp Duration	Frame Duration	Range Resolution	Max Unambiguous Range
77 GHz	4 GHz	2 MHz	250 samples	2	50 μs	50 ms	0.0375	9 m

The experiments were conducted in a small room—3 × 3 m. The subject sat on a chair, and the radar was placed 1 m in front of the subject. The radar was positioned parallel to the chest at a height of about 1 m in the detectable area. The data was collected in binary format. We labeled these samples according to five different respiration classes. The participants were asked to imitate five breathing patterns. Observations were made on each subject with a duration ranging from 5 to 15 s for each class. During data recording, the subjects were not allowed to make any movement to reduce the random body movements that cause noise. The estimated frequency and amplitude will be better if the observation time is larger. However, the observation time is generally limited to the range of 5 to 15 s due to the inherent time-frequency sacrifice.

4.2. Data Collection and Labelling

In this study, we used the breathing waveforms as our data set. Through experiments, we collected 4000 breathing waveforms as the training and testing data. The system randomly divides the data set without following any rules into 80–20% train–test splits for experimental purposes. The collected breathing waveform consists of five classes: normal breathing, deep and quick breathing, deep breathing, quick breathing, and holding the breath.

Before training the data, the pre-processing step is necessary to normalize and eliminate the ambiguous and redundant data from the dataset. In data records, some pieces of data have missing values. To resolve the data, we removed data with missing values from the dataset. We cleaned the noise from the data for better performance and accuracy. The accuracy depends on the input data. We split the breathing waveforms into several data for every 5 s. For each data, we limit the window to 85 steps size. After pre-processing, finally, we had data set with details shown in Table 2.

Table 2. Data set for training and testing.

Class	Training Samples	Testing Samples
Normal breathing	640	160
Deep and quick breathing	640	160
Deep breathing	640	160
Quick breathing	640	160
Holding the breath	640	160
Total	3200	800

The data set was used in the training process to train the classifier model. During the training process, the computer will learn and understand the data to obtain the expected model.

4.3. Experiment and Analysis Results

Before we conducted the experiment, the proposed method was implemented on hardware. The data was collected and labeled as the training datasets. To verify the accuracy of the proposed system, we conducted three experiments for detecting five respiration patterns. The first experiment was conducted without additional feature extraction. The second experiment was conducted using statistical feature extraction, and the third experiment was conducted using MFCC feature extraction.

The statistical feature extraction is used to identify the statistical character of data. In this study, the statistical features were derived from the statistical distribution of the respiratory signal data, such as the mean, median, maximum, variance, standard deviation, absolute deviation, kurtosis, and skewness.

- The mean is the average value of the population.
- The median or middle value is a measure of data centering. If the data is sorted, the observed value is in the middle.
- Maximum describes a greater value than or equal to all values in data.
- Variance presents a square of the average distance between each quantity and mean.
- Standard deviation is used to measure the amount of variation or dispersion of data. The standard deviation describes how far the sample deviates from the mean.
- Absolute deviation represents the absolute difference between each data point and the average. This explains the variability of the data set.
- Kurtosis defines the degree of "tailedness" of a distribution.
- Skewness is known as a measure of slope, which is a number that can indicate whether the curve shape is slanted or not.

Before we trained our data, we plotted it, which has been extracted using feature extraction, into a two-dimensional diagram of linear discriminant analysis (LDA) [94]. The aim is to see the effect of adding MFCC feature extraction. LDA is a classical statistical technique that can reduce the dimensions [94]. With LDA, we can also divide data into several groups (clustering) [94].

Based on the LDA results, Figure 9 describes that MFCC makes the scattering point of one class to be closer and the scattering point for five different classes to be farther. Thus, it shows that MFCC feature extraction helps the classifier in clustering the data. As a comparison, we also show the effect of data extraction using statistical methods. We can see in Figure 9 that the scattering point of the data with the MFCC extraction feature has the least number of overlapping classes.

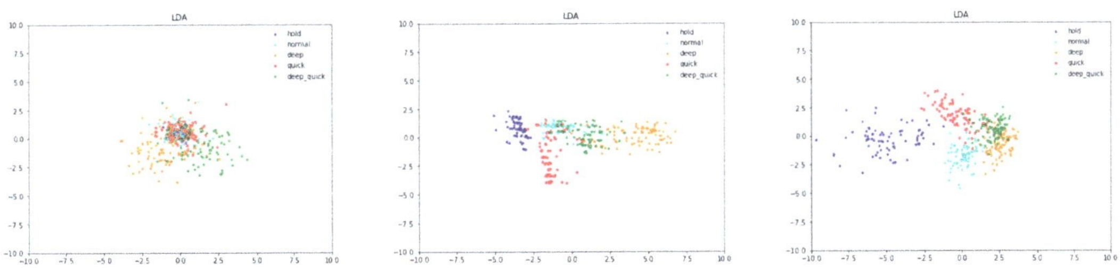

(a) (b) (c)

Figure 9. LDA data scattering point for (**a**) raw data, (**b**) data with statistic feature extraction, and (**c**) data with MFCC feature extraction.

In the next step, the datasets were used to train the XGBoost model. After the entire training phase ends, the resulting model must be tested again using a test set. The evaluation/testing step aims to decide whether the model is good enough or not.

One of the problems in building a learning model is finding the optimal hyperparameter. For example, we need to decide the optimal batch size, the optimal epoch for running a deep-learning model, and the best optimizer for deep-learning models. Many other hyperparameters can be optimized, such as the dropout, number of nodes, number of layers, activation functions, and others. It is time-consuming to use trial and error, trying to change the parameters manually, one by one, to find the best model. One solution to this problem is to use GridSearchCV.

Grid search, as the name implies, looks for parameters in a given "grid"—for example, the number of epochs =—then, we need to decide which of the two values gives the best result. In this study, we used the following parameters:

- *n* estimators: [200 300 400], *n* estimators represent the number of sequential trees modelled in XGBoost.
- Max depth: [3 4 5], max depth means the maximum number of terminal nodes in a tree.
- Learning rate: [0.1, 0.01, 0.001], the learning rate is the learning parameters that control the change value in estimating the prediction. A smaller value causes a stronger model with specific characteristics of the tree. However, lower values will require a larger number of trees to model all relations and do a lot of computation.

The way the Grid Search works is by combining the values inputted in the hyperparameters. An example is when we want to find a combination of hyperparameters A = [1, 2] and B = [3, 4], then the Grid Search will look for all combinations of A and B, namely [1, 3], [1, 4], [2, 3], [2, 4] and choose the best combination based on the value of the highest CV Score. We found the best combinations to obtain higher accuracy. The process was carried out by brute force and reported which parameter has the best accuracy. As we have three parameters with three grids for each, we thus have 27 combinations.

CV, at the end of the word GridSearchCV, stands for cross-validation. This means that our input data will be divided by GridSearchCV into several folds to reduce the bias. In our study, we used five-fold cross-validation. It divided a set of samples randomly into five independent subsets, to do five repetitions for training and testing. For each test, a subset was left for testing and another subset for training. The degree of accuracy is calculated by dividing the total number of correct classifications by the sum of all instances in the initial data.

XGBoost model performance is calculated through a confusion matrix. The confusion matrix presents the amount of data classified correctly and incorrectly. The effectiveness and performance of a machine learning model can be measured by calculating its accuracy. Finally, the result is shown in Figure 10, Figure 11 and Table 3.

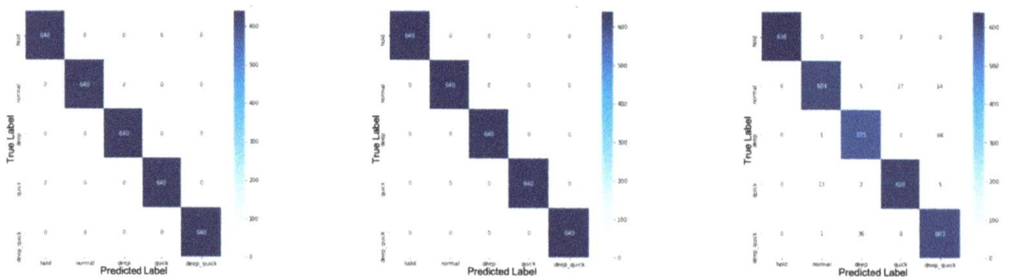

(a) (b) (c)

Figure 10. Confusion matrix for (**a**) raw data, (**b**) data with statistic feature extraction, and (**c**) data with MFCC feature extraction on training stage.

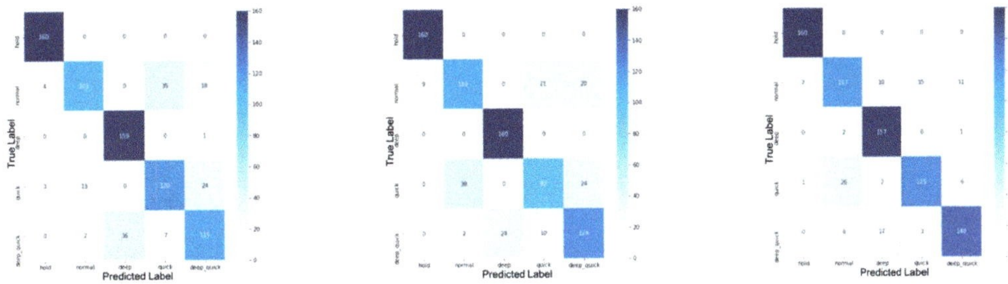

(a) (b) (c)

Figure 11. Confusion matrix for (**a**) raw data, (**b**) data with statistic feature extraction, and (**c**) data with MFCC feature extraction on testing stage.

Table 3. Training and testing accuracy for the raw data set, data set with statistic feature extraction, and with MFCC feature extraction.

Feature Extraction	Training Accuracy	Testing Accuracy
without feature extraction (raw data)	100%	82.125%
statistic	100%	81.375%
MFCC	95%	87.375%

In the confusion matrix, most misclassifications come from predicting deep quick breathing to be deep breathing, normal breathing to be quick breathing, and vice versa. A possible reason is that these two classes have almost the same pattern but are different in the depth of breathing, shown in the amplitude of the waveform. This might happen because the amplitude of the respiratory signal is sensitive to the time window in the normalization process. Besides, the accuracy of the model depends on the input data, whereas chest displacement waves have different variations depending on several factors such as the state of health, location of measurement, variation between people, etc.

Based on the experiments above, we showed that adding MFCC feature extraction gives a better result than without and with statistical feature extraction. Thus, we implemented our proposed system in real time by using MFCC feature extraction.

Let us define X as a disease name. Then, we have four definitions as follows.

From Table 4, there is one important case that needs special attention—false positive. When the system does not detect the patient's disease, in reality the patient is suffering from the disease. This is very dangerous. For example, if the patient has COVID-19 but the system detects that the patient's condition is normal, then the patient will not immediately receive the right treatment. On the other hand, true negatives and false negatives also need attention. If the system detects that a patient is suffering from disease A, but in reality, the patient is suffering from disease B, then the patient will not receive the right treatment. However, if the system detects that the patient suffers from X disease, but the patient is normal, the condition is not dangerous.

Precision is defined as $Precision = \frac{TP}{TP+FP}$. High precision shows that the class can be classified well or have a low FP. Recall is defined as $Recall = \frac{TP}{TP+FN}$. High recall indicates that the class has a low FN. The f1-score is the average of precision and recall that takes these two metrics into $f1\ score = 2\ \frac{precision\ *\ recall}{precision\ +\ recall}$. From the confusion matrix in Figure 11, we thus have the classification report shown in Table 5.

As mentioned before, since a false-positive result is the most dangerous condition, we need to achieve a better precision than recall. To detect deep quick, and quick class, XGBoost with MFCC feature extraction achieves the best precision. However, for the deep class, XGBoost with statistic feature extraction gives the best precision. For the normal and hold class, XGBoost without feature extraction has the best precision.

Table 4. Confusion matrix 2 × 2.

True Positive (TP)	True Negative (TN)
• Prediction: the system detects that the patient suffers from X disease • Reality: the patient suffers from X disease	Prediction: the system detects that the patient suffers from X disease Reality: the patient does not suffer from X disease
False-Positive (FP)	**False-Negative (FN)**
• Prediction: the system does not detect that the patient suffers from X disease • Reality: the patient suffers from X disease	Prediction: the system does not detect that the patient suffers from X disease Reality: the patient does not suffer from X disease

Table 5. Classification report for confusion matrix in Figure 11.

Class.	Raw (without Feature Extraction)			With Statistic Feature Extraction			With MFCC Feature Extraction		
	Precision	Recall	f1-Score	Precision	Recall	f1-Score	Precision	Recall	f1-Score
Normal	0.873	0.644	0.741	0.728	0.688	0.707	0.807	0.731	0.767
Deep quick	0.728	0.719	0.723	0.738	0.775	0.756	0.886	0.875	0.881
Deep	0.815	0.994	0.9	0.87	1	0.93	0.844	0.981	0.908
Quick	0.741	0.75	0.745	0.758	0.606	0.674	0.874	0.781	0.825
Hold	0.958	1	0.979	0.947	1	0.973	0.952	1	0.976

Patients with COVID-19 usually have a quick and short breath at unexpected times. This condition is related to class 4, quick breathing, or short breathing. Thus, if we need to detect patient with COVID-19, it is better to use XGBoost with MFCC feature extraction because it achieves the best precision in detecting quick/short breathing.

We ran the system into a real-time experiment. We conducted five measurements with an object located approximately 1 m in front of the sensor. The results for the detection and classification of breathing waveforms in real time can be seen in Table 6. Table 6 shows the estimated range of the target, chest displacement waveform, estimated breathing rate, and breathing waveform. Five figures in the first left column are the azimuth heat map that shows the range and angle estimation for the object in front of the sensor. It illustrated that the sensor detects 0.1 to 0.5 Hz vibration at approximately 1 m. The figure in the next column shows real-time chest displacement, and the figure in the right column is a real-time breathing waveform.

To clarify whether the breathing waveform was accurate or not, we tried to estimate the respiration rate calculation. The estimated value of the respiration rate was then compared with counting the breathing rate manually. The respiration rate calculation was performed by counting the number of inhalation and exhalation cycles in one minute. The result of the breathing rate is shown in the last two columns of Table 6.

The first experiment was detecting the normal breath, shown in the first row in Table 6. The results show us that the object was detected at about 1.20 m with an angle of 30 degrees. The breathing waveform has a constant breathing waveform and similar pattern during the time. The estimated breathing rate was 20.51 breaths/min.

The second measurement was detecting the deep quick breath, shown in the second row in Table 6. The object was detected at the range of 1.23 m and 30 degrees from the sensor with a breathing rate of 23.44 breath/min. The breathing waveform presents a big amplitude with a higher frequency (higher respiration rate) compared to the normal breathing rate.

The third observation was conducted for a deep breath, shown in the third row in Table 6. The vibration was detected at 1.17 m from the sensor. The detected breathing rate was 17.58 breaths/min. Deep breathing waveform shows a big and large amplitude with a lower respiration rate compared to normal breath.

Table 6. Real-time measurement using TI-IWR 1443 for five breathing classes.

Class	Real-Time Measurement	Breathing Rate	
		Manual	Measured
Normal	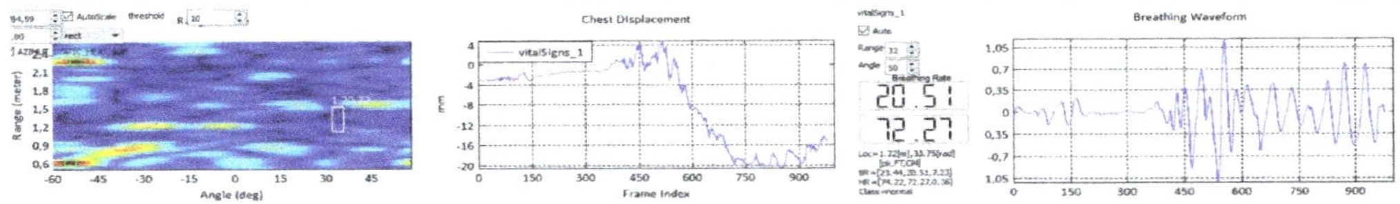	21	20.51
Deep Quick	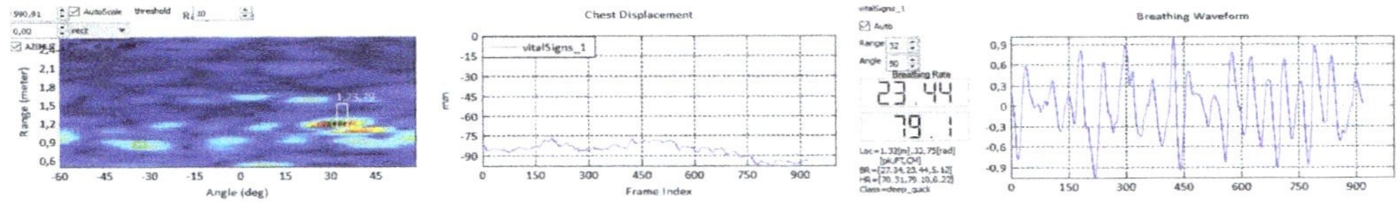	23	23.44
Deep	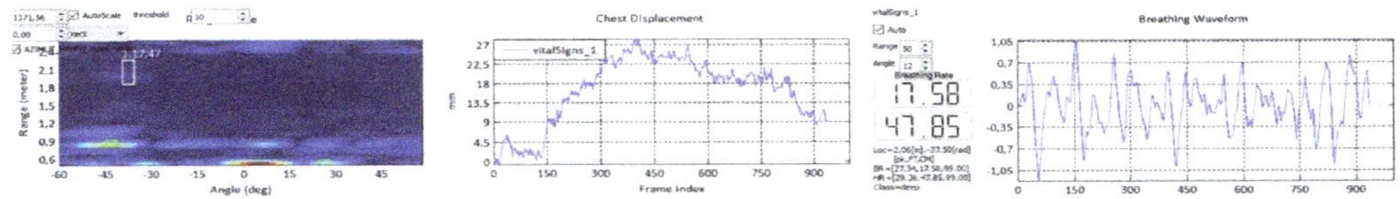	17	17.58
Quick	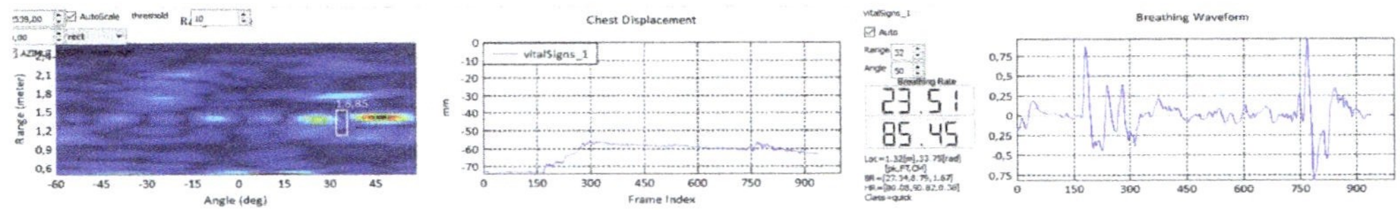	22	23.51
Hold	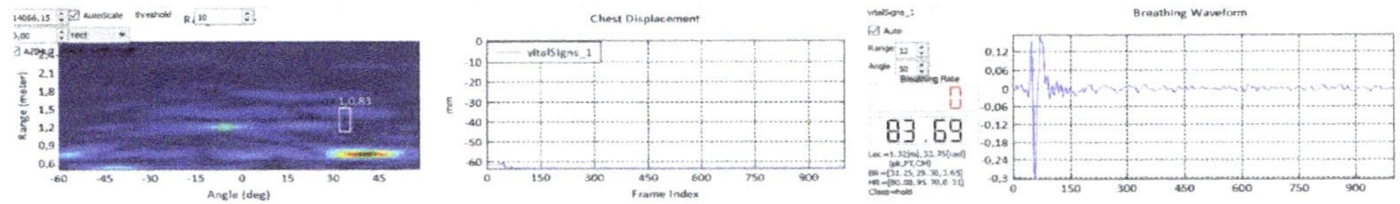	0	0

The fourth experiment detected the quick breath, shown in the fourth row in Table 6. The breathing waveform was detected at 1.88 m from the sensor with a small amplitude and high frequency (high respiration rate). The detected breathing rate was 23.51 breaths/minute.

The last experiment measured the holding breath class. The results show us that the object was detected at about 1.08 m with an angle of 30 degrees. The breathing waveform is almost disappeared, and the amplitudes are close to zero. The estimated breathing rate was 0 breaths/min.

Based on our real-time experiment, Table 6 presents that our real-time implementation can successfully classify five different breathing waveform classes. This shows us that the proposed system can be used to monitor and classify the breathing waveform in real-time. Besides, the breathing rate result shows that our respiration rate has a close value to the manual calculation of the breathing rate, as shown in Table 6.

5. Conclusions

In this paper, we have proposed a non-contact monitoring and classification system for breathing patterns using XGBoost classifier and MFCC feature extraction. Based on the results, the system reached 87.375% accuracy. We also compared the impact of adding MFCC feature extraction to statistical feature extraction and without feature extraction. The results show that the XGBoost classifier with the MFCC feature extraction achieves the best accuracy in classifying five breathing patterns. Thus, we implemented our proposed system in real time by using MFCC feature extraction. Our real-time experiment verifies that our system successfully classifies five different classes of breathing waveform. This shows us that the proposed system can be used to monitor and classify the breathing waveform disorder in real time.

The proposed system will not be a perfect substitute for a professional doctor. It is hoped that this assistance will help practitioners to monitor and analyze the patients. In some cases, the practitioner may make mistakes, pay little attention to the patients, or perform poor report analyses. Thus, it will act as a better solution for now.

In the future, more breathing patterns and classification algorithms will be investigated, and a larger data set will be built. It is hoped that the detection of multiple subjects can be carried out, and the classification model can also be optimized. Since this sensor can be connected to the computer, it also allows us to monitor the breathing waveform with a centralized system. Hence, the supervision of breathing patterns with a centralized system can be developed. In addition, FMCW can also be used to conduct measurements behind interrupted objects such as curtains, walls and others. Therefore, the development of this study is not only useful for the medical field but also for other fields that require detection without physical contact, such as searching for and locating people trapped under rubble. Thus, it would be very helpful for saving lives during a disaster.

Under a controlled environment, all the mentioned methods can work properly. However, monitoring and measuring the breathing pattern in a noisy environment is a challenge that needs to be overcome to make the system stronger and more reliable in the future.

Author Contributions: Conceptualization and methodology, A.T.P. and W.F.H.; software, A.T.P. and W.F.H.; validation, A.T.P. and W.F.H.; formal analysis, A.T.P., D.-B.L. and T.A.; resources, A.T.P., D.-B.L., T.A. and W.F.H.; data, A.T.P., D.-B.L., T.A. and W.F.H.; writing—original draft preparation, A.T.P.; writing—review and editing, A.T.P., D.-B.L., T.A. and W.F.H.; visualization, A.T.P.; supervision, D.-B.L. and T.A. All authors have read and agreed to the published version of the manuscript.

Funding: This work was supported in part by the Ministry of Science and Technology, Taiwan.

Institutional Review Board Statement: Not applicable.

Informed Consent Statement: Not applicable.

Data Availability Statement: Not applicable.

Conflicts of Interest: The authors declare no conflict of interest.

References

1. Spinelli, A.; Pellino, G. COVID-19 pandemic: Perspectives on an unfolding crisis. *Br. J. Surg.* **2020**, *107*, 785–787. [CrossRef] [PubMed]
2. Zheng, Y.Y.; Ma, Y.T.; Zhang, J.Y.; Xie, X. COVID-19 and the cardiovascular system. *Nat. Rev. Cardiol.* **2020**, *17*, 259–260. [CrossRef] [PubMed]
3. Singhal, T. A review of coronavirus disease-2019 (COVID-19). *Indian J. Pediatrics* **2020**, *87*, 281–286. [CrossRef] [PubMed]
4. Dong, D.; Tang, Z.; Wang, S.; Hui, H.; Gong, L.; Lu, Y.; Xue, Z.; Liao, H.; Chen, F.; Yang, F.; et al. The role of imaging in the detection and management of COVID-19: A review. *IEEE Rev. Biomed. Eng.* **2020**, *14*, 16–29. [CrossRef]
5. Cai, J.; Sun, W.; Huang, J.; Gamber, M.; Wu, J.; He, G. Indirect virus transmission in cluster of COVID-19 cases, Wenzhou, China, 2020. *Emerg. Infect. Dis.* **2020**, *26*, 1343–1345. [CrossRef] [PubMed]
6. Jones, N.R.; Qureshi, Z.U.; Temple, R.J.; Larwood, J.P.; Greenhalgh, T.; Bourouiba, L. Two metres or one: What is the evidence for physical distancing in covid-19? *BMJ* **2020**, *370*, m3223. [CrossRef]
7. Salathé, M.; Althaus, C.L.; Neher, R.; Stringhini, S.; Hodcroft, E.; Fellay, J.; Zwahlen, M.; Senti, G.; Battegay, M.; Wilder-Smith, A.; et al. COVID-19 epidemic in Switzerland: On the importance of testing, contact tracing and isolation. *Swiss Med Wkly.* **2020**, *150*, w20225. [CrossRef]
8. Lewnard, J.A.; Lo, N.C. Scientific and ethical basis for social-distancing interventions against COVID-19. *Lancet Infect. Dis.* **2020**, *20*, 631–633. [CrossRef]
9. Pan, L.; Mu, M.; Yang, P.; Sun, Y.; Wang, R.; Yan, J.; Li, P.; Hu, B.; Wang, J.; Hu, C.; et al. Clinical characteristics of COVID-19 patients with digestive symptoms in Hubei, China: A descriptive, cross-sectional, multicenter study. *Am. J. Gastroenterol.* **2020**, *115*. [CrossRef] [PubMed]
10. Su, W.-C.; Juan, P.-H.; Chian, D.-M.; Horng, T.-S.J.; Wen, C.-K.; Wang, F.-K. 2-D Self-Injection-Locked Doppler Radar for Locating Multiple People and Monitoring Their Vital Signs. *IEEE Trans. Microw. Theory Tech.* **2021**, *69*, 1016–1026. [CrossRef]
11. Poyiadji, N.; Shahin, G.; Noujaim, D.; Stone, M.; Patel, S.; Griffith, B. COVID-19—Associated acute hemorrhagic necrotizing encephalopathy: CT and MRI features. *Radiology* **2020**, *296*, E119–E120. [CrossRef]
12. Xu, Z.; Shi, L.; Wang, Y.; Zhang, J.; Huang, L.; Zhang, C.; Liu, S.; Zhao, P.; Liu, H.; Zhu, L.; et al. Pathological findings of COVID-19 associated with acute respiratory distress syndrome. *Lancet Respir. Med.* **2020**, *8*, 420–422. [CrossRef]
13. Wang, Y.; Hu, M.; Li, Q.; Zhang, X.P.; Zhai, G.; Yao, N. Abnormal respiratory patterns classifier may contribute to large-scale screening of people infected with COVID-19 in an accurate and unobtrusive manner. *arXiv* **2020**, arXiv:2002.05534.
14. Massaroni, C.; Lo Presti, D.; Formica, D.; Silvestri, S.; Schena, E. Non-Contact Monitoring of Breathing Pattern and Respiratory Rate via RGB Signal Measurement. *Sensors* **2019**, *19*, 2758. [CrossRef]
15. Cretikos, M.A.; Bellomo, R.; Hillman, K.; Chen, J.; Finfer, S.; Flabouris, A. Respiratory rate: The neglected vital sign. *Med J. Aust.* **2008**, *188*, 657–659. [CrossRef]
16. Hellewell, J.; Abbott, S.; Gimma, A.; Bosse, N.I.; Jarvis, C.I.; Russell, T.W.; Munday, J.D.; Kucharski, A.J.; Edmunds, W.J.; Sun, F.; et al. Feasibility of controlling COVID-19 outbreaks by isolation of cases and contacts. *Lancet Glob. Health* **2020**, *8*, 488–496. [CrossRef]
17. Lee, H.; Kim, B.-H.; Park, J.-K.; Yook, J.-G. A Novel Vital-Sign Sensing Algorithm for Multiple Subjects Based on 24-GHz FMCW Doppler Radar. *Remote Sens.* **2019**, *11*, 1237. [CrossRef]
18. Jin, F.; Zhang, R.; Sengupta, A.; Cao, S.; Hariri, S.; Agarwal, N.K.; Agarwal, S.K. Multiple Patients Behavior Detection in Real-time using mmWave Radar and Deep CNNs. In Proceedings of the 2019 IEEE Radar Conference (RadarConf), Boston, MA, USA, 22–26 April 2019.
19. Cardillo, E.; Caddemi, A. Radar Range-Breathing Separation for the Automatic Detection of Humans in Cluttered Environments. *IEEE Sens. J.* **2020**. [CrossRef]
20. Cardillo, E.; Li, C.; Caddemi, A. Vital Sign Detection and Radar Self-Motion Cancellation Through Clutter Identification. *IEEE Trans. Microw. Theory Tech.* **2021**, *69*, 1932–1942. [CrossRef]
21. Miao, D.; Zhao, H.; Hong, H.; Zhu, X.; Li, C. Doppler radar-based human breathing patterns classification using Support Vector Machine. In Proceedings of the 2017 IEEE Radar Conference (RadarConf), Seattle, WA, USA, 8–12 May 2017.
22. Ji, S.; Wen, H.; Wu, J.; Zhang, Z.; Zhao, K. Systematic Heartbeat Monitoring using a FMCW mm-Wave Radar. In Proceedings of the 2021 IEEE International Conference on Consumer Electronics and Computer Engineering (ICCECE), Guangzhou, China, 15–17 January 2021.
23. Al-Naji, A.; Gibson, K.; Lee, S.-H.; Chahl, J. Monitoring of Cardiorespiratory Signal: Principles of Remote Measurements and Review of Methods. *IEEE Access* **2017**, *5*, 15776–15790. [CrossRef]
24. Harford, M.; Catherall, J.; Gerry, S.; Young, J.D.; Watkinson, P. Availability and performance of image-based, non-contact methods of monitoring heart rate, blood pressure, respiratory rate, and oxygen saturation: A systematic review. *Physiol. Meas.* **2019**, *40*, 06TR01. [CrossRef] [PubMed]
25. Yu, X.; Laurentius, T.; Bollheimer, C.; Leonhardt, S.; Hoog Antink, C. Noncontact Monitoring of Heart Rate and Heart Rate Variability in Geriatric Patients Using Photoplethysmography Imaging. *IEEE J. Biomed. Health Inform.* **2020**, 1. [CrossRef]
26. Kebe, M.; Gadhafi, R.; Mohammad, B.; Sanduleanu, M.; Saleh, H.; Al-Qutayri, M. Human Vital Signs Detection Methods and Potential Using Radars: A Review. *Sensors* **2020**, *20*, 1454. [CrossRef]

27. Fioranelli, F.; Le Kernec, J.; Shah, S.A. Radar for Health Care: Recognizing Human Activities and Monitoring Vital Signs. *IEEE Potentials* **2019**, *38*, 16–23. [CrossRef]
28. Wang, P.; Boufounos, P.; Mansour, H.; Orlik, P.V. Slow-Time MIMO-FMCW Automotive Radar Detection with Imperfect Waveform Separation. In Proceedings of the ICASSP 2020-2020 IEEE International Conference on Acoustics, Speech and Signal. Processing (ICASSP), Barcelona, Spain, 4–8 May 2020.
29. Tang, L.; Meng, H.; Chen, X.; Zhang, J.; Lv, L.; Liu, K. A Novel 3D Imaging Method of FMCW MIMO-SAR. In Proceedings of the 2018 China International SAR Symposium (CISS), Shanghai, China, 10–12 October 2018.
30. Wang, Y.; Wang, W.; Zhou, M.; Ren, A.; Tian, Z. Remote Monitoring of Human Vital Signs Based on 77-GHz mm-Wave FMCW Radar. *Sensors* **2020**, *20*, 2999. [CrossRef] [PubMed]
31. Su, W.-C.; Tang, M.-C.; Arif, R.E.; Horng, T.-S.; Wang, F.-K. Stepped-Frequency Continuous-Wave Radar With Self-Injection-Locking Technology for Monitoring Multiple Human Vital Signs. *IEEE Trans. Microw. Theory Tech.* **2019**, *67*, 5396–5405. [CrossRef]
32. Lee, Y.S.; Pathirana, P.N.; Caelli, T.; Evans, R. Doppler radar in respiratory monitoring: Detection and analysis. In Proceedings of the 2013 International Conference on Control. Automation and Information Sciences (ICCAIS), Nha Trang, Vietnam, 25–28 November 2013.
33. Zito, D.; Pepe, D.; Mincica, M.; Zito, F.; Tognetti, A.; Lanata, A.; De Rossi, D. SoC CMOS UWB Pulse Radar Sensor for Contactless Respiratory Rate Monitoring. *IEEE Trans. Biomed. Circuits Syst.* **2011**, *5*, 503–510. [CrossRef]
34. Chen, T.; Guestrin, C. XGBoost. In Proceedings of the 22nd ACM SIGKDD International Conference on Knowledge Discovery and Data Mining, San Francisco, CA, USA, 13–17 August 2016.
35. Parui, S.; Roshan Bajiya, A.K.; Samanta, D.; Chakravorty, N. Emotion Recognition from EEG Signal using XGBoost Algorithm. In Proceedings of the 2019 IEEE 16th India Council International Conference (INDICON), Rajkot, India, 13–15 December 2019.
36. Sharma, D.; Ali, I. A modified MFCC feature extraction technique for robust speaker recognition. In Proceedings of the 2015 International Conference on Advances in Computing, Communications and Informatics (ICACCI), Kochi, India, 10–13 August 2015.
37. Wang, X.Y. The Improved MFCC Speech Feature Extraction Method and its Application. *Adv. Mater. Res.* **2013**, *756*, 4059–4062. [CrossRef]
38. Kiyokawa, H.; Greenberg, M.; Shirota, K.; Pasterkamp, H. Auditory Detection of Simulated Crackles in Breath Sounds. *Chest* **2001**, *119*, 1886–1892. [CrossRef]
39. Casalino, G.; Castellano, G.; Zaza, G. A mHealth solution for contact-less self-monitoring of blood oxygen saturation. In Proceedings of the 2020 IEEE Symposium on Computers and Communications (ISCC), Rennes, France, 7–10 July 2020.
40. Abdelnasser, H.; Harras, K.A.; Youssef, M. UbiBreathe. In Proceedings of the 16th ACM International Symposium on Mobile Ad Hoc Networking and Computing, Hangzhou, China, 22–25 June 2015.
41. Liu, X.; Cao, J.; Tang, S.; Wen, J.; Guo, P. Contactless Respiration Monitoring Via Off-the-Shelf WiFi Devices. *IEEE Trans. Mob. Comput.* **2016**, *15*, 2466–2479. [CrossRef]
42. Wang, X.; Yang, C.; Mao, S. PhaseBeat: Exploiting CSI Phase Data for Vital Sign Monitoring with Commodity WiFi Devices. In Proceedings of the 2017 IEEE 37th International Conference on Distributed Computing Systems (ICDCS), Atlanta, GA, USA, 5–8 June 2017.
43. Rahman, A.; Lubecke, V.M.; Boric-Lubecke, O.; Prins, J.H.; Sakamoto, T. Doppler Radar Techniques for Accurate Respiration Characterization and Subject Identification. *IEEE J. Emerg. Sel. Top. Circuits Syst.* **2018**, *8*, 350–359. [CrossRef]
44. Ahmad, A.; Roh, J.C.; Wang, D.; Dubey, A. Vital signs monitoring of multiple people using a FMCW millimeter-wave sensor. In Proceedings of the 2018 IEEE Radar Conference (RadarConf18), Oklahoma City, OK, USA, 23–27 April 2018.
45. Hu, W.; Zhao, Z.; Wang, Y.; Zhang, H.; Lin, F. Noncontact Accurate Measurement of Cardiopulmonary Activity Using a Compact Quadrature Doppler Radar Sensor. *IEEE Trans. Biomed. Eng.* **2014**, *61*, 725–735. [CrossRef]
46. Dell'Aversano, A.; Natale, A.; Buonanno, A.; Solimene, R. Through the Wall Breathing Detection by Means of a Doppler Radar and MUSIC Algorithm. *IEEE Sens. Lett.* **2017**, *1*, 1–4. [CrossRef]
47. van Loon, K.; Breteler, M.J.; van Wolfwinkel, L.; Rheineck Leyssius, A.T.; Kossen, S.; Kalkman, C.J.; van Zaane, B.; Peelen, L.M. Wireless non-invasive continuous respiratory monitoring with FMCW radar: A clinical validation study. *J. Clin. Monit. Comput.* **2015**, *30*, 797–805. [CrossRef]
48. He, M.; Nian, Y.; Gong, Y. Novel signal processing method for vital sign monitoring using FMCW radar. *Biomed. Signal Process. Control* **2017**, *33*, 335–345. [CrossRef]
49. Prat, A.; Blanch, S.; Aguasca, A.; Romeu, J.; Broquetas, A. Collimated Beam FMCW Radar for Vital Sign Patient Monitoring. *IEEE Trans. Antennas Propag.* **2019**, *67*, 5073–5080. [CrossRef]
50. Taylor, W.; Abbasi, Q.H.; Dashtipour, K.; Ansari, S.; Shah, S.A.; Khalid, A.; Imran, M.A. A Review of the State of the Art in Non-Contact Sensing for COVID-19. *Sensors* **2020**, *20*, 5665. [CrossRef] [PubMed]
51. AL-Khalidi, F.Q.; Saatchi, R.; Burke, D.; Elphick, H.; Tan, S. Respiration rate monitoring methods: A review. *Pediatric Pulmonol.* **2011**, *46*, 523–529. [CrossRef]
52. Ceniccola, G.D.; Castro, M.G.; Piovacari, S.M.; Horie, L.M.; Corrêa, F.G.; Barrere, A.P.; Toledo, D.O. Current technologies in body composition assessment: Advantages and disadvantages. *Nutrition* **2019**, *62*, 25–31. [CrossRef]
53. Wang, L.; Lin, Z.Q.; Wong, A. COVID-Net: A tailored deep convolutional neural network design for detection of COVID-19 cases from chest X-ray images. *Sci. Rep.* **2020**, *10*, 1–12.

54. Nam, Y.; Kong, Y.; Reyes, B.; Reljin, N.; Chon, K.H. Monitoring of Heart and Breathing Rates Using Dual Cameras on a Smartphone. *PLoS ONE* **2016**, *11*, e0151013. [CrossRef] [PubMed]
55. Bhattacharya, A.; Vaughan, R. Deep Learning Radar Design for Breathing and Fall Detection. *IEEE Sens. J.* **2020**, *20*, 5072–5085. [CrossRef]
56. Barthel, P.; Wensel, R.; Bauer, A.; Muller, A.; Wolf, P.; Ulm, K.; Huster, K.M.; Francis, D.P.; Malik, M.; Schmidt, G. Respiratory rate predicts outcome after acute myocardial infarction: A prospective cohort study. *Eur. Heart J.* **2012**, *34*, 1644–1650. [CrossRef] [PubMed]
57. Silva, T.A.; Silva, L.F.; Muchaluat-Saade, D.C.; Conci, A. A Computational Method to Assist the Diagnosis of Breast Disease Using Dynamic Thermography. *Sensors* **2020**, *20*, 3866. [CrossRef]
58. Lahiri, B.B.; Bagavathiappan, S.; Jayakumar, T.; Philip, J. Medical applications of infrared thermography: A review. *Infrared Phys. Technol.* **2012**, *55*, 221–235. [CrossRef]
59. Qiu, H.; Wu, J.; Hong, L.; Luo, Y.; Song, Q.; Chen, D. Clinical and epidemiological features of 36 children with coronavirus disease 2019 (COVID-19) in Zhejiang, China: An observational cohort study. *Lancet Infect. Dis.* **2020**, *20*, 689–696. [CrossRef]
60. Chen, J.; Qi, T.; Liu, L.; Ling, Y.; Qian, Z.; Li, T.; Li, F.; Xu, Q.; Zhang, Y.; Xu, S.; et al. Clinical progression of patients with COVID-19 in Shanghai, China. *J. Infect.* **2020**, *80*, e1–e6. [CrossRef]
61. Reddi, B.; Fletcher, N. Physics of ultrasound. *Focused Intensive Care Ultrasound* **2019**, 9–16. [CrossRef]
62. Genc, A.; Ryk, M.; Suwała, M.; Zurakowska, T.; Kosiak, W. Ultrasound imaging in the general practitioner's office–A literature review. *J. Ultrason.* **2016**, *16*, 78. [CrossRef]
63. Li, C.; Xiao, Y.; Lin, J. A 5GHz Double-Sideband Radar Sensor Chip in 0.18 μm CMOS for Non-Contact Vital Sign Detection. *IEEE Microw. Wirel. Compon. Lett.* **2008**, *18*, 494–496.
64. Lee, Y.S.; Pathirana, P.N.; Steinfort, C.L.; Caelli, T. Monitoring and Analysis of Respiratory Patterns Using Microwave Doppler Radar. *IEEE J. Transl. Eng. Health Med.* **2014**, *2*, 1–12. [CrossRef] [PubMed]
65. Staderini, E.M. UWB radars in medicine. *IEEE Aerosp. Electron. Syst. Mag.* **2002**, *17*, 13–18. [CrossRef]
66. Immoreev, I. Practical Application of Ultra-Wideband Radars. In Proceedings of the 2006 3rd International Conference on Ultrawideband and Ultrashort Impulse Signals, Sevastopol, Ukraine, 18–22 September 2006.
67. Adib, F.; Mao, H.; Kabelac, Z.; Katabi, D.; Miller, R.C. Smart Homes that Monitor Breathing and Heart Rate. *Proc. 33rd Annu. ACM Conf. Human Factors Comput. Syst.* **2015**, 837–846. [CrossRef]
68. Schleicher, B.; Nasr, I.; Trasser, A.; Schumacher, H. IR-UWB Radar Demonstrator for Ultra-Fine Movement Detection and Vital-Sign Monitoring. *IEEE Trans. Microw. Theory Tech.* **2013**, *61*, 2076–2085. [CrossRef]
69. Li, C.; Lubecke, V.M.; Boric-Lubecke, O.; Lin, J. A Review on Recent Advances in Doppler Radar Sensors for Noncontact Healthcare Monitoring. *IEEE Trans. Microw. Theory Tech.* **2013**, *61*, 2046–2060. [CrossRef]
70. Droitcour, A.; Lubecke, V.; Jenshan, L.; Boric-Lubecke, O. A microwave radio for Doppler radar sensing of vital signs. In Proceedings of the 2001 IEEE MTT-S International Microwave Sympsoium Digest (Cat. No.01CH37157), Phoenix, AZ, USA, 20–24 May 2001.
71. Muehlsteff, J.; Thijs, J.A.J.; Pinter, R. The use of a two channel Doppler radar sensor for the characterization of heart motion phases. In Proceedings of the 2006 International Conference of the IEEE Engineering in Medicine and Biology Society, New York, NY, USA, 30 August–3 September 2006.
72. Pisa, S.; Pittella, E.; Piuzzi, E. A survey of radar systems for medical applications. *IEEE Aerosp. Electron. Syst. Mag.* **2016**, *31*, 64–81. [CrossRef]
73. Tu, J.; Lin, J. Fast Acquisition of Heart Rate in Noncontact Vital Sign Radar Measurement Using Time-Window-Variation Technique. *IEEE Trans. Instrum. Meas.* **2016**, *65*, 112–122. [CrossRef]
74. Hsieh, C.-H.; Chiu, Y.-F.; Shen, Y.-H.; Chu, T.-S.; Huang, Y.-H. A UWB Radar Signal Processing Platform for Real-Time Human Respiratory Feature Extraction Based on Four-Segment Linear Waveform Model. *IEEE Trans. Biomed. Circuits Syst.* **2016**, *10*, 219–230. [CrossRef]
75. Wang, S.; Pohl, A.; Jaeschke, T.; Czaplik, M.; Köny, M.; Leonhardt, S.; Pohl, N. A novel ultra-wideband 80 GHz FMCW radar system for contactless monitoring of vital signs. In Proceedings of the 2015 37th Annual International Conference of the IEEE Engineering in Medicine and Biology Society (EMBC), Milan, Italy, 25–29 August 2015; pp. 4978–4981.
76. Xiong, Y.; Peng, Z.; Gu, C.; Li, S.; Wang, D.; Zhang, W. Differential Enhancement Method for Robust and Accurate Heart Rate Monitoring via Microwave Vital Sign Sensing. *IEEE Trans. Instrum. Meas.* **2020**, *69*, 7108–7118. [CrossRef]
77. Lum, L.C. Hyperventilation Syndromes in Medicine and Psychiatry: A Review. *J. Royal Society Med.* **1987**, *80*, 229–231. [CrossRef]
78. Abnormal-Breathing-Patterns. Available online: https://media.lanecc.edu/users/driscolln/RT127/Softchalk/regulation_of_Breathing/regulation_of_Breathing4.html (accessed on 24 March 2021).
79. Texas Instrument IWR1443. Available online: https://www.ti.com/product/IWR1443 (accessed on 24 March 2021).
80. Brooker, G.M. Understanding millimetre wave FMCW radars. In Proceedings of the 1st International Conference on Sensing Technology, Palmerston North, New Zealand, 21–23 November 2005; pp. 152–157.
81. Itoh, K. Analysis of the phase unwrapping problem. *Appl. Opt.* **1982**, *21*, 2470. [CrossRef]
82. Trouve, E.; Nicolas, J.-M.; Maitre, H. Improving phase unwrapping techniques by the use of local frequency estimates. *IEEE Trans. Geosci. Remote Sens.* **1998**, *36*, 1963–1972. [CrossRef]

83. Kranjec, J.; Beguš, S.; Geršak, G.; Drnovšek, J. Non-contact heart rate and heart rate variability measurements: A review. *Biomed. Signal Process. Control* **2014**, *13*, 102–112. [CrossRef]
84. Smith, S.W. *The Scientist and Engineer's Guide to Digital Signal Processing*; California Technical Pub.: San Diego, CA, USA, 1997.
85. Robertson, N. Design IIR Bandpass Filters. Available online: https://www.dsprelated.com/showarticle/1128.php (accessed on 24 March 2021).
86. Patel, K.; Prasad, R.K. Speech recognition and verification using MFCC & VQ. *Int. J. Emerg. Sci. Eng.* **2013**, *1*, 137–140.
87. Mansour, A.H.; Salh, G.Z.A.; Mohammed, K.A. Voice Recognition using Dynamic Time Warping and Mel-Frequency Cepstral Coefficients Algorithms. *Int. J. Comput. Appl.* **2015**, *116*, 34–41. [CrossRef]
88. Muda, L.; Begam, M.; Elamvazuthi, I. Voice recognition algorithms using mel frequency cepstral coefficient (MFCC) and dynamic time warping (DTW) techniques. *arXiv* **2010**, arXiv:1003.4083.
89. Gupta, D.; Bansal, P.; Choudhary, K. The state of the art of feature extraction techniques in speech recognition. *Speech Lang. Process. Hum. Mach. Commun.* **2018**, 195–207. [CrossRef]
90. Davis, S.T.E.V.E.N.B.; Mermelstein, P.A.U.L. Comparison of Parametric Representations for Monosyllabic Word Recognition in Continuously Spoken Sentences. *Read. Speech Recognit.* **1990**, *28*, 65–74.
91. Brownlee, J. XGBoost for Regression. Available online: https://machinelearningmastery.com/xgboost-for-regression/ (accessed on 24 March 2021).
92. Kasturi, S.N. LightGBM vs XGBOOST: Which Algorithm Win the Race!!! Available online: https://towardsdatascience.com/lightgbm-vs-xgboost-which-algorithm-win-the-race-1ff7dd4917d (accessed on 24 April 2021).
93. Bentéjac, C.; Csörgő, A.; Martínez-Muñoz, G. A comparative analysis of gradient boosting algorithms. *Artif. Intell. Rev.* **2021**, *54*, 1937–1967. [CrossRef]
94. Blei, D.M.; Ng, A.Y.; Jordan, M.I. Latent dirichlet allocation. *J. Mach. Learn. Res.* **2003**, *3*, 993–1022.

sensors

Article

A Baseline for Cross-Database 3D Human Pose Estimation

Michał Rapczyński *, Philipp Werner, Sebastian Handrich and Ayoub Al-Hamadi

Neuro-Information Technology Group, Otto von Guericke University, 39106 Magdeburg, Germany;
Philipp.Werner@ovgu.de (P.W.); Sebastian.Handrich@ovgu.de (S.H.); Ayoub.Al-Hamadi@ovgu.de (A.A.-H.)
* Correspondence: Michal.Rapczynski@ovgu.de

Abstract: Vision-based 3D human pose estimation approaches are typically evaluated on datasets that are limited in diversity regarding many factors, e.g., subjects, poses, cameras, and lighting. However, for real-life applications, it would be desirable to create systems that work under arbitrary conditions ("in-the-wild"). To advance towards this goal, we investigated the commonly used datasets HumanEva-I, Human3.6M, and Panoptic Studio, discussed their biases (that is, their limitations in diversity), and illustrated them in cross-database experiments (for which we used a surrogate for roughly estimating in-the-wild performance). For this purpose, we first harmonized the differing skeleton joint definitions of the datasets, reducing the biases and systematic test errors in cross-database experiments. We further proposed a scale normalization method that significantly improved generalization across camera viewpoints, subjects, and datasets. In additional experiments, we investigated the effect of using more or less cameras, training with multiple datasets, applying a proposed anatomy-based pose validation step, and using OpenPose as the basis for the 3D pose estimation. The experimental results showed the usefulness of the joint harmonization, of the scale normalization, and of augmenting virtual cameras to significantly improve cross-database and in-database generalization. At the same time, the experiments showed that there were dataset biases that could not be compensated and call for new datasets covering more diversity. We discussed our results and promising directions for future work.

Keywords: 3D human pose estimation; deep learning; generalization

Citation: Rapczynski, M.; Werner, P.; Handrich, S.; Al-Hamadi, A. A Baseline for Cross-Database 3D Human Pose Estimation. *Sensors* **2021**, *21*, 3769. https://doi.org/10.3390/s21113769

Academic Editor: Tomasz Krzeszowski

Received: 17 March 2021
Accepted: 24 May 2021
Published: 28 May 2021

1. Introduction

Three-dimensional human body pose estimation is useful for recognizing actions and gestures [1–8], as well as for analyzing human behavior and interaction beyond this [9]. Truly accurate 3D pose estimation requires multiple cameras [10–12], special depth-sensing cameras [13–15], or other active sensors [16–18], because with a regular camera, the distance to an object cannot be measured without knowing the object's actual scale. However, many recent works have shown that 2D images suffice to estimate the 3D pose in a local coordinate system of the body (e.g., with its origin in the human hip). Applications such as the recognition of many actions and gestures do not require the accurate position of the body in the 3D world, so local (also called relative) 3D pose estimation from 2D images can be very useful for them.

Due to the challenges of obtaining accurate 3D ground truths, most prior works used one or two of the few publicly available databases for 2D-image-based 3D pose estimation, such as: Human3.6M [19,20], HumanEva-I and HumanEva-II [21,22], Panoptic Studio [10,11], MPI-INF-3DHP [23], or JTA [24]. All these databases were either recorded in a laboratory (a few sequences of MPI-INF-3DHP were recorded outdoors, but the diversity is still very limited) or synthesized and do not cover the full diversity of possible poses, peoples' appearances, camera characteristics, illuminations, backgrounds, occlusions, etc. However, for real-life applications, it would be desirable to create 3D pose estimation systems that work under arbitrary conditions ("in-the-wild") and are not tuned to the characteristics of a particular limited dataset. Reaching this goal requires much effort, probably including the creation of new datasets. However, one step towards better in-the-wild performance is discussing dataset biases and measuring cross-database performance, that is training with one database and testing with another one [25]. This step was addressed in our paper.

Our key contributions are as follows:

1. We reviewed the literature (Section 2) and discussed biases in the commonly used datasets Human3.6M, HumanEva-I, and Panoptic Studio (Section 3), which we also used for our cross-dataset experiments;
2. We proposed a method for harmonizing the dataset-specific skeleton joint definitions (see Section 4.1). It facilitates cross-dataset experiments and training with multiple datasets while avoiding systematic errors. The source code is available at https://github.com/mihau2/Cross-Data-Pose-Estimation (accessed on 27 May 2021);
3. We proposed a scale normalization method that significantly improves generalization across cameras, subjects, and databases by up to 50% (see Section 4.2). Although normalization is a well-known concept, it has not been consistently used in 3D human pose estimation, especially with the 3D skeletons;
4. We conducted cross-dataset experiments using the method of Martinez et al. [26] (Section 5), showing the negative effect of dataset biases on generalization and the positive impact of the proposed scale normalization. Additional experiments investigated the effect of using more or less cameras (including virtual cameras), training with multiple datasets, applying a proposed anatomy-based pose validation step, and using OpenPose as the basis for the 3D pose estimation. Finally, we discussed our findings, the limitations of our work, and future directions (Section 6).

2. Related Work

Since the work of Shotton et al. [13] and the development of the Kinect sensor, enormous research efforts have been made in the field of human pose estimation. While the work at that time was often based on depth sensors, approaches developed in recent years have focused primarily on estimating the human pose from RGB images. In addition to the high availability of the corresponding sensors, which allows for the generation of extensive datasets in the first place, this is primarily due to the development in the area of deep neural networks, which are very successful in processing visual information. Therefore, all current approaches are based on deep neural networks, but, according to their objectives, can be roughly divided into three categories.

The quantitative results of prior works are summarized in Tables 1–3 for reference.

2.1. 2D Human Pose Estimation

The first class of approaches aims to predict the 2D skeleton joint positions from an RGB input image. In their approach called convolutional pose machines [27], the authors proposed a network architecture of cascading convolutional networks to predict belief maps encoding the 2D joint positions, where each stage refines the prediction of the previous stage. This approach was extended by Newell et al. [28] by replacing the basic convolutional networks with repeated bottom-up, top-down processing networks with intermediate supervision (stacked hourglass) to better consolidate features across all scales and preserve spatial information at multiple resolutions. In [29], the pose estimation problem was split into two stages. A base network with a pyramidal structure aimed to detect the 2D joint positions, while a refinement network explicitly learned to predict the "hard-to-detect" keypoints, i.e., keypoints that were not detected by the base network during the training process. In addition to 2D keypoints, the network in the part affinity field approach [30] learns to predict the orientation and location of several body parts (limbs), resulting in superior keypoint detection. This is particularly helpful when it comes to associating multiple detected joint positions with individuals in multi-person scenarios. This approach was later integrated into the OpenPose framework [31]. In [32], the authors replaced the discrete pixelwise heat map matching with a fully differentiable spatial regression loss.

This led to an improved pose estimation, as the low resolution of the predicted heat maps no longer limited the spatial precision of the detected keypoints. Furthermore, several regularization strategies increasing the prediction accuracy were proposed. Human pose estimation in multi-person scenarios poses a particular challenge. Top-down approaches (e.g., [33]) perform a person detection (bounding boxes) followed by a single-person pose estimation, but typically suffer from partial or even complete overlaps. In contrast, bottom-up approaches [34] first detect all joint positions and then attempt to partition them into corresponding person instances. However, this requires solving an NP-hard partitioning problem. The authors in [35] addressed this problem by simultaneously modeling person detection and joint partitioning as a regression process. For this purpose, the centroid of the associated person was predicted for each pixel of the input image. In [36], the authors first identified similarities among the several approaches for human pose estimation and provided a list of best practices. In their own approach, the authors achieved a state-of-the-art performance by replacing upsample layers with deconvolutional filters and adding optical flow for tracking across multiple images. Whereas all other approaches obtain high-resolution representations by recovering from low-resolution maps using some kind of upscaling networks, Sun et al. [37] proposed HRNet, a network architecture that is able to maintain high-resolution representations throughout all processing steps, leading to superior performance on 2D human pose estimation.

Table 1. Mean per-joint position error (MPJPE) for state-of-the-art approaches on H36M.

Method (Reference)	MPJPE (mm)	Method (Reference)	MPJPE (mm)
Ionescu et al. [20]	162.1	Habibie et al. [38]	65.7
Pavlakos et al. [39]	115.1	Zhou et al. [40]	64.9
Chen and Ramanan [41]	114.2	Sun et al. [42]	64.1
Zhou et al. [43]	113.0	Luo et al. [44]	61.3
Tome et al. [45]	88.4	Rogez et al. [46]	61.2
Martinez et al. [26]	87.3	Nibali et al. [47]	55.4
Pavlakos et al. [48]	75.9	Luvizon et al. [49]	53.2
Wang et al. [50]	71.9	Dabral et al. [51]	52.1
Tekin et al. [52]	69.7	Li et al. [53]	50.9
Chen et al. [54]	68.0	Lin and Lee [55]	46.6
Katircioglu et al. [56]	67.3	Chen et al. [57]	44.1
Benzine et al. [58]	66.4	Wu and Xiao [59]	43.2
Sárándi et al. [60]	65.7	Cheng et al. [61]	42.9

Table 2. Mean per-joint position error (MPJPE) for state-of-the-art approaches on the PAN dataset.

Method (Reference)	MPJPE (mm)
Popa et al. [62]	203.4
Zanfir et al. [63]	153.4
Zanfir et al. [64]	72.1
Benzine et al. [58]	68.5

Table 3. Mean per-joint position error (MPJPE) for state-of-the-art approaches on the HumanEva-I dataset.

Method (Reference)	MPJPE (mm)
Radwan et al. [65]	89.5
Wang et al. [50]	71.3
Yasin et al. [66]	38.9
Moreno-Noguer [67]	26.9
Pavlakos et al. [68]	25.5
Martinez et al. [26]	24.6
Pavlakos et al. [39]	18.3

2.2. 3D Human Pose Estimation from 2D Images

The next class of approaches predicts 3D skeleton joint positions using raw 2D RGB images as the input. Li and Chan [69] used a multitask learning approach to simultaneously train a body part detector and a pose regressor using a fully connected network. In contrast to the direct regression of a pose vector, Pavlakos et al. [68] transferred the idea of the heat map-based 2D pose estimation into the 3D domain and predicted per-joint 3D heat maps using a coarse-to-fine stacked hourglass network, where each voxel contains the probability that the joint is located at this position. Each refinement stage increases the resolution of the z-prediction. Tekin et al. [52] proposed fusing features extracted from the raw image with features extracted from 2D heat maps to obtain a 3D pose regression vector. A similar approach was developed in [40], but instead of deriving features from an already predicted heat map, the authors utilized latent features from the 2D pose regression network. Their end-to-end trainable approach allows for sharing common representations between the 2D and the 3D pose estimation tasks, leading to an improved accuracy. Dabral et al. [51] utilized the same architecture as in [40], but introduced anatomically inspired loss functions, which penalize pose predictions with illegal joint angles and non-symmetric limb lengths. In LCR-Net [46], the pose estimation problem was split into three subtasks: localization, classification, and regression. During localization, candidate boxes and a list of pose proposals are generated using a region proposal network. The proposals are then scored by a classifier and subsequently refined by regressing a set of per-anchor-pose vectors. The subnets share layers so that the complete process can be trained end-to-end. Kanazawa et al. [70] took a slightly different approach. Instead of keypoints, the authors aimed to predict a full 3D mesh by minimizing the reconstruction error. Since the reconstruction loss is highly underconstrained, the authors proposed an adversary training to learn whether a predicted shape is realistic or not. Sun et al. [42] evaluated the performance of the differentiable soft-argmax operation as an alternative to the discrete heat map loss in greater detail and verified its effectiveness. Their approach achieved state-of-the-art results on Human3.6M by splitting the volumetric heat maps into separate x-, y- and z-maps, which allowed for mixed training from both 2D and 3D datasets. Instead of directly dealing with joint coordinates, Luo et al. [44] modeled limb orientations to represent 3D poses. The advantage is that orientations are scale invariant and less dependent on the dataset. Their approach achieved good results on several datasets and generalized well to unseen data. In [49], the authors combined action recognition with human pose estimation. The proposed multitask architecture predicted both local appearance features, as well as keypoint positions, which were then fused to obtain the final action label. The actual pose estimation was based on heat maps and the soft-argmax function. The approach showed state-of-the art results on both pose estimation and action recognition. Another multitask approach was presented by Trumble et al. [71]. It simultaneously estimates 3D human pose and body shape using a symmetric convolutional autoencoder. However, the approach relies on multi-view camera inputs. Approaches that adapt a kinematic skeleton model to the input data typically rely on the detection of corresponding points. This task has been mostly addressed in scenarios where a depth sensor was available. In contrast to this, DensePose [72] maps an input image to the 3D surface of the human body by regressing body part-specific UV coordinates from each RGB input pixel. The approach showed good results, but one has to keep in mind that identifying correspondences is not yet a complete pose estimation due to possible 2D/3D ambiguities and model constraints. All aforementioned approaches learned a direct mapping between the input data and the pose to be estimated. This must be distinguished from approaches that initially learn a latent representation of either the input or the output data [56,73]. In [56], an overcomplete autoencoder network was used to learn a high-dimensional latent pose representation. The input image was then mapped to the latent space, leading to a better modeling of the dependencies among the human joints. In contrast, Rhodin et al. [73] trained a latent representation of the input data by utilizing an autoencoder structure to map one camera view to another. The pose was then regressed from the latent state space.

The approaches showed good, but not the best results. Sárándi et al. [60] demonstrated the effectiveness of data augmentation. By occluding random positions in the RGB image with samples from the Pascal VOC dataset, the mean per-joint position error (MPJPE) can be reduced by up to 20%, making this approach the ECCV pose estimation challenge winner in 2018. The occlusion acts as a regularizer, forcing the network to learn joint positions from several visual cues. The authors used ResNet as the backbone architecture to generate volumetric heat maps. As high-resolution volumetric heat maps are quite memory intensive, the authors of MargiPose [47] proposed to learn three distinct heat maps instead. The maps represent the xy-, xz-, and yz-plane and can be seen as projections of the volumetric heat map. Their approach, which was based on the Inception v4 model, achieved good results and provided a memory-efficient alternative to volumetric heat maps. Habibie et al. [38] contributed by integrating 3D features in the latent space of the learning process. The regressed 3D pose is back-projected to 2D before the loss is computed and thus allows a 3D pose estimation based on 2D datasets. However, there is no explicit supervision of the hidden feature maps that encode the 3D pose cues. A recent work by Wu and Xiao [59] proposed to model the limbs explicitly. Their approach was somewhat similar to OpenPose [31], but extended it to the 3D domain. Next to 2D keypoints from 2D heat maps, the network learned to predict densely-generated limb depth maps. Latent features from the 2D pose estimator and the depth map estimation, as well as 3D specific additional features were then fused to lift the 2D pose to 3D. Their approach significantly outperformed all other methods on the Human3.6M and MPI-INF-3DHP datasets.

2.3. 3D Human Pose Estimation from the 2D Pose

The last class of approaches attempts to predict the 3D pose from an earlier predicted 2D pose, a process typically known as lifting. A big advantage of separating the lifting from the 2D pose estimation is that it can be pre-trained using synthetic poses. Martinez et al. [26] directly regressed 3D poses from 2D poses using only fully connected layers. Their approach achieved excellent results, at least when using 2D ground truth joint positions as the input. In [41], the authors built a huge library of 3D poses and matched it against a detected 2D pose. Using also the stored camera parameters, the best 3D pose was then scaled in a way that it matched the 2D pose. Pavllo et al. [74] exploited temporal information by using dilated temporal convolutions on 2D keypoint sequences. Hossain and Little [75] designed an efficient sequence-to-sequence network taking a sequence of 2D keypoints as the input to predict temporally consistent 3D poses. Their approach achieved state-of-the-art results for every action class of the Human3.6M dataset. While CNNs are suitable for processing grid-like input data (e.g., images), graph convolutional networks (GCNs) can be seen as a generalization of CNNs acting on graphs. In [76], Zhao et al. exploited the hierarchical structure of skeletons by describing both 2D and 3D skeletons as graphs and used CGNs to obtain 3D poses from 2D poses. The aforementioned approaches reported excellent results, in particular when temporal information was used. However, they heavily relied on the quality of the underlying 2D pose estimator. If no 2D ground truth was used, the accuracy was typically similar to approaches that obtained 2D and 3D poses directly from the image.

2.4. Cross-Dataset Generalization

Comprehensive datasets are required in order to train methods for pose estimation. In contrast to 2D pose estimation, reliable 3D pose data cannot be obtained by manually annotating images taken in-the-wild, but are acquired with the help of motion capture systems (e.g., VICON [77], The Captury [78], IMU). This typically limits the acquisition to controlled in-the-lab environments with low variations in terms of subjects, camera view points, backgrounds, occlusions, lighting conditions, etc. This raises the questions how well these approaches (a) perform across multiple controlled datasets and (b) generalize to unconstrained in-the-wild data. Work in this area is still limited. The typical approach is to combine in-the-wild 2D pose data with in-the-lab 3D pose data. Mehta et al. [23] used transfer learning to transfer knowledge from a pre-trained 2D pose network to a 3D pose regression network [23]. They further provided the MPI-INF-3DHP dataset, an augmented in-the-wild 3D pose dataset, by utilizing a marker-less multi-camera system [78] and chroma keying (green screen). The best results on Human3.6M were achieved using transfer learning and including additional data from MPI-INF-3DHP. Zhou et al. [40] mixed 2D and 3D data per batch to learn common representations between 2D and 3D data by computing additional depth regression and anatomical losses for 3D training samples [40]. When additional 2D pose data from the MPII dataset were included, errors on the Human3.6M dataset were reduced by up to 15 mm, and the proportion of correctly estimated joints (PCKs) increased from 37.7% to 69.2% on the MPI-INF-3DHP dataset. This indicated that the constrained setting of Human3.6M is insufficient to generalize to in-the-wild data. The authors also concluded that adding additional 2D data did not improve the accuracy of the 2D pose prediction, but mostly benefited the depth regression via shared feature representations. As mentioned above, Habibie et al. [38] circumvented the problem of missing 3D pose labels by learning both view parameters and 3D poses. The 3D poses were then back-projected to 2D (using a trainable network) before applying the 2D loss. Their approach showed high accuracy and generalized well to in-the-wild scenes. Other approaches attempt to generate 3D labels from existing 2D datasets. Wang et al. [79] achieved this by first mapping a 2D pose to 3D using a "stereo-inspired" neural network and then refined the 3D pose using a geometric searching scheme so that the determined 3D pose matched the 2D pose with pixel accuracy. In [80], which was an updated version of [46], Rogez et al. [46] created pseudo 3D labels for 2D datasets by looking for the 3D pose that best matched a given 2D pose in a large-scale 3D pose database. Further work addressed the problem of missing 3D pose labels by generating synthetic datasets by animating 3D models [81,82] or rendering textured body scans [83]. While rendering may seem promising, both integrating human models in existing images, as well as rendering realistic scenes are not trivial and often require a domain adaption to generalize from synthetic to real images [81,83,84]. Therefore, Rogez and Schmid [85] proposed to build mosaic pictures of real images from 2D pose datasets. While artificial looking, the authors showed that CNNs can be trained on these image and generalize well to real data without the need for any fine-tuning and domain adaption.

While many authors combined multiple training datasets, work on cross-dataset evaluation is still limited. To the best of our knowledge, the very recent work of Wang et al. [86] was the first to systematically examine the differences among existing pose datasets and their effect on cross-database evaluation. However, they focused on systematic differences of camera viewpoints and conducted their experiment with another set of databases, compared to our work.

2.5. Non-Vision-Based Approaches

All approaches listed so far were based on optical sensors, i.e., cameras. We would like to point out to the reader that besides visual methods, other ranges of the electromagnetic spectrum can also be used to estimate the human pose. The major advantage of these approaches is that they are independent of lighting, background, as well as clothing and even allow for person detection and pose estimation through walls and foreground objects. Moreover, privacy issues can be avoided in contrast to camera-based approaches. The most prominent examples are microwaves and millimeter waves. In [16], the authors proposed a radar-based approach (operating in the 5.56–7.25 GHz range) for indoor person location, obtaining a spatial resolution of 8.8 cm. In RFPose [17], the authors utilized radio frequency (RF) signals (20 kHz–300 GHz) and visual information to extract 2D skeleton positions. The approach was later extended to 3D [87], where the authors reported a mean per-joint localization error of 4.2 cm, 4.0 cm, and 4.9 cm for the X-, Y-, and Z-axes, respectively. However, a major disadvantage of this approach is the very specific and high hardware requirements (synchronized 16 + 4 T-shaped antenna array with frequency-modulated continuous waves), which severely limit its possible applications. There are also LIDAR-based approaches (e.g., [18]), but these are usually expensive and power consuming. More recently, WiFi-based approaches were proposed. In [88], Wang et al. [88] developed a human pose estimation system, which reconstructed 2D skeletons from WiFi by mapping the WiFi data to 2D joint heat maps, part affinity fields, and person segmentation masks. The authors reported an average percentage of correctly detected keypoints (PCK) of 78.75% (89.48% for OpenPose [31]). However, their approach performed significantly worse in untrained environments (mPCK = 31.06%). This is a main challenge for all WiFi-based approaches, as WiFi signals exhibit significantly different propagation patterns in different environments. To address this issue and achieve cross-environment generalization, the authors of WiPose [89] proposed to utilize 3D velocity profiles obtained from WiFi signals in order to separate posture-specific features from the static background objects. Their approach achieved an accuracy of up to 2.83 cm (mean per-joint position error), but is currently limited to a single non-moving person.

Camera-based approaches are passive methods, as they capture the ambient light reflected by an object. In contrast, RF-based methods can be considered as active methods, since an illumination signal is actively emitted and interacts with the objects in the scene before being reflected and measured by the receiver. Here, the active illumination signal is often based on appropriately modulated waves or utilizes stochastic patterns. A major drawback of this approach is that the active signal is not necessarily ideal for the specific task, i.e., it is not possible to distinguish between relevant and irrelevant information during the measurement process.

This leads to the idea of *learned sensing* [90], in which the measurement process and the processing of the measurement data are optimized in an overall system. This requires the availability of programmable transmitter hardware whose configuration is determined using machine learning methods in such a way that the emitted illumination signal is optimal for the respective measurement process. This approach has recently been successfully implemented for person recognition, gesture recognition, and human pose estimation tasks. See [91,92] for further details. The idea of *learned sensing* was also applied in the optical domain in order to determine optimal illumination patterns for specific microscopy tasks [93].

For human pose estimation, the learned sensing approach cannot easily be transposed to optical sensors. This is mainly due to the fact that changes in the active illumination signal can be perceived by humans, which is typically undesirable in real-world scenarios. Nevertheless, we suspect that the method can be transferred to approaches that use special (infrared) photodiodes to determine the pose [94]. Furthermore, there may be an application opportunity in multi-camera scenarios. These are often associated with a costly measurement process (high energy consumption, data volume, latency), whereas only a specific part of the measured data is actually required to resolve potentially occurring ambiguities.

3. Datasets

In the following subsections, we describe the three 3D human pose estimation datasets that we used in this article: HumanEva-I, Human 3.6M, and Panoptic. Afterwards, we compare the datasets and discuss dataset biases.

3.1. HumanEva-I (HE1)

In 2006 and 2010, Sigal et al. [21,22] published the HumanEva-I and HumanEva-II datasets to facilitate the quantitative evaluation and comparison of 3D human pose estimation algorithms. We used HumanEva-I, which is larger and more diverse than HumanEva-II. In HumanEva-I, each of four subjects performs six actions (walking, jogging, gesturing, throwing/catching a ball, boxing, and a sequence of several actions) while being recorded with seven cameras. The ground truth positions of the 15 provided skeleton joints were obtained with a motion capture system using reflective markers.

3.2. Human3.6M (H36M)

Ionescu et al. [19,20] collected and published Human3.6M, which is comprised of 3.6 million frames showing diverse body poses of actors performing 15 everyday actions including conversations, eating, greeting, talking on the phone, posing, sitting, smoking, taking photos, waiting, and walking. It total, eleven actors were involved, but they performed individually one after another (i.e., only one person was visible in each video). The data were recorded with four color video cameras and a marker-based motion capture system, providing thirty-two skeleton joint positions.

3.3. Panoptic (Pan)

Aiming at analyzing social interaction, Joo et al. [10,11] recorded the Panoptic Studio dataset. In its current state (Version 1.2), it is comprised of 84 sequences with more than 100 subjects. The sequences are very diverse, among others covering: social games (Haggling, Mafia, and Ultimatum) with up to eight subjects; playing instruments; dancing; playing with toddlers; and covering range of motion. In contrast to the other datasets, there is no categorization or segmentation of the actions (beyond the above-mentioned categories of sequences). To record the dataset, Joo and colleagues built the Panoptic Studio, a special dome with more than 500 cameras in its walls. Using these cameras, Joo et al. [10,11] developed an algorithm for obtaining multi-person 3D skeleton joint ground truths without markers. Their algorithm was based on 2D body pose estimation providing "weak" proposals, triangulation and fusion of the proposals, and temporal refinement.

3.4. Comparison and Dataset Biases

Computer vision datasets are created for quantitatively measuring and comparing the performance of algorithms. However, "are the datasets measuring the right thing, that is, the expected performance on some real-world task?," Torralba and Efros asked in their article on dataset biases [25]. We were interested in the task of relative 3D human body pose estimation in the real world, not only in a specific laboratory. Therefore, we may ask if the error in estimating poses on a specific dataset resembles the expected error in real-world application. Are these datasets representative samples of real-world data or are they biased in some way?

Currently, most "in-the-wild" datasets are collected from the Internet, including datasets commonly used for 2D human body pose estimation [95,96]. Although these datasets are very diverse, they may still suffer from biases compared to the real world, e.g., capture bias (pictures/videos are often taken in similar ways) or selection bias (certain types of images are uploaded or selected for datasets more often) [25].

The datasets of 3D pose estimation are less diverse. They are typically recorded in a laboratory, because (1) multi-view camera systems are state-of-the-art for measuring accurate 3D ground truths and (2) building, installing, and calibrating these systems requires much effort (making it hard to move the systems). All three datasets, HumanEva-I, Human3.6M, and Panoptic, were recorded in such an indoor laboratory with very controlled conditions; see Figure 1 for some example images. The datasets differ in size and diversity, as summarized in Table 4. Compared to in-the-wild data, the three datasets suffer from several biases:

- Lighting: The recordings are homogeneously lit, typically without any overexposed or strongly shadowed areas. Further, there is no variation in lighting color and color temperature. Real-world data are often more challenging, e.g., consider an outdoor scene with unilateral sunlight or a nightclub scene with colored and moving lighting;
- Background: The backgrounds are static and homogeneous. Real-world data often include cluttered and changing backgrounds, which may challenge the computer vision algorithms more;
- Occlusion: In real-world data, people are often partially occluded by their own body parts, other people, furniture, or other objects; or parts of the body are outside the image. Self-occlusion is covered in all three databases. Human3.6M is comprised

of more self-occlusions than the other datasets (and also some occlusions by chairs), because it includes many occlusion-causing actions such as sitting, lying down, or bending down. Occlusions by other people are common in Panoptic's multi-person sequences. Additionally, parts of the bodies are quite frequently outside of the cameras' field of view in Panoptic;

- Subject appearance: Human3.6M and especially HumanEva-I suffer from a low number of subjects, which restricts variability in body shapes, clothing, hair, age, ethnicity, skin color, etc. Although Panoptic includes many more and quite diverse subjects, it may still not sufficiently cover the huge diversity of real-world human appearances;
- Cameras: In-the-wild data are recorded from different viewpoints with varying resolutions, noise, motion blur, fields of view, depths of field, white-balance, camera-to-subject distance, etc. Within the three databases, only the viewpoint is varied systematically, and the other factors are mostly constant. With more than 500 cameras, Panoptic is the most diverse regarding viewpoint (also using three types of cameras). In contrast to the others, it also includes high-angle and low-angle views (down- and up-looking cameras). If only a few cameras are used, as in Human3.6M and HumanEva-I, there may be a bias in the body poses, because people tend to turn towards one of the cameras (also see [86] on this issue);
- Actions and poses: HumanEva-I and Human3.6M are comprised of the acted behavior of several action categories, whereas the instructions in Human3.6M allowed quite free interpretation and performance. Further, the actions and poses in Human3.6M are much more diverse than in HumanEva-I, including many everyday activities and non-upright poses such as sitting, lying down, or bending down (compared to only upright poses in HumanEva-I). However, some of the acted behavior in Human3.6M used imaginary objects and interaction partners, which may cause subtle behavioral biases compared to natural interaction. Panoptic captured natural behavior in real social interactions of multiple people and interactions with real objects such as musical instruments. Thus, it should more closely resemble real-world behavior;
- Annotated skeleton joints: The labels of the datasets, the ground truth joints provided, differ among the datasets in their number and meaning. Most obviously, the head, neck, and hip joints were defined differently by the dataset creators. In Section 4.1, we discuss this issue in detail and propose a way to handle it.

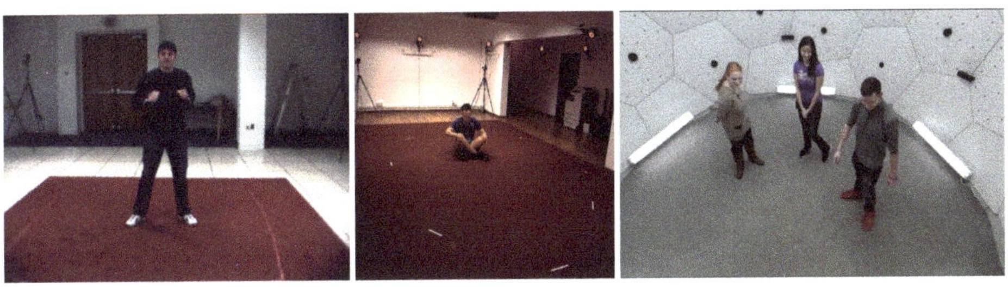

Figure 1. Example images of the HumanEva-I (**left**), the Human3.6M (**middle**), and the Panoptic databases (**right**).

Table 4. Quantitative comparison of the datasets.

	HumanEva-I	Human3.6M	Panoptic
Subjects	4	11	>100
Actions	6	15	many
Multi-person	-	-	✓
Recording duration	10 min	298 min	689 min
Cameras	7	4	>500
Total frames	0.26 M	3.6 M	>500 M
Skeleton joints	15	32	19

Although all the datasets have been and still are very useful to advance the state-of-the-art, we expect that many of these datasets' biases will degrade real-world performance in 3D human pose estimation. As all the datasets were sampled from the real world, we used training and testing with different databases as a surrogate for roughly estimating the expected in-the-wild performance. Such cross-database evaluation is a common practice or a targeted goal in many other domains of computer vision [25,97–102].

Some of the biases, such as lighting, background, as well as subject ethnicity, clothing, and hair, only affect the images, but not the position of body joints. A limited diversity in these factors may be acceptable in 3D pose estimation datasets, because it is no problem for training a geometry-based approach that estimates the 3D pose from the 2D joint positions, given the used 2D pose estimation model has been trained with a sufficiently diverse 2D pose estimation dataset. Other factors, especially cameras and poses, heavily influence the position of body joints and must be covered in great diversity in both 2D and 3D pose datasets.

4. Methods

4.1. Joint Harmonization

As mentioned before, the skeleton joint positions provided in the datasets differed in their number and definition. To be able to conduct cross-dataset experiments, we selected a common set of 15 joints based on HumanEva-I. One keypoint was the central hip joint, which is the origin of the local body coordinate system, i.e., it is always at (0, 0, 0). Thus, we excluded it from the training and error evaluation. The remaining 14 joints are listed in Table 5. The first problem we faced was that there was no head keypoint in Panoptic, because this has not been annotated in the MS COCO dataset [96], which is used for training OpenPose [27,31] and other 2D pose estimators. However, there are MS COCO keypoints for the left and right ear. We calculated the center of gravity of these two points (Number 17 and 18 in Panoptic and OpenPose) in order to get a keypoint at the center of the head.

Table 5. Our joint definitions for the different datasets and OpenPose. The numbers in the table correspond to the joint number in the datasets' original joint definition. The joints marked with * were repositioned in the harmonization process.

Joint	HumanEva-I	Human3.6M	Panoptic	OpenPose
R Hip	1 *	1 *	12	9
R Knee	2	2	13	10
R Ankle	3	3	14	11
L Hip	4 *	6 *	6	12
L Knee	5	7	7	13
L Ankle	6	8	8	14
Neck	7	13 *	0	1
Head	8 *	15 *	(17 + 18)/2	(17 + 18)/2
L Shoulder	9	17	3	5
L Elbow	10	18	4	6
L Hand	11	19	5	7
R Shoulder	12	25	9	2
R Elbow	13	26	10	3
R Hand	14	27	11	4

After this step, we had keypoints for all the joints listed in Table 5. However, there were still obvious differences in some of the joints' relative placements, as illustrated in Figure 2a,c,e. The different skeleton joint definitions introduced systematic errors into the cross-dataset experiments. To counter these effects, we harmonized the joint positions, using Panoptic (and thus the MS COCO-based joints) as the reference. This facilitated combining the 3D pose estimation with MS COCO-based 2D pose estimators, which is a promising research direction, and comparing future results with ours.

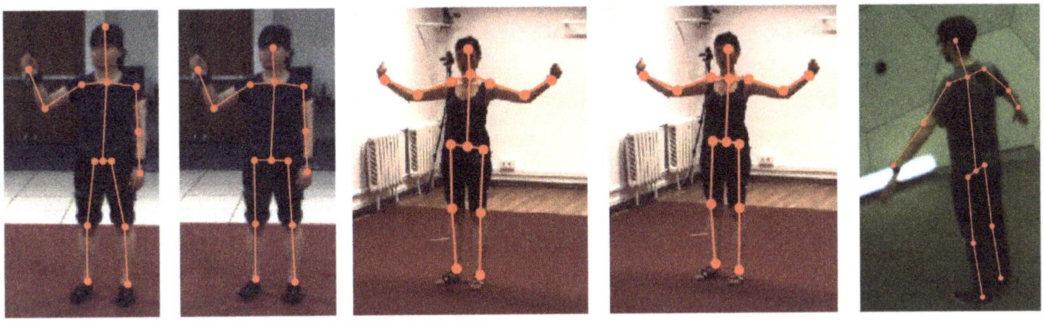

(a) orig. joints, HE1 (b) harm. joints, HE1 (c) orig. joints, H36M (d) harm. joints, H36M (e) orig. joints, PAN

Figure 2. Examples showing the skeleton of HumanEva-I (HE1) and Human3.6M (H36M) before ("original" = orig.) and after the harmonization (harm.) of the head, neck, and hip joints. Panoptic (PAN) was used as the reference for harmonization.

We adjusted the obvious differences in the head, neck, and hip positions in the HumanEva-I (HE1) and Human3.6M (H36M) datasets: (1) the head joint was moved to in between the ears in both HE1 and H36M; (2) the neck was placed between the shoulders in the H36M; and (3) the hip width (distance between the left and right hip keypoints) was expanded in HE1 and reduced in H36M.

To be more precise, the positions of the head and hip joints of the HE1 dataset were harmonized as follows: To move the head closer to the neck, we multiplied the direction vector between the neck and the head joint by a factor of 0.636. We calculated the factor from the ratio of the means of the neck-to-head length of the HE1 (316.1 mm) and the Panoptic (PAN) test datasets (201.1 mm). The hip joint distance was increased from the common center point by a factor of 2.13, again based on scaling the direction vector by the ratio of the mean distances (PAN 205.2 mm, HE1 96.3 mm). Figure 2a,b illustrates the effect of our adjustment.

The joint harmonization of the H36M dataset changed the position of the neck, head, and hip joints. The neck joint, which was defined at a higher position than in the other datasets, was moved to the center between the shoulder joints. To move the head point closer to the neck, we multiplied the direction vector between the repositioned neck joint and the head joint by a factor of 0.885. The factor was calculated from the ratio of the means of the neck-to-head length of the H36M (227.3 mm) and the PAN test datasets (201.1 mm). The hip joint distance was reduced from the common center point to 0.775 of the original value, based on the mean distances (PAN 205.2 mm, H36M 264.9 mm). Figure 2c,d illustrates the effect of the adjustment.

We provided the Python source code for harmonizing the joints at https://github.com/mihau2/Cross-Data-Pose-Estimation/ (accessed on 27 May 2021).

4.2. Scale Normalization

People differ in their heights and limb lengths. On the one hand, this is a problem for 2D-image-based 3D pose estimation, because, in the general case, the real height and limb lengths of a person (as well as the distance from the camera) cannot be measured from a single 2D image; therefore, accurate estimation of 3D joint positions is only possible up to an unknown scaling factor. Nevertheless, most state-of-the-art methods train their relative pose estimation models in a way that forces them to implicitly predict this scale, because they train the models to predict 3D joint coordinates, which implicitly contain the overall scale and the body proportions. This imposes a burden that encourages the models to learn dataset-specific heuristics, such as the height of individual subjects, the mean height of the subjects, the characteristics of the camera used, or the expected height/depth depending on the position and/or size of the person in the image. We expect that this way of training worsens generalization to in-the-wild data and in cross-dataset evaluations. On the other hand, knowing the scaling factor (the real height and limb lengths of the person) is not necessary for many applications that only require relative joint positions. Normalizing the joint positions from absolute to relative coordinates is common practice. We went a step further and proposed to normalize the scale of the skeletons, in order to remove the (often unnecessary) burden of predicting the scale and to improve the cross-dataset performance.

The absolute joint coordinates $\mathbf{p}_i$ of each pose sample were normalized individually based on the skeleton's relative joint positions in relation to the center hip point $\mathbf{p}_0$, which was in the origin of the local coordinate system. We quantified the scale s by calculating the mean of the Euclidean distances between the origin and all N joint positions:

$$s = \frac{1}{N}\sum_{i=1}^{N}\|\mathbf{p}_i - \mathbf{p}_0\| \tag{1}$$

Afterwards, we resized the skeleton by dividing all joint position coordinates by the scale, yielding a normalized scale of 1. The normalized joint positions $\hat{\mathbf{p}}_i$ were calculated as follows:

$$\hat{\mathbf{p}}_i = \frac{1}{s}(\mathbf{p}_i - \mathbf{p}_0) \tag{2}$$

This normalized all poses to a similar coordinate scale. This transformation was applied individually in both the 3D target data (with $\mathbf{p}_i \in \mathbb{R}^3$) and the 2D input data (with $\mathbf{p}_i \in \mathbb{R}^2$).

4.3. Baseline Model and Training

We performed our experiments with the "Baseline" neural network architecture proposed by Martinez et al. [26]. We decided to use an existing method rather than developing a completely new approach, because the focus of our work was on cross-dataset evaluation and proposing improvements that can be applied in many contexts. The approach by Martinez et al. did not rely on images, but mapped 2D skeleton joint positions to relative 3D joint positions, which is also called "lifting". Thus, it can be combined with any existing or future 2D body pose estimation method. This way, the results can benefit from advances in 2D pose estimation, which are faster than in 3D pose estimation, because in-the-wild 2D pose datasets are much easier to create than their counterparts with 3D ground truths. Other advantages include: (1) The approach is independent of image-related issues, such as lighting, background, and several aspects of subject appearance, which are covered with great diversity in 2D pose datasets. By decoupling the 2D pose estimation from the "lifting", we avoided overfitting the 3D pose estimation to the quite restricted diversity of the 3D pose datasets regarding lighting, background, and subject appearance. (2) The approach allowed augmenting the training data by creating synthetic poses and virtual cameras, which can massively increase the variability of the available data and lead to better generalization. (3) No images were needed, so additional sources of training data may be exploited, such as motion capture data recorded in sports, biomechanics, or entertainment. (4) The source code is available. Therefore, it is easy to reproduce the results, apply the method with other data, and start advancing the approach.

The architecture by Martinez et al. [26] was a deep neural network consisting of fully connected layers and using batch normalization, ReLU, dropout, and residual connections. The first layer maps the 2D coordinates ($2n = 28$ dimensions) to a 1024-dimensional space. It is followed by two residual blocks, each including two fully connected layers. Finally, there is another linear layer that maps the 1024-dimensional space to the $3n = 42$-dimensional 3D coordinate output.

The model, training, and testing were implemented in the TensorFlow2 deep learning framework using the Keras API. The networks were trained with the Adam optimizer, minimizing the mean squared error loss function. The training set was separated into training and validation data with a 90/10% split, and the training data were shuffled before each epoch. We used a batch size of 512 and a dropout rate of 0.5. The training of each neural network started with a learning rate of 10^{-3}, which was reduced during the training by a factor of 0.5 if the loss on the validation set did not decrease for 3 epochs. The training was stopped if the validation loss did not decrease for 10 epochs or the learning rate was reduced below 10^{-6}. The model with the lowest validation loss was saved for testing.

4.4. Anatomical Pose Validation

We proposed an optional pose validation step, which assessed the predicted poses using the constraints of human anatomy. The human body is usually symmetrical regarding the length of the left and right extremities and has, according to Pietak et al. [103], stable ratios regarding the lengths of the upper and lower limbs with little variation between individuals.

For every pose, the ratios of each upper and lower extremity, as well as its left and right counterpart were calculated. The ratios were measured as the difference in length in %, based on the shorter of the two compared limbs. Therefore, a ratio of 2:1 and 1:2 would both result in a difference of +100%. If one of the 8 calculated ratios was greater than 100%, the pose was rejected by the validation.

The effect of this approach is analyzed in Section 5.7. All the other experiments were conducted without applying this validation step, because it led to the exclusion of rejected poses from the error calculation and thus limited the comparability of the error measures (which may be based on different subsets of the data).

4.5. Use of Datasets

Each dataset was split into training and test data based on the sessions/subjects and cameras, as illustrated in Figure 3. No single camera or session was used for both the test and training set. Although parts of the datasets were unused, we selected this way of splitting because our focus was to measure the generalization across subjects, camera viewpoints, and datasets rather than reaching the highest in-dataset performance. The cameras were assigned to the test set and the reduced and full training set, as illustrated in Figure 4 and detailed below. Further, because the number of cameras was quite low in Human3.6M (only 4), we generated synthetic camera views as described in Section 4.5.2. After the main split, the training data was further randomly split into 90%, which were used as the actual training set by the optimizer, and 10%, which were used as the validation set.

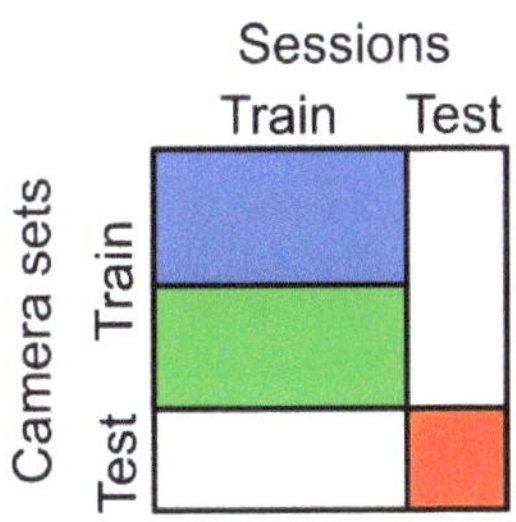

Figure 3. Separation of the training and test sets for the subject sessions and camera sets (blue: reduced training camera set; blue and green: full training camera set; red: test camera set; white: unused data).

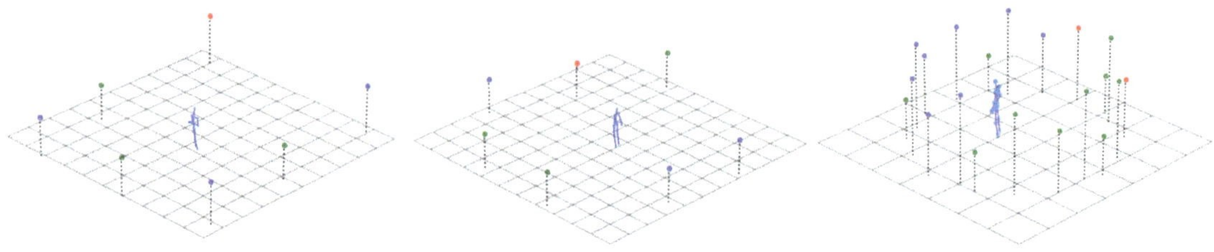

(a) HE1 (b) H36M (c) PAN

Figure 4. Camera positions and an example pose for the used datasets (grid at z = 0 with 1 meter cell size; blue: reduced training camera set; blue and green: full training camera set; red: test camera set).

4.5.1. Dataset Split Details

For HumanEva-I (HE1), Subjects 0 and 1 were used as the training set and Subject 2 as the test set. The camera "BW1" was used for the test set. We used the other black-and-white and color cameras for the training set. The reduced camera set only contained the black-and-white cameras. For the evaluation using the OpenPose 2D joint positions, we used all videos of Subject 2 that contained the corresponding video and motion capture data. This reduced the OpenPose test dataset, in comparison to the standard test set, because only a subset of the sessions included both motion capture and video files.

For Human3.6M (H36M), Subjects 1, 5, 6, 7, and 8 were used for the training set and Subjects 9 and 11 for testing. We used Camera 3 as the test camera. Cameras 0, 1, and 2 were the reduced camera training set. The full camera training set contained Cameras 0, 1, and 2 and their modified synthetic copies (see Section 4.5.2). For the evaluation using the generated OpenPose 2D joint estimations, we used all videos of Subjects 9 and 11.

For Panoptic (PAN), the *Range of Motions* sessions (sequence names: 171026_pose1, 171026_pose2, 171026_pose3, 171204_pose1, 171204_pose2, 171204_pose3, 171204_pose4, 171204_pose5, 171204_pose6) were used for testing and all other sessions for training. Of each panel, we only used VGA Camera No. 1. The cameras on Panels 9 and 10 were used for testing. The cameras on Panels 1–8 were the reduced camera training set, and the cameras on Panels 1–8 and 11–20 were the full training set.

Table 6 shows the resulting sample sizes for the different databases and successfully mapped OpenPose (OP) samples.

Table 6. Sample sizes in thousands of poses.

	Training Set		**Testing Set**
	Reduced	**Full**	
HE1	113	225	17.8
H36M	1169	2312	137.7
PAN	4131	9809	292.0
HE1 (OP)	-	-	1.7
H36M (OP)	-	-	52.1

4.5.2. Virtual Camera Augmentation

The H36M dataset was recorded with only 4 cameras. In order to make the ratio of training and test cameras in the databases more similar, three more camera were added. For this purpose, we virtually copied the training Cameras 0, 1, and 2 by rotating their extrinsic camera parameters by 90° around the world coordinate center in the middle of the recording space without changing the intrinsic camera parameters. This can be seen in Figure 4b, where the blue points represent the original training camera positions, and the newly created cameras are shown in green.

4.6. Implementation Details

First, the original 3D pose data in the world coordinate space were loaded. If a pose contained a non-valid joint position, usually (0, 0, 0), the pose was discarded. Further, we used the jointwise confidence score provided in the PAN dataset to remove unreliable data. If the score of any of the 14 used joints was <0.1, we discarded the corresponding pose. Next, the 3D pose data were transformed to the camera coordinates for each camera of the used set.

If *scale normalization* was applied, the scale of the 3D joint positions was normalized to a mean distance of 1 from the center of the hips. After that, the pose was repositioned to the camera coordinates (0, 0, 50) for the projection step. All pose transformations described in this section used the center of the hips as the reference point.

In the next step, the 3D pose was projected onto the camera 2D image plane including the distortion parameters. Poses with at least one joint outside the projected camera image (1000×1000 px in H36M and 640×480 px in HE1 and PAN) were discarded. This was necessary due to the nonlinear components in the distortion model, which could result in extremely outlying projected points in the image plane if the 3D joint positions were not in the original image frame for which the distortion parameters were calibrated. Next, an additional pose validation step was performed. The limb lengths (for left and right: upper arm, lower arm, upper leg, lower leg, shoulder-to-neck; also the hip width and neck-to-head distance) were calculated once using the original joint descriptions for every database on its complete training set. From that data, the mean length μ and standard deviation σ of the noted limbs were determined. Irregular poses, where at least one limb length deviated more than 3σ from μ, were discarded in the full and reduced training set. These two data validation checks were only done for the projection with the original joint descriptions and without the use of *scale normalization*, to keep the differently processed datasets comparable. The validity status for each sample was saved and applied if the *scale normalization* and/or the harmonized joint descriptions were used, to ensure the same subset of samples was used regardless of the preprocessing steps.

If the *scale normalization* was applied, then the 2D joints of the pose were also normalized to a mean distance of 1 to the center of the hips, and the pose was moved to (0, 0). The 2D poses in the image coordinates were used as training inputs, and the 3D poses in world coordinates, moved to (0, 0, 0), were used as training targets. The data were normalized to a mean of 0 and a standard deviation of 1 for every net input and output channel.

For the evaluation of a model, the 2D inputs of the test dataset were normalized with the models' normalization values calculated on the training set, and the resulting prediction outputs were denormalized analogously. If *scale normalization* was used, the output pose was first scaled up to the scale of the ground truth pose before calculating the joint errors.

5. Results

In this section, we summarize the results of our cross-dataset and in-dataset evaluation. All experiments were repeated five times, that is each reported result was the average performance of five independently trained models. The error was calculated as the mean of the sum of all joint Euclidean distances between the output and the corresponding ground truth pose in mm.

We calculated two error types for the evaluation: The first was a no-alignment error, where the data of the predicted output pose were not post-processed and directly compared to the relative 3D ground truth pose, with the center of the hips at (0, 0, 0). The *no-alignment* error was used for most of the results. Second, we calculated the *Procrustes* error, where the output pose was moved, scaled, and rotated, minimizing the joint distances between the prediction and ground truth. Some *Procrustes* error values are presented in Table 12 for comparison with the no-alignment errors. The other *Procrustes* error tables for the presented data can be found in the Supplemental Materials.

If not explicitly mentioned otherwise, the results reported in the following were obtained with harmonized joints and the full camera set.

The prediction speed on the trained models was tested using an NVIDIA GeForce RTX 2080 TI graphics card. A batch with a size of 256 samples was calculated in around 30 milliseconds, which would result in 8533 pose estimations per second. The proposed model can therefore calculate 3D poses from 2D points in real time.

5.1. Joint Harmonization

Table 7 shows the mean and standard deviation of the errors for the evaluation over all datasets with and without joint harmonization. All entries in a row share the same training database; those in a column share the same test database. On the main diagonal are the in-database errors, which were significantly lower than the cross-database error (off the main diagonal). This difference showed the presence of dataset biases and their negative effect on cross-dataset generalization

The joint harmonization improved the results significantly from an overall mean error of 133.7 mm to 120.0 mm ($p = 0.040$, paired t-test). The impact differed among the individual training and test dataset combinations. As to be expected, the estimation error was mainly reduced in the cross-database results, where it was decreased by up to −29%. The greatest effect can be seen for HE1, which was the smallest dataset and whose joint definition deviated most from those of the other datasets.

The high absolute errors of the models trained with the HE1 were especially prominent in the ankle and knee joints. The errors can be attributed to the low diversity of poses in HE1, which did not include wide arm movements and no non-standing poses, which however were very common in H36M and PAN.

Table 7. Errors with original vs. harmonized joints (no-alignment errors in mm, mean ± std. deviation).

Training Data	Test Data HE1	 H36M	 PAN
		original joints (mean 133.7)	
HE1	95.9 ± 2.9	299.7 ± 9.5	148.8 ± 4.9
H36M	142.1 ± 3.9	67.6 ± 0.6	95.1 ± 3.2
PAN	166.7 ± 2.4	143.6 ± 1.2	43.9 ± 0.3
		harmonized joints (mean 120.0)	
HE1	91.7 ± 1.9	254.1 ± 5.8	125.4 ± 4.3
H36M	141.7 ± 3.8	67.0 ± 0.6	98.3 ± 2.2
PAN	117.8 ± 2.4	140.4 ± 1.3	43.7 ± 0.2
		mean error change	
HE1	−4.3%	−15.2%	−15.7%
H36M	−0.2%	−0.9%	3.4%
PAN	−29.3%	−2.3%	−0.6%

One-sided paired-sample t-test $p = 0.040$.

5.2. Number of Cameras

We compared the estimation error for the full camera set with a reduced camera set. For this purpose, the amount of used cameras was halved. Details about the used camera sets and their placement can be found in Section 4.5 and Figure 4.

Table 8 shows the results. The use of more cameras, and therefore more viewpoints and pose samples, changed the individual testing errors by in between 6.8% and −28.8%. Overall, the mean error decreased from 132.6 mm to 120.0 mm, which was a statistically significant difference ($p = 0.031$ in a one-sided paired t-test). The increase in the number of cameras had a positive impact on the testing results when training with the HE1 or H36M dataset, which both only had three camera views in the reduced camera set, with changes in the error of −5.1% up to −28.8%.

Table 8. Errors with the reduced vs. the full camera set (no-alignment errors in mm, mean ± std. deviation).

Training Data	Test Data HE1	H36M	PAN
		reduced camera set (mean 132.6)	
HE1	96.6 ± 1.9	270.6 ± 11.5	176.2 ± 5.4
H36M	166.4 ± 7.7	75.1 ± 0.4	105.2 ± 5.5
PAN	129.5 ± 2.1	131.4 ± 0.7	42.2 ± 0.3
		full camera set (mean 120.0)	
HE1	91.7 ± 1.9	254.1 ± 5.8	125.4 ± 4.3
H36M	141.7 ± 3.8	67.0 ± 0.6	98.3 ± 2.2
PAN	117.8 ± 2.4	140.4 ± 1.3	43.7 ± 0.2
		mean error change	
HE1	−5.1%	−6.1%	−28.8%
H36M	−14.8%	−10.7%	−6.6%
PAN	−9.0%	6.8%	3.4%

One-sided paired-sample t-test $p = 0.031$.

5.3. Scale Normalization

Table 9 shows the mean error and the standard deviation for the evaluation with and without scale normalization. Scale normalization significantly decreased the pose estimation error, from on average 120.0 mm to 90.1 mm ($p = 0.015$ in a one-sided sample-paired t-test). For the in-database evaluation, the error decreased between −13.2% and −24.6%. Cross-database testing resulted in even bigger reductions up to −42.9%.

The results of the models trained on HE1 and H36M and tested on the PAN dataset showed less improvement or even a worse result when using scale normalization. This error increase can be attributed to the test samples with a low camera viewing angle, which was not contained in the HE1 and H36M datasets.

Table 9. Error with and without scale normalization (no-alignment errors in mm, mean ± std. deviation).

Training Data	Test Data HE1	H36M	PAN
		no scale normalization (mean 120.0)	
HE1	91.7 ± 1.9	254.1 ± 5.8	125.4 ± 4.3
H36M	141.7 ± 3.8	67.0 ± 0.6	98.3 ± 2.2
PAN	117.8 ± 2.4	140.4 ± 1.3	43.7 ± 0.2
		with scale normalization (mean 90.1)	
HE1	69.2 ± 0.7	170.3 ± 4.0	152.7 ± 2.7
H36M	86.0 ± 1.2	55.2 ± 0.5	89.2 ± 0.7
PAN	67.3 ± 1.0	83.1 ± 0.6	37.9 ± 0.4
		mean error change	
HE1	−24.6%	−33.0%	21.8%
H36M	−39.3%	−17.7%	−9.3%
PAN	−42.9%	−40.8%	−13.2%

One-sided paired-sample t-test $p = 0.015$.

Interestingly, the scale normalization error when training on PAN and testing on HE1 decreased below the in-database error of HE1. The training set of PAN was larger and more diverse than that of HE1, which helped the cross-dataset generalization outperform the in-dataset generalization in this case.

Figure 5 shows the jointwise errors with and without scale normalization of only the cross-database evaluation as a box plot. The median error decreased for all joints, most for the leg joints. Most of the high-error outliers occurring with the original representation disappeared when using scale normalization.

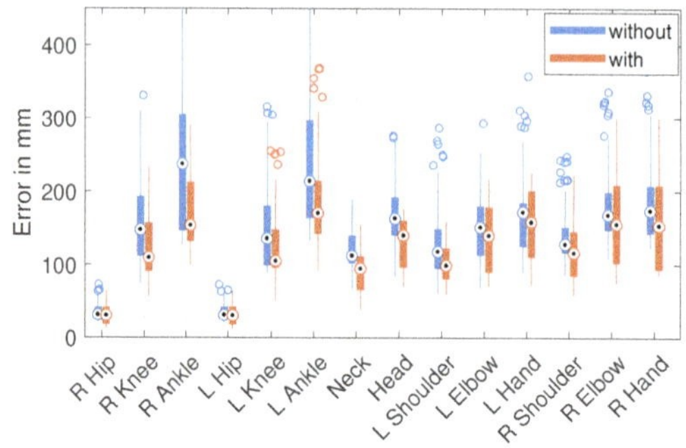

Figure 5. Jointwise error of all cross-dataset results with and without scale normalization. Box plot with median (circle with dot), 1'st/3'rd quartile (bottom/top of thick bar), and outliers (circles, default settings of MATLAB 2017b).

In order to illustrate the need for scale normalization, we calculated the scale differences without the normalization as the ratio of the Frobenius norms, of all joints, of the predicted and ground truth poses, after moving the centroid of the poses to (0, 0, 0). Table 10 shows the scale differences for the test sets. Several systematic prediction errors of up to 16% can be seen in the scale ratios, especially in the cross-dataset experiments when testing on HE1 and H36M. We found an interesting in-database result (main diagonal) with HE1: The scale of the predicted HE1 test poses (0.89) was significantly smaller than the absolute scale of the ground truth, while the PAN (1.0) and H36M (0.99) model predicted their own scale from their training data with greater accuracy.

Table 10. Mean scale ratios between ground truth and prediction without scale normalization.

Training Data		Test Data	
	HE1	**H36M**	**PAN**
		scale error (full cam set)	
HE1	0.89	1.09	0.97
H36M	0.90	0.99	1.01
PAN	0.84	0.90	1.00

This difference in the size of the ground truth poses can be attributed to the distances between the cameras and the recorded subjects. The camera positions in the HE1 were set up in a rectangle of around 8 × 9 m with a capture space of 2 × 3 m in the center of that. Our randomly chosen test camera was at one of the corners and therefore one of the most distant cameras in the dataset. The H36M dataset had its cameras in a 5 × 10 m setup and used a capture space of 3 × 4 m. We virtually copied and rotated the three training cameras so that the cameras were positioned close to circularly around the subjects. The PAN dataset had a capture space with diameter of 5 m in which the subjects could act freely, but due to the curvature of the dome and the constraint that the pose had to be fully captured in the camera view, only a limited range of distances could be used for training.

Therefore, the positioning of the cameras and the capture spaces led to different distances from the recorded subjects and systematic differences in the pose scale in the training data. Figure 6 shows the relative distribution of all joint-to-camera distances for some of the training and test datasets. It can be seen that the training poses of all datasets and testing poses of HE1 differed strongly in the distance to the cameras, which probably resulted in the failure to predict the true scale of the presented 2D pose. Other factors that can lead to this effect are the camera field-of-view/focal length, the camera resolution, and systematic biases in the body size of the subjects. The presence and effect of such dataset biases illustrate the importance of scale normalization for improving the cross-dataset and in-the-wild performance of 3D pose estimation.

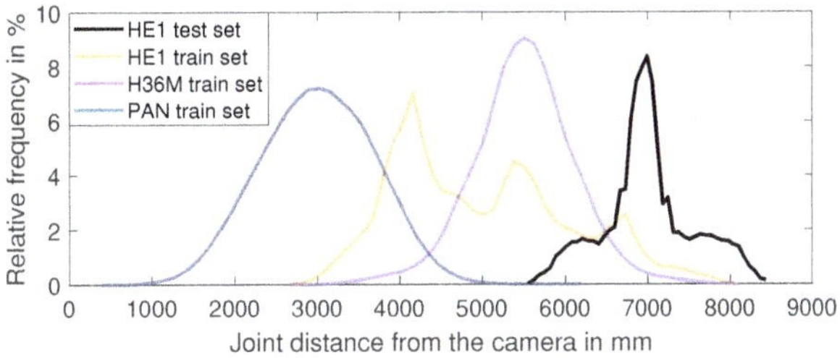

Figure 6. Relative distribution of all joint-to-camera distances for the HE1 test set and all training sets (full camera set).

5.4. Multi-Database Training

In order to improve the cross-dataset generalization, we tried to increase the diversity in the training data by combining datasets. We used a leave-one-out approach for the training and testing, that is we always left out one database for cross-database testing and used the other two for training. The training sets were combined by concatenating the data (and new normalization parameters for the nets inputs and outputs were derived).

Table 11 shows the generalization errors with and without scale normalization. Scale normalization improved the pose estimation error in the multi-database training, with a value of $p = 0.003$ in a paired t-test. For cross-database training and test cases, the error decreased between −0.6% and −50.1%. Similar to the single-database training in Table 9, the effect was bigger on the HE1 and H36M test set than on PAN. The error for the cases, in which the model was tested on one of the training databases, decreased between −10.8% and −41.5%.

In Table 11, single-database training results are added for easier comparison to multi-database. When testing on the HE1 dataset, combining H36M and PAN for training improved the cross-database results from 67.3 mm (PAN only) to 64.9 mm, which was significantly below using HE1 for training (69.2 mm). Further, combining HE1 and PAN for training reduced the error slightly below using PAN only. Apart from that, the multi-database training did not reduce the test errors in comparison to single-database training; Training with the bigger dataset alone achieved a similar or slightly better result than training with the combination of two datasets.

Table 11. Error of multi-database training with and without scale normalization (no-alignment errors in mm, mean ± std. deviation).

Training Data	Test Data HE1	H36M	PAN
		no scale normalization (mean 103.0)	
H36M + PAN	130.2 ± 2.9	103.8 ± 1.6	43.0 ± 0.3
HE1 + PAN	115.0 ± 1.3	143.0 ± 2.2	45.5 ± 1.6
HE1 + H36M	135.5 ± 1.2	75.1 ± 1.1	103.7 ± 4.3
		with scale normalization (mean 69.0)	
H36M + PAN	64.9 ± 0.5	63.0 ± 0.4	38.3 ± 0.3
HE1 + PAN	67.2 ± 0.7	83.2 ± 0.8	38.3 ± 0.7
HE1 + H36M	100.4 ± 1.2	62.6 ± 0.8	103.0 ± 2.1
HE1	69.2 ± 0.7	170.3 ± 4.0	152.7 ± 2.7
H36M	86.0 ± 1.2	55.2 ± 0.5	89.2 ± 0.7
PAN	67.3 ± 1.0	83.1 ± 0.6	37.9 ± 0.4
		mean error change	
H36M + PAN	−50.1%	−39.3%	−10.8%
HE1 + PAN	−41.5%	−41.8%	−15.9%
HE1 + H36M	−25.8%	−16.6%	−0.6%

One-sided paired-sample *t*-test $p = 0.003$.

5.5. OpenPose Evaluation

In order to test the generalization of the 3D pose estimation with a widely used 2D pose estimator, we conducted experiments with OpenPose [31].

First, the test set videos of the HE1 and the H36M datasets were processed with OpenPose. The videos of the Panoptic database could not be obtained on several occasions, due to availability issues with the host file server. The obtained 2D joint coordinates were used as inputs for the trained models to predict 3D joint positions, which were compared to the ground truth pose data. Note that the models were not fine-tuned with points provided by OpenPose. A noticeable difference between the two OpenPose datasets was the underlying image quality. While the HE1 was recorded at 640 × 480 px, the H36M dataset had a higher resolution of 1000 × 1000 px and better image quality. The video frames and motion-capture joint poses were synchronized for the OpenPose evaluation. The synchronization was manually corrected for the HE1 with an offset of 10 frames. The 3D pose evaluation error for every frame was calculated to the timewise closest motion-capture pose if that corresponding pose was valid.

Table 12 shows the test results for the OpenPose (OP) data and, for better comparison, the standard evaluation results. The errors are given for the *no-alignment* case and after the *Procrustes* alignment. As in the previous sections, the test results were generally better when training and testing with the same dataset, except for HE1. When testing on HE1 and HE1 (OP), the cross-database training on PAN outperformed the in-database training on HE1 in both the no-alignment and Procrustes error. On HE1 (OP) with Procrustes error, also, cross-dataset training with H36M performed better than in-dataset training with HE1.

Table 12. Error (no alignment vs. Procrustes) with OpenPose 2D joints (OP) and ground truth joint projection, with scale normalization (errors in mm, mean ± std. dev.).

Training Data	Evaluation Data				
	HE1 (OP)	**H36M (OP)**	**HE1**	**H36M**	**PAN**
			no alignment		
HE1	138.3 ± 1.3	184.4 ± 4.0	69.2 ± 0.7	170.3 ± 4.0	152.7 ± 2.7
H36M	151.3 ± 1.7	108.6 ± 0.7	86.0 ± 1.2	55.2 ± 0.5	89.2 ± 0.7
PAN	126.1 ± 1.2	130.8 ± 1.1	67.3 ± 1.0	83.1 ± 0.6	37.9 ± 0.4
			Procrustes alignment		
HE1	105.8 ± 0.6	109.5 ± 1.3	57.8 ± 0.7	105.0 ± 2.3	104.6 ± 1.9
H36M	103.1 ± 0.8	65.6 ± 0.6	61.4 ± 0.6	41.5 ± 0.2	48.6 ± 0.9
PAN	93.4 ± 0.7	71.9 ± 0.5	55.0 ± 0.8	55.4 ± 0.3	28.2 ± 0.3
			mean error change		
HE1	−23.5%	−40.6%	−16.4%	−38.3%	−31.5%
H36M	−31.8%	−39.6%	−28.6%	−24.8%	−45.5%
PAN	−26.0%	−45.0%	−18.3%	−33.3%	−25.6%

The *no alignment* error for the H36M (OP) test dataset was, excluding the HE1-trained models, consistently around 50 mm higher compared to the projected H36M data. This increase was evenly distributed over most of the joints, with the exception of the hip joints, for the models trained on both the H36M itself and the PAN datasets. The models trained on HE1 achieved an error reduction on certain joints (R knee, R ankle) and increased in the others, which was probably due to the lack of training data and the higher error rates to begin with. Similar effects can be seen for the results of the HE1 (OP) dataset, where the error increase was also distributed over all joints for all test cases, with slightly lower error increases for the hip, neck, and shoulder joints.

The *Procrustes* calculation minimized the errors in the scaling, rotation, and positioning of the skeleton. Therefore, the errors were smaller than without this alignment step in all cases. Analogous to the *no alignment* error, the results for the testing on the H36M (OP) dataset were, excluding the HE1 trained models, consistently around 20 mm higher compared to the projected H36M data. The results for the HE1 (OP) dataset were around 40 mm higher than for the projected HE1 data. For the projected test datasets, the error reductions for the same training and test database cases were between −16.4% and −25.6%, and the the cross-database results improved by up to −45.5% The absolute errors for the pose estimation were reduced to a range between 28 mm and 61 mm using the bigger training datasets (H36M, PAN) and 105 mm for the smaller HE1.

The *Procrustes* error changes of the individual joints are shown in Figure 7. It can be seen that rotation and repositioning during the *Procrustes* optimization increased the error in the hip joints, but decreased the error for all other joints. The effect increased with the distance to the skeletal root between the hips, because the joints further away from the center tended to have a greater impact on the Procrustes distance and minimization.

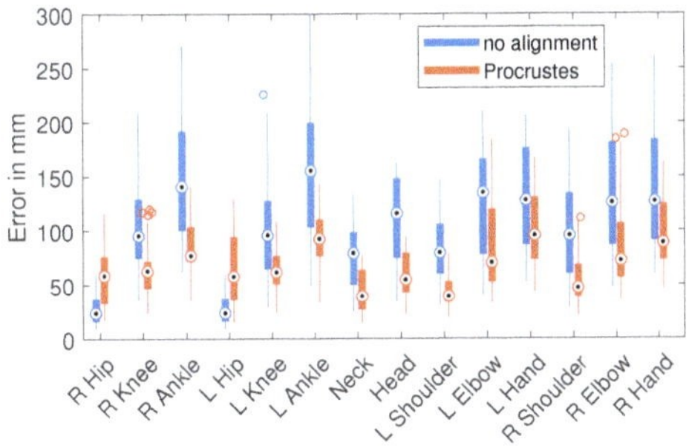

Figure 7. Jointwise no-alignment and Procrustes error of all single-database training results. Box plot with median (circle with dot), 1'st/3'rd quartile (bottom/top of thick bar), and outliers (circles, default settings of MATLAB 2017b).

5.6. Rotation Errors

Due to the separation of the cameras into training and test sets, the test camera viewpoints were not used for the training and were novel to the models. This often led to skeleton predictions with rotation errors. We calculated the rotation error from the Procrustes alignment as the magnitude of the minimal rotation in 3D space needed to minimize the joint distances between the ground truth and prediction. Table 13 shows the rotation errors for the OpenPose and projected test sets, using both camera sets and with or without scale normalization.

Table 13. Mean rotation corrections of the *Procrustes* alignment for different camera sets and scale normalization.

Training Data	Test Data				
	HE1 (OP)	H36M (OP)	HE1	H36M	PAN
	rotation error (reduced cam set, no scale norm)				
HE1	23.2°	35.2°	9.3°	31.3°	20.2°
H36M	28.5°	11.2°	22.7°	8.2°	11.0°
PAN	22.3°	18.6°	15.3°	17.1°	4.2°
	rotation error (full cam set, no scale norm)				
HE1	21.3°	35.1°	10.1°	30.6°	13.6°
H36M	26.7°	10.9°	18.1°	7.2°	10.2°
PAN	19.8°	18.6°	12.1°	18.7°	4.1°
	rotation error (full cam set, using scale norm)				
HE1	24.8°	24.3°	8.9°	20.9°	24.1°
H36M	18.1°	12.5°	8.8°	6.5°	9.3°
PAN	18.9°	14.8°	7.7°	8.9°	4.7°

The rotation error generally decreased with the addition of new camera positions, which we saw when comparing the first part of the table (reduced cam set) with the second part (full cam set). The effect was especially strong (−6.6°) when training on HE1 and testing on PAN.

The additional use of scale normalization decreased the error for most of the combinations even further, up to −9.7° and −9.3° for the HE1 and H36M cross-dataset evaluations. The PAN-trained models also had better rotation accuracy with the cross-database test results decreasing from −4.4° to −9.8°. The effects were smaller (−1.2° HE1 and −0.7° H36M) or even slightly worse (+0.6° PAN) for in-database training and testing. An outlying increase of the rotation error can be seen for the HE1 trained models, when evaluated on the PAN dataset. This was probably due to the introduction of bigger camera-to-pose view angles by the repositioning of the poses before the 3D to 2D projection.

5.7. Anatomical Pose Validation

Table 14 compares the pose estimation results with and without the anatomical pose validation that we proposed in Section 4.4. The validation approach successfully identified many wrongly estimated poses, which was revealed by the decreasing error in all tested database combinations.

For the testing on the projected ground truth data (HE1, H36M, and PAN), the decreases were smaller for in-database, with decreases from −0.5% to −2.5%. Bigger improvements can be seen in the results for the cross-database testing. The error rates decreased here from −1.3% to −9.3%.

The biggest impact of the pose validation was on the models trained with the HE1 dataset. It had the biggest error reduction, and up to 20.7% of the poses were rejected, while the rate for the other datasets was between 0.3% and 3.3%. Many poses that occurred in H36M and PAN test data were not part of the small HE1 dataset, e.g., HE1 only contained upright poses and only a limited range of arm and leg movements. Training with this dataset resulted in poor generalization to completely unseen poses, leading to many anatomically impossible skeletons. The other two datasets reached a lower cross-database pose rejection in the range from 1.1% to 3.3%, which showed better generalization.

All datasets had high pose rejection rates on the HE1 (OP) testing set. This effect was not present for the H36M (OP) dataset, where only the HE1-trained models showed a higher pose rejection rate, which was similar to the rate for the ground truth projection H36M dataset. This correlated with the low sample size of the training data and poor video quality of the HE1 dataset, which led to higher pose errors for all models.

Table 14. Error with and without anatomical pose validation, with scale normalization (no-alignment errors in mm, mean ± std. deviation).

Training Data	Test Data HE1 (OP)	H36M (OP)	HE1	H36M	PAN
			no validation		
HE1	138.3 ± 1.3	184.4 ± 4.0	69.2 ± 0.7	170.3 ± 4.0	152.7 ± 2.7
H36M	151.3 ± 1.7	108.6 ± 0.7	86.0 ± 1.2	55.2 ± 0.5	89.2 ± 0.7
PAN	126.1 ± 1.2	130.8 ± 1.1	67.3 ± 1.0	83.1 ± 0.6	37.9 ± 0.4
			using validation		
HE1	125.8 ± 2.4	166.1 ± 4.3	67.5 ± 0.8	155.4 ± 6.6	138.6 ± 2.3
H36M	142.4 ± 1.9	108.3 ± 0.7	84.9 ± 1.3	54.9 ± 0.6	88.9 ± 0.7
PAN	113.9 ± 1.1	130.4 ± 1.0	65.9 ± 1.0	81.8 ± 0.5	37.6 ± 0.4
			mean error change		
HE1	−9.0%	−9.9%	−2.5%	−8.8%	−9.3%
H36M	−5.8%	−0.3%	−1.3%	−0.5%	−0.3%
PAN	−9.7%	−0.3%	−2.1%	−1.6%	−0.7%
			rate of rejected poses		
HE1	15.8%	15.7%	1.8%	13.8%	20.7%
H36M	12.2%	1.3%	1.1%	0.3%	2.2%
PAN	15.6%	2.9%	1.3%	3.3%	0.6%

6. Discussion

In this article, we conducted cross-dataset experiments and discussed dataset biases as a step towards better cross-database generalization and in-the-wild performance of 3D human pose estimation systems.

The used datasets, HumanEva-I, Human3.6M, and Panoptic datasets, differed in their ground truth skeleton joint definitions, which impeded using these datasets together. Thus, we proposed a joint harmonization approach that facilitated cross-dataset experiments and reduced the biases among the datasets. In-the-wild performance would benefit from unifying the ground truth of additional datasets. However, a limitation of our approach was that it needed to be parameterized manually for each new dataset. For future works, it may be promising to develop generalized, automatic, and more accurate harmonization methods for post-processing existing datasets and to agree on a standardized skeleton joint model for collecting new datasets.

We analyzed the impact of the number of camera viewpoints used for the training. For databases with a small number of cameras such as H36M and HE1, adding more cameras improved the pose estimation significantly for in-database and cross-database evaluation. This showed that a certain coverage of viewpoints was needed for good generalization. With approaches that lift 2D poses to 3D poses, such as the one of Martinez et al. [26], datasets may be augmented by projecting the 3D ground truth to new virtual cameras, as was tested on the H36M dataset, improving the evaluation error up to −14%.

Many prior works expected the pose estimation model to learn the correct 3D scale from single-image 2D data, which is an impossible task in the general case. This imposed a burden that encouraged the models to learn dataset-specific heuristics and, as a consequence, to overfit to the dataset. We showed that the used databases were biased regarding the parameters, positions, and distances of the used cameras, which resulted in systematic scale errors in the output of the trained poses. Our proposed *scale normalization* step reduced the pose estimation error on the test datasets significantly, in 17 of 18 test cases and in the best case by more than −50% (see Tables 9 and 11). We investigated the one case in which the scale normalization decreased the performance. In this case, the repositioning of test poses in the preprocessing step increased the relative rotation of the pose to the camera, which led to higher prediction errors because these rotations were not present in the training dataset. However, this weakness could be compensated by augmenting the training dataset using virtual cameras with additional viewing angles, as mentioned above. Another limitation of the presented scale normalization approach is that the effects of camera distortions cannot be trained, because the position and scale are normalized in both 2D and 3D space. This error was not relevant in comparison to other factors in our experiments, but could become an issue for cameras with a very wide field of view.

Several of the dataset biases could be compensated in future works by adding virtual cameras, as described in Section 4.5.2, with various camera elevations, angles, and distances. We see this as a promising and more general augmentation approach for all available pose datasets. This approach could generate more training data for camera-to-subject distances and view angles with variation of the extrinsic camera calibration parameters. More camera types can also be added by variation of the intrinsic camera parameters, such as the focal length, to create data with different angles-of-view and enable better generalization. Additionally, this idea may be used with arbitrary motion capture data, including data for which no images are available, but probably requires advancing the proposed harmonization approach as mentioned above.

The presented anatomical pose validation achieved a high rate of pose rejection for the small HE1 dataset, catching malformed poses originally not contained in the training dataset. It also identified many invalid poses predicted with OpenPose from low-quality video. Most of the rejected poses had big shifts in the depth component (distance from the camera) of one or multiple joints, probably because there was no similar pose in the training set. Next to such a validation step, a promising alternative direction for all future works would be to include anatomical constraints in the model training to avoid such errors in the first place, e.g., as proposed in [40,51].

The evaluated multi-dataset training could not consistently improve the results compared to single database training, probably due to the big differences in the sample size of the used datasets (by a factor of approximately 10 to 40). A combination of databases is probably most beneficial if the datasets contain different poses and motions that can add new information, while more camera viewing angles may be created artificially, as stated above.

The prior work that is most similar to our work was published recently by Wang et al. [86]. They systematically examined the differences among existing pose datasets and their effect on cross-database evaluation. However, compared to our work, they focused on the systematic differences of camera viewpoints and conducted their experiment with another set of databases. Quantitative comparison to other works is difficult, because our evaluation protocol was designed for cross-dataset experiments that have not been published before. Nevertheless, the improvement by methodological advancements can be measured in comparison with the approach of Martinez et al. [26], which we used as the starting point. Table 15 shows that the proposed modifications (joint harmonization, scale normalization, and virtual camera augmentation (tested when training with H36M)) improved generalization across subjects, camera viewpoints, and datasets. The proposed anatomical pose validation (APV) reduced the error further. Joint harmonization, scale normalization, and APV can be applied with other 3D pose estimation approaches, and we see this as a promising direction for improving generalization. Virtual camera augmentation can be applied for all 2D to 3D pose lifting approaches, which may easily benefit from motion capture data and synthesized data and avoid overfitting to image-related dataset biases.

Table 15. No alignment errors of the proposed method compared with Martinez et al. [26]. The proposed method extended Martinez et al. [26] by joint harmonization, scale normalization, some virtual camera augmentation, and, optionally, anatomical pose validation (APV).

Training Data →	**HE1**			**H36M**			**PAN**			
Test Data →	**HE1**	**H36M**	**PAN**	**HE1**	**H36M**	**PAN**	**HE1**	**H36M**	**PAN**	**Mean**
Martinez et al. [26]	95.9	299.7	148.8	146.0	78.7	107.8	166.7	143.6	43.9	136.8
Proposed	69.2	170.3	152.7	86.0	55.2	89.2	67.3	83.1	37.9	90.1
Proposed + APV	67.5	155.4	138.6	84.9	54.9	88.9	65.9	81.8	37.6	86.2

As a promising direction for improving the cross-database performance (and testing of the proposed approaches), we suggest a multi-task training combining in-the-wild 2D datasets with 3D datasets, integrating a pretrained 2D-to-3D pose lifting network. Further, a logical advancement of our work is evaluating cross-database performance with additional datasets, especially new "in-the-wild" datasets, in order to gain additional insights about dataset biases and about how to improve 3D pose estimation so that it works well on arbitrary data.

Supplementary Materials: The following are available online at https://www.mdpi.com/article/10.3390/s21113769/s1, Table S1: Errors with original vs. harmonized joints, corresponding to Table 7, Table S2: Errors with reduced vs. full camera set, corresponding to Table 8, Table S3: Error with and without scale normalization, corresponding to Table 9, Table S4: Error of multi-database training with and without scale normalization, corresponding to Table 11.

Author Contributions: Conceptualization, M.R., P.W., and S.H.; methodology, M.R., P.W., and S.H.; software, M.R. and P.W.; validation, M.R. and P.W.; formal analysis, M.R., P.W., and S.H.; investigation, M.R., P.W., and S.H.; resources, M.R., P.W., S.H., and A.A.-H.; data curation, M.R. and S.H.; writing—original draft preparation, M.R., P.W., and S.H.; writing—review and editing, M.R., P.W., S.H., and A.A.-H.; visualization, M.R.; supervision, A.A.-H.; project administration, A.A.-H.; funding acquisition, A.A.-H. All authors read and agreed to the published version of the manuscript.

Funding: This work was funded by the German Federal Ministry of Education and Research (BMBF) under Grant Nos. 03ZZ0470 (HuBA), 03ZZ0448L (RoboAssist), and 03ZZ04X02B (RoboLab) within the Zwanzig20 Alliance 3Dsensation. The responsibility for the content lies solely with the authors.

Institutional Review Board Statement: Ethical review and approval were waived for this study, due to the use of public databases, which were conducted according to the guidelines of the Declaration of Helsinki and approved by the relevant review boards. We complied with the terms of use of the databases regarding the publication of data.

Informed Consent Statement: According to the documentation of the used public databases, informed consent was obtained from all subjects involved.

Data Availability Statement: The source code for this paper is available at http://added.later (accessed on 27 May 2021). The original baseline model can be found at https://github.com/una-dinosauria/3d-pose-baseline (accessed on 27 May 2021). The Human3.6M dataset can be obtained at vision.imar.ro/human3.6m/ (accessed on 27 May 2021). The Panoptic dataset can be obtained at http://domedb.perception.cs.cmu.edu/ (accessed on 27 May 2021). The Human Eva Dataset can be obtained at http://humaneva.is.tue.mpg.de/ (accessed on 27 May 2021).

Conflicts of Interest: The authors declare no conflict of interest.

References

1. Lo Presti, L.; La Cascia, M. 3D skeleton-based human action classification: A survey. *Pattern Recognit.* **2016**, *53*, 130–147. [CrossRef]
2. Handrich, S.; Rashid, O.; Al-Hamadi, A., Non-intrusive Gesture Recognition in Real Companion Environments. In *Companion Technology: A Paradigm Shift in Human-Technology Interaction*; Biundo, S., Wendemuth, A., Eds.; Springer International Publishing: Cham, Switzerland, 2017; pp. 321–343. [CrossRef]
3. Shi, L.; Zhang, Y.; Cheng, J.; Lu, H. Two-Stream Adaptive Graph Convolutional Networks for Skeleton-Based Action Recognition. In Proceedings of the IEEE/CVF Conference on Computer Vision and Pattern Recognition (CVPR), Long Beach, CA USA, 16–20 June 2019.
4. Yan, S.; Xiong, Y.; Lin, D. Spatial Temporal Graph Convolutional Networks for Skeleton-Based Action Recognition. In Proceedings of the 32nd AAAI Conference on Artificial Intelligence (AAAI), New Orleans, LA, USA, 2–7 February 2018; pp. 7444–7452.
5. Zhang, X.; Xu, C.; Tao, D. Context Aware Graph Convolution for Skeleton-Based Action Recognition. In Proceedings of the IEEE/CVF Conference on Computer Vision and Pattern Recognition (CVPR), Seattle, WA, USA, 14–19 June 2020.
6. Liu, Z.; Zhang, H.; Chen, Z.; Wang, Z.; Ouyang, W. Disentangling and Unifying Graph Convolutions for Skeleton-Based Action Recognition. In Proceedings of the 2020 IEEE/CVF Conference on Computer Vision and Pattern Recognition (CVPR), Seattle, WA, USA, 14–19 June 2020.
7. Cheng, K.; Zhang, Y.; He, X.; Chen, W.; Cheng, J.; Lu, H. Skeleton-Based Action Recognition With Shift Graph Convolutional Network. In Proceedings of the IEEE/CVF Conference on Computer Vision and Pattern Recognition (CVPR), Seattle, WA, USA, 14–19 June 2020.
8. Li, C.; Zhang, X.; Liao, L.; Jin, L.; Yang, W. Skeleton-Based Gesture Recognition Using Several Fully Connected Layers with Path Signature Features and Temporal Transformer Module. *Proc. AAAI Conf. Artif. Intell.* **2019**, *33*, 8585–8593. [CrossRef]
9. Joo, H.; Simon, T.; Cikara, M.; Sheikh, Y. Towards Social Artificial Intelligence: Nonverbal Social Signal Prediction in a Triadic Interaction. In Proceedings of the IEEE/CVF Conference on Computer Vision and Pattern Recognition (CVPR), Long Beach, CA, USA, 16–20 June 2019.
10. Joo, H.; Liu, H.; Tan, L.; Gui, L.; Nabbe, B.; Matthews, I.; Kanade, T.; Nobuhara, S.; Sheikh, Y. Panoptic Studio: A Massively Multiview System for Social Motion Capture. In Proceedings of the IEEE International Conference on Computer Vision (ICCV), Santiago, Chile, 7–13 December 2015.
11. Joo, H.; Simon, T.; Li, X.; Liu, H.; Tan, L.; Gui, L.; Banerjee, S.; Godisart, T.; Nabbe, B.; Matthews, I.; et al. Panoptic Studio: A Massively Multiview System for Social Interaction Capture. *IEEE Trans. Pattern Anal. Mach. Intell.* **2019**, *41*, 190–204. [CrossRef] [PubMed]
12. Iskakov, K.; Burkov, E.; Lempitsky, V.; Malkov, Y. Learnable Triangulation of Human Pose. In Proceedings of the IEEE/CVF International Conference on Computer Vision (ICCV), Seoul, Korea, 27 October–3 November 2019.
13. Shotton, J.; Fitzgibbon, A.; Cook, M.; Sharp, T.; Finocchio, M.; Moore, R.; Kipman, A.; Blake, A. Real-time human pose recognition in parts from single depth images. In Proceedings of the The 24th IEEE Conference on Computer Vision and Pattern Recognition (CVPR 2011), Colorado Springs, CO, USA, 20–25 June 2011; pp. 1297–1304.
14. Handrich, S.; Al-Hamadi, A. Localizing body joints from single depth images using geodetic distances and random tree walk. In Proceedings of the 2017 IEEE International Conference on Image Processing (ICIP), Beijing, China, 17–20 September 2017; pp. 146–150. [CrossRef]

15. Handrich, S.; Waxweiler, P.; Werner, P.; Al-Hamadi, A. 3D Human Pose Estimation Using Stochastic Optimization in Real Time. In Proceedings of the 2018 25th IEEE International Conference on Image Processing (ICIP), Athens, Greece, 7–10 October 2018; pp. 555–559.
16. Adib, F.; Kabelac, Z.; Katabi, D.; Miller, R.C. 3D Tracking via Body Radio Reflections. In Proceedings of the 11th USENIX Conference on Networked Systems Design and Implementation, NSDI'14, Seattle, WA, USA, 2–4 April 2014; USENIX Association: Berkeley, CA, USA, 2014; pp. 317–329.
17. Zhao, M.; Li, T.; Alsheikh, M.A.; Tian, Y.; Zhao, H.; Torralba, A.; Katabi, D. Through-Wall Human Pose Estimation Using Radio Signals. In Proceedings of the 2018 IEEE/CVF Conference on Computer Vision and Pattern Recognition (CVPR), Salt Lake City, UT, USA, 18–23 June 2018; pp. 7356–7365. [CrossRef]
18. Wang, Z.; Liu, Y.; Liao, Q.; Ye, H.; Liu, M.; Wang, L. Characterization of a RS-LiDAR for 3D Perception. In Proceedings of the 2018 IEEE 8th Annual International Conference on CYBER Technology in Automation, Control, and Intelligent Systems (CYBER), Tianjin, China, 18–23 July 2018; pp. 564–569. [CrossRef]
19. Ionescu, C.; Li, F.; Sminchisescu, C. Latent Structured Models for Human Pose Estimation. In Proceedings of the International Conference on Computer Vision (ICCV), Barcelona, Spain, 6–13 November 2011.
20. Ionescu, C.; Papava, D.; Olaru, V.; Sminchisescu, C. Human3.6M: Large Scale Datasets and Predictive Methods for 3D Human Sensing in Natural Environments. *IEEE Trans. Pattern Anal. Mach. Intell.* **2014**, *36*, 1325–1339. [CrossRef] [PubMed]
21. Sigal, L.; Black, M.J. *HumanEva: Synchronized Video and Motion Capture Dataset for Evaluation of Articulated Human Motion*; Technical Report; Brown University: Providence, RI, USA, 2006.
22. Sigal, L.; Balan, A.O.; Black, M.J. HumanEva: Synchronized video and motion capture dataset and baseline algorithm for evaluation of articulated human motion. *Int. J. Comput. Vis.* **2010**, *87*, 4–27. [CrossRef]
23. Mehta, D.; Rhodin, H.; Casas, D.; Fua, P.; Sotnychenko, O.; Xu, W.; Theobalt, C. Monocular 3D Human Pose Estimation in the Wild Using Improved CNN Supervision. In Proceedings of the 2017 International Conference on 3D Vision (3DV), Verona, Italy, 10–12 October 2017. [CrossRef]
24. Fabbri, M.; Lanzi, F.; Calderara, S.; Palazzi, A.; Vezzani, R.; Cucchiara, R. Learning to Detect and Track Visible and Occluded Body Joints in a Virtual World. In Proceedings of the European Conference on Computer Vision (ECCV), Munich, Germany, 8–14 September 2018.
25. Torralba, A.; Efros, A.A. Unbiased look at dataset bias. In Proceedings of the IEEE Computer Society Conference on Computer Vision and Pattern Recognition (CVPR), Colorado Springs, CO, USA, 20–25 June 2011; pp. 1521–1528. [CrossRef]
26. Martinez, J.; Hossain, R.; Romero, J.; Little, J.J. A simple yet effective baseline for 3d human pose estimation. In Proceedings of the IEEE International Conference on Computer Vision (ICCV), Venice, Italy, 22–29 October 2017. [CrossRef]
27. Wei, S.E.; Ramakrishna, V.; Kanade, T.; Sheikh, Y. Convolutional Pose Machines. In Proceedings of the IEEE Conference on Computer Vision and Pattern Recognition (CVPR), Las Vegas, NV, USA, 27–30 June 2016. [CrossRef]
28. Newell, A.; Yang, K.; Deng, J. Stacked Hourglass Networks for Human Pose Estimation. In Proceedings of the European Conference on Computer Vision (ECCV), Amsterdam, The Netherlands, 8–16 October 2016.
29. Chen, Y.; Wang, Z.; Peng, Y.; Zhang, Z.; Yu, G.; Sun, J. Cascaded Pyramid Network for Multi-Person Pose Estimation. In Proceedings of the IEEE Computer Society Conference on Computer Vision and Pattern Recognition (CVPR), Salt Lake City, UT, USA, 18–23 June 2018; pp. 7103–7112.
30. Cao, Z.; Simon, T.; Wei, S.E.; Sheikh, Y. Realtime Multi-Person 2D Pose Estimation using Part Affinity Fields. In Proceedings of the 2017 IEEE Conference on Computer Vision and Pattern Recognition (CVPR), Honolulu, HI, USA, 21–26 July 2017. [CrossRef]
31. Cao, Z.; Hidalgo Martinez, G.; Simon, T.; Wei, S.; Sheikh, Y.A. OpenPose: Realtime Multi-Person 2D Pose Estimation using Part Affinity Fields. *IEEE Trans. Pattern Anal. Mach. Intell.* **2019**, *43*, 172–186. [CrossRef]
32. Nibali, A.; He, Z.; Morgan, S.; Prendergast, L. Numerical Coordinate Regression with Convolutional Neural Networks. *arXiv* **2019**, arXiv:1801.07372.
33. Papandreou, G.; Zhu, T.; Kanazawa, N.; Toshev, A.; Tompson, J.; Bregler, C.; Murphy, K. Towards Accurate Multi-person Pose Estimation in the Wild. In Proceedings of the 30th IEEE Conference on Computer Vision and Pattern Recognition (CVPR), Honolulu, HI, USA, 21–26 July 2017; pp. 3711–3719.
34. Pishchulin, L.; Insafutdinov, E.; Tang, S.; Andres, B.; Andriluka, M.; Gehler, P.; Schiele, B. DeepCut: Joint Subset Partition and Labeling for Multi Person Pose Estimation. In Proceedings of the IEEE Computer Society Conference on Computer Vision and Pattern Recognition (CVPR), Boston, MA, USA, 7–12 June 2015; pp. 4929–4937.
35. Nie, X.; Feng, J.; Xing, J.; Yan, S. Pose Partition Networks for Multi-Person Pose Estimation. In Proceedings of the European Conference on Computer Vision (ECCV), Munich, Germany, 8–14 September 2018; pp. 684–699.
36. Xiao, B.; Wu, H.; Wei, Y. Simple baselines for human pose estimation and tracking. In Proceedings of the European Conference on Computer Vision (ECCV), Munich, Germany, 8–14 September 2018; pp. 466–481. [CrossRef]
37. Sun, K.; Xiao, B.; Liu, D.; Wang, J. Deep High-Resolution Representation Learning for Human Pose Estimation. In Proceedings of the IEEE Computer Society Conference on Computer Vision and Pattern Recognition (CVPR), Long Beach, CA, USA, 15–20 June 2019; pp. 5686–5696.
38. Habibie, I.; Xu, W.; Mehta, D.; Pons-Moll, G.; Theobalt, C. In the Wild Human Pose Estimation Using Explicit 2D Features and Intermediate 3D Representations. In Proceedings of the IEEE Computer Society Conference on Computer Vision and Pattern Recognition (CVPR), Long Beach, CA, USA, 15–20 June 2019; pp. 10897–10906.

39. Pavlakos, G.; Zhou, X.; Daniilidis, K. Ordinal Depth Supervision for 3D Human Pose Estimation. In Proceedings of the 2018 IEEE/CVF Conference on Computer Vision and Pattern Recognition (CVPR), Salt Lake City, UT, USA, 18–23 June 2018; pp. 7307–7316. [CrossRef]
40. Zhou, X.; Huang, Q.; Sun, X.; Xue, X.; Wei, Y. Towards 3D Human Pose Estimation in the Wild: A Weakly-supervised Approach. In Proceedings of the IEEE International Conference on Computer Vision (ICCV), Venice, Italy, 22–29 October 2017; pp. 398–407.
41. Chen, C.H.; Ramanan, D. 3D human pose estimation = 2D pose estimation + matching. In Proceedings of the 30th IEEE Conference on Computer Vision and Pattern Recognition (CVPR), Honolulu, HI, USA, 21–26 July 2017; pp. 5759–5767. [CrossRef]
42. Sun, X.; Xiao, B.; Wei, F.; Liang, S.; Wei, Y. Integral human pose regression. In Proceedings of the European Conference on Computer Vision (ECCV), Munich, Germany, 8–14 September 2018; pp. 529–545. [CrossRef]
43. Zhou, X.; Zhu, M.; Leonardos, S.; Derpanis, K.; Daniilidis, K. Sparseness Meets Deepness: 3D Human Pose Estimation from Monocular Video. In Proceedings of the IEEE Conference on Computer Vision and Pattern Recognition (CVPR), Boston, MA, USA, 7–12 June 2015; pp. 4966–4975.
44. Luo, C.; Chu, X.; Yuille, A. OriNet: A Fully Convolutional Network for 3D Human Pose Estimation. In Proceedings of the British Machine Vision Conference BMVC, Newcastle, UK, 3–6 September 2018.
45. Tome, D.; Russell, C.; Agapito, L. Lifting from the Deep: Convolutional 3D Pose Estimation from a Single Image. In Proceedings of the 30th IEEE Conference on Computer Vision and Pattern Recognition (CVPR), Honolulu, HI, USA, 21–26 July 2017; pp. 2500–2509.
46. Rogez, G.; Weinzaepfel, P.; Schmid, C. LCR-Net: Localization-Classification-Regression for Human Pose. In Proceedings of the 30th IEEE Conference on Computer Vision and Pattern Recognition (CVPR), Honolulu, HI, USA, 21–26 July 2017.
47. Nibali, A.; He, Z.; Morgan, S.; Prendergast, L. 3D Human Pose Estimation with 2D Marginal Heatmaps. In Proceedings of the 2019 IEEE Winter Conference on Applications of Computer Vision (WACV), Waikoloa, HI, USA, 7–11 January 2019.
48. Pavlakos, G.; Zhu, L.; Zhou, X.; Daniilidis, K. Learning to Estimate 3D Human Pose and Shape from a Single Color Image. In Proceedings of the 2018 IEEE/CVF Conference on Computer Vision and Pattern Recognition (CVPR), Salt Lake City, UT, USA, 18–23 June 2018; pp. 459–468. [CrossRef]
49. Luvizon, D.C.; Picard, D.; Tabia, H. 2D/3D Pose Estimation and Action Recognition Using Multitask Deep Learning. In Proceedings of the 2018 IEEE/CVF Conference on Computer Vision and Pattern Recognition (CVPR), Salt Lake City, UT, USA, 18–23 June 2018.
50. Wang, C.; Wang, Y.; Lin, Z.; Yuille, A.L.; Gao, W. Robust Estimation of 3D Human Poses from a Single Image. In Proceedings of the 2014 IEEE Conference on Computer Vision and Pattern Recognition (CVPR), Columbus, OH, USA, 23–28 June 2014; pp. 2369–2376. [CrossRef]
51. Dabral, R.; Mundhada, A.; Kusupati, U.; Afaque, S.; Sharma, A.; Jain, A. Learning 3D Human Pose from Structure and Motion. In Proceedings of the European Conference on Computer Vision (ECCV), Munich, Germany, 8–14 September 2018.
52. Tekin, B.; Márquez-Neila, P.; Salzmann, M.; Fua, P. Learning to Fuse 2D and 3D Image Cues for Monocular Body Pose Estimation. In Proceedings of the 2017 IEEE International Conference on Computer Vision (ICCV), Venice, Italy, 22–29 October 2017.
53. Li, S.; Ke, L.; Pratama, K.; Tai, Y.W.; Tang, C.K.; Cheng, K.T. Cascaded Deep Monocular 3D Human Pose Estimation With Evolutionary Training Data. In Proceedings of the 2020 IEEE/CVF Conference on Computer Vision and Pattern Recognition (CVPR), Seattle, WA, USA, 14–19 June 2020; pp. 6172–6182. [CrossRef]
54. Chen, C.H.; Tyagi, A.; Agrawal, A.; Drover, D.; Rohith, M.V.; Stojanov, S.; Rehg, J.M. Unsupervised 3D Pose Estimation With Geometric Self-Supervision. In Proceedings of the IEEE Computer Society Conference on Computer Vision and Pattern Recognition (CVPR), Long Beach, CA, USA, 15–20 June 2019; pp. 5707–5717. [CrossRef]
55. Lin, J.; Lee, G.H. Trajectory Space Factorization for Deep Video-Based 3D Human Pose Estimation. In Proceedings of the British Machine Vision Conference (BMVC), Cardiff, UK, 9–12 September 2019.
56. Katircioglu, I.; Tekin, B.; Salzmann, M.; Lepetit, V.; Fua, P. Learning Latent Representations of 3D Human Pose with Deep Neural Networks. *Int. J. Comput. Vis.* **2018**, *126*, 1326–1341. [CrossRef]
57. Chen, T.; Fang, C.; Shen, X.; Zhu, Y.; Chen, Z.; Luo, J. Anatomy-aware 3D Human Pose Estimation with Bone-based Pose Decomposition. *IEEE Trans. Circuits Syst. Video Technol.* **2021**. [CrossRef]
58. Benzine, A.; Luvison, B.; Pham, Q.C.; Achard, C. Single-shot 3D multi-person pose estimation in complex images. *Pattern Recognit.* **2021**, *112*, 107534. [CrossRef]
59. Wu, H.; Xiao, B. 3D Human Pose Estimation via Explicit Compositional Depth Maps. In Proceedings of the AAAI Conference on Artificial Intelligence, New York, NY, USA, 7–12 February 2020; pp. 12378–12385. [CrossRef]
60. Sárándi, I.; Linder, T.; Arras, K.O.; Leibe, B. Synthetic Occlusion Augmentation with Volumetric Heatmaps for the 2018 ECCV PoseTrack Challenge on 3D Human Pose Estimation. *arXiv* **2018**, arXiv:1809.04987v3.
61. Cheng, Y.; Yang, B.; Wang, B.; Wending, Y.; Tan, R. Occlusion-Aware Networks for 3D Human Pose Estimation in Video. In Proceedings of the IEEE/CVF International Conference on Computer Vision (ICCV), Seoul, Korea, 27 October–3 November 2019; pp. 723–732. [CrossRef]
62. Popa, A.I.; Zanfir, M.; Sminchisescu, C. Deep Multitask Architecture for Integrated 2D and 3D Human Sensing. In Proceedings of the 30th IEEE Conference on Computer Vision and Pattern Recognition, CVPR, Honolulu, HI, USA, 21–26 July 2017; pp. 4714–4723. [CrossRef]

63. Zanfir, A.; Marinoiu, E.; Sminchisescu, C. Monocular 3D Pose and Shape Estimation of Multiple People in Natural Scenes—The Importance of Multiple Scene Constraints. In Proceedings of the IEEE Conference on Computer Vision and Pattern Recognition (CVPR), Salt Lake City, UT, USA, 18–23 June 2018.
64. Zanfir, A.; Marinoiu, E.; Zanfir, M.; Popa, A.I.; Sminchisescu, C. Deep Network for the Integrated 3D Sensing of Multiple People in Natural Images. In *Advances in Neural Information Processing Systems*; Bengio, S., Wallach, H., Larochelle, H., Grauman, K., Cesa-Bianchi, N., Garnett, R., Eds.; Curran Associates, Inc.: Red Hook, NY, USA, 2018; Volume 31.
65. Radwan, I.; Dhall, A.; Goecke, R. Monocular Image 3D Human Pose Estimation under Self-Occlusion. In Proceedings of the 2013 IEEE International Conference on Computer Vision (ICCV), Sydney, Australia, 8–12 April 2013; pp. 1888–1895. [CrossRef]
66. Yasin, H.; Iqbal, U.; Kruger, B.; Weber, A.; Gall, J. A Dual-Source Approach for 3D Pose Estimation from a Single Image. In Proceedings of the 2016 IEEE Conference on Computer Vision and Pattern Recognition (CVPR), Las Vegas, USA, 27–30 June 2016; Volume 172, pp. 4948–4956. [CrossRef]
67. Moreno-Noguer, F. 3D Human Pose Estimation from a Single Image via Distance Matrix Regression. In Proceedings of the 30th IEEE Conference on Computer Vision and Pattern Recognition (CVPR), Honolulu, HI, USA, 21–26 July 2017; pp. 1561–1570.
68. Pavlakos, G.; Zhou, X.; Derpanis, K.G.; Daniilidis, K. Coarse-to-Fine Volumetric Prediction for Single-Image 3D Human Pose. In Proceedings of the 30th IEEE Conference on Computer Vision and Pattern Recognition (CVPR), Honolulu, HI, USA, 21–26 July 2017.
69. Li, S.; Chan, A.B. 3D human pose estimation from monocular images with deep convolutional neural network. In *Asian Conference on Computer Vision*; Springer: Cham, Switzerland; Singapore, 2014; pp. 332–347 [CrossRef]
70. Kanazawa, A.; Black, M.J.; Jacobs, D.W.; Malik, J. End-to-end Recovery of Human Shape and Pose. In Proceedings of the 30th IEEE Conference on Computer Vision and Pattern Recognition (CVPR), Honolulu, HI, USA, 21–26 July 2017. [CrossRef]
71. Trumble, M.; Gilbert, A.; Hilton, A.; Collomosse, J. Deep autoencoder for combined human pose estimation and body model upscaling. In Proceedings of the European Conference on Computer Vision ECCV, Munich, Germany, 8–14 September 2018; pp. 784–800. [CrossRef]
72. Güler, R.A.; Neverova, N.; Kokkinos, I. DensePose: Dense Human Pose Estimation In The Wild. In Proceedings of the IEEE Conference on Computer Vision and Pattern Recognition (CVPR), Salt Lake City, UT, USA, 18–23 June 2018.
73. Rhodin, H.; Salzmann, M.; Fua, P. Unsupervised geometry-aware representation for 3D human pose estimation. In Proceedings of the European Conference on Computer Vision ECCV, Munich, Germany, 8–14 September 2018; pp. 765–782. [CrossRef]
74. Pavllo, D.; Feichtenhofer, C.; Grangier, D.; Auli, M. 3D human pose estimation in video with temporal convolutions and semi-supervised training. In Proceedings of the IEEE Computer Society Conference on Computer Vision and Pattern Recognition (CVPR), Salt Lake City, UT, USA, 18–22 June 2018; pp. 7745–7754.
75. Hossain, M.R.I.; Little, J.J. Exploiting temporal information for 3D human pose estimation. In Proceedings of the European Conference on Computer Vision ECCV, Munich, Germany, 8–14 September 2018; pp. 68–84. [CrossRef]
76. Zhao, L.; Peng, X.; Tian, Y.; Kapadia, M.; Metaxas, D.N. Semantic Graph Convolutional Networks for 3D Human Pose Regression. In Proceedings of the IEEE Computer Society Conference on Computer Vision and Pattern Recognition (CVPR), Long Beach, CA, USA, 15–20 June 2019; pp. 3420–3430. [CrossRef]
77. Vicon. Available online: https://ien.vicon.eu (accessed on 27 May 2021).
78. The Captury. Available online: https://captury.com (accessed on 27 May 2021).
79. Wang, L.; Chen, Y.; Guo, Z.; Qian, K.; Lin, M.; Li, H.; Ren, J.S. Generalizing monocular 3D human pose estimation in-the-wild. In Proceedings of the 2019 International Conference on Computer Vision Workshop (ICCVW), Seoul, Korea, 27–28 October 2019; pp. 4024–4033. [CrossRef]
80. Rogez, G.; Weinzaepfel, P.; Schmid, C. LCR-Net++: Multi-Person 2D and 3D Pose Detection in Natural Images. *IEEE Trans. Pattern Anal. Mach. Intell.* **2020**, *42*, 1146–1161. [CrossRef] [PubMed]
81. Chen, W.; Wang, H.; Li, Y.; Su, H.; Wang, Z.; Tu, C.; Lischinski, D.; Cohen-Or, D.; Chen, B. Synthesizing Training Images for Boosting Human 3D Pose Estimation. In Proceedings of the 2016 4th International Conference on 3D Vision 2016, Stanford, CA, USA, 25–28 October 2016; pp. 479–488.
82. de Souza, C.R.; Gaidon, A.; Cabon, Y.; Peña, A.M.L. Procedural Generation of Videos to Train Deep Action Recognition Networks. In Proceedings of the 30th IEEE Conference on Computer Vision and Pattern Recognition, CVPR, Honolulu, HI, USA, 21–26 July 2017; pp. 2594–2604.
83. Varol, G.; Romero, J.; Martin, X.; Mahmood, N.; Black, M.J.; Laptev, I.; Schmid, C. Learning from Synthetic Humans. In Proceedings of the 30th IEEE Conference on Computer Vision and Pattern Recognition, CVPR, Honolulu, HI, USA, 21–26 July 2017; pp. 4627–4635. [CrossRef]
84. Peng, X.; Sun, B.; Ali, K.; Saenko, K. Learning Deep Object Detectors from 3D Models. In Proceedings of the 2015 IEEE International Conference on Computer Vision (ICCV), Las Condes, Chile, 7–13 December 2015.
85. Rogez, G.; Schmid, C. Image-based Synthesis for Deep 3D Human Pose Estimation. *Int. J. Comput. Vis.* **2018**, *126*, 993–1008. [CrossRef]
86. Wang, Z.; Shin, D.; Fowlkes, C.C. Predicting Camera Viewpoint Improves Cross-dataset Generalization for 3D Human Pose Estimation. In Proceedings of the European Conference on Computer Vision (ECCV), Glasgow, UK, 23–28 August 2020.

87. Zhao, M.; Tian, Y.; Zhao, H.; Alsheikh, M.A.; Li, T.; Hristov, R.; Kabelac, Z.; Katabi, D.; Torralba, A. RF-based 3D skeletons. In Proceedings of the 2018 Conference of the ACM Special Interest Group on Data Communication, Budapest, Hungary, 20–25 August 2018; pp. 267–281. [CrossRef]
88. Wang, F.; Zhou, S.; Panev, S.; Han, J.; Huang, D. Person-in-WiFi: Fine-Grained Person Perception Using WiFi. In Proceedings of the IEEE/CVF International Conference on Computer Vision (ICCV), Seoul, Korea, 27 October–3 November 2019; pp. 5451–5460. [CrossRef]
89. Jiang, W.; Xue, H.; Miao, C.; Wang, S.; Lin, S.; Tian, C.; Murali, S.; Hu, H.; Sun, Z.; Su, L. Towards 3D human pose construction using wifi. In Proceedings of the 26th Annual International Conference on Mobile Computing and Networking, New York, NY, USA, 21–25 September 2020; pp. 1–14. [CrossRef]
90. Hougne, P.; Imani, M.F.; Diebold, A.V.; Horstmeyer, R.; Smith, D.R. Learned Integrated Sensing Pipeline: Reconfigurable Metasurface Transceivers as Trainable Physical Layer in an Artificial Neural Network. *Adv. Sci.* **2020**, *7*, 1901913. [CrossRef]
91. Li, L.; Shuang, Y.; Ma, Q.; Li, H.; Zhao, H.; Wei, M.; Liu, C.; Hao, C.; Qiu, C.W.; Cui, T.J. Intelligent metasurface imager and recognizer. *Light. Sci. Appl.* **2019**, *8*, 2047–7538. [CrossRef] [PubMed]
92. Li, H.Y.; Zhao, H.T.; Wei, M.L.; Ruan, H.X.; Shuang, Y.; Cui, T.J.; del Hougne, P.; Li, L. Intelligent Electromagnetic Sensing with Learnable Data Acquisition and Processing. *Patterns* **2020**, *1*, 100006. [CrossRef] [PubMed]
93. Kim, K.; Konda, P.C.; Cooke, C.L.; Appel, R.; Horstmeyer, R. Multi-element microscope optimization by a learned sensing network with composite physical layers. *Opt. Lett.* **2020**, *45*, 5684. [CrossRef] [PubMed]
94. Li, T.; Liu, Q.; Zhou, X. Practical Human Sensing in the Light. In Proceedings of the 14th Annual International Conference on Mobile Systems, Applications, and Services, MobiSys'16, Singapore, 26–30 June 2016; pp. 71–84. [CrossRef]
95. Andriluka, M.; Pishchulin, L.; Gehler, P.; Schiele, B. 2D human pose estimation: New benchmark and state-of-the-art analysis. In Proceedings of the 2014 IEEE Conference on Computer Vision and Pattern Recognition (CVPR), Columbus, OH, USA, 23–28 June 2014; pp. 3686–3693. [CrossRef]
96. Lin, T.Y.; Maire, M.; Belongie, S.; Hays, J.; Perona, P.; Ramanan, D.; Dollár, P.; Zitnick, C.L. Microsoft COCO: Common objects in context. In *European Conference on Computer Vision (ECCV)*; Springer: Zurich, Switzerland, 2014 ; Volume 8693 LNCS, pp. 740–755. [CrossRef]
97. Werner, P.; Saxen, F.; Al-Hamadi, A. Handling Data Imbalance in Automatic Facial Action Intensity Estimation. In Proceedings of the British Machine Vision Conference (BMVC), Swansea, UK, 7–10 September 2015; pp. 124.1–124.12. [CrossRef]
98. Zhu, Y.; Long, Y.; Guan, Y.; Newsam, S.; Shao, L. Towards Universal Representation for Unseen Action Recognition. In Proceedings of the IEEE Conference on Computer Vision and Pattern Recognition (CVPR), Salt Lake City, UT, USA, 18–23 June 2018.
99. Othman, E.; Werner, P.; Saxen, F.; Al-Hamadi, A.; Walter, S. Cross-database evaluation of pain recognition from facial video. In Proceedings of the International Symposium on Image and Signal Processing and Analysis (ISPA), Dubrovnik, Croatia, 23–25 September 2019; pp. 181–186. [CrossRef]
100. Werner, P.; Lopez-Martinez, D.; Walter, S.; Al-Hamadi, A.; Gruss, S.; Picard, R. Automatic Recognition Methods Supporting Pain Assessment: A Survey. *IEEE Trans. Affect. Comput.* **2019**. [CrossRef]
101. Li, S.; Deng, W. Deep Facial Expression Recognition: A Survey. *IEEE Trans. Affect. Comput.* **2020**, *3045*, 1–20. [CrossRef]
102. Wang, M.; Dong, W. Deep Face Recognition: A Survey. *arXiv* **2020**, arXiv:1804.06655.
103. Pietak, A.; Ma, S.; Beck, C.W.; Stringer, M.D. Fundamental ratios and logarithmic periodicity in human limb bones. *J. Anat.* **2013**, *222*, 526–537. [CrossRef] [PubMed]

Short Biography of Authors

Michal Rapczynski received his B.Sc. and M.Sc. degree at the Otto von Guericke University Magdeburg, Germany. Since 2013, he is a Researcher and Ph.D. candidate in the Neuro-Information Technology Group at Otto von Guericke University Magdeburg. His research focuses on computer vision, image processing, machine learning and biomedical signal processing.

Philipp Werner received his Masters degree (Dipl.-Ing.-Inf.) in computer science from the Otto-von-Guericke University Magdeburg, Germany, in 2011. Since then he has been working as a Research Assistant and Ph.D. candidate in the Neuro-Information Technology group of the Otto von Guericke University. His research focuses on pain recognition, facial expression recognition, human behavior recognition, computer vision, pattern recognition, and deep learning. Since 2018 he has been a research team leader at the Neuro-Information Technology Group of the Otto von Guericke University Magdeburg, Germany. He has authored and co-authored more than 40 articles, which have been cited more than 700 times. See http://philipp-werner.info for more details.

Sebastian Handrich received his B.S. and M.S. Degree in electrical engineering from the University of Magdeburg, Germany in 2008. After working as a research assistant at the University of Oldenburg in the field of biological psychology, he is currently working on his Ph.D. in electrical engineering and information technology at the University of Magdeburg. His research focuses on human pose estimation, facial expression analysis, affective computing and human machine interaction.

Ayoub Al-Hamadi received the Ph.D. degree in technical computer science, in 2001, and the Habilitation degree in artificial intelligence and the Venia Legendi degree in pattern recognition and image processing from Otto von Guericke University Magdeburg, Germany, in 2010. He is Professor and the Head of the Neuro-Information Technology Department (NIT), Otto-von-Guericke University Magdeburg. He is the author of more than 350 papers in peer-reviewed international journals, conferences, and books. His research interests include computer vision, pattern recognition, artificial intelligence, and human-roboter interaction. See http://www.iikt.ovgu.de/al_hamadi.html for more details.

sensors

Article

Design of a Plantar Pressure Insole Measuring System Based on Modular Photoelectric Pressure Sensor Unit

Bin Ren * and Jianwei Liu

Shanghai Key Laboratory of Intelligent Manufacturing and Robotics, School of Mechatronic Engineering and Automation, Shanghai University, Shanghai 200444, China; jianweiliu@shu.edu.cn
* Correspondence: binren@i.shu.edu.cn

Abstract: Accurately perceiving and predicting the parameters related to human walking is very important for man–machine coupled cooperative control systems such as exoskeletons and power prostheses. Plantar pressure data is rich in human gait and posture information and is an essential source of reference information as the input of the exoskeleton control system. Therefore, the proper design of the pressure sensing insole and validation is a big challenge considering the requirements such as convenience, reliability, no interference and so on. In this research, we developed a low-cost modular sensing unit based on the principle of photoelectric sensing and designed a plantar pressure sensing insole to achieve the purpose of sensing human walking gait and posture information. On the one hand, the sensor unit is made of economy-friendly commercial flexible circuits and elastic silicone, and the mechanical and electrical characteristics of the modular sensor unit are evaluated by a self-developed pressure-related calibration system. The calibration results show that the modular sensor based on the photoelectric sensing principle has fast response and negligible hysteresis. On the other hand, we analyzed the area where the plantar pressure is densely distributed. One benefit of the modular sensing unit design is that it is rather convenient to fabricate different insole solutions, so we fabricated and compared several pressure-sensitive insole solutions in this preliminary study. During the dynamic locomotion experiments of wearing the pressure-sensing insole, the time series signal of each sensor unit was collected and analyzed. The results show that the pressure sensing insole based on the photoelectric effect can sense the distribution of the plantar pressure by capturing the deformation of the insole caused by the foot contact during locomotion, and provide reliable gait information for wearable applications.

Keywords: optical sensing principle; modular sensing unit; plantar pressure measurement; gait parameters

Citation: Ren, B.; Liu, J. Design of a Plantar Pressure Insole Measuring System Based on Modular Photoelectric Pressure Sensor Unit. *Sensors* **2021**, *21*, 3780. https://doi.org/10.3390/s21113780

Academic Editors: Tomasz Krzeszowski, Adam Świtoński, Michal Kepski and Carlos Tavares Calafate

Received: 14 May 2021
Accepted: 27 May 2021
Published: 29 May 2021

1. Introduction

In recent years, much research on lower limb exoskeleton robots has been carried out [1–4] to help with human activities and enhance the functions of the human body. Among many exoskeleton/prosthetic assist devices, the primary task is to provide the wearer with assistance in walking motions. The detection and sensing of data information related to human motion is the basis of and key to the compliance control of the lower limb wearable device [5,6]. The human wearer is the controlling center of an exoskeleton system. The real-time information on the human body is the primary source of the exoskeleton man-machine coupling control system, which accurately senses and predicts the state of human walking. An exoskeleton controller can detect the intent of the motion and control the corresponding parts of the drive module by sensors. At the same time, the comparison between the human body and the exoskeleton motion is analyzed to provide feedback to ensure that the exoskeleton can respond to human action quickly and accurately. It is also vital to provide a safety guarantee for the human body in the human–machine coupling system.

The interactive contact between the feet and the ground is the most intuitive manifestation of human motion dynamic information. Plantar pressure data contains abundant human gait and posture information [7]. In the initial stage, the fixed system [8,9] (such as motion capture system, force measurement platform system, etc.) is used to provide a simple and effective way to explore the basic biomechanical laws walking process. However, these non-mobile systems can only be used in limited space and usually have expensive construction and maintenance costs [7]. In practical applications, when people need to wear mobile assistive devices to cope with various environments or terrains in an outdoor environment, the motion perception system requires efficiency and portability. Thus, pressure-sensitive insoles/socks provide a better trade-off. They usually use flexible materials as their medium (such as silicone [10], fabric [11], composite materials [12], etc.), employing different sensing principles (for example, piezoresistive, capacitive, piezoelectric, etc.) for the portable wearable plantar pressure measurement system to collect information on the movement of the portable robot. Powerful technical support is provided in the wearable application.

There are some commercialized sensing insoles based on different sensing principles. The F-Scan system (Tekscan®, South Boston, MA, USA) [13] uses FSR (Force-sensing resistors) sensors, the ParoTech system (Paromed®, Neubauer, Germany) [14] uses piezoresistive sensors and the Pedar system (Novel® GmbH, Munich, Germany) [15] uses an embedded capacitive sensor. In addition to the commercial insole design, researchers are still trying to innovate in structural layout and processing algorithms. Liu et al. [16] designed a pressure-sensitive foot for the lower extremity exoskeleton. The pressure-sensitive foot can measure plantar pressure to sense the contact with the ground and reflect the wearer's behavioral intentions. Lim et al. [17] compared the three flexible pressure sensors of FSR, FlexiForce and capacitive sensors. They chose the FlexiForce sensor to design the pressure insole and detect the gait phase based on the threshold segmentation method of the pressure center. Wu et al. [18] used an insole made of three FSR sensors to detect four gait sub-phases. Chen et al. [19] used FlexiForce sensors to design a pair of insoles with eight sensors to identify walking patterns. Zhang et al. [11] developed a simple, low-cost and highly integrated insole based only on fabric for measuring plantar pressure, the principle of which mainly relies on the capacitive mechanism. However, as emphasized in the paper [20], it is precisely because of the light, thin and soft characteristics of these sensing units that they will produce unpredictable distortion and deformation on the contact surface, making the sensing response unable to be accurately estimated. What is more, this type of sensor usually needs to go through an additional modulation circuit to amplify the signal.

In addition to those plantar pressure-sensing insole solutions based on membrane-based sensor units mentioned above, the research teams tried to develop their pressure-sensing insoles to provide more reliable plantar pressure information sensing solutions. Park et al. [21] showed a novel use of high-sensitivity crack-based strain sensors to make plantar pressure insoles. The technical solution based on photoelectric induction has attracted the attention of many researchers. Leal et al. [12,22] grasped the characteristics of polymer optical fiber (POF), such as lightweight, anti-magnetic and electrical isolation, and initially designed and integrated four POF of the sensing unit [12] to monitor the ground reaction force during the gait, and the follow-up research work [22], combined with the advantages of 3D printing technology rapid prototyping, developed customizable pressure sensing insoles and increased the number of POF sensors to 15. The research team from Santa Ana [20,23,24] used the light-emitting unit and the photosensitive unit arranged on the same side and realized the sensing of pressure to electric signal with the help of an elastic rubber cover covering the sensor element, and applied this technology to the outside, in the interactive signal perception between the bones and the human body in the ring. These technical solutions paved the way for the research on a more stable and comprehensive plantar pressure sensing system.

In this paper, we mainly completed the following work: firstly, a modular pressure sensor based on photoelectric sensing technology is properly designed and fabricated. The components of the sensor are from commercially available materials. Its structural design is novel, and no additional signal amplifier is needed in the sensing acquisition circuit to capture the sensing signal. In the manuscript, we introduced how to use easily accessible, low-cost manufacturing methods and materials to make such a modular sensor so that other researchers can easily reproduce technology and further carry out related research work. Secondly, the designed modular pressure sensor is implemented on a specially designed programmable control calibration instrument [25,26]. The mechanical and electrical characteristic evaluation experiment proved that the modular sensor has specific applicability in pressure sensing. Thirdly, based on the analysis of the pressure distribution area, two different sensor layout schemes were specified, and the modular pressure sensor was integrated into the pressure sensing insole. The performance of two insole solutions were compared in the preliminary experiment. Finally, combined with the dynamic walking experiment, the performance of the manufactured pressure sensing insole in the application of collecting plantar pressure was explored, and the results showed that the insole system could monitor in real-time plantar pressure and provide reliable gait-related parameters, which provides potential value in wearable walking robot equipment, exoskeleton, power prosthetics and other applications.

2. Materials and Methods

2.1. Modular Pressure Sensing Unit

2.1.1. Sensing Principle

The sensor technology used in this research mainly relies on the photoelectric effect. That is, the photoresistor exhibits different resistance characteristics under different ambient light intensities. The resistance of the photoresistor decreases as the incident light (visible light) increases. Under normal conditions, its resistance can reach 10,000 to 10 million ohms while, under photosensitive conditions (such as 100 Lux), its resistance is only a few hundred to a few thousand ohms.

Generally, when the walking foot touches the ground, the sole exerts a force on the ground through the insole and the shoehorn. During this period, the insole undergoes a certain degree of deformation in the direction perpendicular to the contact surface. We hope to use this tiny deformation relationship to induce the induction between the photodiode and the photoresistor. In other words, during the deformation process, the distance between the light emitter (light-emitting diode) and the light receiver (photoresistor) changes to cause a change in light, which in turn triggers a change in the resistance of the photoresistor, as shown in Figure 1. We use photoelectric technology to capture the slight deformation of the insole caused by plantar pressure. It is crucial to design an appropriate design structure to integrate the light-emitting diode and photoresistor into a narrow space with the limited thickness of the insole and provide a light-transmitting medium with suitable material properties.

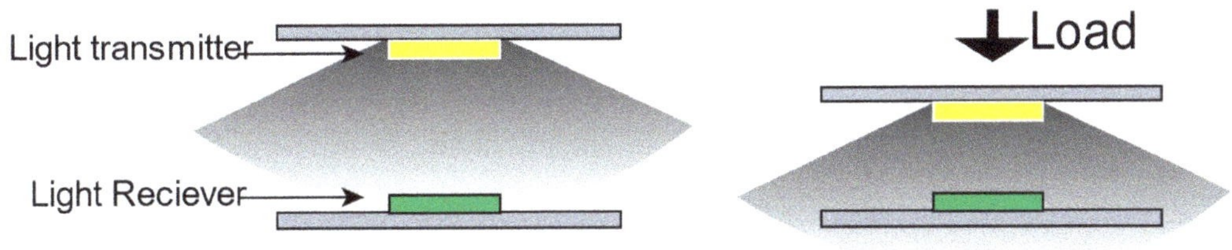

Figure 1. Photoelectric sensing principle diagram.

2.1.2. Design and Manufacturing

Compared with the overall layout, the modular sensing solutions are convenient for customizing. Therefore, we designed a modular sensing unit based on the sensing principle introduced above, adjusting the sensor layout according to different foot sizes to embed pressure sensing insoles. The designed modular sensing unit is mainly composed of three parts: (1) a flexible circuit board containing a photodiode and a photoresistor; (2) an elastic light-transmitting silica gel medium used to absorb the applied pressure and recover when the pressure is removed; (3) some necessary electrical connections.

(a) Flexible circuit board

Optical transmitters and optical receivers play an essential role in optical sensing technology. In this article, we introduce a low-cost method to use this technology. The light-emitting element and the photosensitive element are obtained by modifying the commercial LED strip (Telesky, Shenzhen, China), as shown in the figure. The LED strip is based on a flexible printed circuit board (FPC), a photodiode powered by 5 V and a corresponding current limiting resistor. Each light-emitting diode is independently powered and can normally work when the anode and cathode are connected to a 5 V power supply. It is worth noting that the applicable model of the photodiode is 5050 (5 mm × 5 mm), which means that the LED footprint can fully accommodate the 1206 SMD surface mount package. Therefore, according to the positive and negative polarity, we chose a 1206 SMD surface mount package type photoresistor, replacing the lamp beads in the LED light strip. At the same time, we soldered a signal wire from the voltage divider circuit node and led it out. After using a jumper wire to connect the LED light bar and the photoresistor bar, we used hot melt glue to cover the soldering point to improve its reliability, as shown in Figure 2.

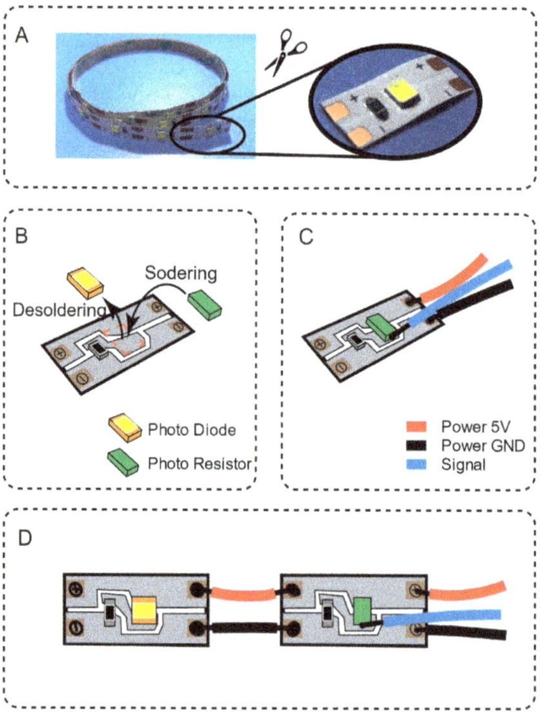

Figure 2. Fabrication of flexible circuit board inside the sensing unit. (**A**) cut out a single LED unit from the light strip; (**B**) replace the LED lamp beads with a photoresistor; (**C**) weld the power supply and signal wires; (**D**) weld the LED strip and the photoresistor strip together.

(b) Flexible light-transmitting medium

The pressure sensing unit needs to tolerate a specific pressure range and return to the original state when the applied pressure is released. As an inexpensive material, organic silicon materials are widely used in the design of many flexible sensors. This article uses semi-transparent silica gel (Beijing Hibas Technology Co., Ltd., Beijing, China) as the primary elastic material and is mixed with corresponding plasticizers to catalyze the solidification process of silica gel. In the initial prototype design, we found that only silica gel was used to prepare a light guide medium with higher hardness, which made the signal not obvious enough for us in the plantar pressure range of interest. To optimize the design of the sensor unit, we have included a softener (dimethazone). By mixing different proportions of silica gel and softener, we finally determined the appropriate ratio of the mixture as the elastomer medium, and its weight ratio is silica gel: softener = 4:1.

(c) Integration process

After preparing the circuit and the elastic medium, we used Autodesk Fusion 360 modelling software to design multiple molds for casting and used FDM3D printers to prepare the molds. As shown in Figure 3, one of the molds is used to make the sensor unit's silicone shell baffle (thickness 1 mm). The other mold is used to integrate the entire sensor unit (including the silicone shell baffle, circuit and elastic medium). Figure 3B shows the process of fixing the sensing unit circuit in the grooves of the two silicone shell baffles and, finally, integrating it into a whole with the cured silicone medium. The final size of the sensor is a square flexible sensing unit which is 20 mm in width, 20 mm in length and 7 mm in height. One single sensing unit weighs 2~2.2 $\times$ 10^{-3} kg (silica gel density 700 kg/m^3). Figure 4 shows the manufactured sensor with and without power supply status.

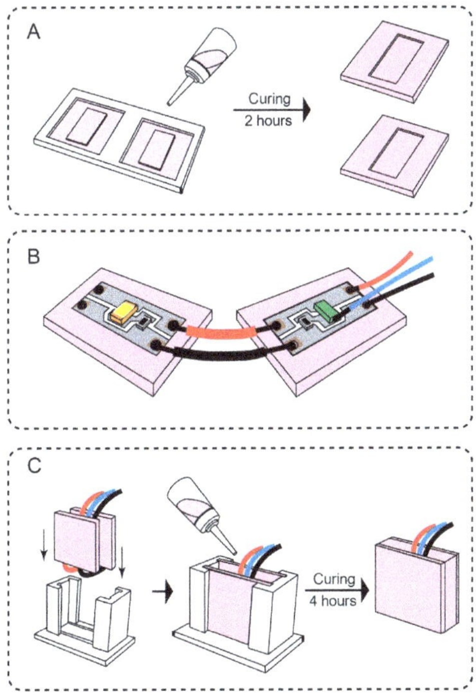

Figure 3. The integration process of the flexible module of the sensing unit. (**A**) Silica gel baffle pouring (**B**) The sensor unit is fixed on the silica gel baffle (**C**) Put into the mold and add the silica gel mixture to wait for solidification.

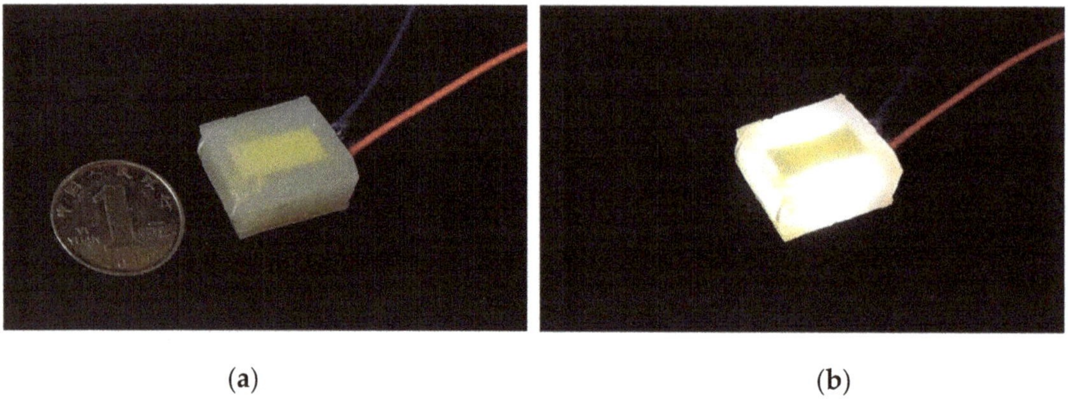

Figure 4. The manufactured sensing unit. (a) unit without power (b) unit with power.

2.1.3. Characteristic Analysis System

We analyzed the mechanical and electrical characteristics of the manufactured modular sensing unit and calibrated the mapping relationship between the quasi-static pressure and the output signal. It is necessary to conduct a characteristic evaluation experiment on the sensing unit before being integrated into the insole. In the paper [19], the author provides a low-cost calibration method that allows researchers to carry out the calibration test of pressure-related sensing units without using expensive calibration equipment. Here, we introduce an improved version. The calibration analysis system is shown in Figure 5. First, we established a calibration instrument according to the process described in [19] to measure the load force and deformation during the static load test. The calibration instrument is a microsystem composed of three parts: (1) HX711 force measurement unit, on both sides of which are bolted 3D printed rigid plastic (PLA) boards; (2) HX711 amplifier circuit module; (3) an Arduino NANO microcontroller for collecting and recording data from the measuring instrument; secondly, we replaced the original printing platform of the FDM3D printer with a load cell. At the same time, the print head of the printer was replaced with a corresponding contact pressure head according to different test purposes. The loading and unloading can be designed by writing G-code control codes for the 3D printer test. In other words, compared to manually adding weight, the mechanical frame of the 3D printer makes the process more controllable.

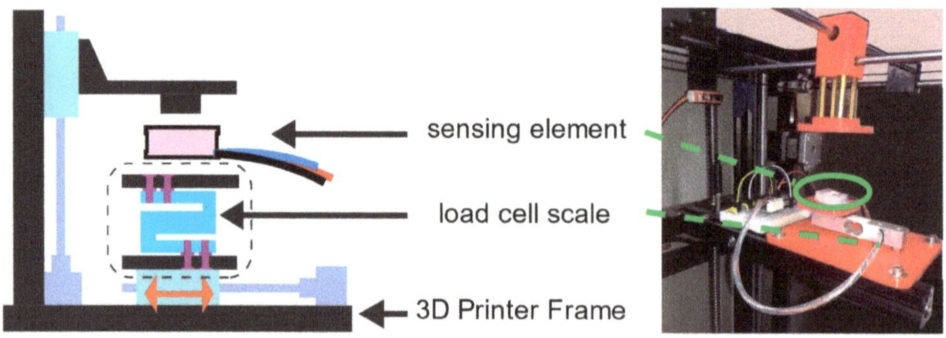

Figure 5. Characteristic analysis system.

2.2. Insole Solution

2.2.1. The Layout of Insole

This part mainly introduces the design and production of flexible pressure sensing insoles based on the photoelectric pressure sensing unit designed above. The most challenging problem is that the layout of the sensing unit in the pressure sensing insole needs to fully consider the plantar pressure distribution. From the intuitive impression, due to the irregular surface of the sole and the dynamically changing contact position, the pressure is not evenly distributed on all the surfaces of the insole. For example, the pressure on the inside of the foot arch is slight, while the heel and forefoot areas have greater pressure. Figure 6 is a diagram of plantar pressure distribution in the standing state from reference [13]. From the heat diagram, it can be observed that the plantar pressure is mainly distributed in the heel, forefoot and toes, among which the force in the toe area is mainly located on the thumb. Therefore, placing sensors in these locations can provide more relevant data on plantar pressure.

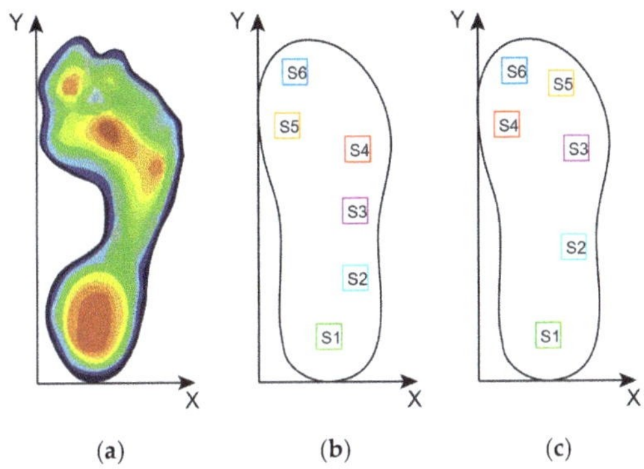

Figure 6. The plantar pressure distribution. (**a**) typical plantar pressure distribution in a standing position; (**b**) first insole layout solution; (**c**) second insole layout solution.

Because the difference in the foot size varies among different people, and the layout of the pressure-sensing insole does not have a proper guideline, two preliminary steps were made based on the author's foot size. A flexible pressure insole solution, as shown in the figure: the layout of the sensor mainly refers to the aforementioned plantar pressure distribution. In the first solution, sensors are placed in six places: the first toe, the third toe, the first metatarsal, the fifth metatarsal, the outside of the arch of the foot and the heel. In the second solution, sensors are placed in six places: the first toe, the first metatarsal, the fifth metatarsal, two outside of the foot's arch and the heel. "S1" in Figure 6b, c represents sensor 1, "S2" represents sensor 2, and so on.

2.2.2. Insole Manufacturing

The manufacturing process of the insole is as follows: first, we designed and 3D printed the casting mold for the insole (right foot), which is size 43 according to Chinese standards; secondly, the sensor unit was fixed in the insole casting mold according to the corresponding position of the two insole layout solutions (six sensors for each solution); all the wires are guided to the outlet at the heel of the casting mold and fixed in the free position where the sensor does not interfere with each other; after the mold outlet is closed with a 3D printed lid, the silica gel mixed with the same proportion softener (dimethazone) was poured into the casting mold—it should be noted that the cavity height of the insole casing mold is 0.5 mm higher than the height of the single sensing unit which ensures the surface of the insole after casting is as flat as possible; finally, the mixed silica gel takes 4 h to solidify, and all the power supply wires are welded into a bus for external power supply. The weight of the two insole solutions are 136 g and 132 g, respectively. Figure 7 shows the two plantar pressure-sensing insole solutions in the non-powered state and the powered state.

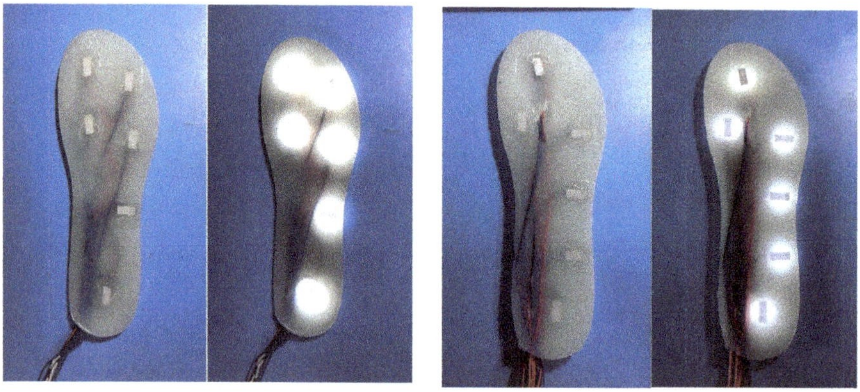

(a) (b) (c) (d)

Figure 7. Two insole solutions for the right foot insole. (**a**) Insole solution No. 1 without power (**b**) Insole solution No. 1 with power (**c**) Insole solution No. 2 without power (**d**) Insole solution No. 2 with power.

2.2.3. Electrical System

To measure the sensor's signal and record the data, we designed the circuit system according to the system framework shown in Figure 8. The circuit system is mainly used for sensor signal acquisition, data preprocessing and data storage, including a microcontroller module, a data storage module and a power supply device. Since the voltage divider circuit of the sensing unit has been integrated inside the sensing unit, there is no need to use additional modulation circuits and operational amplifiers to process the signal. The signal channels from the pressure sensing insole (6 per foot, 12 on both feet) are connected to the input port of the 16-channel multiplexer module (HC4067, NXP). Under the control of the microcontroller (Arduino UNO, Ivrea, Italy), the multiplexer traverses all the connection channels in turn, and transmits the collected analogue signal to the analogue input port of the microcontroller and passes the built-in ADC (analogue-to-digital converter), which converts the voltage signal into a digital signal for storage. The sensor signal data is recorded in a file on the SD card for offline analysis and evaluation of the performance of the plantar pressure-sensitive insole. To improve the overall ease of use, we designed an Arduino UNO expansion integrated circuit board to integrate all the above modules into the expansion circuit board, as shown in Figure 8. The entire system can be powered by a DC voltage source of 5 V~12 V for power supply, such as a polymer lithium battery.

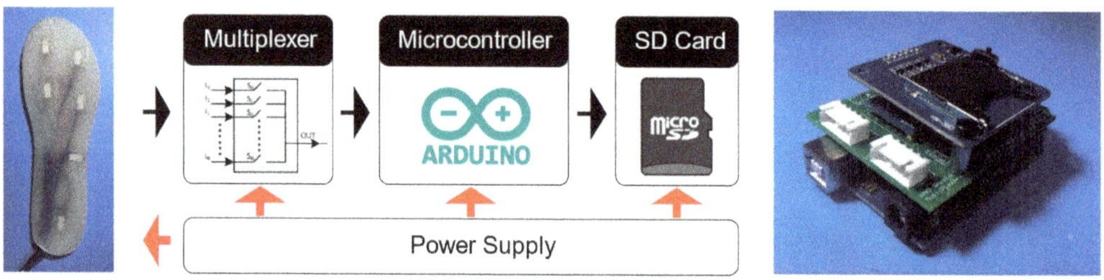

(a) (b)

Figure 8. Circuit system diagram (**a**) and Arduino UNO expansion integrated circuit board (**b**).

2.2.4. Plantar Pressure Center

COP is widely used in the study of plantar pressure-related parameters, especially the division of the gait phase, so it can be used as the most direct evaluation parameter for verifying pressure-sensing insoles. During exercise, due to the movement of the body's center of gravity, the plantar pressure center shows a periodic trend. It moves from the heel to the toe in a single foot and switches back and forth between the two feet. Therefore, based on our design, we refer to the method introduced in the paper [11] to calculate the COP. The COP is divided into CoP_x along the inner and outer directions and CoP_y in the front and rear directions. The calculation method is shown in Equations (1) and (2):

$$CoP_x = \frac{\sum_{i=1}^{6} X_i \cdot P_i}{\sum_{n=1}^{6} P_i}, \tag{1}$$

$$CoP_y = \frac{\sum_{i=1}^{6} Y_i \cdot P_i}{\sum_{n=1}^{6} P_i} \tag{2}$$

where X_i and Y_i represent the position of the sensing unit along with the medial/lateral directions and front/rear direction, respectively, as shown in Figure 7. P_i represents the signal value of the *i*th sensing unit. It is worth noticing that the plantar pressure center only exists in the standing stage of the leg. Therefore, we define that the center of pressure during the swing stage is located at (0,0) to distinguish the standing and swing phases.

3. Experiments and Results

3.1. Sensor Characteristic

Using the calibration analysis system introduced above, we can easily carry out the characteristic analysis experiment of a single sensor. The characteristic analysis experiment is mainly to anchor the center of the sensor unit perpendicular to the static load test of the pressure sensing surface. The static load test is defined as a step of 0.025 mm perpendicular to the sensor's surface and then staying for 3 s to have enough time for stable measurement. The maximum distance is 1 mm (accounting for 14.3% of the thickness of the sensing unit). After the loading process, the unloading process is completed according to the same stepping distance and dwell time until the indenter leaves the surface of the sensing unit. Results of stiffness (force-strain response), sensitivity (resistance-force response) and hysteresis characteristics of a batch of six sensing elements are analyzed.

As shown in Figure 9a, all sensing units exhibit certain mechanical hysteresis characteristics in terms of mechanical characteristics. According to the quantification method of mechanical hysteresis characteristics in the paper [5], that is, through calculation, the ratio of the area enclosed by the loading range and the horizontal axis to the area enclosed by the unloading range and the horizontal axis in the curve is used to quantify the mechanical hysteresis characteristics. The mechanical hysteresis coefficients of each sensor are 0.928, 0.937, 0.947, 0.935, 0.921 and 0.933, respectively. We can observe that the mechanical characteristics of the pressure sensing unit are relatively consistent, which are mainly related to the characteristics of the silicone elastomer inside the sensing unit.

As shown in Figure 9b, there is a certain linear relationship between the sensing signal and the load in terms of electrical characteristics. With the help of MATLAB's cftool toolbox, we chose to use a polynomial to fit the curve. Here, the relationship curve between signal response and loading force is fitted with a two-order polynomial ($\mathrm{F(s)} = a_0 + a_1 s + a_2 s^2$ where S represents the loading force, and F represents the output response signal). The fitting results are shown in Table 1. It can be found from Figure 10 that the electrical hysteresis characteristic is almost negligible during loading and unloading. Combining Figure 10 and Table 1, we can observe that the initial sensing signals (a_0) of the six sensing units under no-load are more or less different. However, from the 0-60N load range result, the sensor's sensing range is relatively close.

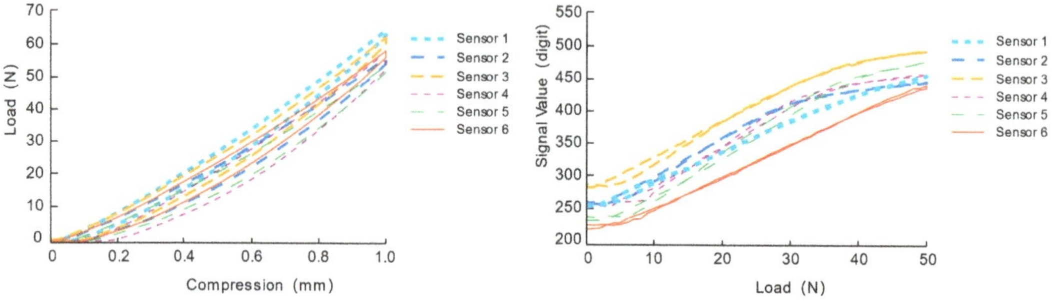

(**a**) (**b**)

Figure 9. The mechanical and electrical characteristics. (**a**) The mechanical characteristics (**b**) The electrical characteristics.

Table 1. Fitting Result of Sensing Unit Electrical Characteristic.

Sensor	Fitted Coefficients			Fitting Effect		0–60 N Load RanΔS
	a_0	a_1	a_2	RMSE	R^2	
1#	261.2	9.867	−0.105	1.074	0.997	161.3
2#	234.5	5.421	−0.011	2.395	0.992	176.4
3#	281.5	6.517	−0.051	2.753	0.998	169.7
4#	264.7	8.714	−0.091	2.711	0.998	173.4
5#	240.1	8.098	−0.058	2.475	0.996	185.9
6#	225.5	4.974	−0.027	1.713	0.993	178.3

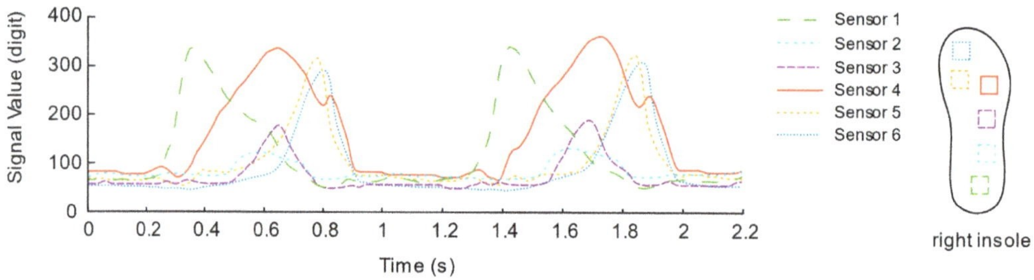

Figure 10. Signal Curve of the first insole solution.

3.2. Gait Data Collection

3.2.1. Comparison of Insole Solutions

We evaluate the performance of the two designed induction insole solutions. Preliminary test experiments were carried out on both. The author of this article (male, age 24, height 1.79 m, weight 76.6 kg, shoe size 43) wears two kinds of insoles in indoor corridors according to walking habits, and conducts the following tests: from a natural standing state to a waking state with an average pace, then staying still for a few seconds in the end.

In the experiment, we found that the induction signal has an abnormally negative value. Through observation, we found that the abnormal phenomenon is mainly caused by the asymmetry of the flexible circuit board area in the sensor unit, which causes the internal optical path of the sensor unit to shift when pressure is applied to the circuit around the sensor unit. At this time, the optical path is deflected. The most direct effect of the shift is that the light intensity is reduced compared to the case where the light path is directly facing, which leads to abnormal negative values of the induced signal in the experimental results. Therefore, this is an inevitable feature in the design principle of light-sensitive pressure sensing based on commercial LED light strips in this study. That is, the sensing signal at the center of the sensing unit will be interfered with by the surrounding pressure, but it will be affected by the center of the sensing unit. When the pressure is positive, the pressure signal is in line with theoretical expectations. In response to this phenomenon, we used the linear rectification activation function (Rectified Linear Unit, ReLU) in the neural network to preprocess the signal of the sensing unit and use the negative signal caused by the pressure around the sensing unit under the function of the function. It is filtered, and only the positive part of the sensing signal is retained. The expression equation can be described as:

$$f(x) = \begin{cases} 0 & if(x \leq 0) \\ x & if(x > 0) \end{cases}, \tag{3}$$

After the rectification activation function is processed, as shown in Figures 10 and 11, the signal curves are drawn from a piece of data intercepted from the walking experiments of the two insole schemes. In contrast, the data curve of solution No. 1 is "messy", and the mess is mainly reflected in the insufficient regularity of the sensor signal fluctuations located in the toe area.

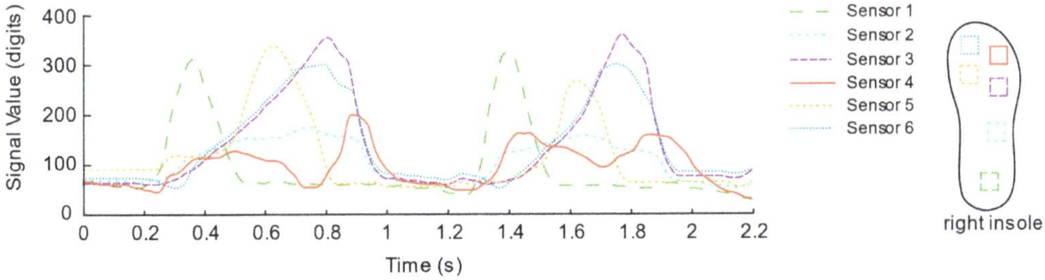

Figure 11. Signal Curve of the second insole solution.

To further compare the quality of the two solutions, the pressure centers are calculated by Equations (1) and (2), as shown in Figures 12 and 13. From the CoP_y of the curve of solution No. 1, we can observe that the plantar pressure center during the stance phase is disturbed suddenly and, then, the CoP_y of the curve of the solution No. 2 can better reflect that the pressure center is standing. The tendency of the phase is to move from the heel to the toe, therefore the layout of the second solution is regarded as a better sensor layout.

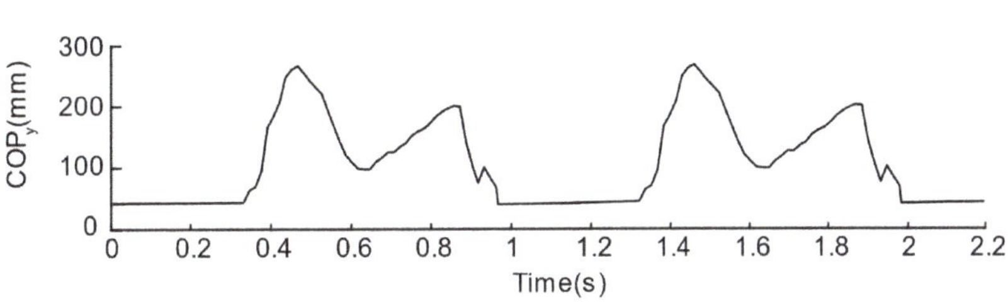

Figure 12. COP values of the first insole solution.

3.2.2. Biped Gait Data Collection

We completed the left foot insole production according to the same production process as the right foot insole. Since the circuit system design reserved up to 16 sensing channels for both feet, one only needs to open the left foot sensing channel in the acquisition program to acquire all the sensing channels of the feet (a total of 12 sensing units). The equipment used for the bipedal gait data collection experiment is shown in Figure 14, including a pair of versatile 43 size shoes and a pair of self-designed two-point photoelectric pressure sensor insoles. Before collecting data, the wearer uses a nylon bayonet to fix the wire on the back of the lower limbs, in the manner shown in Figure 14, to avoid the influence of wire swinging on the usual walking movement during walking. The nylon bayonet and cable tie are fixed on the wearer's waist, and the power bank can be placed in the wearer's trouser pocket after power is supplied to achieve the minimum hindrance to walking. The gait data of a subject was collected using the device. The subject has never suffered from any disease that hinders walking posture. The experiment process is also from a natural standing state to a waking state with an average pace followed by staying still for a few seconds.

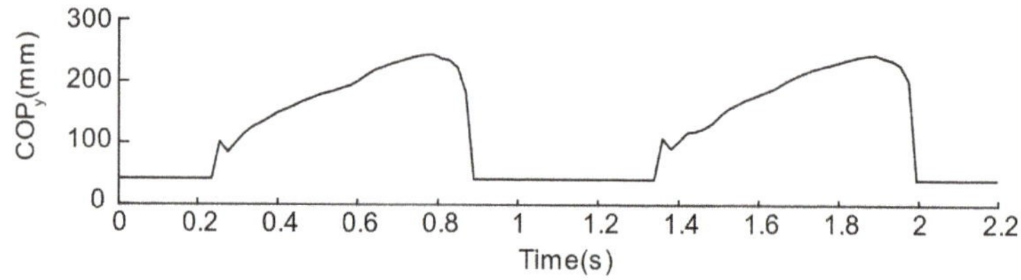

Figure 13. COP values of the second insole solution.

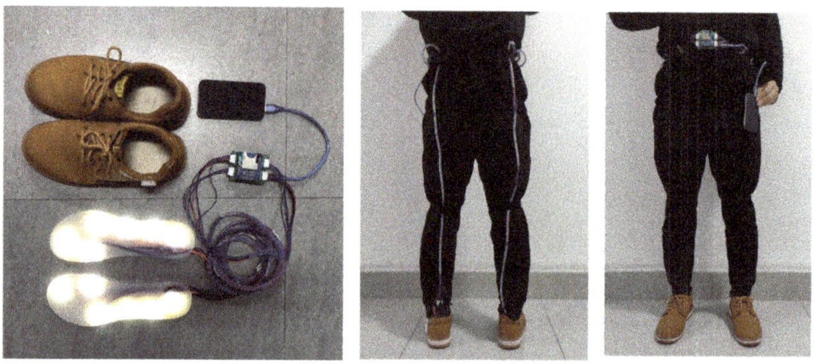

(a) (b) (c)

Figure 14. The experimental setup. (**a**) equipment used (**b**) sensing system setup back view (**c**) sensing system setup front view.

Figures 15 and 16 describes the dynamic bipedal walking experiment data. Observing the data, each sensor presents a periodic "rest state" and "active state". During the "active state", the sensor signals reach their respective peaks in succession. During the "resting state", all sensors returned to their lower levels, which is consistent with our intuitive impression of the phase of standing support and swing phase during the complete gait cycle of a single leg.

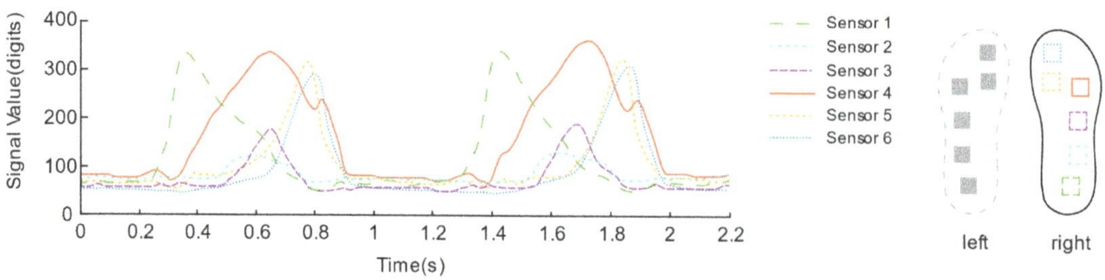

Figure 15. Right foot plantar pressure sensor signal during the bipedal walking experiment.

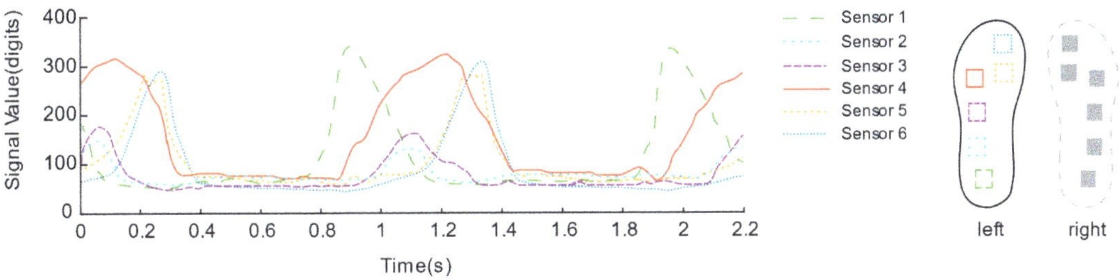

Figure 16. Left foot plantar pressure sensor signal during the bipedal walking experiment.

In addition, it can be estimated that the overall walking frequency is about 31 steps per minute from the periodically changing curve. Focusing on the data of each sensor of the insole, we can observe that, during the initial period from the "rest state" to the "active state", the No. 1 sensor unit located on the heel first senses the pressure and quickly rises to the peak; then, as the pressure of the No. 1 sensor gradually decreases, the pressure of the No. 2 sensors located on the outside of the arch of the foot also change; secondly, the No. 3, No. 4 and No. 5 sensors located on the forefoot and the toes were in the same interval, reaching their respective peaks over a long time; finally, all sensor signals return to a lower level of resting state. Connecting the pressure curves of the left and right feet, we can see that in the short period when the pressure of the right foot is about to enter a lower level, the heel pressure of the left foot has already been generated and rapidly increased to the peak. Similarly, the pressure of the left foot is about to enter a higher level. During the short period of low level, the heel pressure of the right foot has also begun to reach its peak rapidly, which also reveals the "bipedal standing phase" that is not easily noticeable when walking. Both feet are in contact with the ground during walking.

4. Discussion

The content shown in this research is mainly focused on the integration of photoelectric sensing technology into the pressure sensing unit to achieve the purpose of sensing plantar pressure information, including the analysis of the photoelectric sensing principle and technology, and the introduction of the technical method realization process. We conducted a test and result analysis of sensor unit characteristics, and performance evaluation during actual use. Here, we will discuss and analyze the results obtained, and make reasonable assumptions and note prospects for further research work.

In this study, in terms of the design of the modular sensing unit, the modular-type sensor provides flexibility for the layout of the pressure sensing insole solution. This modular sensor unit is designed ingeniously and economically. The production of the sensor unit can be completed by using some materials that can be easily purchased from the market. However, as far as the manufacturing method is concerned, the method provided in this article is only for small batches and is a hand-made method, so there are certain defects in the stability and repeatability of the sensor characteristics, which can be found in the sensor characterization section. The result analysis shows that in further research work, if one wants to obtain a more stable and reliable modular sensor unit, it is needed to improve the existing processing technology and manufacturing equipment, and choosing a mature engineering technology may be able to solve this problem. On the other hand, it is worth mentioning that our research introduced a compromised pressure analysis instrument, which can carry out pressure-related calibration test work by building a simple force plate and transforming a desktop-level 3D printer. The design of a programmable calibration instrument derived from the analysis of the mechanical and electrical characteristics of the sensor in this study can also be useful for research teams with limited experimental conditions, that is, who cannot obtain equipment with higher precision and more comprehensive functions.

As for the layout of the pressure sensing insole, in some research work, high-resolution intensive plantar pressure sensing is settled as the research target, such as the research work from [20,23,24]. However, from the perspective of practical application, high resolution means a sharp increase in the number of sensor units. The consequence is that more complex computing processors to deal with the hypermultiplet-channel signal and larger power supply units are required to maintain long-term data recording. However, the plantar pressure distribution is continuous, which means there is information redundancy in the same area. The modular sensor unit can be used to lay out the area with main pressure characteristics (such as the pressure sensing insole layout solutions in this article) to reduce the density of the sensing element. In fact, by adjusting the layout position of the sensing unit, the stability and accuracy of the acquisition of plantar pressure gait data can be improved in a controllable manner as exhibited in this research. On the one hand, the experimental results show that, due to the structural configuration of the sensing unit itself, there exists a phenomenon of "negative pressure" around the sensing unit. We analyzed the cause of this phenomenon and used the rectification activation function to preprocess the negative signal. On the other hand, under normal circumstances, the pressure sensor can better monitor the changes in plantar pressure in the position where the force is more extensive. The sensor unit located in the toe area may not be selected as the pressure sensor unit due to the need to control the stability–sensation area. We mainly focused on the analysis of the results of COP in the front and rear directions, which also showed a specific rule; that is, during the period when the observed leg is in the standing support phase, the movement trend of the center of pressure is to shift from the heel position to the toe position gradually. From this information, the following stages of gait can be preliminarily observed:

(1) Swing stage: the sole hardly exerts a force on the insole, and the total pressure is in a stable state and lower than the standing state.
(2) Heel contact stage: the heel touches the ground and bears weight, and the pressure on the heel area increases significantly.
(3) Intermediate stance phase: the heel no longer bears the same pressure as the heel contact phase, and part of the pressure is transferred to the front foot.
(4) Toe off stage: the body's center of gravity is almost moved to the other side of the body, the heel is off the ground, and the pressure is mainly concentrated on the forefoot.

However, this article only shows a hand-made/manual operation of modifying the sensor layout. In view of the increasing application of machine learning and other technologies in the engineering field, if the sensor layout can be used as an optimization goal, optimization methods based on machine learning can be employed to perform a much better sensing unit layout for a wider range of plantar pressure distribution information. We believe that the optimization strategy using artificial intelligence algorithms will bring reference significance for better design layout.

What is more, when discussing this research work from the perspective of signal transmission, the current design work adopts the method of transmitting the signal through the cable to obtain the pressure signal. It needs to be admitted that although, in our design, the cables are carefully routed according to the path that does not hinder the movement as much as possible, it is inevitable that there are annoying obstacles such as position displacement and winding during the actual locomotion. Wires will negatively affect the reliability and complexity of wearable devices (usually tied to the human body). Therefore, in further research work, wireless communication protocols (such as WIFI, Bluetooth) are urgently desired for signal transmission. The development of a wireless transmission version of the plantar pressure measurement system can get rid of the annoying winding that hinders movement and improve the integration of the entire sensor measurement system, providing a more convenient interface for further integration into wearable applications.

5. Conclusions

The design and application of a simple and reliable plantar pressure data acquisition device is very important for wearable human body assist equipment such as exoskeletons and power prosthesis. In this research, we conducted a novel plantar pressure sensing insole based on photoelectric sensing technology. The innovation of this modular sensing unit focuses on sensing principles, structural design and elastic materials. We introduced how to use low-cost manufacturing methods and materials to fabricate this modular sensor so that other researchers can easily replicate and further develop related research work. The designed modular pressure sensor realizes calibration analysis on a self-designed programmable calibration instrument. The electromechanical performance evaluation experiment proves that the modular pressure sensor has special applicability in pressure sensing. Subsequently, based on the analysis of plantar pressure, we proposed and compared different pressure-sensitive insole layout solutions, and integrated the modular pressure sensor into the pressure-sensitive insole. The dynamic locomotion results showed that the pressure-sensing insole based on the photoelectric effect can capture the deformation of the insole caused by plantar pressure during walking, sense the distribution of plantar pressure and provide reliable gait-related parameters with no interference to the wearer. This research provides a reliable gait information acquisition device for wearable applications such as powered exoskeletons, prosthesis and orthotics.

Author Contributions: Conceptualization, B.R. and J.L.; methodology, B.R. and J.L.; software, B.R. and J.L.; validation, B.R.; formal analysis, J.L.; investigation, B.R. and J.L.; resources, B.R.; data curation, B.R.; writing—original draft preparation, J.L.; writing—review and editing, B.R; visualization, J.L.; supervision, B.R.; project administration, B.R. All authors have read and agreed to the published version of the manuscript.

Funding: This research was funded by the National Natural Science Foundation of China (Grant No. 51775325); National Key Research and Development Program of China (Grant No. 2018YFB1309200); Young Eastern Scholars Program of Shanghai (Grant No. QD2016033).

Data Availability Statement: The datasets generated and/or analysed during the current study are available from the corresponding author on reasonable request.

Acknowledgments: The authors would like to thank the anonymous reviewers for their constructive comments that will help us to improve the quality of this manuscript.

Conflicts of Interest: The authors declare no conflict of interest.

References

1. Chu, A.; Kazerooni, H.; Zoss, A. On the biomimetic design of the berkeley lower extremity exoskeleton (BLEEX). In Proceedings of the 2005 IEEE International Conference on Robotics and Automation, Barcelona, Spain, 18–22 April 2005; pp. 4345–4352.
2. Kawamoto, H.; Sankai, Y. Power assist method based on phase sequence and muscle force condition for HAL. *Adv. Robot.* **2005**, *19*, 717–734. [CrossRef]
3. Kazerooni, H.; Steger, R. The Berkeley lower extremity exoskeleton. *J. Dyn. Syst. Meas. Control* **2006**, *128*, 14–25. [CrossRef]
4. Urry, S. Plantar pressure-measurement sensors. *Meas. Sci. Technol.* **1999**, *10*, R16. [CrossRef]
5. Anam, K.; Al-Jumaily, A.A. Active exoskeleton control systems: State of the art. *Procedia Eng.* **2012**, *41*, 988–994. [CrossRef]
6. Kazerooni, H.; Racine, J.-L.; Huang, L.; Steger, R. On the control of the berkeley lower extremity exoskeleton (BLEEX). In Proceedings of the 2005 IEEE International Conference on Robotics and Automation, Barcelona, Spain, 18–22 April 2005; pp. 4353–4360.
7. Razak, A.H.A.; Zayegh, A.; Begg, R.K.; Wahab, Y. Foot plantar pressure measurement system: A review. *Sensors* **2012**, *12*, 9884–9912. [CrossRef] [PubMed]
8. Cavanagh, P.R.; Lafortune, M.A. Ground reaction forces in distance running. *J. Biomech.* **1980**, *13*, 397–406. [CrossRef]
9. Mengüç, Y.; Park, Y.-L.; Pei, H.; Vogt, D.; Aubin, P.M.; Winchell, E.; Fluke, L.; Stirling, L.; Wood, R.J.; Walsh, C.J. Wearable soft sensing suit for human gait measurement. *Int. J. Robot. Res.* **2014**, *33*, 1748–1764. [CrossRef]
10. Zhu, Y.; Zhang, G.; Xu, W.; Zhao, J. Flexible force-sensing system for wearable exoskeleton using liquid pressure detection. *Sens. Mater.* **2018**, *30*, 1655–1664. [CrossRef]
11. Zhang, Q.; Wang, Y.L.; Xia, Y.; Wu, X.; Chen, X.D. A low-cost and highly integrated sensing insole for plantar pressure measurement. *Sens. Bio Sens. Res.* **2019**, *26*, 100298. [CrossRef]

12. Leal, A.G.; Frizera, A.; Avellar, L.M.; Marques, C.; Pontes, M.J. Polymer Optical Fiber for In-Shoe Monitoring of Ground Reaction Forces During the Gait. *IEEE Sens. J.* **2018**, *18*, 2362–2368. [CrossRef]
13. Tekscan. Tekscan® F-Scan64® System 2021. Available online: https://www.tekscan.com/introducing-f-scan64 (accessed on 21 May 2021).
14. Paromed. ParoTech System 2021. Available online: https://www.paromed.com.au/our-products/ (accessed on 21 May 2021).
15. GmbH N. Pedar System 2021. Available online: https://www.novel.de/products/pedar/ (accessed on 21 May 2021).
16. Liu, J.; Li, H.; Chen, W.; Wang, J. A novel design of pressure sensing foot for lower limb exoskeleton. In Proceedings of the 2013 IEEE 8th Conference on Industrial Electronics and Applications (ICIEA), Melbourne, VIC, Australia, 19–21 June 2013; pp. 1517–1520.
17. Lim, D.-H.; Kim, W.-S.; Kim, H.-J.; Han, C.-S. Development of real-time gait phase detection system for a lower extremity exoskeleton robot. *Int. J. Precis. Eng. Manuf.* **2017**, *18*, 681–687. [CrossRef]
18. Wu, G.; Wang, C.; Wu, X.; Wang, Z.; Ma, Y.; Zhang, T. Gait phase prediction for lower limb exoskeleton robots. In Proceedings of the 2016 IEEE International Conference on Information and Automation (ICIA), Ningbo, China, 1–3 August 2016; pp. 19–24.
19. Chen, B.; Wang, X.; Huang, Y.; Wei, K.; Wang, Q. A foot-wearable interface for locomotion mode recognition based on discrete contact force distribution. *Mechatronics* **2015**, *32*, 12–21. [CrossRef]
20. Crea, S.; Donati, M.; De Rossi, S.M.; Oddo, C.M.; Vitiello, N. A wireless flexible sensorized insole for gait analysis. *Sensors* **2014**, *14*, 1073–1093. [CrossRef] [PubMed]
21. Park, J.; Kim, M.; Hong, I.; Kim, T.; Kang, D. Foot Plantar Pressure Measurement System Using Highly Sensitive Crack-Based Sensor. *Sensors* **2019**, *19*, 5504. [CrossRef]
22. Leal-Junior, A.G.; Diaz, C.R.; Marques, C.; Pontes, M.J.; Frizera, A. 3D-printed POF insole: Development and applications of a low-cost, highly customizable device for plantar pressure and ground reaction forces monitoring. *Opt. Laser Technol.* **2019**, *116*, 256–264. [CrossRef]
23. Donati, M.; Vitiello, N.; De Rossi, S.M.M.; Lenzi, T.; Crea, S.; Persichetti, A.; Giovacchini, F.; Koopman, B.; Podobnik, J.; Munih, M. A flexible sensor technology for the distributed measurement of interaction pressure. *Sensors* **2013**, *13*, 1021–1045. [CrossRef] [PubMed]
24. Martini, E.; Fiumalbi, T.; Dell'Agnello, F.; Ivanic, Z.; Munih, M.; Vitiello, N.; Crea, S. Pressure-Sensitive Insoles for Real-Time Gait-Related Applications. *Sensors* **2020**, *20*, 1448. [CrossRef] [PubMed]
25. Tahir, A.M.; Chowdhury, M.E.; Khandakar, A.; Al-Hamouz, S.; Abdalla, M.; Awadallah, S.; Reaz, M.B.I.; Al-Emadi, N. A systematic approach to the design and characterization of a smart insole for detecting vertical ground reaction force (vGRF) in gait analysis. *Sensors* **2020**, *20*, 957. [CrossRef] [PubMed]
26. Chen, D.L.; Cai, Y.; Huang, M.C. Customizable Pressure Sensor Array: Design and Evaluation. *IEEE Sens. J.* **2018**, *18*, 6337–6344. [CrossRef]

sensors

MDPI

Article

Attention-Based 3D Human Pose Sequence Refinement Network

Do-Yeop Kim and Ju-Yong Chang *

Department of Electronics and Communication Engineering, Kwangwoon University, Seoul 01897, Korea; dyub1@kw.ac.kr
* Correspondence: jychang@kw.ac.kr; Tel.: +82-2-940-5136

Abstract: Three-dimensional human mesh reconstruction from a single video has made much progress in recent years due to the advances in deep learning. However, previous methods still often reconstruct temporally noisy pose and mesh sequences given in-the-wild video data. To address this problem, we propose a human pose refinement network (HPR-Net) based on a non-local attention mechanism. The pipeline of the proposed framework consists of a weight-regression module, a weighted-averaging module, and a skinned multi-person linear (SMPL) module. First, the weight-regression module creates pose affinity weights from a 3D human pose sequence represented in a unit quaternion form. Next, the weighted-averaging module generates a refined 3D pose sequence by performing temporal weighted averaging using the generated affinity weights. Finally, the refined pose sequence is converted into a human mesh sequence using the SMPL module. HPR-Net is a simple but effective post-processing network that can substantially improve the accuracy and temporal smoothness of 3D human mesh sequences obtained from an input video by existing human mesh reconstruction methods. Our experiments show that the noisy results of the existing methods are consistently improved using the proposed method on various real datasets. Notably, our proposed method reduces the pose and acceleration errors of VIBE, the existing state-of-the-art human mesh reconstruction method, by 1.4% and 66.5%, respectively, on the 3DPW dataset.

Keywords: 3D human mesh reconstruction; 3D human pose estimation; deep neural network

Citation: Kim, D.-Y.; Chang, J.-Y. Attention-Based 3D Human Pose Sequence Refinement Network. *Sensors* **2021**, *21*, 4572. https://doi.org/10.3390/s21134572

Academic Editors: Tomasz Krzeszowski, Adam Świtoński, Michal Kepski and Carlos Tavares Calafate

Received: 13 June 2021
Accepted: 1 July 2021
Published: 3 July 2021

1. Introduction

Three-dimensional human pose estimation is an important and actively studied problem in computer vision. Various methods have been proposed to generate successful pose estimation results on the basis of deep learning. These methods have been used to address the problem of reconstructing 3D human pose from a single RGB image or video obtained from a monocular camera. Recently, methods for estimating dense 3D mesh beyond sparse 3D joints have been proposed on the basis of a statistical shape model for human body. However, reconstructing 3D human poses from RGB images accurately remains a difficult problem.

Recent methods for 3D human mesh reconstruction extract features from input images on the basis of deep learning and directly regress the pose and identity parameters of a statistical shape model, such as a skinned multi-person linear model (SMPL) [1], from the extracted features. However, in the case of an image including occlusion or an unseen pose that is not included in training data, the network has difficulty in estimating the correct pose. Methods of estimating temporally coherent pose sequences from input videos have shown moderate performance [2–4]. However, the above problems still prevent the existing methods to reconstruct the correct pose in some frames and generate noisy human motion. For example, the top row of Figure 1 shows the results obtained by VIBE [3], a state-of-the-art method for reconstructing 3D human mesh from video. In the 3rd frame, VIBE fails to estimate the correct pose, resulting in temporally noisy results. Our study focusses on this problem, namely temporally coherent human pose estimation from input video. Specifically, we propose a *human pose refinement network* (i.e., HPR-Net) that can refine the noisy human pose sequence reconstructed by existing methods. The bottom row of Figure 1 shows the improved results through our HPR-Net.

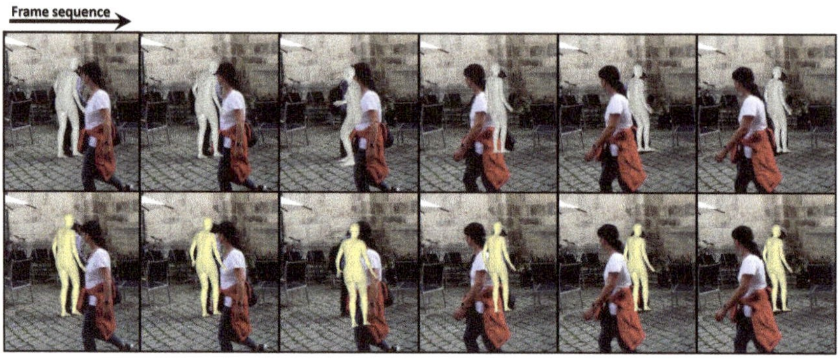

Figure 1. This figure shows a 3D human mesh sequence estimated by VIBE (**top row**) and its refined result by our proposed method (**bottom row**). In the 3rd frame, VIBE fails to estimate the correct pose of the target person due to severe occlusion. Our method effectively refines the incorrectly estimated results.

Weighted averaging is a simple but effective method that has been widely used for refinement of signals, including images. The basic idea of this paper to refine the noisy 3D human pose sequence is based on weighted averaging. However, applying weighted averaging to human pose refinement is not trivial due to the following two problems. The first problem is how to determine weights for weighted averaging. To accomplish this task, we learn a module that generates optimal weights on the basis of large-scale data. Specifically, we define a weight as affinity between two 3D poses. We propose a non-local attention-based weight regression module that can consider long-range interactions to compute this affinity. The proposed module is supervised to output weights that can reconstruct a ground-truth pose sequence from a noisy pose sequence estimated by existing human pose estimation methods.

Our human pose refinement method relies on SMPL, where 3D human pose is represented as a set of 3D rotations of joints. However, 3D rotation and 3D pose, including the rotation, cannot be regarded as a vector defined in Euclidean space. Thus, weighted averaging cannot be applied directly to 3D human poses. Performing weighted averaging for 3D rotation requires a complex optimization process [5]. To alleviate this problem, we use Gramkow's study [6], which proves that the mean of unit quaternions is a quadratic approximation of the mean of 3D rotations. Specifically, we first represent the 3D rotation constituting the 3D human pose as a unit quaternion and then perform weighted averaging on the 3D human pose sequence represented as a sequence of unit quaternions. This weighted averaging based on the unit quaternion can be represented as a simple algebraic equation without an optimization process and can be included in our network for learning.

Suppose that SMPL-based human mesh reconstruction methods estimate the 3D pose and identity parameter sequence. Our proposed system consisting of a weight-regression module, a weighted-averaging module, and an SMPL module performs pose refinement for a noisy 3D pose sequence through the following process. To refine a pose of a frame, the weight-regression module first generates weights for the poses in a window of a predefined size around that frame. Next, the weighted-averaging module outputs an improved 3D pose by applying weighted averaging on the basis of the generated weights to the poses inside the window. Finally, the SMPL module generates human meshes and 3D joints from the improved 3D pose parameters. This process is repeated for all frames to reconstruct the refined 3D pose and mesh sequence. An overview of the proposed method is shown in Figure 2.

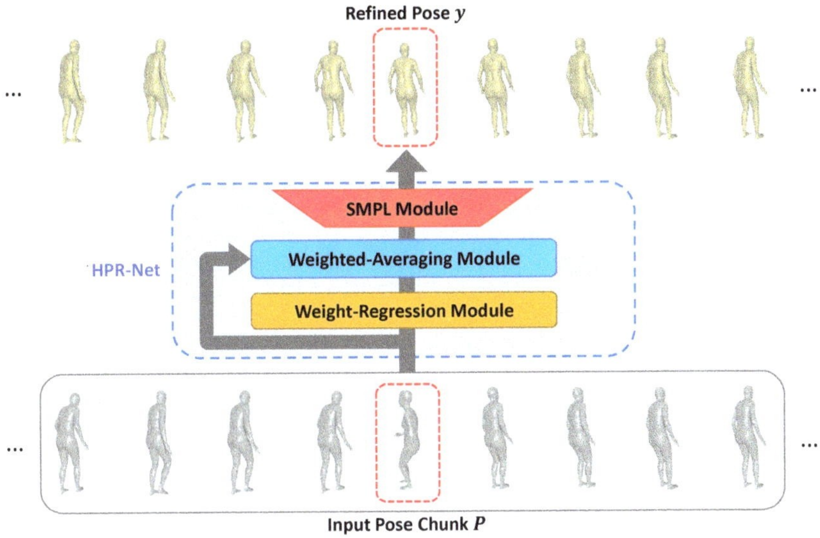

Figure 2. Overall framework of the proposed method. The input to our model is a noisy 3D human pose sequence estimated by existing 3D human pose estimation methods. Our proposed HPR-Net refines the noisy 3D human pose sequence and generates a refined human pose sequence.

The contributions of this paper can be summarized as follows:

- We propose a novel method to refine a 3D human pose sequence consisting of 3D rotations of joints. The proposed method performs human pose refinement independently from existing 3D human pose estimation methods. It can be applied to the results of any existing method in a model-agnostic manner and is easy to use.
- The proposed method is based on a simple but effective weighted-averaging operation and generates interpretable affinity weights using a non-local attention mechanism.
- In accordance with our experimental results, the proposed method consistently improves the 3D pose estimation and mesh reconstruction performance (i.e., accuracy and smoothness of output sequences) of existing methods for various real datasets.

2. Related Work

Human mesh reconstruction. Many recent 3D human mesh reconstruction methods directly regress the parameters of statistical shape models, such as SMPL [1]. These methods can be broadly classified into a single image-based approach [7–10] and a video-based approach [2–4].

The single image-based approaches reconstruct 3D human mesh from a monocular image. Bogo et al. [7] proposed a method that estimates 2D joints from an input image on the basis of a pretrained 2D joint regression network and optimizes an energy function to fit SMPL to the regressed 2D joints. Pavlakos et al. [8] extended [7] to optimize an improved energy function to fit the SMPL-X model to the regressed 2D full-body joints for holistic body modeling. A variational autoencoder(VAE)-based pose prior for valid pose parameter regression was proposed for optimization. Kanazawa et al. [9] proposed a model that directly maps features extracted by a deep network from a single image to SMPL parameters. In their method, an adversarial prior for the estimated parameters was proposed and learned to help obtain a realistic human mesh. Kolotouros et al. [10] combined the optimization-based method and the regression-based method in an end-to-end manner. The SMPL parameter estimated from a single image is used as an initial parameter, which is iteratively optimized through the method of [7]. The optimized parameter is used as a pseudo ground-truth for learning the regressor to construct a self-improving framework.

The video-based approaches reconstruct the 3D human mesh sequence from a video. Kanazawa et al. [2] proposed a temporal convolutional network that reconstructs the SMPL model from an image sequence. This method is supervised to predict SMPL models in the nearby few frames to learn information about human motion better. It can estimate past and future meshes from a single image through a hallucinator. Kocabas et al. [3] proposed a method that reconstructs an SMPL model sequence from a feature sequence computed using bidirectional gated recurrent unit from an input video. To compensate for the lack of 3D annotated data, this method performs weak supervision with various 2D datasets and adversarial training using large-scale motion datasets, resulting in successful human mesh reconstruction performance. Luo et al. [4] tried to solve the jittering problem from the inference results of existing methods for video data. This method reconstructs coarse motion by learning a VAE-based motion prior and then performs refinement for each frame's pose. Thus, the smoothness of the output SMPL sequence is improved. Despite these recent advances in 3D human mesh reconstruction, most methods still produce erroneous poses or jittered motions due to unseen poses or occlusions from input images or videos acquired in an uncontrolled environment. Our work can substantially improve the accuracy and smoothness of human mesh sequences reconstructed by existing methods.

Non-local attention. Non-local attention was proposed to model long-range dependency in natural language processing [11,12] and computer vision [13–15]. Vaswani et al. [12] proposed the transformer, which is a framework using only attention mechanisms to overcome the limitations of existing recurrent models for natural language processing tasks and successfully solves the long-range dependency problem. Recently, the transformer architecture is known to improve image recognition performance and is actively used for various computer vision tasks [16–20]. Wang et al. [13] attempted to model the long-range dependency in image features using non-local operations proposed in [21]. For this, a non-local block based on attention mechanisms was proposed. On this basis, the method in [12] can be regarded as a special case of non-local neural networks. In the study of Cao et al. [14], the position-wise attention map of [13] was analyzed qualitatively, and most of the attention maps of each position have similar attention aspects. On this Basis, a more efficient non-local attention block was proposed. Woo et al. [15] proposed a method that extracts new features by successively applying channel attention and spatial attention to input features. This method shows a stronger representation power than features based on existing fully convolutional baselines. Our method generates a temporal non-local attention map inspired by [13,21]. The generated attention weights suppress features that are useless for refinement and strengthen helpful features. Our method can refine noisy pose parameter sequences through this attention mechanism.

Human pose refinement. The goal of human pose refinement studies is mainly to refine an estimated sparse joint set. Existing pose refinement methods are included as part of the joint regression network or used as a post-processing module for inference results. Newell et al. [22] proposed a network in which several hourglass modules are stacked. Hourglass module repeats top-down and bottom-up processing, extracts features at various scales, and is trained with intermediate supervision. Each stage module generates a heatmap, which is used as input to the next stage module for refinement. Chen et al. [23] proposed a cascaded pyramid network that combines GlobalNet, a Resnet-based pyramid network, and RefineNet, which refines the heatmap generated by GlobalNet. RefineNet considers all features obtained from each step of the pyramid to find occluded joint positions that are difficult to estimate. Moon et al. [24] proposed a model-agnostic refinement model based on the error distribution of 2D pose estimation models investigated in Ronchi et al.'s work [25]. This method is independent on the pose estimation model because it does not work in an end-to-end manner, and pose estimation performance can be improved for various existing approaches. Mall et al. [26] proposed a method to refine noisy motion capture data. The proposed network consisting of linear layers and bidirectional long short-term memories regresses the standard deviation of a Gaussian kernel to improve the pose of a current frame. The proposed method obtains a denoised pose using the Gaussian kernel obtained through this network to calculate a temporally weighted sum for an input noisy pose sequence. In [26], 3D human pose is represented in the form of 126 joint angles, and the weighted sum is computed for this joint angle sequence. Our work provides a more reliable basis for computation in non-Euclidean space where 3D rotation actually exists. While the values of weights are limited by the Gaussian kernel in [26], they are not in our method.

Several methods have been proposed to refine the SMPL pose parameter [9,10]. Kanazawa et al. [9] proposed a regressor that performs iterative refinement to estimate the SMPL parameter. Kolotouros et al. [10] presented a method that refines the estimated SMPL parameter through an optimization process. In [9,10], the refinement process is included in the model, which outputs SMPL identity and pose parameters directly from an input image. Our refinement method is independent of the pose estimation model and can be applied to the results of any method for estimating the SMPL pose parameter sequence regardless of their network structure. Our work is the first to propose a post-processing method for SMPL pose parameter refinement, and the proposed method is simple but works effectively.

3. Proposed Method

This section provides detailed descriptions of each module constituting our proposed HPR-Net. As presented in Section 1, we propose HPR-Net that generates a refined 3D human pose sequence from a noisy 3D human pose sequence estimated by other methods, such as VIBE. As shown in Figure 2, HPR-Net refines a noisy 3D pose of a target frame from input 3D poses that consist of all 3D poses within a window of size N (N is an odd number) centered on the target frame. We term these input 3D poses as a *pose chunk*. HPR-Net consists of a weight-regression module, a weighted-averaging module, and an SMPL module. Each module is explained in the following subsections. We first introduce the SMPL module to explain what 3D human model is used, how the 3D human pose is defined in the SMPL model, and why this module is needed in our framework. The weight-regression module consists of 1D convolution layers and generates an N-dimensional weight vector using non-local self-attention mechanism from an input pose chunk. The weighted-averaging module outputs a refined 3D pose by weighted averaging with the input pose chunk and weights from the weight-regression module. The above procedure is applied to the noisy 3D pose sequence with a sliding window manner, so we can obtain the refined 3D human pose sequence.

3.1. SMPL Module

SMPL is a 3D statistical shape model used to represent a human body and includes low-dimensional parameters to control the body shape. The parameter set included in the SMPL model consists of an identity parameter $\boldsymbol{\beta} \in \mathbb{R}^{10}$ and a pose parameter $\boldsymbol{\theta} \in \mathbb{R}^{72}$. The pose parameter represents the relative 3D rotations of 24 joints in an axis-angle form. This parameter controls the 3D pose of the human body represented by the SMPL model. From the given identity and pose parameters, the SMPL module generates a 3D human mesh model in a differentiable manner. The vertices $M \in \mathbb{R}^{3\times 6890}$ of the generated mesh model are multiplied with a pretrained linear regression matrix included in the SMPL model, so that 24 joints $X_{smpl,3d} \in \mathbb{R}^{3\times 24}$ can be additionally obtained. In HPR-Net, the SMPL module computes the human mesh model and the 3D joints from the refined pose parameters using our weighted-averaging module and the identity parameters estimated by existing methods. We can compare the 3D joints from the refined mesh generated by the SMPL module with its ground-truth to compute a loss function for learning and error for evaluating the proposed method. We also use the joint set $X_{3d} \in \mathbb{R}^{3\times 14}$ obtained by converting $X_{smpl,3d}$ into 14 joints compatible with the joint definition of Human3.6M [27] for learning and evaluation.

3.2. Weight-Regression Module

Network structure. The weight-regression module of HPR-Net generates weights for pose refinement of the target pose from an input noisy pose chunk. Figure 3 shows the detailed structure of the weight-regression module consisting of 1D temporal convolution layer with a kernel size of 3, layer normalization [28], rectified linear unit activation, and self-attention layer. Suppose that $\Phi = \{\boldsymbol{\beta}_i, \boldsymbol{\theta}_i\}_{i=0}^{N-1}$, which is a chunk of length N for the noisy SMPL parameter sequence, is given. Here, $\boldsymbol{\beta}_i \in \mathbb{R}^{10}$ and $\boldsymbol{\theta}_i \in \mathbb{R}^{72}$ are the identity and pose parameters in the i-th frame, respectively. The pose parameter $\boldsymbol{\theta}_i$ represents the 3D rotations for 24 SMPL joints represented in an axis-angle form. We first convert $\boldsymbol{\theta}_i$ to pose parameter $\boldsymbol{p}_i \in \mathbb{R}^{96}$ in a unit quaternion form. We apply frame-wise positional encoding to the unit quaternion pose chunk $P = [\boldsymbol{p}_0, \ldots, \boldsymbol{p}_{N-1}] \in \mathbb{R}^{96\times N}$, similar to [12], before feeding it into the network. Specifically, to inject positional information into P, we concatenate a relative position index vector $[-\lfloor\frac{N}{2}\rfloor, \ldots, -1, 0, 1, \ldots, \lfloor\frac{N}{2}\rfloor]$ with P to construct $\tilde{P} \in \mathbb{R}^{97\times N}$ and feed the concatenated tensor into the weight-regression module. The weight-regression module first computes temporal feature $H = [\boldsymbol{h}_0, \boldsymbol{h}_1, \ldots, \boldsymbol{h}_{N-1}] \in \mathbb{R}^{24\times N}$ from $\tilde{P}$ through three 1D temporal convolution layers. $\boldsymbol{h}_i \in \mathbb{R}^{24}$ represents the temporal feature of the i-th frame. A pose affinity vector $\boldsymbol{w} \in \mathbb{R}^N$ is generated through a non-local self-attention mechanism [12,13] as follows:

$$\boldsymbol{w} = \text{Softmax}(H^T \cdot \boldsymbol{h}_{\lfloor\frac{N}{2}\rfloor}). \quad (1)$$

Similar to the existing self-attention-based methods, our weight-regression module simply uses matrix multiplication and softmax operation to construct pose affinity vector $\boldsymbol{w} = [w_0, w_1, \ldots, w_{N-1}]^T$, where $w_i \in \mathbb{R}$ represents how the pose of i-th frame affects the computation of the refined pose at the $\lfloor\frac{N}{2}\rfloor$-th (i.e., center) frame. As our HPR-Net refines the center frame's pose $\boldsymbol{p}_{\lfloor\frac{N}{2}\rfloor}$ from P, we choose the center frame's feature $\boldsymbol{h}_{\lfloor\frac{N}{2}\rfloor}$ from H to compare the feature with all other features within the chunk.

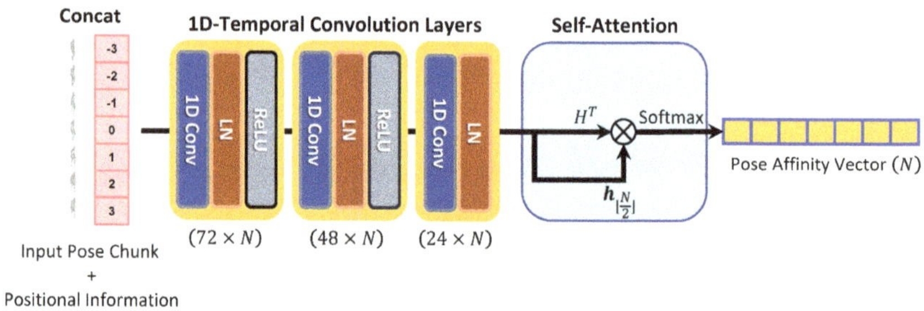

Figure 3. Detailed pipeline of the weight-regression module. ⊗ represents matrix multiplication. First, the weight-regression module concatenates positional information to an input pose chunk. Second, the positional encoded input chunk is fed into the weight-regression module that consists of three 1D temporal convolution layers. Finally, pose affinity vector is generated from the output temporal feature of the convolution layers.

Why do we use LayerNorm? From our experiments, we observed that the use of layer normalization after the convolution layer shows higher performance than the commonly used batch normalization [29]. In our method, the 3D pose in an input pose chunk consists of 3D rotations, and this 3D rotation is represented in a unit quaternion form that is geometrically on a 4D unit sphere. Layer normalization helps to learn the weight-regression module by enforcing the features extracted through the convolution layer to be on the unit sphere.

3.3. Weighted-Averaging Module

Pose refinement by weighted averaging. Using $\boldsymbol{w}$ generated by the weight-regression module, we perform weighted averaging on the input pose chunk P and obtain the refined pose $\boldsymbol{y} \in \mathbb{R}^{96}$ as follows. Figure 4 shows the detailed structure of the weighted-averaging module. Weighted averaging cannot be directly applied to 3D rotations because they are defined in non-Euclidean space. Therefore, we obtain a second-order approximation of optimal rotation averaging by performing weighted averaging based on unit quaternion following Gramkow's work [6]. By weighted averaging, we first obtain $\tilde{\boldsymbol{y}}$ as follows:

$$\tilde{\boldsymbol{y}} = \sum_{i=0}^{N-1} w_i \boldsymbol{p}_i, \tag{2}$$

where w_i is the i-th component of vector $\boldsymbol{w}$ and represents the contribution of $\boldsymbol{p}_i$ to weighted averaging. However, $\tilde{\boldsymbol{y}} = [\tilde{\boldsymbol{q}}_0^T, \tilde{\boldsymbol{q}}_1^T, \ldots, \tilde{\boldsymbol{q}}_{23}^T]^T$ cannot be guaranteed to consist of unit quaternions. Therefore, we additionally perform normalization to make the 3D rotations $\tilde{\boldsymbol{q}}_j$ belonging to $\tilde{\boldsymbol{y}}$ into a unit quaternion form using $\boldsymbol{q}_j = \tilde{\boldsymbol{q}}_j / \|\tilde{\boldsymbol{q}}_j\|$. The weighted-averaging module outputs the refined 3D pose $\boldsymbol{y}$ consisting of unit quaternions as follows:

$$\boldsymbol{y} = [\boldsymbol{q}_0^T, \boldsymbol{q}_1^T, \ldots, \boldsymbol{q}_{23}^T]^T \in \mathbb{R}^{96}, \tag{3}$$

where $\boldsymbol{q}_j$ denotes the 3D rotation of the j-th SMPL joint.

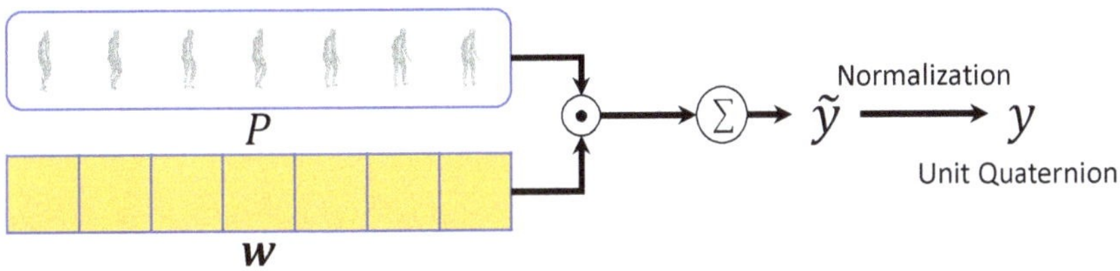

Figure 4. Detailed pipeline of the weighted-averaging module. $\odot$ represents element-wise multiplication with broadcasting. Σ represents summation for across time dimension. Input pose vectors P are multiplied with pose affinity weights $\boldsymbol{w}$ which are generated by the weight-regression module. Then weighted pose vectors are added to output a refined pose vector $\tilde{\boldsymbol{y}}$. To ensure that the refined pose parameters consist of unit quaternions, we additionally normalize $\tilde{\boldsymbol{y}}$ to output a valid pose vector $\boldsymbol{y}$.

Loss functions. The refined 3D human pose $\boldsymbol{y}$ is converted to an axis-angle form and then fed into the SMPL module along with the identity parameter $\boldsymbol{\beta}$ estimated by other methods to generate the refined mesh $\hat{M}$ and 3D joints $\hat{X}_{3d} = [\hat{\boldsymbol{x}}_{3d,1}, \ldots, \hat{\boldsymbol{x}}_{3d,14}]$. The joint loss function L_{joint} for learning the proposed network is defined as follows:

$$L_{\text{joint}} = \frac{1}{J} \sum_{j=1}^{J} \|\hat{\boldsymbol{x}}_{3d,j} - \boldsymbol{x}_{3d,j}\|_1, \tag{4}$$

where $J = 14$ is the number of joints, and $\hat{\boldsymbol{x}}_{3d,j}$ and $\boldsymbol{x}_{3d,j}$ denote the estimated and ground-truth coordinate vectors of the j-th joint, respectively. L_{joint} is defined as L1 loss and we supervise only $\hat{X}_{3d}$ that is generated from $\hat{M}$ by the SMPL module.

4. Experimental Results

4.1. Datasets and Evaluation Metrics

We use Human3.6M [27] and 3DPW [30] for training and evaluation. Human3.6M is a large-scale dataset obtained from an indoor environment, and has been used in many existing 3D human pose estimation methods. The Human3.6M dataset consists of videos, where 11 subjects perform 15 actions and includes 2D and 3D joint annotations for each frame. Image data in Human3.6M were captured in four camera views. 3DPW is a dataset obtained in an in-the-wild environment. The 3DPW dataset includes 60 videos and is divided into train, validation, and test sets. The three sets consist of 24, 12, and 24 videos, respectively. The 3DPW dataset provides 2D and 3D joint annotations and SMPL parameter annotations.

We split each dataset into training and test data. In Human3.6M, we use 5 subjects (1, 5, 6, 7, 8) as training data and 2 subjects (9, 11) as test data following the convention of previous studies [3,4,10]. For the 3DPW dataset, we use the train and validation sets as training data and the test set as test data. For convenience of training and evaluation, we apply VIBE to the training data of Human3.6M and 3DPW datasets and store the estimated SMPL parameters offline. The saved results are used as input for training the proposed network. We apply SPIN [10], VIBE [3], and MEVA [4] to the test data of each dataset, store the estimated SMPL parameters offline, and use them as input for evaluation. At the training stage, we train the proposed HPR-Net using all the training data of each dataset. We then evaluate the proposed method by applying HPR-Net to the test data of each dataset and report the performance quantitatively and qualitatively.

To evaluate the performance of the proposed method, we report MPJPE, PA-MPJPE, MPVE, and acceleration error. MPJPE and PA-MPJPE are metrics used to evaluate joint position error. MPJPE calculates the average 3D joint distance (mm). PA-MPJPE calculates the average 3D joint distance (mm) after performing Procrustes alignment [31] on the estimated and ground-truth joint sets. MPVE calculates the average position error (mm) of the vertices of the generated SMPL mesh. Acceleration error [2] is a metric for evaluating the temporal smoothness (mm/s^2) of the estimated pose sequence. Acceleration vectors are computed for the 3D joint sequence, and the acceleration error is calculated as the average difference between the estimated and ground-truth acceleration vectors.

4.2. Implementation Details

HPR-Net is trained end-to-end, and the input pose chunk in the training process is determined by random shuffling at each iteration. Zero padding is applied to the temporal 1D convolution of the weight-regression module. We set the length N of the input pose chunk for training the HPR-Net to 17 and calculate the loss function for the joints of the refined mesh corresponding to the center frame. We use Adam [32] as the optimizer of the network. The learning rate is set to 10^{-4}. We do not decay the learning rate during training. The batch size and the number of epochs are set to 64 and 20, respectively. In each epoch, 1000 iterations are performed. The learning rate, batch size, and number of epochs are determined through simple greedy search using the validation set of 3DPW. Pytorch [33] was used to implement the proposed method, which was trained with a single Nvidia RTX3090 GPU. In the evaluation process, the input pose chunk is not randomly determined and is fed into the network in the order of the frames of the evaluation video. We refine the input video except for 16 frames (i.e., 8 frames each at the beginning and end of the video). The pose chunk of length 17 is fed into HPR-Net in a sliding window manner with stride 1.

4.3. Ablation Study

In ablation experiments, we report how the hyperparameters and component changes affect the performance of HPR-Net. We use VIBE's pose sequence estimation result as our HPR-Net's input chunk. We set the length of the input pose chunk to 17 in all experiments, except for the pose chunk length ablation experiment. In ablation experiments, HPR-Net is evaluated on 3DPW test set.

Pose chunk length. To determine the optimal length N of the pose chunk, we perform training using various lengths and analyze the results. Table 1 shows the performance in accordance with the length of the input chunk. HPR-Net shows the best performance with length 17, except for PA-MPJPE. Thus, we set the pose chunk length to 17.

Table 1. Performance comparison of HPR-Net according to different pose chunk length. Bold values indicate best results.

Length	MPJPE ↓	PA-MPJPE ↓	MPVE ↓	Accel-Error ↓
9	82.14	51.82	98.25	7.31
17	**81.10**	51.26	**97.13**	**6.94**
33	82.11	51.63	98.26	18.36
65	81.23	**50.97**	97.24	8.19
129	81.81	51.35	97.89	11.69

Various loss combinations. In the proposed method, only 3D joints are supervised to train HPR-Net using the joint loss function in Equation (4). To justify this condition, we conduct an experiment to investigate how various combinations of loss functions affect the performance of HPR-Net. Specifically, we perform direct supervision with the joint loss function L_{joint} and losses that can be defined using the outputs of HPR-Net. The mesh loss function L_{mesh} and the pose loss function L_{pose} are additionally defined as follows:

$$L_{\text{mesh}} = \frac{1}{6890} \sum_{v=1}^{6890} \|\hat{\boldsymbol{m}}_v - \boldsymbol{m}_v\|_1, \tag{5}$$

$$L_{\text{pose}} = \frac{1}{24} \sum_{j=1}^{24} \|\hat{R}_j - R_j\|_F^2. \tag{6}$$

The mesh loss function L_{mesh} is defined as L1 loss, where $\hat{\boldsymbol{m}}_v$ and $\boldsymbol{m}_v$ denote the estimated and ground-truth coordinate vectors for the v-th vertex, respectively. The pose loss function L_{pose} is for the pose parameters, including 3D rotations, where $\hat{R}_j$ and $R_j \in \mathbb{R}^{3\times 3}$ denote the estimated and ground-truth rotation matrices for the j-th joint, respectively. Frobenius norm for their difference represents the distance (i.e., chordal distance [5]) between two 3D rotations in non-Euclidean space. The total loss function L for this ablation experiment is defined as follows:

$$L = \lambda_j L_{\text{joint}} + \lambda_m L_{\text{mesh}} + \lambda_p L_{\text{pose}}, \tag{7}$$

where λ_j, λ_m, and λ_p denote the weights that determine the strength of each loss.

Table 2 shows the performance of HPR-Net in accordance with the weights of L. HPR-Net shows the highest performance, except for PA-MPJPE when only the joint loss function L_{joint} is used. Using the pose loss function L_{pose} leads to performance degradation (1st, 4th, 6th, 7th rows). Supervising the mesh vertices shows lower PA-MPJPE (2nd row) than only using L_{joint} (3rd row). We use Human3.6M and 3DPW datasets for training. However, the Human3.6M dataset does not include SMPL annotations. Thus, only the 3DPW dataset is used to supervise the network when we calculate L_{mesh} and L_{pose}. The size of 3DPW training data is smaller than that of Human3.6M. However, the experimental result from supervising with only L_{mesh} shows the highest PA-MPJPE and second highest performance on other metrics. If more datasets containing SMPL annotations are available, then the use of the mesh loss function will lead to further performance improvements.

Table 2. Performance comparison of HPR-Net according to various combinations of loss functions. (✓= 1.0, blank = 0.0). Bold values indicate best results.

λ_j	λ_m	λ_p	MPJPE ↓	PA-MPJPE ↓	MPVE ↓	Accel-Error ↓
		✓	85.60	55.03	102.07	12.39
	✓		81.29	**51.04**	97.33	7.72
✓			**81.10**	51.26	**97.13**	**6.94**
✓		✓	84.37	54.12	100.76	10.38
✓	✓		81.46	51.16	97.48	9.79
	✓	✓	85.64	55.20	102.12	12.11
✓	✓	✓	83.76	53.39	100.06	14.12

Positional encoding. Most of the non-local attention-based methods inject positional information into their input. HPR-Net performs positional encoding, which helps to distinguish the pose of each frame in input pose chunk. We investigate the effect of positional encoding and its method on the performance of HPR-Net. Table 3 shows the performance of HPR-Net in accordance with the positional encoding method. For the experiment, we train and evaluate with three different models, one without positional encoding (*None*), one with sinusoidal positional encoding according to [12] (*Sinusoidal*), and one with positional encoding used in the proposed method (*Ours*). When positional encoding is not used, HPR-Net shows decreased PA-MPJPE performance compared with VIBE, but the other metrics are improved. Using the sinusoidal positional encoding shows improved results and best performance on PA-MPJPE. Our encoding method shows slightly lower PA-MPJPE compared with the sinusoidal positional encoding, but the best performance on the other metrics.

Table 3. Comparison of refinement performance of HPR-Net according to positional encoding method. Bold values indicate best results.

Methods	MPJPE ↓	PA-MPJPE ↓	MPVE ↓	Accel-Error ↓
None	81.63	52.00	97.72	6.97
Sinusoidal	81.53	**51.15**	97.58	8.42
Ours	**81.10**	51.26	**97.13**	**6.94**

Layer normalization. The weight-regression module is composed of simple 1D temporal convolution layers. Layer normalization is adopted as the feature normalization layer of the proposed weight-regression module. To justify the use of layer normalization for HPR-Net, we trained three models, one without feature normalization, one using batch normalization, and one using layer normalization. Table 4 shows the performance comparison in accordance with the normalization method used in HPR-Net. When layer normalization is used, HPR-Net achieves the best performance in all metrics compared with other methods. From the result, layer normalization helps the learning of the weight-regression module.

Table 4. Comparison of refinement performance of HPR-Net according to feature normalization method. Bold values indicate best results.

Methods	MPJPE ↓	PA-MPJPE ↓	MPVE ↓	Accel-Error ↓
None	82.12	51.84	98.17	7.76
BatchNorm	82.66	52.07	98.81	12.93
LayerNorm	**81.10**	**51.26**	**97.13**	**6.94**

4.4. *Refinement on State-of-the-Art Methods*

We evaluate the performance of applying HPR-Net to state-of-the-art methods [3,4,10] for different datasets [27,30]. Tables 5 and 6 report the performance of existing methods and their refinement performance by HPR-Net on each evaluation dataset. Existing methods are re-evaluated using publicly provided pretrained models. HPR-Net achieves performance improvement in all metrics for all methods on 3DPW and Human3.6M datasets. HPR-Net considerably improves the acceleration error in every experiments. We trained our HPR-Net with the pose estimation result by VIBE as input. However, HPR-Net consistently improves other methods (i.e., SPIN and MEVA). These results show our HPR-Net's generalization capability for other methods.

Table 5. HPR-Net's pose refinement performance for various existing methods on 3DPW test data. Bold values indicate performance improvements.

Methods	MPJPE ↓	PA-MPJPE ↓	MPVE ↓	Accel-Error ↓
VIBE	82.28	51.72	98.42	20.69
VIBE + HPR-Net	**81.10**	**51.26**	**97.13**	**6.94**
SPIN	102.46	60.05	129.22	29.78
SPIN + HPR-Net	**100.95**	**59.30**	**127.58**	**8.19**
MEVA	85.81	53.54	102.18	14.37
MEVA + HPR-Net	**85.43**	**53.50**	**101.79**	**6.63**

Table 6. HPR-Net's pose refinement performance for various existing methods on Human3.6M test data. Bold values indicate performance improvements.

Methods	MPJPE ↓	PA-MPJPE ↓	Accel-Error ↓
VIBE	78.35	53.58	9.76
VIBE + HPR-Net	**77.77**	**53.17**	**2.13**
SPIN	68.22	46.16	14.21
SPIN + HPR-Net	**67.35**	**45.53**	**2.74**
MEVA	73.64	48.48	7.22
MEVA + HPR-Net	**73.06**	**48.06**	**1.83**

4.5. Comparison with Other Pose Refinement Methods

The pose parameter sequence can be refined in several methods. We compare HPR-Net with other methods in improving the pose sequence. Table 7 shows the quantitative improvement results of SLERP, Gaussian-filtering-based method (HPR-Gaussian), direct-regression-based method (HPR-DR), and HPR-Net. All the methods are evaluated on 3DPW test set. SLERP calculates the interpolated unit quaternion between two unit quaternions. MEVA uses SLERP to further smoothen their output pose parameter sequence. We test SLERP to evaluate its refinement performance and compare it with our HPR-Net's performance. HPR-Gaussian regresses standard deviations to create optimal joint-wise Gaussian kernels. We implement the HPR-Gaussian model by modifying the structure of the weight-regression module in HPR-Net. We only change the kernel size of the 3rd temporal 1D convolution layer of the weight-regression module to N and set the number of channels to 24. HPR-Gaussian's weight-regression module creates 24 joint-wise standard deviations, where the 24 Gaussian kernels with kernel size N are created. Each kernel is used for Gaussian filtering for the 3D rotation of each of the 24 joints. Specifically, weighted averaging of 3D rotations along the temporal axis is performed using the values of the kernels as weights. HPR-DR directly regresses the refined pose of the center frame from the input pose chunk. To implement HPR-DR, our proposed HPR-Net is modified as follows. We change the number of channels and kernel size in the last 1D convolution layer of the weight-regression module to 96 and N, respectively, so that the modified network (i.e., HPR-DR) generates a 96D vector. This vector is converted into a refined pose vector consisting of unit quaternions through normalization.

Table 7. Comparison of refinement performance between HPR-Net and other pose sequence refinement methods on the 3DPW dataset. Bold values indicate best results.

Methods	MPJPE ↓	PA-MPJPE ↓	MPVE ↓	Accel-Error ↓
SLERP	82.72	52.13	99.88	12.38
HPR-Gaussian	82.15	51.58	98.30	18.04
HPR-DR	183.01	102.79	223.20	14.28
HPR-Net	**81.10**	**51.26**	**97.13**	**6.94**

From the quantitative improvement results of each method in Table 7, we observe that SLERP does not improve the performance, except for the acceleration error. Acceleration is defined as the second derivative of the joint position and is very sensitive to the small noise in the refined pose sequence. Since SLERP performs weighted averaging for interpolation between two poses, it is effective in reducing the small noise and the acceleration error. HPR-Gaussian improves VIBE quantitatively. However, the performance gain for the acceleration error is smaller than SLERP because HPR-Gaussian over-smooths the pose sequence. HPR-DR fails to refine the results of VIBE. It is because the size of the training data, which is not large enough to train the direct regression model, leads to overfitting. Our HPR-Net adaptively adjusts the shape of the kernel to prevent over-smoothing and outperforms the other methods in all metrics, especially the acceleration error. Experimental results show that our HPR-Net is superior to the other human pose refinement methods.

4.6. Network Design Based on Non-Local Attention

Our proposed HPR-Net is based on non-local attention. Transformer [12] is a representative method and has the non-local attention-based structure. Our HPR-Net's network structure is similar to that of the Transformer's non-local self-attention module, but HPR-Net does not include components, such as multi-head attention and linear projection. To explore how these components affect our model, we compare our HPR-Net with two HPR-Net variants with a multi-head attention structure (MHA) and a single-head attention structure (SHA). The details of each structure are shown in Figure 5.

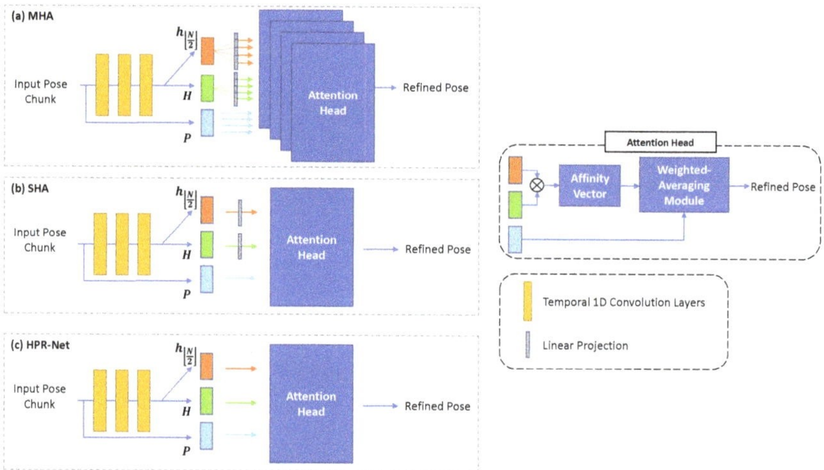

Figure 5. Detailed pipelines of multi-head structure (**a**), linear projection structure (**b**), and our proposed HPR-Net's structure (**c**) for network design experiment. We did not apply linear projection to input pose chunk P in (**a**–**c**), because it should be averaged with affinity weights. Attention head contains affinity vector generation by self-attention and weighted-averaging processes.

Table 8 shows the VIBE refinement performance of MHA, SHA, and our HPR-Net on the 3DPW dataset. All the experimented structures show acceleration error improvement. MHA shows higher MPJPE, PA-MPJPE, and MPVE than SHA. However, the two models fail to improve MPJPE, PA-MPJPE, and MPVE compared with VIBE. Unlike the two structures, HPR-Net improves the performance in all metrics and achieves the lowest acceleration error. The difference between our method and the two structures are that the features obtained from the convolution layers are not linearly projected, and the MHA is not used in the proposed HPR-Net. From the results, the linear projection layer seems to cause performance degradation by confusing to generate an appropriate affinity weight vector from input pose information. The MHA seems to result in overfitting by complicating the network structure more than necessary. Our network has a simpler structure and performs better. HPR-Net is more optimal in solving our problem than the commonly used non-local attention structure.

Table 8. Comparison of refinement performance according to network design of HPR-Net. Bold values indicate best results.

Methods	MPJPE ↓	PA-MPJPE ↓	MPVE ↓	Accel-Error ↓
MHA	84.00	53.20	99.94	7.71
SHA	84.13	53.52	100.44	7.49
HPR-Net	**81.10**	**51.26**	**97.13**	**6.94**

4.7. Qualitative Results

Acceleration error improvement. HPR-Net consistently shows a significant improvement in acceleration error across all methods and datasets on the basis of quantitative results. We present qualitative improvement results using a graph. Figure 6 shows the acceleration error of VIBE, SPIN, MEVA and their refined results after applying HPR-Net to each method. The acceleration errors are calculated for every three consecutive frames from a video of 3DPW. Compared with existing methods' result, HPR-Net effectively improves the acceleration errors for all methods. In particular, the acceleration error is significantly reduced in frames with high peaks where the errors are noticeable.

Refinement result. We present the qualitative results to show that HPR-Net substantially refines a 3D human pose sequence estimated by existing methods. Figures 7 and 8 show the refined results for VIBE and SPIN, respectively. For each example in Figures 7 and 8, the top, middle, and bottom rows show the input image sequence, the estimation result by the existing method, and the refinement result by the proposed HPR-Net, respectively. We do not report the qualitative result for MEVA, because the SMPL estimation results by MEVA's official code are projected incorrectly in the image. In the topmost example of Figure 7, a pedestrian causes occlusion. Thus, the pose of the target subject is incorrectly estimated. HPR-Net refines the results by reconstructing the appropriate pose using the information of nearby frames. In the top-left example of Figure 8, SPIN predicts the global orientation incorrectly due to challenging illumination. This incorrect global orientation is well refined in the result of HPR-Net. From the other results, HPR-Net refines the incorrect estimations of arms and legs.

4.8. Discussion

In accordance with our experimental results, the refinement of the human pose sequence estimated by existing methods can be achieved through a data-driven approach on the basis of a large-scale dataset and a deep neural network. To realize this, the proposed HPR-Net adaptively performs weighted averaging, a well-known framework for noise reduction, on input data, therefore consistently improving the human pose estimation performance of existing state-of-the-art methods. The pose refinement by HPR-Net is performed independently of the existing human pose estimation method. This modularity can be a benefit of our approach because it makes the use and analysis of the proposed method easy. However, HPR-Net has a limitation of depending on the pose estimation results of existing methods. Combining HPR-Net with the existing pose estimation network and learning it in an end-to-end manner may bring additional performance improvement. We plan to continue our research to investigate the end-to-end approach and overcome the limitations of the proposed method.

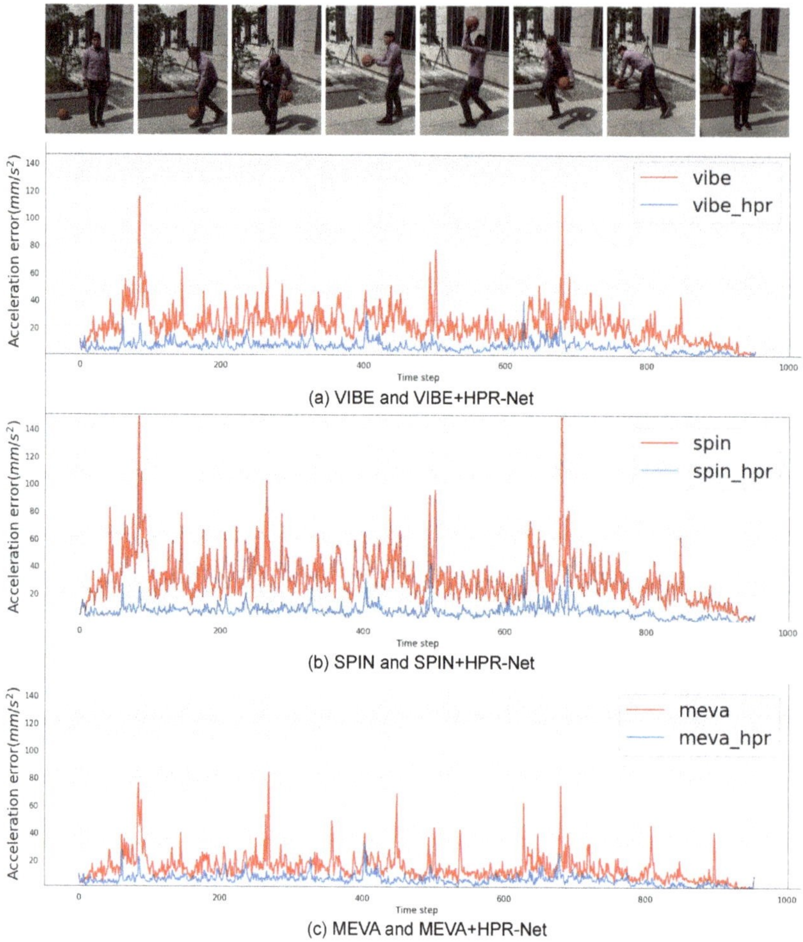

Figure 6. Comparison of acceleration error between HPR-Net and previous methods (VIBE, SPIN, and MEVA). HPR-Net effectively suppresses acceleration error for all methods, even there are very high peaks of acceleration error.

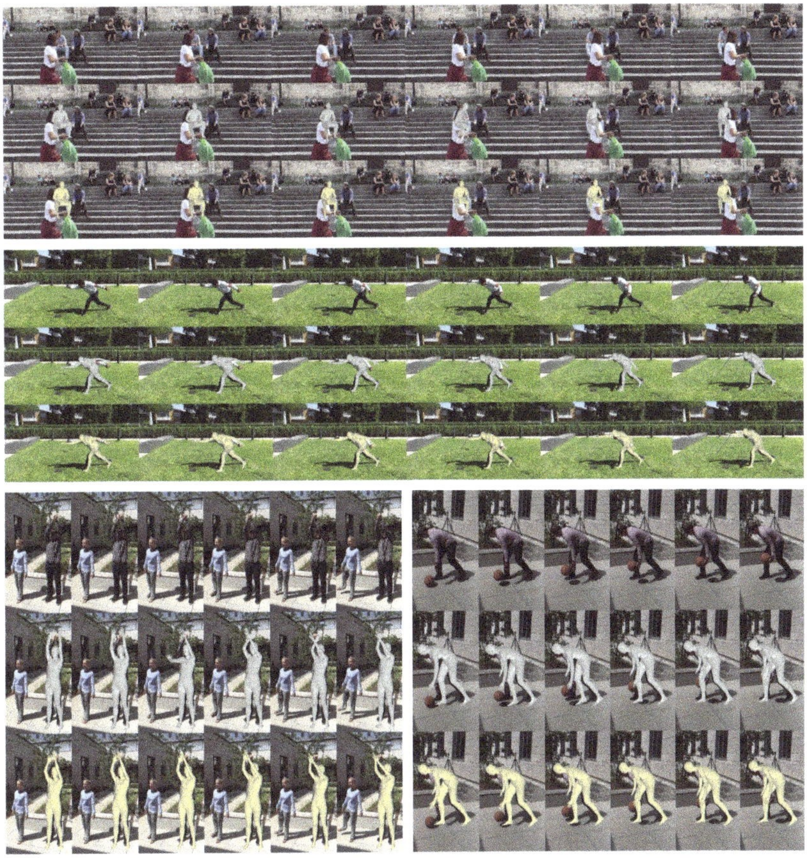

Figure 7. Input images (**top**) and reconstruction results of VIBE (**middle**, gray SMPL mesh) and HPR-Net (**bottom**, yellow SMPL mesh) on the 3DPW dataset.

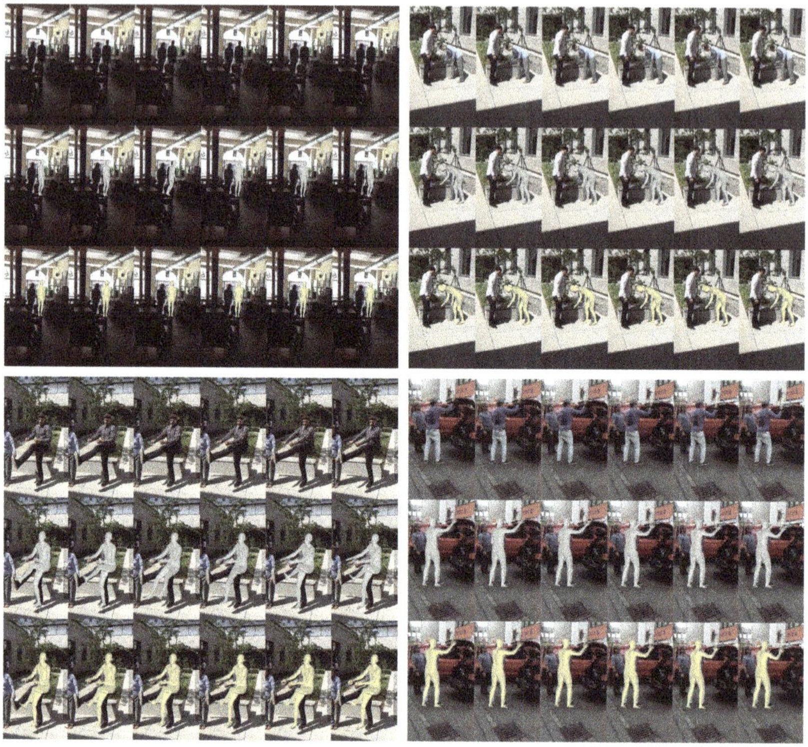

Figure 8. Input images (**top**) and reconstruction results of SPIN (**middle**, gray SMPL mesh) and HPR-Net (**bottom**, yellow SMPL mesh) on the 3DPW dataset.

5. Conclusions

We propose HPR-Net to refine the noisy 3D human pose parameter sequence. HPR-Net improves the accuracy and temporal smoothness of the 3D human pose sequence through a simple non-local attention-based weighted averaging for a noisy pose parameter chunk represented in a unit-quaternion form. We report quantitatively and qualitatively that the proposed method can improve 3D human reconstruction performance for various real datasets, such as Human3.6M and 3DPW. From the experiments for improving the results of various existing methods such as SPIN, VIBE, and MEVA, a consistent performance improvement is observed regardless of the method used to estimate the input human pose sequence. This finding shows that our method works in a model-agnostic manner. The superiority of HPR-Net is confirmed by comparing it with other approaches that can refine 3D human pose parameters.

Author Contributions: Conceptualization, D.-Y.K. and J.-Y.C.; methodology, D.-Y.K.; software, D.-Y.K.; validation, D.-Y.K.; formal analysis, D.-Y.K.; investigation, D.-Y.K.; resources, J.-Y.C.; data curation, D.-Y.K.; writing—original draft preparation, D.-Y.K.; writing—review and editing, J.-Y.C.; visualization, D.-Y.K.; supervision, J.-Y.C.; project administration, J.-Y.C.; funding acquisition, J.-Y.C. All authors have read and agreed to the published version of the manuscript.

Funding: This work was supported by Samsung Research Funding Center of Samsung Electronics (No. SRFC-IT1901-06) and by the National Research Foundation of Korea(NRF) grant funded by the Korea government(MSIT) (No. 2019R1C1C1008462). The present research has been conducted by the Research Grant of Kwangwoon University in 2021.

Conflicts of Interest: The authors declare no conflict of interest.

References

1. Loper, M.; Mahmood, N.; Romero, J.; Pons-Moll, G.; Black, M.J. SMPL: A skinned multi-person linear model. *ACM Trans. Graph. (TOG)* **2015**, *34*, 1–16. [CrossRef]
2. Kanazawa, A.; Zhang, J.Y.; Felsen, P.; Malik, J. Learning 3D human dynamics from video. In Proceedings of the IEEE/CVF Conference on Computer Vision and Pattern Recognition, Long Beach, CA, USA, 15–20 June 2019; pp. 5614–5623.
3. Kocabas, M.; Athanasiou, N.; Black, M.J. VIBE: Video inference for human body pose and shape estimation. In Proceedings of the IEEE/CVF Conference on Computer Vision and Pattern Recognition, Seattle, WA, USA, 13–19 June 2020; pp. 5253–5263.
4. Luo, Z.; Golestaneh, S.A.; Kitani, K.M. 3D human motion estimation via motion compression and refinement. In Proceedings of the Asian Conference on Computer Vision, Kyoto, Japan, 30 November–4 December 2020.
5. Hartley, R.; Trumpf, J.; Dai, Y.; Li, H. Rotation averaging. *Int. J. Comput. Vis.* **2013**, *103*, 267–305. [CrossRef]
6. Gramkow, C. On averaging rotations. *J. Math. Imaging Vis.* **2001**, *15*, 7–16. [CrossRef]
7. Bogo, F.; Kanazawa, A.; Lassner, C.; Gehler, P.; Romero, J.; Black, M.J. Keep it SMPL: Automatic estimation of 3D human pose and shape from a single image. In Proceedings of the European Conference on Computer Vision, Amsterdam, The Netherlands, 8–16 October 2016; Springer: Berlin/Heisenberg, Germany, 2016; pp. 561–578.
8. Pavlakos, G.; Choutas, V.; Ghorbani, N.; Bolkart, T.; Osman, A.A.; Tzionas, D.; Black, M.J. Expressive body capture: 3D hands, face, and body from a single image. In Proceedings of the IEEE/CVF Conference on Computer Vision and Pattern Recognition, Long Beach, CA, USA, 15–20 June 2019; pp. 10975–10985.
9. Kanazawa, A.; Black, M.J.; Jacobs, D.W.; Malik, J. End-to-end recovery of human shape and pose. In Proceedings of the IEEE Conference on Computer Vision and Pattern Recognition, Salt Lake City, UT, USA, 18–23 June 2018; pp. 7122–7131.
10. Kolotouros, N.; Pavlakos, G.; Black, M.J.; Daniilidis, K. Learning to reconstruct 3D human pose and shape via model-fitting in the loop. In Proceedings of the IEEE/CVF International Conference on Computer Vision, Seoul, Korea, 27–28 October 2019; pp. 2252–2261.
11. Bahdanau, D.; Cho, K.; Bengio, Y. Neural machine translation by jointly learning to align and translate. *arXiv* **2014**, arXiv:1409.0473.
12. Vaswani, A.; Shazeer, N.; Parmar, N.; Uszkoreit, J.; Jones, L.; Gomez, A.N.; Kaiser, L.; Polosukhin, I. Attention is all you need. *arXiv* **2017**, arXiv:1706.03762.
13. Wang, X.; Girshick, R.; Gupta, A.; He, K. Non-local neural networks. In Proceedings of the IEEE Conference on Computer Vision and Pattern Recognition, Salt Lake City, UT, USA, 18–23 June 2018; pp. 7794–7803.
14. Cao, Y.; Xu, J.; Lin, S.; Wei, F.; Hu, H. Gcnet: Non-local networks meet squeeze-excitation networks and beyond. In Proceedings of the IEEE/CVF International Conference on Computer Vision Workshops, Seoul, Korea, 27–28 October 2019.
15. Woo, S.; Park, J.; Lee, J.Y.; Kweon, I.S. CBAM: Convolutional block attention module. In Proceedings of the European Conference on Computer Vision, Munich, Germany, 8–14 September 2018; pp. 3–19.
16. Zhu, X.; Su, W.; Lu, L.; Li, B.; Wang, X.; Dai, J. Deformable DETR: Deformable transformers for end-to-end object detection. *arXiv* **2020**, arXiv:2010.04159.
17. Carion, N.; Massa, F.; Synnaeve, G.; Usunier, N.; Kirillov, A.; Zagoruyko, S. End-to-end object detection with transformers. In Proceedings of the European Conference on Computer Vision, Glasgow, UK, 23–28 August 2020; Springer: Berlin/Heisenberg, Germany, 2020; pp. 213–229.
18. Chen, H.; Wang, Y.; Guo, T.; Xu, C.; Deng, Y.; Liu, Z.; Ma, S.; Xu, C.; Xu, C.; Gao, W. Pre-trained image processing transformer. *arXiv* **2020**, arXiv:2012.00364.
19. Chen, M.; Radford, A.; Child, R.; Wu, J.; Jun, H.; Luan, D.; Sutskever, I. Generative pretraining from pixels. In Proceedings of the International Conference on Machine Learning, PMLR, Online, 13–18 July 2020; pp. 1691–1703.
20. Dosovitskiy, A.; Beyer, L.; Kolesnikov, A.; Weissenborn, D.; Zhai, X.; Unterthiner, T.; Dehghani, M.; Minderer, M.; Heigold, G.; Gelly, S.; et al. An image is worth 16 × 16 words: Transformers for image recognition at scale. *arXiv* **2020**, arXiv:2010.11929.
21. Buades, A.; Coll, B.; Morel, J.M. A non-local algorithm for image denoising. In Proceedings of the IEEE Computer Society Conference on Computer Vision and Pattern Recognition, San Diego, CA, USA, 20–25 June 2005; Volume 2, pp. 60–65.
22. Newell, A.; Yang, K.; Deng, J. Stacked hourglass networks for human pose estimation. In Proceedings of the European Conference on Computer Vision, Amsterdam, The Netherlands, 8–16 October 2016; Springer: Berlin/Heisenberg, Germany, 2016; pp. 483–499.
23. Chen, Y.; Wang, Z.; Peng, Y.; Zhang, Z.; Yu, G.; Sun, J. Cascaded pyramid network for multi-person pose estimation. In Proceedings of the IEEE Conference on Computer Vision and Pattern Recognition, Salt Lake City, UT, USA, 18–23 June 2018; pp. 7103–7112.
24. Moon, G.; Chang, J.Y.; Lee, K.M. Posefix: Model-agnostic general human pose refinement network. In Proceedings of the IEEE/CVF Conference on Computer Vision and Pattern Recognition, Long Beach, CA, USA, 15–20 June 2019; pp. 7773–7781.
25. Ruggero Ronchi, M.; Perona, P. Benchmarking and error diagnosis in multi-instance pose estimation. In Proceedings of the IEEE International Conference on Computer Vision, Venice, Italy, 22–29 October 2017; pp. 369–378.
26. Mall, U.; Lal, G.R.; Chaudhuri, S.; Chaudhuri, P. A deep recurrent framework for cleaning motion capture data. *arXiv* **2017**, arXiv:1712.03380.
27. Ionescu, C.; Papava, D.; Olaru, V.; Sminchisescu, C. Human3. 6m: Large scale datasets and predictive methods for 3D human sensing in natural environments. *IEEE Trans. Pattern Anal. Mach. Intell.* **2013**, *36*, 1325–1339. [CrossRef] [PubMed]
28. Ba, J.L.; Kiros, J.R.; Hinton, G.E. Layer normalization. *arXiv* **2016**, arXiv:1607.06450.

29. Ioffe, S.; Szegedy, C. Batch normalization: Accelerating deep network training by reducing internal covariate shift. In Proceedings of the International Conference on Machine Learning, PMLR, Lille, France, 7–9 July 2015; pp. 448–456.
30. von Marcard, T.; Henschel, R.; Black, M.J.; Rosenhahn, B.; Pons-Moll, G. Recovering accurate 3D human pose in the wild using imus and a moving camera. In Proceedings of the European Conference on Computer Vision, Munich, Germany, 8–14 September 2018; pp. 601–617.
31. Gower, J.C. Generalized procrustes analysis. *Psychometrika* **1975**, *40*, 33–51. [CrossRef]
32. Kingma, D.P.; Ba, J. Adam: A method for stochastic optimization. *arXiv* **2014**, arXiv:1412.6980.
33. Paszke, A.; Gross, S.; Massa, F.; Lerer, A.; Bradbury, J.; Chanan, G.; Killeen, T.; Lin, Z.; Gimelshein, N.; Antiga, L.; et al. PyTorch: An imperative style, high-performance deep learning library. In *Advances in Neural Information Processing Systems 32*; Wallach, H., Larochelle, H., Beygelzimer, A., d'Alché-Buc, F., Fox, E., Garnett, R., Eds.; Curran Associates, Inc.: Red Hook, NY, USA, 2019; pp. 8024–8035.

sensors

MDPI

Article

Gap Reconstruction in Optical Motion Capture Sequences Using Neural Networks

Przemysław Skurowski [1,*] and Magdalena Pawlyta [1,2]

[1] Department of Graphics, Computer Vision and Digital Systems, Faculty of Automatic Control, Electronics and Computer Science, Silesian University of Technology, Akademicka 16, 44-100 Gliwice, Poland; Magdalena.Pawlyta@polsl.pl
[2] Polish-Japanese Academy of Information Technology, Koszykowa 86, 02-008 Warsaw, Poland
* Correspondence: przemyslaw.skurowski@polsl.pl; Tel.: +48-32-237-2151

Abstract: Optical motion capture is a mature contemporary technique for the acquisition of motion data; alas, it is non-error-free. Due to technical limitations and occlusions of markers, gaps might occur in such recordings. The article reviews various neural network architectures applied to the gap-filling problem in motion capture sequences within the FBM framework providing a representation of body kinematic structure. The results are compared with interpolation and matrix completion methods. We found out that, for longer sequences, simple linear feedforward neural networks can outperform the other, sophisticated architectures, but these outcomes might be affected by the small amount of data availabe for training. We were also able to identify that the acceleration and monotonicity of input sequence are the parameters that have a notable impact on the obtained results.

Keywords: motion capture; neural networks; reconstruction; gap filling; FFNN; LSTM; BILSTM; GRU

Citation: Skurowski, P.; Pawlyta, M. Gap Reconstruction in Optical Motion Capture Sequences Using Neural Networks. *Sensors* **2021**, *21*, 6115. https://doi.org/10.3390/s21186115

Academic Editors: Tomasz Krzeszowski, Adam Świtoński, Michal Kepski and Carlos Tavares Calafate

Received: 30 July 2021
Accepted: 8 September 2021
Published: 12 September 2021

1. Introduction

Motion capture (mocap) [1,2], in recent years, has become a mature technology that has an important role in many application areas. Its main application is in computer graphics, where it is applied in gaming and movie FX for the generation of realistic-looking character animation. Other prominent applications areas are biomechanics [3], sports [4], medical sciences (involving biomechanical [5] and the other branches, i.e., neurology [6]), and rehabilitation [7].

Optical motion capture (OMC) relies on the visual tracking and triangulation of active or retro-reflective passive markers. Assuming a rigid body model, successive positions of markers (trajectories) are used in further stages of processing to drive an associated skeleton, which is used as a key model for the animation of human-like or animal characters.

OMC is commonly considered the most reliable mocap technology; it is sometimes called the 'gold standard', as it outperforms the other mocap technologies. However, the process of acquiring marker locations is not error-free. Noise, which is immanent in any measurement system, has been studied in numerous works [8,9], which suggests it is not just simple additive Gaussian process. The noise types present in OMC systems were identified in [10]; these are red, pink, white, blue-violet, and Markov–Gaussian-correlated noises; however, they are not a big issue for the mocap operators since they have rather low amplitudes and can be quite efficiently filtered out. The most annoying errors come from marker observation issues. They occur due to marker occlusion and the marker leaving the scene, and result in a lack of the recorded data-gaps that are typically represented as not a number (NaN) values.

The presence of gaps is common and results in everyday praxis, which requires painstaking visual trajectory examination and manual trajectory editing by operators. This can be assisted by software support for trajectory reconstruction.

In this work, we propose a marker-wise approach that addresses the trajectory reconstruction problem. We analyze the usability of various neural network architectures applied to regressive tasks. The regression/prediction exploits inter-marker correlations between markers placed on the same body parts. Therefore, we employed a functional body mesh structure (FBM) [11] as a framework to model the kinematic structure of the subject. I Thisan be calculated ad-hoc for any articulated subject or rigid objects, so we do not need a skeleton model.

The article is organized as follows: in Section 2, we disclose the background for the article—mocap pipeline with sources of distortion and former works on the distortions in optical mocap systems; Section 3 describes the proposed method, with its rationales and design considerations, and experiment plan. In the Section 4 we provide results, and a discussion and interpretation of results. Section 5 summarizes the article.

2. Background

2.1. Optical Motion Capture Pipeline

Optical motion capture systems track the markers—usually passive retro-reflective spheres in near-infrared images (NIR) images. The basic pipeline is shown in Figure 1. The markers are observed by several geometrically calibrated NIR cameras. The visual wavelengths cut-off, and, hence, the images, contain just white dots, which are matched between the views and triangulated, so the outcome of the early stage of mocap is a time series containing Cartesian coordinates of all markers. An actor and/or object wears a sufficient number of markers to represent body segments—marker layout usually follows a predefined layout standard. The body segments are represented by a predefined mesh, which identifies the body segments and is a marker-wise representation of body structure. Finally, mocap recording takes the form of a skeleton angle time series, which represents the mocap sequence as orientations (angles) in joints and a single Cartesian coordinate for body root (pelvis usually).

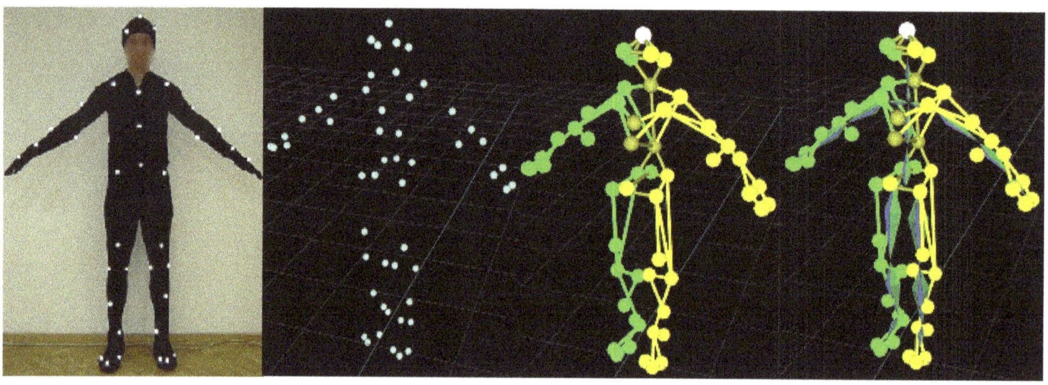

(a) (b) (c) (d)

Figure 1. Stages of the motion capture pipeline: actor (**a**); registered markers (**b**); body mesh (**c**); mesh matched skeleton (**d**).

2.2. Functional Body Mesh

Functional body mesh (FBM) is a authors' original contribution, that forms a framework for marker-wise mocap data processing, which incorporates also the kinematic structure of a represented object. The FBM structure is not given in advance, but it can be inferred based on the articulated object representative motions [11]. For human actors it resembles standard meshes, but it can be applied for virtually any vertebrates. It assumes the body is divided into rigid segments (submeshes), which are organized into a tree structure. The model represents the hierarchy of subjects' kinematic structure, reflecting bonds between body segments, where every segment is a local rigid body model—usually based on an underlying bone.

The rigid segments maintain the distance between the markers and, additionally, for each child segment, one representative marker is assumed within the parent one, which is also assumed to maintain a constant distance from the child markers. The typical FBM for the human actor is shown in Figure 2b as a tree. The segments and constituent markers are located in nodes, whereas the parent marker is denoted on the parent–child edge.

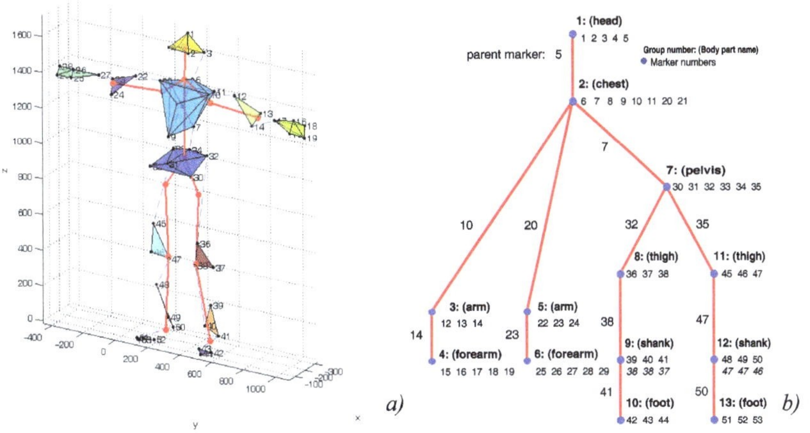

Figure 2. Outline of the body model (**a**), and corresponding parts hierarchy annotated with parents and siblings (**b**).

2.3. Previous Works

Gap filling is a classical problem frequently addressed in research on mocap technologies. It was in numerous works, which proposed various approaches. The existing methods can be divided into three main groups—skeleton-based, marker-wise, and coordinate-based.

A classical skeleton-based method was proposed by Herda et al. [12], they estimate skeleton motion and regenerate markers on the body envelope. Aristidou and Lanesby [13] proposed the other method based on a similar concept, where the skeleton is a source for constraints in inverse kinematics estimation of marker location. Also, Perepichka et al. [14] combined IK of skeleton model with deep NN to detect erroneously located markers and to place them on a probable trajectory. All aforementioned approaches require either to have a predefined skeleton or to infer the skeleton as the entry step of an algorithm.

The skeleton-free methods consider information from markers only, usually acknowledging the whole sequence as a single multivariable (matrix), thus losing the kinematic structure of the represented actor. They rely on various concepts, starting from the simple interpolating methods [15–17]. The proposal by Liu and McMillan [18] employed 'local' (neighboring markers) low-dimensional least squares models combined with PCA for missing marker reconstruction. A significant group of gap reconstruction proposals is based on the low-rank matrix completion methods. They employ various mathematical tools (e.g., matrix factorization with SVD) for the missing data completion, relying on inter marker correlations. Among the others, these methods are described in the following works [19,20]. Another approach is somewhat related: it is a fusion of several regressions and interpolation methods, which was proposed in [21].

Predicting markers (or joint) position is another concept that is the basis of gap-filling techniques. One such concept is a predictive model by Piazza et al. [22], which decomposes the motion into linear and circular and finds momentary predictors by curve fitting. More sophisticated dynamical models based on the Kalman filter (KF) are commonly applied. Wy and Boulanger [23] proposed a KF with velocity constraints; however, this achieved moderate success due to drift. A KF with an expectation-maximization algorithm was also used in two related approaches by Li et al.—DynaMMo [24], and BoLeRO [25] (the latter is actually Dynammo with bone length constraints). Another approach was proposed by Burke and Lanesby [26], who applied dimensionality reduction by PCA and then Kalman smoothing for the reconstruction of missing markers.

Another group of methods is dictionary-based. These algorithms recover the trajectories using a dictionary created from previously recorded sequences. They result in satisfactory outcomes as long the specific motion is in the database. They are represented by the works of Wang et al. [27], Aristidou et al. [28], and Zhang and van de Panne [29].

Finally, neural networks are another group of methods used in marker trajectory reconstruction. The task can be described as a sequence-to-sequence regression problem, whereas NN applied for regression has been recognized since the early 1990s in the work of Hornik [30]; hence, NN seems to be a natural choice for the task. Surprisingly, however, they become popular quite late. In the work of Fragkiadaki et al. [31], an encoder–recurrent-decoder (ERD) was proposed, employing long-short term memory (LSTM) as a recurrent layer. A similar approach (ERD) was proposed by Harvey et al. [32] for in-between motion generation on the basis of asmall amount of keyframes. Mall et al. [33] modified the ERD and proposed an encoder–bidirectional-filter (EBF) based on the bidirectional LSTM (BILSTM). In the work of Kucharenko et al. [34], a classical two-layer LSTM and window-based feed-forward NN (FFNN) were employed. A variant of ResNet is applied by Holden [35] to reconstruct marker positions from noisy data as a ttrajectory reconstruction task. A set of extensions to the plain LSTM were proposed by Ji et al. [36]; they introduced attention (a weighting mechanism) and LS-derived spatial constraints, which result in an improvement in performance. Convolution auto-encoders was proposed by Kaufmann et al. [37].

3. Materials and Methods

3.1. Proposed Regression Approach

The proposed approach involves employing various neural networks architectures for the regression task. These are FFNN and three variants of contemporary recursive neural networks—gated recurrent unit (GRU), long-short-term memory (LSTM), and bidirectional LSTM (BILSTM). In our proposal, these methods predict trajectories of lost markers on the basis of a local dataset—the trajectories of neighboring markers.

The proposed utilization procedure of NN differs from the scenario that is typically employed in machine learning. We do not feed the NNs with a massive amount of training sequences in advance to form a predictive model. Instead, we consider each sequence separately and try to reconstruct the gaps in individual motion trajectory on the basis of its own data only. This makes sense as long as the marker motion is correlated and most of the sequence is correct and representative enough. This is the same as for the other common regression methods, starting with the least squares. Therefore, the testing data are the whole 'lost' segment (gap), whereas the training is the remaining part of the trajectory. Depending on the gap sizes, and sequence length used in the experiment, the testing can be between 0.6% (for short gaps and long sequences) and up to 57.1% (for long gaps in short sequences).

The selection of such a non-typical approach requires a justification. It is likely that training the NN models for prediction of marker position in a conventional way, using a massive dataset of mocap sequences, would be able to generalize enough to adjust to different body sizes and motions. However, it will be tightly coupled with the marker configuration, not to mention the other actors, such as animals. The other issue is obtaining such a large amount of data. Despite our direct access to the lab resources, this is still quite a cumbersome task, since we believe these might be not enough, especially as the resources available online from various other labs are hardly usable, since they employ different marker setups.

The forecasting of timeseries is a typical problem addressed by RNNs [38]. Usually, numerous training and testing sequences allow for a prediction of the future states of the modelled system (e.g., power consumption or remaining useful life of devices). A more similar situation, where RNNs are also applied, is forecasting the time series for problems lacking massive training data (e.g., COVID-19 [39]). An analysis of LSTM architectures for similar cases is presented in [40]. However, in these works, the forecast of future values is based on the past values. What makes our case a bit different is the fact that we usually have to predict the value in-the-middle, so the past and future values are available.

3.1.1. Feed Forward Neural Network

FFNN is the simplest neural network architecture. In this architecture, the information flows in one direction, as its structure forms an acyclic directed graph. The neurons are modeled in the nodes with activation functions (usually sigmoid) using the weighted sum of inputs. These networks are typically organized into layers, where the output from the previous layer becomes an input to a successive one. This architecture of networks is employed for regression and classification tasks, either alone or as final stages in a larger structures (such as modern deep NN). The architecture of the NN that we employed is shown in Figure 3. The basic equation (output) of a single—k-th artificial neuron is given as:

$$y_k(x) = f\left(\sum_j w_{jk}x_j + b\right), \tag{1}$$

where x_j is j-th input, w_{kj} is j-th input weight, b—a bias value, f—is transfer (activation) function. Transfer function depends on the layer purpose; these are typically a sigmoid for hidden layers, threshold, linear, or softmax for final layers (for regression and classification problems, respectively), or others.

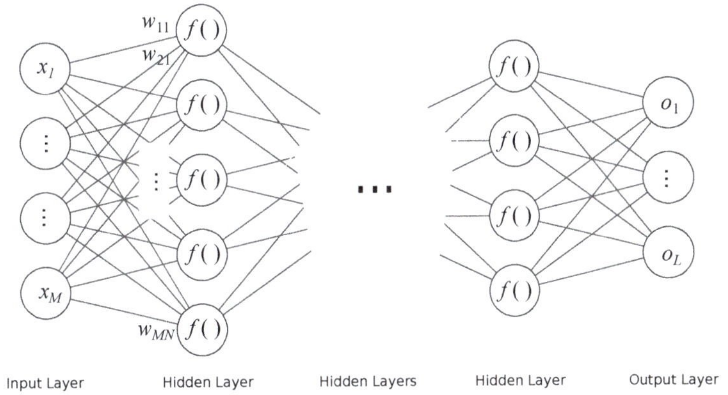

Figure 3. Schematic of FFNN.

3.1.2. Recurrent Neural Networks

Recurrent neural networks (RNN) are the types of architecture that employ cycles in NN structure; this allows for the consideration of current input value as well as preserving the previous inputs and internal states of NN in memory (and future ones for bidirectional architecture). Such an approach allows for NN to deal with timed processes and to recognize process dynamics, not just static values—it applies to such tasks as a signal prediction or recognition of sequences. Regarding the applicability, aside from classic problem dichotomy (classification and regression), RNN results might need another task differentiation. One must decide whether the task is a sequence-to-one or sequence-to-sequence problem, so the network has to return either a single result for the whole sequence or a single result for each data tuple in sequence. The prediction/regression task is a sequence-to-sequence problem, as demonstrated with RNNs in Figure 4 in different variants—both folded and unfolded, uni- and bi-directional.

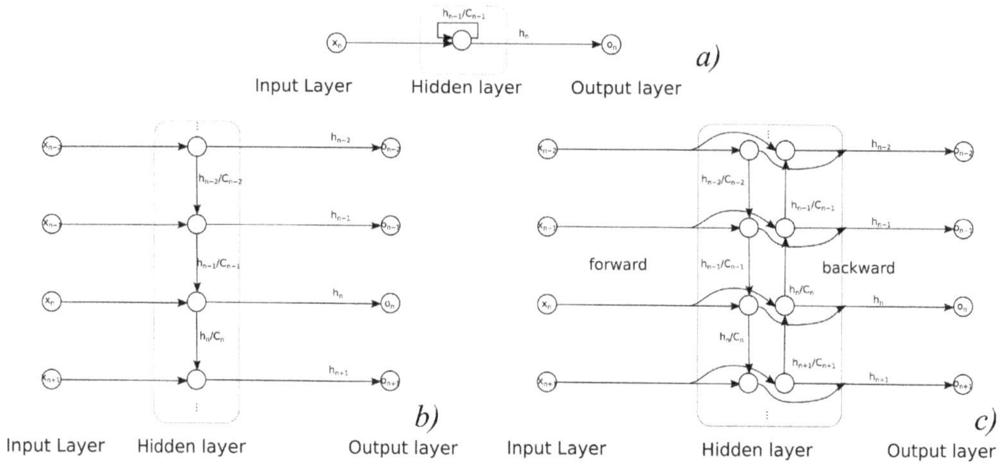

Figure 4. Usage of recurrent NNs in sequence to sequence task: (**a**) folded, (**b**) unfolded unidirectional variant, (**c**) unfolded bidirectional variant.

At present two types of neuron are predominantly applied in RNN–long short term memory (LSTM) and gated recurrent unit (GRU), of which the former is also applied in bidirectional variant (BILSTM). They evolved from a plain RNN called 'vanilla', and they prevent vanishing gradient problems when back-propagating errors in the learning process. Their detailed designs are unfolded in Figure 5. These cell types rely on the input information and information from previous time steps, and those previous states are represented in various ways. GRU passes an output (hidden signal h) between the steps, whereas LSTM also passes a h and internal cell state C. These values are interpreted as memory—h as short term, and C as long term. Their activation function is typical sigmoid, which is modeled with a hyperbolic tangent (tanh), but there are additional elements present in the cell. The contributing components, such as input or previous values, are subject to *'gating'*—their share is controlled by Hadamard product (element-wise product denoted as $\odot$ or $\otimes$ in diagram) with 0–1 sigmoid function $\sigma(x) = \frac{1}{1+e^{-x}}$. The individual σ values are obtained by weighted input and state values.

Figure 5. LSTM (**left**) and GRU (**right**) neurons in detail.

In more detail, in LSTM, we pass two variables h, C and have three gates—forget, input and output. They govern how much of the respective contribution passes to further processing. The forget gate (f_t) decides how much of the past cell internal state (C_{t-1}) is to be kept; the input gate (i_t) controls how much new contribution $\tilde{C}_t$ caused by input (x_t) annd taken into the current cell state (C_t). Finally, the output gate (o_t) controls what part of activation is based on the cell internal state; (C_t) is taken as cell output (h_t). The equations are as follows:

$$f_t = \sigma(W_f \cdot [x_t, h_{t-1}] + b_f), \tag{2}$$
$$i_t = \sigma(W_i \cdot [x_t, h_{t-1}] + b_f), \tag{3}$$
$$\tilde{C}_t = \tanh(W_c \cdot [x_t, h_{t-1}] + b_c), \tag{4}$$
$$C_t = f_t \odot C_{t-1} + i_t \odot \tilde{C}_t, \tag{5}$$
$$o_t = \sigma(W_o \cdot [x_t, h_{t-1}] + b_f), \tag{6}$$
$$h_t = o_t \odot \tanh(C_t). \tag{7}$$

The detailed schematic of GRU is a bit simpler. Only one signal, hidden (layer output) value (h for hi), is passed between steps. There are two gates present—the reset gate (r_t), which controls how much past output (h_{t-1}) contributes to the overall cell activation, and the update gate (u_t), which controls how much current activation ($\tilde{h}_t$) contributes to the final cell output.The above are described by the following equations:

$$u_t = \sigma(W_u \cdot [x_t, h_{t-1}] + b_u), \tag{8}$$
$$r_t = \sigma(W_u \cdot [x_t, h_{t-1}] + b_u), \tag{9}$$
$$\tilde{h}_t = \tanh(W_h \cdot [x_t, r_t \odot h_{t-1}] + b_h), \tag{10}$$
$$h_t = (1 - u_t) \odot h_{t-1} + u_t \odot \tilde{h}_t. \tag{11}$$

3.1.3. Employed Reconstruction Methods

We compared the performance of five architectures of NN—two variants of FFNN and three RNN-FCs based on GRU, LSTM, and BILSTM; the outline of the latter is depicted in Figure 6. The detailed structures and hyperparameters of NNs were established empirically, since there are no strict rules or guidelines. Usually, this requires simulating, with parameters sweeping the domain of feasible numbers of layers and neurons [41]. We shared this approach and reviewed the performance of NN using the test data.

- $FFNN_{lin}$, with 1 hidden fully connected (FC) layer—containing 8 linear neurons;
- $FFNN_{tanh}$, with 1 hidden FC layer—containing 8 sigmoidal neurons;
- LSTM followed by 1 FC layer containing 8 sigmoidal neurons;
- GRU followed by 1 FC layer containing 8 sigmoidal neurons;
- BILSTM followed by 1 FC layer containing 8 sigmoidal neurons.

The output is three valued x, y, z vectors, containing reconstructed marker coordinates.

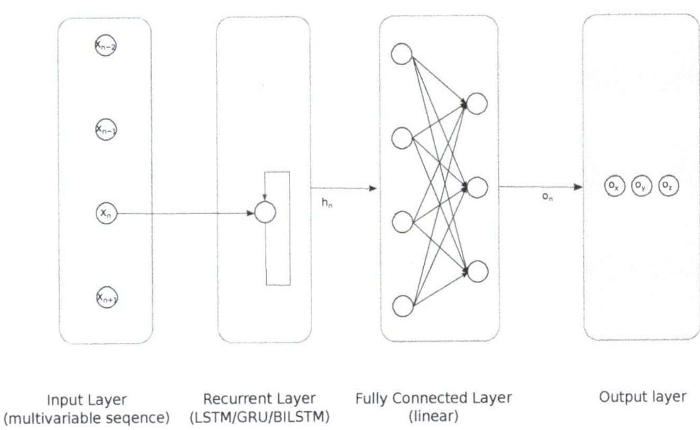

Figure 6. Proposed RNN-FC architecture for the regression task.

3.1.4. Implementation Details

The training process was performed using 600 epochs, with the SGDM solver running on the GPU. It involved the whole input sequence with gaps excluded. There was a single instance of sequence in the batch. The sequence parts containing gaps were used as the test data; the remainder was used for training—therefore, the relative size of test part varies between 0.6% and 57.1%. The other parameters are:

- Initial Learn Rate: 0.01;
- Learn Rate Drop Factor: 0.9;
- Learn Rate Drop Period: 10;
- Gradient Threshold 0.7;
- Momentum: 0.8.

We also applied z-score normalization for the input and target data.

Additionally, for comparison, we used a pool of other methods, which should provide nice results for short-term gaps. These are interpolations: linear, spline, modified Akima (makima), piecewise cubic hermite interpolating polynomial (pchip), and the low-rank matrix completion method (mSVD0). All but linear interpolation methods are actually variants of piecewise Hermite cubic polynomial interpolations, which differ in the details of how they compute interpolant slopes. Spline is a generic method, whereas pchip tries to preserve shape, and makima avoids overshooting. However, mSVD [42] is an iterative method decomposing motion capture data with SVD and neglecting the least significant part of the basis transformed signal, reconstructing the original data with replacing missing values using reconstructed ones. The procedure finishes when convergence is reached. We implemented the algorithm, as outlined in [24].

The implementation of methods and experiments was carried out in Matlab 2021a using its implementations of numerical methods and deep learning toolbox.

3.2. Input Data Preparation

Constructing the predictor for certain markers, we obtained the locations from all the sibling markers and a single parent one, as they are organized within an FBM structure. For j-th marker ($X_j = [x_j, y_j, z_j]$), we consider parent (X_p) and sibling markers ($X_{s1}, \ldots, X_{sL}$). To form an input vector, we take two of their values—one for the current moment and with one sample lag. The other variants with more lags or values raised to the higher powers were considered, but after preliminary tests, we neglected them since they did not improve performance.

Each input vector T, for the moment n, is quite long and is assembled of certain parts, as given below:

$$T(n,*) = \begin{bmatrix} \overbrace{x_p(n), y_p(n), z_p(n), x_p(n-1), y_p(n-1), z_p(n-1)}^{\text{current and former values of parent marker } (p)}, \\ \overbrace{x_{s1}(n), y_{s1}(n), z_{s1}(n), x_{s1}(n-1), y_{s1}(n-1), z_{s1}(n-1)}^{\text{current and former value of first sibling } s_1}, \\ \vdots \\ \underbrace{x_{sL}(n), y_{sL}(n), z_{sL}(n), x_{sL}(n-1), y_{sL}(n-1), z_{sL}(n-1)}_{\text{current and former value of last sibling } s_L} \end{bmatrix}. \tag{12}$$

Finally, the input and output data are z-score standardized—zero centered and standard deviation scaled to 1, since such a step notably improves the final results.

3.3. Test Dataset

For testing purposes, we used a dataset (Table 1) acquired for professional purposes in the motion-capture laboratory. The ground truth sequences were obtained at the PJAIT human motion laboratory using the industrial-grade Vicon MX system. The system capture volume was 9 m × 5 m × 3 m. To minimize the impact of external interference such as infrared interference from sunlight or vibrations, all windows were permanently darkened and cameras were mounted on scaffolding instead of tripods. The system was equipped with 30 NIR cameras manufactured by Vicon: MX-T40, Bonita10, Vantage V5—wth 10 pieces of each kind.

During the recording, we employed a standard animation pipeline, where data were obtained with Vicon Blade software using a 53-marker setup. The trajectories were acquired at 100 Hz and, by default, they were processed in a standard, industrial-quality way, which includes manual data reviewing, cleaning and denoising, so they can be considered distortion-free.

Several parameters for the test sequences are also presented in Table 2. We selected these parameters as one could consider them to potentially describe prediction difficulty. They are various, and based on different concepts such as information theory, statistics, kinematics, and dynamics, but all characterize the variability in the Mocap signal. They are usually the average value per marker, except for standard deviation (std dev), which reports value per coordinate.

Table 1. List of mocap sequence scenarios used for the testing.

No.	Name	Scenario	Duration	Difficulty
1	Static	Actor stands in the middle of scene, looking around and shifting from one foot to another, freely swinging arms	32 s	varied motions
2	Walking	Actor stands still at the edge of the scene, then walks straight for 6 m, then stands still	7 s	low dynamics, easy
3	Running	Actor stands in the middle of scene, then goes backwards to the edge of the scene and runs for 6 m, then goes backwards to the middle of the scene	16 s	moderate dynamics
4	Sitting	Actor stands in the middle of scene, then sits on a stool, and, after a few seconds, stands again	15 s	occlusions
5	Boxing	Actor stands in the middle of scene, and performs some fast boxing punches	14 s	high dynamics
6	Falling	Actor stands on 0.5 m elevation in the middle of scene, the walks to edge of platform, then falls on the mattress, lies for 2 s and stands	16 s	high dynamics, occlusions

Two non-obvious measures are enumerated: monotonicity and complexity. The monotonicity indicates, on average, the extent to which the coordinate is monotonic. For this purpose, we employed an average Spearman rank correlation, which can be described as follows:

$$monotonicity = \frac{1}{M} \sum_{m=1}^{M} \operatorname{corr}(\operatorname{rank}(X_i), 1 \ldots N), \tag{13}$$

where X_m is mth coordinate, M is number of coordinates, N is sequence length.

Complexity, on the other hand, is how we estimate the variability of poses in the sequence. For that purpose, we employed PCA, which identifies eigenposes as a new basis for the sequence. The corresponding eigenvalues describe how much of the overall variance is described by each of the eigenposes. Therefore, we decided to take the remainder of the fraction of variance described by the sum of the five largest eigenvalues (λ_i) as a term describing how complex (or rather simple) the sequence is—the simpler the sequence, the more variance is described, with a few eigenposes. Therefore, our complexity measure is simply given as:

$$complexity = 1 - \sum_{i=1}^{5} \lambda_i / \sum_{i=1}^{M} \lambda_i, \tag{14}$$

where M is a number of coordinates.

Table 2. Input sequence characteristics.

No	Entropy ($H(X)$) [Bits/Mark.]	Stddev (σ_X) [mm/Coordinate]	Velocity ($\frac{\partial X}{\partial t}$) [m/s/Mark.]	Acc. ($\frac{\partial^2 X}{\partial t^2}$) [m/s^2/Mark.]	Jerk ($\frac{\partial^3 X}{\partial t^3}$) [m/s^3/Mark.]	Monotonicity [-]	Complexity [-]
1	12.697	129.705	0.208	1.561	64.817	0.352	0.027
2	13.943	941.123	0.773	6.476	829.271	0.582	0.000
3	15.710	982.342	0.895	6.176	643.337	0.379	0.001
4	10.231	135.356	0.190	2.863	452.142	0.347	0.016
5	11.356	121.094	0.259	3.557	507.975	0.323	0.023
6	14.152	601.140	0.589	6.703	799.039	0.745	0.007

3.4. Quality Evaluation

The natural criterion for the reconstruction task is root mean square error (RMSE), which, in our case, is calculated only for the time and marker, where the gaps occur:

$$\text{RMSE} = \sqrt{\frac{1}{|W|}\sum_{i\in W}(\hat{X}_i - X_i)^2}, \tag{15}$$

where W is a gap map, logically indexing locations of gaps, $\hat{X}$ is a reconstructed coordinate, X is the original coordinate.

Additionally, we calculated RMSEs for individual gaps. Local RMSE is a variant of the above formula, and simply given as:

$$\text{RMSE}_\text{k} = \sqrt{\frac{1}{|w_k|}\sum_{i\in w_k}(\hat{X}_i - X_i)^2}, \tag{16}$$

where $w_k \subset W$ is a single gap map logically indexing the location of k-th gap, $\hat{X}$ is reconstructed coordinate, X is original coordinate. $RMSE_k$ is intended to reveal variability in reconstruction capabilities; hence, we used it to obtain statistical descriptors—mean, median, mode, and quartiles and interquartile range.

A more complex evaluation of regression models can be based on infromation criteria. These quality measures incorporate squared error and a number of tunable parameters, as they were designed by searching for a tradeoff between the number of tunable parameters and the obtained error. The two most popular ones are Bayesian Information Criterion (BIC) and Akaike Information Criterion (AIC). BIC is calculated as:

$$\text{BIC} = n\log(MSE) + p\log(n), \tag{17}$$

whereas AIC formula is as follows:

$$\text{AIC} = n\log(MSE) + 2p, \tag{18}$$

where: mean squared error MSE $=$ RMSE^2, n is a number of testing data, p is a number of tunable parameters.

3.5. Experimental Protocol

During the experiments, we simulated gap occurrence in perfectly reconstructed source sequences. We simulated gaps of different average lengths—10, 20, 50, 100 and 200 samples (0.1, 0.2, 0.5, 1, and 2 s, respectively). The assumed gap sizes were chosen to represent situations of various levels of difficulty, from short-and-simple to difficult ones, when gaps are long. For every gap length, we performed 100 simulation iterations, where the training and testing data do not intermix between simulation runs. The steps performed in every iteration are as follows:

1. We introduce two gaps of assumed length (on average) to the random markers at random moments; actual values are stored as testing data;
2. The model is trained using the remaining part of the sequence (all but gaps);
3. We reconstruct (predict) the gaps using the pool of methods;
4. The resulting values are stored for evaluation.

We report the results as RMSE and descriptive statistical descriptors for $RMSE_k$ for every considered reconstruction technique. Additionally, we verified the correlation between RMSE and the variability descriptors for sequences. It is intended to reveal what are the sources of difficulties in predicting the marker trajectories.

Gap Generation Procedure

The procedure of gap contamination, which was employed, introduces distortions into the sequences in a controlled way. The parameter characterizing the experiment is an average-length number of occurrences of gaps. the sequence of operations distorting the signal is as follows: at first, we draw moments to contaminate, then select a random marker. The duration of distortions and intervals is a Poisson process, an average length of distortion set-up according to the considered gap length in the experiment, whereas the interval length results from the sequence length and number of intervals, which, for two gaps per sequence, are three—ahead of the first gap, in-between, and after the second gap.

4. Results and Discussion

The section comprises two parts. First, we present RMSE results; they illustrate the performance of each of the considered gap reconstruction methods. The second part is the interpretation of results, searching for the aspects of Mocap sequence that might affect the resulting performance.

4.1. Gap Reconstruction Efficiency

The detailed numerical values are presented in Table 3 for the first sequence as an example. In the table, we also emphasize the best result for each measurement of gap size. Forclarity, the numerical outcomes of the experiment are only presented in this chapter with representative examples. To see the complete set of results in the tabular form, please refer to Appendix A. The complete results for the gap reconstruction are also demonstrated in a visual form in Figure 7. Additionally, the zoomed variant of the fragments of the plot (dash square annotated) for gaps 10–50 are presented in Figure 8.

The first observation, regarding the performance measures, is the fact that the results are very coherent, regardless of which measure was used. This is shown in Figure 7, where all the symbols coherently denote statistical descriptors scale. It is also clearly visible in the values emphasized in Table 3, where all measures but one (mode) indicate the same best (smallest) results. Hence, we can use a single quality measure; in our case, we assumed RMSE for further analysis.

Analyzing the results for several sequences, various observations regarding the performance of the considered methods can be noted. These are listed below:

- It can be seen that, for the short gaps, interpolation methods outperform any of the NN-based methods.
- For gaps that are 50 samples long, the results become less obvious and NN results are no worse or (usually) better than interpolation methods.
- Linear FFNN usually performed better than any other methods (including non-linear $FFNN_{tanh}$), for gaps of 50 samples or longer, for most of the sequences.
- In very rare cases of short-gap cases, RNNs performed better than $FFNN_{lin}$, but, in general, simpler $FFNN_{lin}$ outperformed more complex NN models.
- There are two situations when the $FFNN_{lin}$, performed no better or worse than interpolation methods (walking and falling). This occurred for sequences with larger monotonicity values in Table 2. They have also increased velocity/acceleration/jerk

values; the 'running' sequence has similar values for these, but $FFNN_{lin}$ perform the best in this case, so the kinematic/dynamic parameters should not be considered.

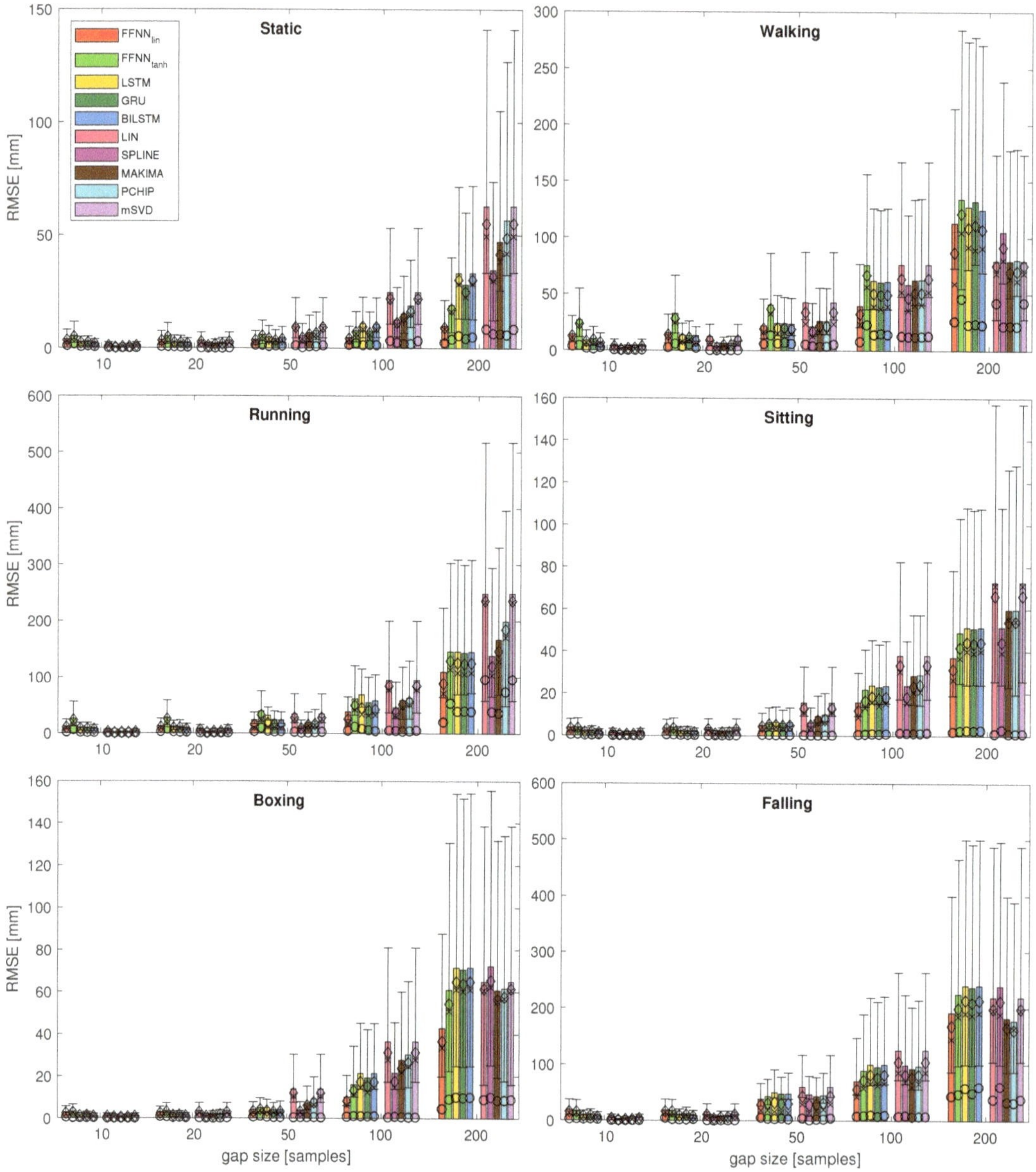

Figure 7. Results for most of the quality measures for all the test sequences. Bars denote $RMSE$; for $RMSE_k$: $\diamond$ denotes mean value, $\times$ denotes median, $\circ$ denotes mode, whiskers indicate IQR; standard deviation is not depicted here; dash-outlined areas are zoomed in Figure 8.

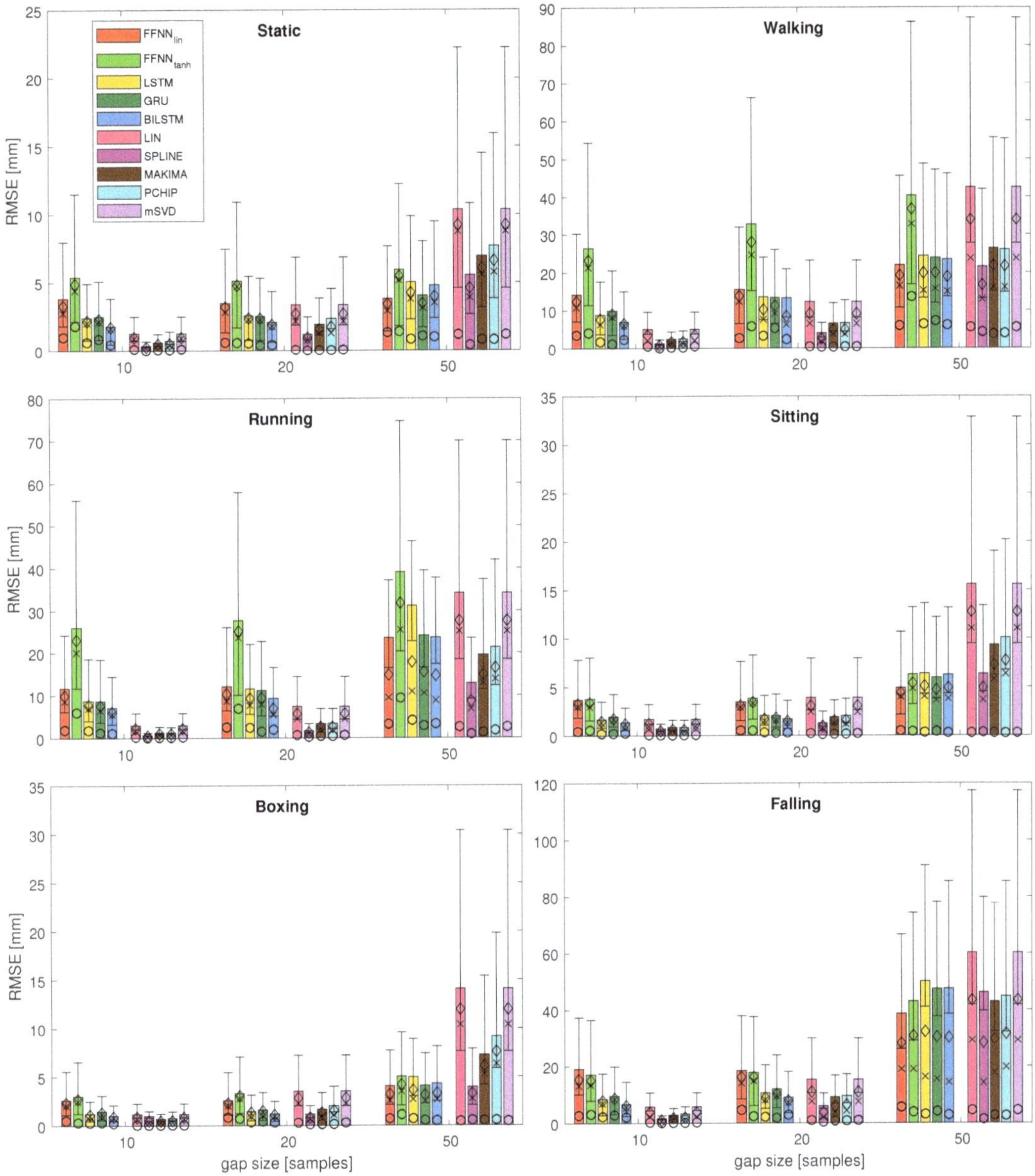

Figure 8. Results of the most of the quality measures for all the test sequences—zoomed variant for gaps 10, 20, and 50. Bars denote $RMSE$; for $RMSE_k$: $\diamond$ denotes mean value, $\times$ denotes median, $\circ$ denotes mode, whiskers indicate IQR; standard deviation is not depicted here.

Table 3. Quality measures for the static (No. 1) sequence.

Len		$FFNN_{lin}$	$FFNN_{tanh}$	LSTM	GRU	BILSTM	LIN	SPLINE	MAKIMA	PCHIP	mSVD
10	RMSE	3.830	5.375	2.410	2.494	1.801	1.267	**0.348**	0.610	0.737	1.267
	mean($RMSE_k$)	3.280	4.869	2.175	2.290	1.708	0.971	**0.243**	0.468	0.512	0.971
	median($RMSE_k$)	2.746	4.399	2.035	2.120	1.614	0.893	**0.205**	0.406	0.391	0.893
	mode($RMSE_k$)	0.993	1.821	0.626	0.861	0.455	0.099	**0.000**	0.045	0.036	0.099
	stddev($RMSE_k$)	1.893	2.209	0.939	0.989	0.573	0.695	**0.216**	0.336	0.458	0.695
	iqr($RMSE_k$)	2.123	2.905	0.881	0.901	0.684	0.692	**0.235**	0.370	0.434	0.692
20	RMSE	3.474	5.114	2.559	2.527	2.082	3.366	**1.191**	1.914	2.354	3.366
	mean($RMSE_k$)	3.187	4.775	2.371	2.351	1.903	2.694	**0.933**	1.525	1.738	2.694
	median($RMSE_k$)	2.828	4.709	2.274	2.235	1.779	2.147	**0.764**	1.251	1.287	2.147
	mode($RMSE_k$)	0.605	0.584	0.540	0.381	0.415	0.052	**0.005**	0.026	0.023	0.052
	stddev($RMSE_k$)	1.442	1.871	0.891	0.898	0.826	1.831	**0.664**	1.045	1.483	1.831
	iqr($RMSE_k$)	1.841	2.394	1.103	1.013	0.813	1.983	**0.866**	1.173	1.437	1.983
50	RMSE	**3.813**	5.910	5.001	4.041	4.777	10.363	5.517	6.928	7.677	10.363
	mean($RMSE_k$)	**3.401**	5.434	4.233	3.445	3.958	9.207	4.572	6.027	6.573	9.207
	median($RMSE_k$)	**2.906**	5.154	3.776	3.118	3.496	8.733	3.888	5.512	5.733	8.733
	mode($RMSE_k$)	1.326	1.393	0.831	1.066	1.000	1.169	**0.400**	0.800	0.793	1.169
	stddev($RMSE_k$)	**1.688**	2.168	2.430	1.921	2.448	4.464	2.852	3.174	3.764	4.464
	iqr($RMSE_k$)	**1.421**	2.216	2.169	1.642	2.282	6.078	2.418	3.770	4.373	6.078
100	RMSE	**4.759**	7.805	10.798	7.678	10.716	24.634	12.548	15.231	18.746	24.634
	mean($RMSE_k$)	**4.233**	7.134	9.460	6.721	9.302	21.812	11.236	13.587	16.108	21.812
	median($RMSE_k$)	**3.658**	6.329	8.333	5.953	8.198	21.129	10.345	12.875	14.785	21.129
	mode($RMSE_k$)	1.517	2.252	**1.377**	1.465	1.400	3.266	2.546	1.986	1.937	3.266
	stddev($RMSE_k$)	**2.132**	3.143	5.114	3.692	5.230	11.305	5.472	6.825	9.556	11.305
	iqr($RMSE_k$)	**2.215**	3.473	5.650	4.217	5.700	14.536	6.850	8.029	11.019	14.536
200	RMSE	**9.959**	18.970	33.147	27.987	33.104	62.786	34.481	47.259	56.570	62.786
	mean($RMSE_k$)	**9.062**	17.303	30.204	24.837	30.135	55.099	31.616	41.676	48.789	55.099
	median($RMSE_k$)	**8.683**	16.200	28.352	22.655	28.462	49.641	29.914	38.410	42.155	49.641
	mode($RMSE_k$)	**2.404**	3.973	5.523	4.263	5.010	8.510	6.518	6.459	6.033	8.510
	stddev($RMSE_k$)	**4.013**	7.631	13.450	12.743	13.503	29.934	13.511	22.022	28.463	29.934
	iqr($RMSE_k$)	**5.084**	9.413	18.231	16.895	18.436	48.864	17.125	36.315	46.222	48.864

Looking at the results of various NN architectures, it might be surprising that the sophisticated RNNs often returned worse results than relatively simple FFNN, especially for relatively long gaps. Conversely, one might expect that RNNs would outperform other methods, since they would be able to model longer-term dependencies in the motion. Presumably, the source of such a result is in the limited amount of training data, which, depending on the length of the source file, varies between hundreds and thousands of registered coordinates. Therefore, solvers are unable to find actually good values for a massive amount of parameters—see Table 4 for the formulas and numbers of learnable parameters for an exemplary case when input comprises 30 values—coordinates of four siblings and a parent at current and previous frames.

An obvious solution to such an issue would be increasing the training data. We could achieve this by employing very long recordings or by using numerous recordings. In the former, it would be difficult to achieve long enough recordings; the latter is different from the case which we try to address, where we only obtain a fresh mocap recording and reconstruct it with the minimal model given by FBM. Training the predictive model in advance with a massive amount of data is, of course, an interesting solution, but would cost the generality. For every marker configuration, a separate set of predicting NNs would need to be trained, so the result would only be practical for standardized body models.

Considering the length of the training sequences, its contribution to the final results seems far less important than other factors, at least within the range of considered cases. The analysis of its influence is illustrated in Figure 9. Since the MSE results are entangled, we employed two additional information criterions, Akaike Information Criterion (AIC), Bayesian Information Criterion (BIC), which disentangle the results by accounting for the number of trainable parameters. For every sequence and every NN model, we obtain a series of five results, which decrease, as the training sequence grows longer when we have shorter gaps (i.e., the annotated quintuple in the Figure). Analyzing the results in Figure 9, it is most convenient to observe this in the AIC/BIC plots since, for each model, the number of parameters remains the same (Table 4), so we can easily compare the results of the testing sequences. The zoomed versions (to the right) reveal differences at appropriate scales for the RNN results.

Lookng at the reults, we observe that, regardless the length of the training sequence, the MSE (AIC/BIC) of the NN model remains at the same order of magnitude—this is clearly visible in the Figure, where we have very similar values for each gap size for variable sequences (represented as different marker shapes) for each of the NN types (represented by a color). The most notable reduction in the error is probably observed with the increased sequence length, when the sequence (Seq. 1—static) is several folds longer than the others. However, we cannot observe this difference for shorter sequences in our data, with notably different lengths (e.g., walking—running). The quality of prediction could be likely improved if the recordings were longer, but, in everyday praxis, the length of the motion caputre sequences is only minutes, so one should not expect the results for RNN data to be notably improved compared to those for FFNN.

The observations hold for both FFNN models and all RNNs. These ambiguous outcomes confirm the results shown in [40], where the quality of results does not depend on the length of the training data in a straightforward way.

Table 4. List of mocap sequence scenarios used for the testing.

NN Type	Number of Learnable Parameters	Value for Exemplary Case
FFNN:	$hiddenLayerSize \times inputvectorSize + hiddenLayerSize$ $+3 \times hiddenLayerSize + 3$	275
LSTM:	$4 \times hiddenRecurrentNeurons \times inputvectorSize$ $+4 \times hiddenRecurrentNeurons \times$ $hiddenRecurrentNeurons$ $+4 \times hiddenRecurrentNeurons$ $+3 \times hiddenRecurrentNeurons + 3$	22,023
GRU:	$3 \times hiddenRecurrentNeurons \times inputvectorSize$ $+3 \times hiddenRecurrentNeurons \times$ $hiddenRecurrentNeurons$ $+3 \times hiddenRecurrentNeurons$ $+3 \times hiddenRecurrentNeurons + 3$	16,563
BILSTM:	$8 \times hiddenRecurrentNeurons \times inputvectorSize$ $+8 \times hiddenRecurrentNeurons \times$ $hiddenRecurrentNeurons$ $+8 \times hiddenRecurrentNeurons$ $+3 \times 2 \times hiddenRecurrentNeurons + 3$	47,043

4.2. Motion Factors Affecting Performance

In this section, we try to identify the correlation in which features (parameters) of the input sequences relate to the performance of gap-filling methods. The results presented here are concise; we only present and discuss the most conclusive results. The complete tables containing correlation values for all gap sizes are presented in Appendix B.

Foremost, a generalized view into the correlation between gap-filling outcomes and input sequence characteristics is given in Table 5. It contains Pearson correlation coefficients (CC) between RMSE and input sequence characteristic parameters; the values are Pearson CCs, averaged across all the considered gap sizes. Additionally, for the interpretation of the results, in Table 6, we provide CCs between RMSE and the descriptive parameters for the whole sequences for all the test recordings.

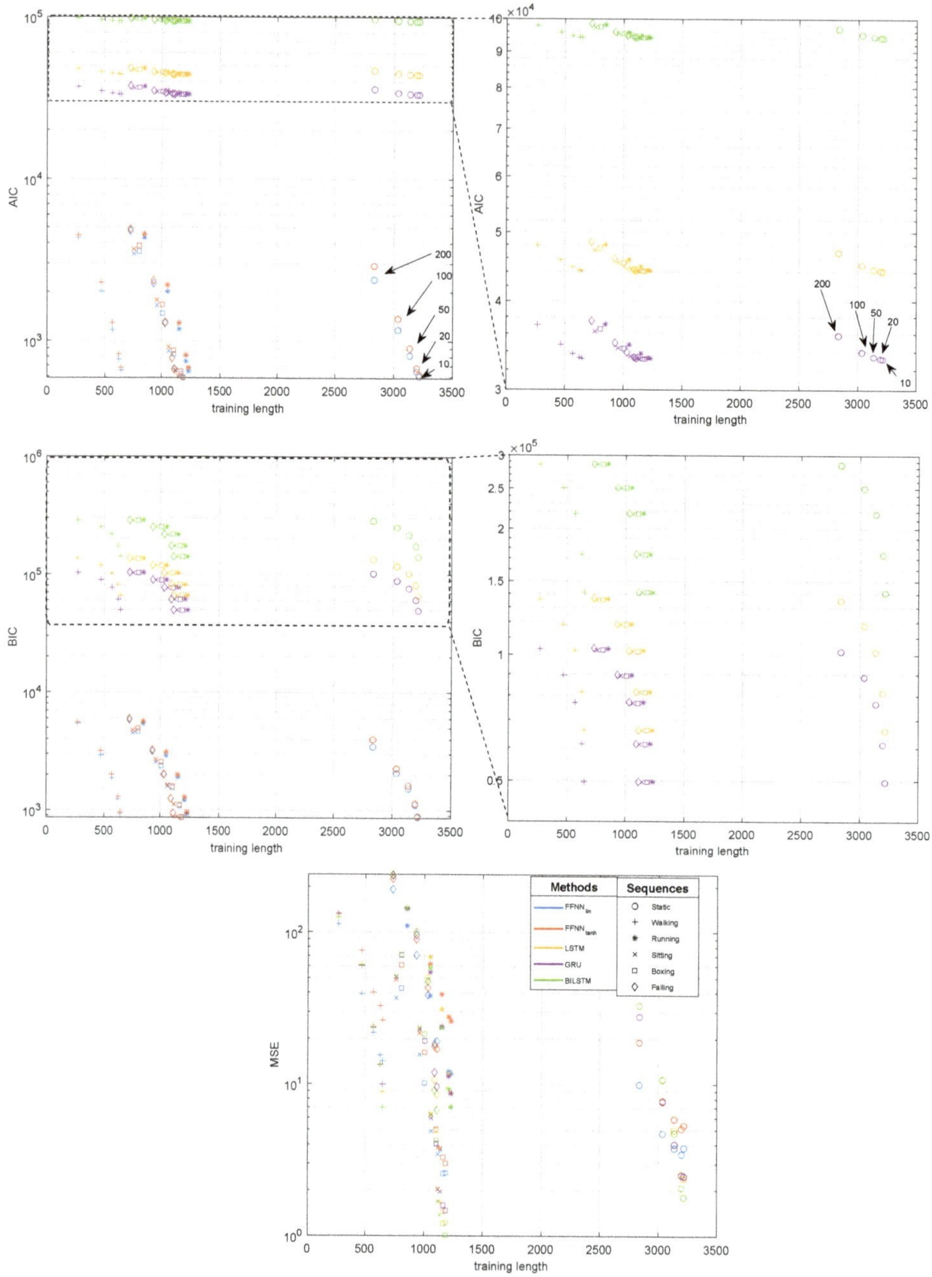

Figure 9. Influence of training sequence length on the quality of obtained results for NN methods: Akaike Information Criterion (AIC), Bayesian Information Criterion (BIC) and MSE.

Knowing that correlation, as a statistical measure, makes little sense for a sparse dataset, we treat it as a kind of measurement of co-linearity between the measures. However, for part of the parameters, the (high) correlation values are connected, with quite satisfactory low p-values; these are given in Appendix B.

Table 5. Correlation between RMSE and sequence parameters (averaged for all gap sizes).

	$FFNN_{lin}$	$FFNN_{tanh}$	LSTM	GRU	BILSTM	LIN	SPLINE	MAKIMA	PCHIP	mSVD
Entropy	0.708	0.793	0.775	0.736	0.735	0.680	0.486	0.624	0.630	0.680
Stddev	0.741	0.892	0.805	0.781	0.778	0.706	0.517	0.653	0.631	0.706
Velocity	0.744	0.886	0.813	0.784	0.781	0.713	0.521	0.656	0.640	0.713
Acceleration	0.905	0.912	0.903	0.907	0.890	0.854	0.791	0.844	0.818	0.854
Jerk	0.803	0.794	0.777	0.799	0.779	0.753	0.758	0.763	0.725	0.753
Monotonicity	0.900	0.713	0.798	0.847	0.819	0.824	0.926	0.888	0.862	0.824
Complexity	−0.779	−0.886	−0.815	−0.804	−0.794	−0.742	−0.589	−0.702	−0.670	−0.742

Table 6. Correlation between sequence parameters.

	Entropy	Stddev	Velocity	Acceleration	Jerk	Monotonicity	Complexity
Entropy	1.000	0.869	0.898	0.730	0.459	0.465	−0.712
Stddev	0.869	1.000	0.992	0.879	0.732	0.501	−0.949
Velocity	0.898	0.992	1.000	0.890	0.731	0.477	−0.929
Acceleration	0.730	0.879	0.890	1.000	0.941	0.735	−0.913
Jerk	0.459	0.732	0.731	0.941	1.000	0.695	−0.847
Monotonicity	0.465	0.501	0.477	0.735	0.695	1.000	−0.560
Complexity	−0.712	−0.949	−0.929	−0.913	−0.847	−0.560	1.000
				p-values			
Entropy	1.000	0.025	0.015	0.100	0.360	0.353	0.112
Stddev	0.025	1.000	0.000	0.021	0.098	0.311	0.004
Velocity	0.015	0.000	1.000	0.017	0.099	0.338	0.007
Acceleration	0.100	0.021	0.017	1.000	0.005	0.096	0.011
Jerk	0.360	0.098	0.099	0.005	1.000	0.125	0.033
Monotonicity	0.353	0.311	0.338	0.096	0.125	1.000	0.248
Complexity	0.112	0.004	0.007	0.011	0.033	0.248	1.000

Looking into the results in Table 5, we observe that all the considered sequence parameters are related, to some extent, to RMSE. However, for all the gap-filling methods, we identified two key parameters that have higher CCs than the others. These are acceleration and monotonicity, which seem to be promising candidate measures for describing the susceptibility of sequences to the employed reconstruction methods.

Regarding inter-parameter correlations in Table 6, we can observe that most of the measures are correlated with each other. This is expected, since kinematic/dynamic parameters are connected with the location of the markers over time, so values such as entropy, position standard deviation, velocity, acceleration, and jerk are correlated (for the derivatives, the smaller the difference in the derivative order, the higher the CCs).

On the other hand, the two less typical measures, monotonicity and complexity, are different; therefore, their correlation with the other measures is less predictable. Complexity appeared to have a notable negative correlation with most of the typical measures. Monotonicity, on the other hand, is more interesting. Since it is only moderately correlated with remaining measures, it still has quite a high CC, with RMSEs for all the gap reconstruction methods. Therefore, we can suppose this describes an aspect of the sequence that is independent of the other measures, which is related to susceptibility to the gap reconstruction procedures.

5. Summary

In this article, we addressed the issue of filling the gaps that occurred in the mocap signal. We considered this to be a regressive problem and reviewed the results of several NN-based regressors, which were compared with several interpolation and low-rank matrix completion (mSVD) methods.

Generally, in the case of short gaps, the interpolation methods returned the best results, but since the gaps became longer, part of the NNs gained an advantage. We reviewed five variants of neural networks. Surprisingly, the tests revealed that simple linear FFNNs, using momentary (current and previous sample) and local (from neighboring markers) coordinates as input data, outperformed quite advanced recurrent NNs for the longer gaps. For the shorter gaps, RNNs offered better results, but all the NNs were outperformed by interpolations. The boundary between 'long' and 'short' terms are gaps of 50 samples long. Finally, we were able to identify which factors of the input mocap sequence influence the reconstruction errors.

The approach to the NNs given here does not incorporate skeletal information. Instead, the kinematic structure is based on the FBM framework and all the predictions are performed with the local data, as obtained from FBM. Currently, none of the analyzed approaches considered body constraints such as limb length or size, but we can easily obtain such information from the FBM model. We plan to apply this as an additional processing stage in the future. In the future, we plan to test more sophisticated NN architectures, such as combined LSTM convolution, or averaged multiregressions.

Supplementary Materials: The following are available at https://www.mdpi.com/1424-8220/21/18/6115/s1, The motion capture sequences.

Author Contributions: conceptualization, P.S.; methodology, P.S., M.P.; software, P.S., M.P.; investigation, P.S.; resources, M.P.; data curation, M.P.; writing—original draft preparation, P.S.; writing—review and editing, P.S., M.P.; visualization, P.S. All authors have read and agreed to the published version of the manuscript.

Funding: The research described in the paper was performed within the statutory project of the Department of Graphics, Computer Vision and Digital Systems at the Silesian University of Technology, Gliwice (RAU-6, 2021). APC were covered from statutory research funds. M.P. was supported by grant no WND-RPSL.01.02.00-24-00AC/19-011 funded by under the Regional Operational Programme of the Silesia Voivodeship in the years 2014–2020.

Institutional Review Board Statement: Not applicable.

Informed Consent Statement: Not applicable.

Data Availability Statement: The motion capture sequences are provided as Supplementary Files accompanying the article.

Acknowledgments: The research was supported with motion data by Human Motion Laboratory of Polish-Japanese Academy of Information Technology.

Conflicts of Interest: The authors declare no conflict of interest. The funders had no role in the design of the study; in the collection, analyses, or interpretation of data; in the writing of the manuscript, or in the decision to publish the results.

Abbreviations

The following abbreviations are used in this manuscript:

BILSTM	bidirectional LSTM
CC	correlation coefficient
FC	fully connected
FBM	functional body mesh
FFNN	feed forward neural network
GRU	gated recurrent unit
HML	Human Motion Laboratory
IK	inverse kinematics
KF	Kalman filter
LS	least squares
LSTM	long-short term memory
Mocap	MOtion CAPture
MSE	Mean Square Error
NARX-NN	nonlinear autoregressive exogenous neural network

NaN	not a number
NN	neural network
OMC	optical motion capture
PCA	principal component analysis
PJAIT	Polish-Japanese Academy of Information Technology
RMSE	root mean squared error
RNN	recurrent neural network
STDDEV	standard deviation
SVD	singular value decomposition

Appendix A. Performance Results for All Sequences

Table A1. Quality measures for the walking (No. 2) sequence.

Len		$FFNN_{lin}$	$FFNN_{tanh}$	LSTM	GRU	BILSTM	LIN	SPLINE	MAKIMA	PCHIP	mSVD
10	RMSE	14.222	26.428	8.844	9.932	7.004	5.088	1.287	2.464	2.507	5.088
	mean($RMSE_k$)	12.398	23.213	7.659	9.014	6.495	3.442	0.810	1.621	1.697	3.442
	median($RMSE_k$)	10.865	21.290	6.327	8.262	5.956	2.051	0.511	1.087	1.180	2.051
	mode($RMSE_k$)	3.499	4.068	1.755	1.140	2.344	0.536	0.056	0.237	0.239	0.536
	stddev($RMSE_k$)	6.930	12.645	4.634	4.744	3.371	3.505	0.938	1.773	1.788	3.505
	iqr($RMSE_k$)	8.986	12.914	3.644	4.297	3.444	3.180	0.652	1.293	1.334	3.180
20	RMSE	15.490	32.802	13.491	13.382	13.303	12.274	4.031	6.590	6.619	12.274
	mean($RMSE_k$)	13.743	27.978	10.155	11.171	8.396	9.071	2.591	4.798	4.904	9.071
	median($RMSE_k$)	12.334	24.575	7.568	9.116	6.209	6.508	1.823	3.767	3.728	6.508
	mode($RMSE_k$)	2.654	5.774	3.242	5.247	2.352	0.401	0.314	0.316	0.382	0.401
	stddev($RMSE_k$)	6.723	16.042	8.161	6.609	8.827	8.020	2.828	4.290	4.175	8.020
	iqr($RMSE_k$)	7.454	15.726	4.545	5.667	2.491	6.791	1.571	3.308	3.921	6.791
50	RMSE	21.907	40.375	24.343	23.833	23.434	42.517	21.474	26.332	25.995	42.517
	mean($RMSE_k$)	19.168	36.769	19.788	19.867	18.831	33.944	16.673	21.757	21.607	33.944
	median($RMSE_k$)	16.432	32.752	15.196	15.655	14.926	23.652	12.952	16.134	15.996	23.652
	mode($RMSE_k$)	5.905	13.574	6.336	7.173	6.100	5.500	4.293	3.782	3.921	5.500
	stddev($RMSE_k$)	10.486	16.289	13.174	12.408	13.037	25.484	12.659	14.438	13.993	25.484
	iqr($RMSE_k$)	12.421	22.207	13.308	11.413	12.903	29.918	12.189	17.991	18.129	29.918
100	RMSE	39.346	75.817	61.641	60.420	60.823	76.058	58.357	62.302	62.419	76.058
	mean($RMSE_k$)	32.287	66.701	50.195	49.019	49.453	63.445	46.476	50.803	50.693	63.445
	median($RMSE_k$)	23.318	56.329	38.960	37.001	39.074	51.683	35.447	40.065	40.418	51.683
	mode($RMSE_k$)	8.122	22.940	14.125	15.094	14.334	12.943	12.407	12.074	12.493	12.943
	stddev($RMSE_k$)	22.397	35.709	35.107	34.707	34.685	41.564	34.503	35.371	35.700	41.564
	iqr($RMSE_k$)	18.933	41.446	39.727	40.427	40.813	63.062	39.062	49.784	50.440	63.062
200	RMSE	112.933	134.121	127.416	132.150	124.566	79.741	105.237	79.407	80.457	79.741
	mean($RMSE_k$)	87.084	121.229	108.733	111.164	107.192	75.307	91.826	69.585	70.031	75.307
	median($RMSE_k$)	59.288	104.710	91.987	89.523	91.019	68.567	80.427	63.559	61.704	68.567
	mode($RMSE_k$)	26.007	46.150	23.032	23.675	22.813	42.408	21.841	21.984	21.602	42.408
	stddev($RMSE_k$)	71.160	57.197	66.944	71.470	63.693	26.502	53.401	39.296	40.746	26.502
	iqr($RMSE_k$)	61.864	71.116	90.839	90.285	90.685	42.057	88.500	65.862	66.873	42.057

Table A2. Quality measures for the running (No. 3) sequence.

Len		$FFNN_{lin}$	$FFNN_{tanh}$	LSTM	GRU	BILSTM	LIN	SPLINE	MAKIMA	PCHIP	mSVD
10	RMSE	11.702	25.988	8.748	8.666	7.066	3.001	0.701	1.291	1.259	3.001
	mean($RMSE_k$)	9.939	23.049	7.675	7.581	6.105	2.221	0.476	0.985	0.942	2.221
	median($RMSE_k$)	8.661	20.122	6.973	6.485	5.540	1.743	0.346	0.831	0.720	1.743
	mode($RMSE_k$)	1.933	6.022	1.838	1.236	1.106	0.234	0.079	0.149	0.151	0.234
	stddev($RMSE_k$)	5.919	11.837	4.214	4.245	3.797	1.714	0.439	0.692	0.691	1.714
	iqr($RMSE_k$)	7.005	15.692	5.106	4.850	3.513	1.835	0.286	0.835	0.799	1.835

Table A2. *Cont.*

Len		FFNN$_{lin}$	FFNN$_{tanh}$	LSTM	GRU	BILSTM	LIN	SPLINE	MAKIMA	PCHIP	mSVD
20	RMSE	12.141	27.729	11.594	11.232	9.321	7.397	1.742	3.401	3.439	7.397
	mean(RMSE$_k$)	10.331	25.124	9.324	9.440	6.919	5.676	1.274	2.601	2.589	5.676
	median(RMSE$_k$)	8.695	23.641	7.664	7.948	5.424	4.496	0.968	1.988	1.853	4.496
	mode(RMSE$_k$)	2.547	6.946	2.438	1.512	1.953	0.661	0.237	0.453	0.438	0.661
	stddev(RMSE$_k$)	6.215	11.425	6.552	5.753	5.787	4.017	1.010	1.889	2.021	4.017
	iqr(RMSE$_k$)	8.168	12.490	4.481	5.442	3.111	3.995	1.017	2.154	2.061	3.995
50	RMSE	23.573	39.084	31.147	24.057	23.597	34.144	12.857	19.473	21.328	34.144
	mean(RMSE$_k$)	14.767	31.801	17.835	15.504	14.637	27.624	8.608	14.842	16.431	27.624
	median(RMSE$_k$)	9.523	25.412	10.904	10.501	8.853	25.122	6.834	12.894	13.844	25.122
	mode(RMSE$_k$)	3.229	9.379	4.119	2.888	3.306	2.559	0.896	1.291	1.737	2.559
	stddev(RMSE$_k$)	18.345	22.596	25.456	18.049	18.231	18.865	8.914	11.837	12.760	18.865
	iqr(RMSE$_k$)	6.432	16.838	6.719	7.811	7.903	20.224	6.920	9.883	11.590	20.224
100	RMSE	38.173	61.656	68.606	54.639	58.223	94.347	45.740	58.606	62.724	94.347
	mean(RMSE$_k$)	25.165	49.288	44.780	40.344	42.251	83.854	37.303	51.072	55.958	83.854
	median(RMSE$_k$)	18.493	41.944	33.811	31.168	32.177	77.220	32.103	46.438	50.903	77.220
	mode(RMSE$_k$)	4.901	11.780	8.178	5.555	4.181	4.989	4.549	3.554	3.884	4.989
	stddev(RMSE$_k$)	27.594	35.231	50.041	35.158	38.271	41.350	25.286	27.575	27.272	41.350
	iqr(RMSE$_k$)	13.060	29.863	24.844	24.922	25.449	47.512	25.816	26.432	29.725	47.512
200	RMSE	110.196	145.641	145.387	143.360	145.050	248.552	138.231	167.249	199.417	248.552
	mean(RMSE$_k$)	88.708	129.262	125.767	123.634	125.213	235.787	119.848	146.780	185.085	235.787
	median(RMSE$_k$)	70.845	113.902	108.387	105.181	107.987	233.618	103.952	128.657	171.109	233.618
	mode(RMSE$_k$)	20.092	53.434	39.113	39.722	38.728	96.336	38.444	36.027	74.145	96.336
	stddev(RMSE$_k$)	63.969	65.135	70.990	70.695	71.285	73.293	66.963	77.021	70.628	73.293
	iqr(RMSE$_k$)	67.200	73.343	87.747	82.080	89.947	77.986	83.010	64.869	47.085	77.986

Table A3. Quality measures for the sitting (No. 4) sequence.

Len		FFNN$_{lin}$	FFNN$_{tanh}$	LSTM	GRU	BILSTM	LIN	SPLINE	MAKIMA	PCHIP	mSVD
10	RMSE	3.701	3.792	1.664	1.954	1.373	1.697	0.711	0.841	0.839	1.697
	mean(RMSE$_k$)	3.272	3.386	1.463	1.737	1.210	1.218	0.478	0.617	0.606	1.218
	median(RMSE$_k$)	2.996	2.987	1.351	1.682	1.108	0.948	0.339	0.475	0.429	0.948
	mode(RMSE$_k$)	0.437	0.558	0.197	0.212	0.249	0.072	0.059	0.041	0.043	0.072
	stddev(RMSE$_k$)	1.896	1.767	0.806	0.846	0.642	1.094	0.483	0.530	0.537	1.094
	iqr(RMSE$_k$)	2.282	2.025	0.991	1.301	0.702	1.049	0.260	0.467	0.480	1.049
20	RMSE	3.464	3.829	2.060	2.025	1.688	3.902	1.285	1.904	2.029	3.902
	mean(RMSE$_k$)	3.106	3.429	1.708	1.797	1.475	3.057	0.942	1.515	1.559	3.057
	median(RMSE$_k$)	2.911	3.319	1.519	1.572	1.318	2.434	0.739	1.230	1.169	2.434
	mode(RMSE$_k$)	0.522	0.497	0.300	0.240	0.271	0.211	0.126	0.155	0.161	0.211
	stddev(RMSE$_k$)	1.577	1.750	1.122	0.962	0.812	2.415	0.838	1.153	1.311	2.415
	iqr(RMSE$_k$)	2.233	2.263	1.038	1.069	0.934	2.762	0.781	0.979	0.995	2.762
20	RMSE	4.901	6.291	6.392	5.952	6.255	15.596	6.334	9.332	10.056	15.596
	mean(RMSE$_k$)	4.383	5.355	5.064	4.697	4.895	12.767	4.902	7.260	7.710	12.767
	median(RMSE$_k$)	3.982	4.831	4.007	3.623	3.803	11.036	3.652	5.788	6.343	11.036
	mode(RMSE$_k$)	0.482	0.417	0.313	0.422	0.277	0.267	0.332	0.267	0.240	0.267
	stddev(RMSE$_k$)	2.276	3.254	3.793	3.568	3.778	8.741	3.880	5.667	6.265	8.741
	iqr(RMSE$_k$)	2.978	3.833	5.160	4.098	4.999	11.116	5.269	6.546	6.801	11.116
20	RMSE	15.716	21.780	23.727	23.023	23.575	38.083	23.547	28.358	28.813	38.083
	mean(RMSE$_k$)	11.904	16.468	18.440	17.539	18.222	33.439	18.245	23.435	24.033	33.439
	median(RMSE$_k$)	8.596	13.132	15.903	14.109	15.147	30.517	15.467	20.365	20.691	30.517
	mode(RMSE$_k$)	0.643	0.711	0.927	0.743	0.950	1.324	1.170	1.139	1.121	1.324
	stddev(RMSE$_k$)	9.980	13.839	14.495	14.484	14.524	17.840	14.459	15.569	15.542	17.840
	iqr(RMSE$_k$)	7.816	11.087	13.380	12.476	13.201	23.419	13.054	15.405	14.280	23.419

Table A3. *Cont.*

Len		$FFNN_{lin}$	$FFNN_{tanh}$	LSTM	GRU	BILSTM	LIN	SPLINE	MAKIMA	PCHIP	mSVD
20	RMSE	37.101	48.909	51.388	50.842	51.274	72.745	51.478	59.839	59.857	72.745
	mean($RMSE_k$)	31.439	41.811	44.331	43.711	44.219	66.280	44.321	54.030	54.156	66.280
	median($RMSE_k$)	26.422	36.792	40.178	39.257	40.099	71.201	39.395	55.235	54.311	71.201
	mode($RMSE_k$)	1.783	2.342	2.592	2.372	2.558	0.972	2.819	0.875	0.912	0.972
	stddev($RMSE_k$)	20.198	25.924	26.514	26.496	26.480	30.443	26.659	26.183	26.001	30.443
	iqr($RMSE_k$)	22.947	30.188	29.510	29.617	29.241	37.209	29.316	29.572	28.094	37.209

Table A4. Quality measures for the boxing (No. 5) sequence.

Len		$FFNN_{lin}$	$FFNN_{tanh}$	LSTM	GRU	BILSTM	LIN	SPLINE	MAKIMA	PCHIP	mSVD
10	RMSE	2.603	3.006	1.217	1.467	1.008	1.175	0.986	0.668	0.735	1.175
	mean($RMSE_k$)	2.321	2.697	1.087	1.316	0.885	0.848	0.484	0.461	0.507	0.848
	median($RMSE_k$)	2.036	2.476	1.001	1.173	0.783	0.666	0.276	0.317	0.322	0.666
	mode($RMSE_k$)	0.505	0.309	0.270	0.303	0.218	0.036	0.043	0.035	0.034	0.036
	stddev($RMSE_k$)	1.174	1.354	0.521	0.613	0.456	0.712	0.765	0.420	0.473	0.712
	iqr($RMSE_k$)	1.449	1.769	0.504	0.705	0.542	0.709	0.307	0.341	0.490	0.709
20	RMSE	2.581	3.298	1.446	1.591	1.200	3.534	1.157	1.648	2.021	3.534
	mean($RMSE_k$)	2.295	3.030	1.341	1.458	1.070	2.818	0.797	1.309	1.519	2.818
	median($RMSE_k$)	2.022	2.780	1.242	1.353	0.934	2.282	0.608	0.983	1.071	2.282
	mode($RMSE_k$)	0.826	0.700	0.326	0.402	0.303	0.273	0.106	0.125	0.126	0.273
	stddev($RMSE_k$)	1.161	1.308	0.541	0.606	0.549	1.965	0.819	0.930	1.249	1.965
	iqr($RMSE_k$)	1.415	1.704	0.736	0.732	0.494	2.153	0.491	1.038	1.333	2.153
50	RMSE	4.045	5.038	4.965	4.067	4.295	14.095	3.956	7.248	9.171	14.095
	mean($RMSE_k$)	3.211	4.183	3.609	3.109	3.306	11.957	3.262	6.083	7.562	11.957
	median($RMSE_k$)	2.546	3.503	2.661	2.500	2.634	10.384	2.736	5.271	6.318	10.384
	mode($RMSE_k$)	0.699	1.102	0.699	0.538	0.480	0.444	0.542	0.513	0.546	0.444
	stddev($RMSE_k$)	2.460	2.788	3.404	2.614	2.747	7.236	2.235	3.802	4.994	7.236
	iqr($RMSE_k$)	1.743	1.595	1.968	1.540	2.062	9.821	2.059	5.074	7.302	9.821
100	RMSE	10.134	16.216	21.424	19.275	21.386	36.436	21.538	27.723	30.374	36.436
	mean($RMSE_k$)	8.175	13.241	17.438	15.384	17.357	31.336	17.608	23.779	26.421	31.336
	median($RMSE_k$)	6.398	11.337	14.702	12.285	14.627	27.834	14.823	22.008	24.825	27.834
	mode($RMSE_k$)	0.864	1.156	1.085	0.973	1.090	0.514	0.912	0.632	0.490	0.514
	stddev($RMSE_k$)	5.837	9.220	12.123	11.372	12.169	18.465	12.075	14.128	14.876	18.465
	iqr($RMSE_k$)	6.261	12.033	16.415	16.672	16.414	25.577	16.361	18.637	19.008	25.577
200	RMSE	42.833	60.847	71.465	70.625	71.514	64.829	72.201	60.721	61.704	64.829
	mean($RMSE_k$)	36.693	54.330	64.743	63.732	64.805	61.507	65.477	56.493	57.666	61.507
	median($RMSE_k$)	33.631	50.764	61.017	60.170	61.057	60.782	62.218	55.492	57.030	60.782
	mode($RMSE_k$)	4.592	9.116	10.042	9.788	9.974	8.998	10.077	8.616	8.609	8.998
	stddev($RMSE_k$)	21.768	26.954	29.620	29.819	29.609	20.171	29.798	22.097	21.740	20.171
	iqr($RMSE_k$)	21.992	31.408	36.039	35.205	36.075	24.945	36.485	29.814	28.505	24.945

Table A5. Quality measures for the falling (No. 6) sequence.

Len		$FFNN_{lin}$	$FFNN_{tanh}$	LSTM	GRU	BILSTM	LIN	SPLINE	MAKIMA	PCHIP	mSVD
10	RMSE	19.193	17.106	8.537	9.585	6.720	5.763	1.601	2.872	3.365	5.763
	mean($RMSE_k$)	15.455	15.022	7.818	8.772	6.166	3.827	0.994	1.851	1.968	3.827
	median($RMSE_k$)	13.186	13.571	6.947	8.341	5.616	2.359	0.618	1.107	1.145	2.359
	mode($RMSE_k$)	2.760	3.139	2.310	2.880	2.110	0.244	0.105	0.145	0.149	0.244
	stddev($RMSE_k$)	11.270	8.163	3.494	3.852	2.555	4.023	1.138	2.039	2.551	4.023
	iqr($RMSE_k$)	9.203	10.174	3.101	4.009	3.520	3.723	0.789	1.795	1.813	3.723

Table A5. *Cont.*

Len		FFNN$_{lin}$	FFNN$_{tanh}$	LSTM	GRU	BILSTM	LIN	SPLINE	MAKIMA	PCHIP	mSVD
20	RMSE	18.496	17.762	10.664	11.914	9.057	15.278	6.073	9.213	9.596	15.278
	mean(RMSE$_k$)	16.206	16.199	8.940	10.261	7.897	10.937	3.694	5.981	6.392	10.937
	median(RMSE$_k$)	14.108	14.897	8.130	9.530	7.106	7.613	2.089	3.687	4.319	7.613
	mode(RMSE$_k$)	4.659	2.388	2.143	1.822	2.821	0.953	0.339	0.756	0.832	0.953
	stddev(RMSE$_k$)	8.511	7.184	6.211	5.915	4.455	9.596	4.383	6.169	6.567	9.596
	iqr(RMSE$_k$)	9.496	8.219	4.321	5.520	3.642	10.133	3.402	5.298	5.154	10.133
50	RMSE	38.618	43.058	50.077	47.367	47.474	60.232	46.220	42.945	44.782	60.232
	mean(RMSE$_k$)	28.149	30.795	32.292	30.356	30.213	43.423	28.314	29.543	31.603	43.423
	median(RMSE$_k$)	18.927	18.873	16.214	15.491	14.312	29.262	14.172	17.724	19.705	29.262
	mode(RMSE$_k$)	5.585	3.883	3.061	4.112	2.789	4.507	1.345	2.710	2.615	4.507
	stddev(RMSE$_k$)	25.916	29.395	37.233	35.345	35.587	42.053	35.660	31.333	31.914	42.053
	iqr(RMSE$_k$)	15.417	17.126	31.871	21.094	29.043	38.828	27.009	24.239	28.168	38.828
100	RMSE	70.671	89.650	100.005	95.523	100.282	125.495	98.770	92.878	97.573	125.495
	mean(RMSE$_k$)	55.641	72.172	81.983	76.503	81.814	104.667	81.794	76.620	81.261	104.667
	median(RMSE$_k$)	42.728	57.532	66.468	62.277	66.311	86.119	68.990	59.710	66.757	86.119
	mode(RMSE$_k$)	7.967	8.688	10.247	7.912	9.749	7.809	9.146	6.892	7.449	7.809
	stddev(RMSE$_k$)	43.593	53.283	57.286	57.268	58.033	69.796	55.712	52.857	54.185	69.796
	iqr(RMSE$_k$)	52.533	72.029	82.980	85.218	86.618	85.060	92.529	71.859	75.211	85.060
200	RMSE	192.371	224.989	240.459	237.068	240.104	219.332	238.962	182.390	177.973	219.332
	mean(RMSE$_k$)	168.542	199.701	214.118	209.626	213.731	198.998	212.497	165.908	161.330	198.998
	median(RMSE$_k$)	145.399	185.636	190.954	187.446	191.565	196.704	189.560	169.458	163.676	196.704
	mode(RMSE$_k$)	43.924	47.226	58.636	49.156	60.128	38.386	60.706	33.396	32.207	38.386
	stddev(RMSE$_k$)	92.157	103.406	108.703	110.129	108.684	94.898	108.542	77.915	77.491	94.898
	iqr(RMSE$_k$)	102.432	114.007	119.186	116.515	119.440	153.708	117.857	121.289	120.480	153.708

Appendix B. Correlations between RMSE an Sequence Parameters

Table A6. Correlation between RMSE and entropy of input sequence.

Len	FFNN$_{lin}$	FFNN$_{tanh}$	LSTM	GRU	BILSTM	LIN	SPLINE	MAKIMA	PCHIP	mSVD
10	0.741	0.878	0.890	0.849	0.890	0.614	0.261	0.552	0.520	0.614
20	0.760	0.827	0.852	0.842	0.790	0.608	0.466	0.550	0.533	0.608
50	0.744	0.851	0.740	0.678	0.670	0.660	0.503	0.603	0.608	0.660
100	0.639	0.719	0.724	0.649	0.662	0.742	0.576	0.661	0.679	0.742
200	0.658	0.691	0.667	0.665	0.664	0.777	0.626	0.756	0.812	0.777
10	0.092	0.021	0.017	0.033	0.017	0.195	0.617	0.256	0.290	0.195
20	0.080	0.042	0.031	0.036	0.061	0.200	0.352	0.258	0.276	0.200
50	0.090	0.032	0.093	0.139	0.146	0.153	0.309	0.205	0.200	0.153
100	0.172	0.108	0.104	0.163	0.152	0.091	0.231	0.153	0.138	0.091
200	0.155	0.129	0.148	0.150	0.150	0.069	0.184	0.082	0.050	0.069

Table A7. Correlation between RMSE and standard deviation of input sequence.

Len	FFNN$_{lin}$	FFNN$_{tanh}$	LSTM	GRU	BILSTM	LIN	SPLINE	MAKIMA	PCHIP	mSVD
10	0.775	0.997	0.950	0.928	0.956	0.729	0.419	0.668	0.586	0.729
20	0.823	0.986	0.969	0.943	0.948	0.688	0.505	0.595	0.556	0.688
50	0.736	0.924	0.703	0.661	0.645	0.718	0.479	0.627	0.614	0.718
100	0.673	0.833	0.755	0.708	0.698	0.747	0.611	0.718	0.707	0.747
200	0.696	0.719	0.649	0.667	0.641	0.648	0.570	0.659	0.694	0.648

Table A7. *Cont.*

Len	$FFNN_{lin}$	$FFNN_{tanh}$	LSTM	GRU	BILSTM	LIN	SPLINE	MAKIMA	PCHIP	mSVD
10	0.070	0.000	0.004	0.007	0.003	0.100	0.408	0.147	0.222	0.100
20	0.044	0.000	0.001	0.005	0.004	0.131	0.307	0.213	0.252	0.131
50	0.095	0.008	0.119	0.153	0.167	0.108	0.336	0.183	0.195	0.108
100	0.143	0.040	0.083	0.115	0.123	0.088	0.198	0.108	0.116	0.088
200	0.125	0.107	0.163	0.148	0.170	0.164	0.237	0.155	0.126	0.164

Table A8. Correlation between RMSE and velocity of input sequence.

Len	$FFNN_{lin}$	$FFNN_{tanh}$	LSTM	GRU	BILSTM	LIN	SPLINE	MAKIMA	PCHIP	mSVD
10	0.768	0.983	0.943	0.915	0.950	0.701	0.419	0.640	0.564	0.701
20	0.812	0.962	0.950	0.927	0.916	0.669	0.486	0.576	0.540	0.669
50	0.749	0.921	0.724	0.672	0.657	0.716	0.478	0.624	0.619	0.716
100	0.681	0.825	0.772	0.715	0.709	0.771	0.615	0.728	0.723	0.771
200	0.712	0.742	0.679	0.694	0.673	0.710	0.609	0.714	0.755	0.710
10	0.074	0.000	0.005	0.011	0.004	0.121	0.409	0.172	0.243	0.121
20	0.050	0.002	0.004	0.008	0.010	0.146	0.328	0.231	0.269	0.146
50	0.087	0.009	0.104	0.143	0.157	0.110	0.338	0.186	0.190	0.110
100	0.136	0.043	0.072	0.111	0.115	0.072	0.194	0.101	0.104	0.072
200	0.112	0.091	0.138	0.126	0.143	0.114	0.199	0.111	0.083	0.114

Table A9. Correlation between RMSE and acceleration of input sequence.

Len	$FFNN_{lin}$	$FFNN_{tanh}$	LSTM	GRU	BILSTM	LIN	SPLINE	MAKIMA	PCHIP	mSVD
10	0.901	0.867	0.917	0.928	0.922	0.879	0.775	0.853	0.806	0.879
20	0.923	0.853	0.916	0.932	0.894	0.870	0.754	0.806	0.779	0.870
50	0.896	0.952	0.870	0.858	0.846	0.909	0.740	0.845	0.844	0.909
100	0.886	0.960	0.928	0.916	0.907	0.914	0.858	0.926	0.916	0.914
200	0.918	0.929	0.884	0.901	0.879	0.699	0.830	0.789	0.745	0.699
10	0.014	0.025	0.010	0.008	0.009	0.021	0.070	0.031	0.053	0.021
20	0.009	0.031	0.010	0.007	0.016	0.024	0.083	0.053	0.068	0.024
50	0.016	0.003	0.024	0.029	0.034	0.012	0.093	0.034	0.034	0.012
100	0.019	0.002	0.008	0.010	0.012	0.011	0.029	0.008	0.010	0.011
200	0.010	0.007	0.019	0.014	0.021	0.122	0.041	0.062	0.089	0.122

Table A10. Correlation between RMSE and jerk of input sequence.

Len	$FFNN_{lin}$	$FFNN_{tanh}$	LSTM	GRU	BILSTM	LIN	SPLINE	MAKIMA	PCHIP	mSVD
10	0.784	0.711	0.752	0.785	0.760	0.823	0.861	0.818	0.765	0.823
20	0.811	0.723	0.778	0.798	0.784	0.810	0.720	0.750	0.720	0.810
50	0.772	0.813	0.736	0.749	0.737	0.833	0.674	0.770	0.766	0.833
100	0.806	0.881	0.826	0.846	0.827	0.797	0.800	0.855	0.830	0.797
200	0.843	0.843	0.791	0.816	0.785	0.502	0.736	0.625	0.546	0.502
10	0.065	0.113	0.084	0.064	0.080	0.044	0.028	0.047	0.076	0.044
20	0.050	0.104	0.068	0.057	0.065	0.051	0.107	0.086	0.106	0.051
50	0.072	0.049	0.095	0.086	0.095	0.040	0.142	0.073	0.076	0.040
100	0.053	0.020	0.043	0.034	0.042	0.057	0.056	0.030	0.041	0.057
200	0.035	0.035	0.061	0.048	0.064	0.311	0.095	0.185	0.262	0.311

Table A11. Correlation between RMSE and monotonicity of input sequence.

Len	$FFNN_{lin}$	$FFNN_{tanh}$	LSTM	GRU	BILSTM	LIN	SPLINE	MAKIMA	PCHIP	mSVD
10	0.918	0.533	0.722	0.781	0.709	0.952	0.866	0.971	0.993	0.952
20	0.898	0.529	0.694	0.759	0.703	0.971	0.999	0.993	0.996	0.971
50	0.883	0.774	0.857	0.914	0.918	0.937	0.974	0.965	0.953	0.937
100	0.908	0.873	0.853	0.908	0.904	0.817	0.951	0.897	0.890	0.817
200	0.892	0.858	0.866	0.871	0.862	0.441	0.842	0.612	0.476	0.441
10	0.010	0.276	0.106	0.067	0.115	0.003	0.026	0.001	0.000	0.003
20	0.015	0.281	0.126	0.080	0.119	0.001	0.000	0.000	0.000	0.001
50	0.020	0.071	0.029	0.011	0.010	0.006	0.001	0.002	0.003	0.006
100	0.012	0.023	0.031	0.012	0.013	0.047	0.003	0.015	0.018	0.047
200	0.017	0.029	0.026	0.024	0.027	0.381	0.036	0.196	0.340	0.381

Table A12. Correlation between RMSE and complexity of input sequence.

Len	$FFNN_{lin}$	$FFNN_{tanh}$	LSTM	GRU	BILSTM	LIN	SPLINE	MAKIMA	PCHIP	mSVD
10	−0.795	−0.937	−0.913	−0.906	−0.922	−0.781	−0.532	−0.729	−0.645	−0.781
20	−0.837	−0.931	−0.936	−0.920	−0.919	−0.733	−0.568	−0.644	−0.599	−0.733
50	−0.763	−0.914	−0.730	−0.703	−0.687	−0.770	−0.544	−0.685	−0.670	−0.770
100	−0.744	−0.878	−0.802	−0.775	−0.758	−0.787	−0.682	−0.780	−0.759	−0.787
200	−0.754	−0.769	−0.692	−0.714	−0.685	−0.637	−0.618	−0.673	−0.675	−0.637
10	0.059	0.006	0.011	0.013	0.009	0.067	0.278	0.100	0.167	0.067
20	0.038	0.007	0.006	0.009	0.010	0.097	0.239	0.167	0.209	0.097
50	0.078	0.011	0.099	0.119	0.131	0.074	0.265	0.134	0.145	0.074
100	0.090	0.021	0.055	0.070	0.081	0.063	0.135	0.067	0.080	0.063
200	0.083	0.074	0.128	0.111	0.133	0.174	0.191	0.143	0.141	0.174

References

1. Kitagawa, M.; Windsor, B. *MoCap for Artists: Workflow and Techniques for Motion Capture*; Elsevier: Amsterdam, The Netherlands; Focal Press: Boston, MA, USA, 2008.
2. Menache, A. *Understanding Motion Capture for Computer Animation*, 2nd ed.; Morgan Kaufmann: Burlington, MA, USA, 2011.
3. Mündermann, L.; Corazza, S.; Andriacchi, T.P. The evolution of methods for the capture of human movement leading to markerless motion capture for biomechanical applications. *J. Neuroeng. Rehabil.* **2006**, *3*, 6. [CrossRef] [PubMed]
4. Szczęsna, A.; Błaszczyszyn, M.; Pawlyta, M. Optical motion capture dataset of selected techniques in beginner and advanced Kyokushin karate athletes. *Sci. Data* **2021**, *8*, 13. [CrossRef] [PubMed]
5. Świtoński, A.; Mucha, R.; Danowski, D.; Mucha, M.; Polański, A.; Cieślar, G.; Wojciechowski, K.; Sieroń, A. Diagnosis of the motion pathologies based on a reduced kinematical data of a gait. *PrzegląD Elektrotechniczny* **2011**, *87*, 173–176.
6. Lachor, M.; Świtoński, A.; Boczarska-Jedynak, M.; Kwiek, S.; Wojciechowski, K.; Polański, A. The Analysis of Correlation between MOCAP-Based and UPDRS-Based Evaluation of Gait in Parkinson's Disease Patients. In *Brain Informatics and Health*; Ślęzak, D., Tan, A.H., Peters, J.F., Schwabe, L., Eds.; Number 8609 in Lecture Notes in Computer Science; Springer International Publishing: Cham, Switzerland, 2014; pp. 335–344. [CrossRef]
7. Josinski, H.; Świtoński, A.; Stawarz, M.; Mucha, R.; Wojciechowski, K. Evaluation of rehabilitation progress of patients with osteoarthritis of the hip, osteoarthritis of the spine or after stroke using gait indices. *Przegląd Elektrotechniczny* **2013**, *89*, 279–282.
8. Windolf, M.; Götzen, N.; Morlock, M. Systematic accuracy and precision analysis of video motion capturing systems—Exemplified on the Vicon-460 system. *J. Biomech.* **2008**, *41*, 2776–2780. [CrossRef]
9. Jensenius, A.; Nymoen, K.; Skogstad, S.; Voldsund, A. A Study of the Noise-Level in Two Infrared Marker-Based Motion Capture Systems. In Proceedings of the 9th Sound and Music Computing Conference, SMC 2012, Copenhagen, Denmark, 11–14 July 2012; pp. 258–263.
10. Skurowski, P.; Pawlyta, M. On the Noise Complexity in an Optical Motion Capture Facility. *Sensors* **2019**, *19*, 4435. [CrossRef] [PubMed]
11. Skurowski, P.; Pawlyta, M. Functional Body Mesh Representation, A Simplified Kinematic Model, Its Inference and Applications. *Appl. Math. Inf. Sci.* **2016**, *10*, 71–82. [CrossRef]
12. Herda, L.; Fua, P.; Plankers, R.; Boulic, R.; Thalmann, D. Skeleton-based motion capture for robust reconstruction of human motion. In Proceedings of the Proceedings Computer Animation 2000, Philadelphia, PA, USA, 3–5 May 2000; pp. 77–83. ISSN: 1087-4844. [CrossRef]

13. Aristidou, A.; Lasenby, J. Real-time marker prediction and CoR estimation in optical motion capture. *Vis. Comput.* **2013**, *29*, 7–26. [CrossRef]
14. Perepichka, M.; Holden, D.; Mudur, S.P.; Popa, T. Robust Marker Trajectory Repair for MOCAP using Kinematic Reference. In *Motion, Interaction and Games*; Association for Computing Machinery: New York, NY, USA, 2019; MIG'19, pp. 1–10. [CrossRef]
15. Lee, J.; Shin, S.Y. A hierarchical approach to interactive motion editing for human-like figures. In Proceedings of the 26th Annual Conference on Computer Graphics and Interactive Techniques, Los Angeles, CA, USA, 8–13 August 1999; ACM Press/Addison-Wesley Publishing Co.: New York, NY, USA, 1999; pp. 39–48. [CrossRef]
16. Howarth, S.J.; Callaghan, J.P. Quantitative assessment of the accuracy for three interpolation techniques in kinematic analysis of human movement. *Comput. Methods Biomech. Biomed. Eng.* **2010**, *13*, 847–855. [CrossRef]
17. Reda, H.E.A.; Benaoumeur, I.; Kamel, B.; Zoubir, A.F. MoCap systems and hand movement reconstruction using cubic spline. In Proceedings of the 2018 5th International Conference on Control, Decision and Information Technologies (CoDIT), Thessaloniki, Greece, 10–13 April 2018; pp. 1–5. [CrossRef]
18. Liu, G.; McMillan, L. Estimation of missing markers in human motion capture. *Vis. Comput.* **2006**, *22*, 721–728. [CrossRef]
19. Lai, R.Y.Q.; Yuen, P.C.; Lee, K.K.W. Motion Capture Data Completion and Denoising by Singular Value Thresholding. In *Eurographics 2011—Short Papers*; Avis, N., Lefebvre, S., Eds.; The Eurographics Association: Geneve, Switzerland, 2011. [CrossRef]
20. Gløersen, Ø.; Federolf, P. Predicting Missing Marker Trajectories in Human Motion Data Using Marker Intercorrelations. *PLoS ONE* **2016**, *11*, e0152616. [CrossRef]
21. Tits, M.; Tilmanne, J.; Dutoit, T. Robust and automatic motion-capture data recovery using soft skeleton constraints and model averaging. *PLoS ONE* **2018**, *13*, e0199744. [CrossRef]
22. Piazza, T.; Lundström, J.; Kunz, A.; Fjeld, M. Predicting Missing Markers in Real-Time Optical Motion Capture. In *Modelling the Physiological Human*; Magnenat-Thalmann, N., Ed.; Number 5903 in Lecture Notes in Computer Science; Springer: Berlin/Heidelberg, Germany, 2009; pp. 125–136.
23. Wu, Q.; Boulanger, P. Real-Time Estimation of Missing Markers for Reconstruction of Human Motion. In Proceedings of the 2011 XIII Symposium on Virtual Reality, Uberlandia, Brazil, 23–26 May 2011; pp. 161–168. [CrossRef]
24. Li, L.; McCann, J.; Pollard, N.S.; Faloutsos, C. DynaMMo: Mining and summarization of coevolving sequences with missing values. In *Proceedings of the 15th ACM SIGKDD International Conference on Knowledge Discovery and Data Mining*; Association for Computing Machinery: New York, NY, USA, 2009; pp. 507–516. [CrossRef]
25. Li, L.; McCann, J.; Pollard, N.; Faloutsos, C. BoLeRO: A Principled Technique for Including Bone Length Constraints in Motion Capture Occlusion Filling. In *Proceedings of the 2010 ACM SIGGRAPH/Eurographics Symposium on Computer Animation*; Eurographics Association: Aire-la-Ville, Switzerland, 2010; pp. 179–188.
26. Burke, M.; Lasenby, J. Estimating missing marker positions using low dimensional Kalman smoothing. *J. Biomech.* **2016**, *49*, 1854–1858. [CrossRef]
27. Wang, Z.; Liu, S.; Qian, R.; Jiang, T.; Yang, X.; Zhang, J.J. Human motion data refinement unitizing structural sparsity and spatial-temporal information. In Proceedings of the IEEE 13th International Conference on Signal Processing (ICSP), Chengdu, China, 6–10 November 2017; pp. 975–982.
28. Aristidou, A.; Cohen-Or, D.; Hodgins, J.K.; Shamir, A. Self-similarity Analysis for Motion Capture Cleaning. *Comput. Graph. Forum* **2018**, *37*, 297–309. [CrossRef]
29. Zhang, X.; van de Panne, M. Data-driven autocompletion for keyframe animation. In Proceedings of the 11th Annual International Conference on Motion, Interaction, and Games, New York, NY, USA, 8–10 November 2018; Association for Computing Machinery: New York, NY, USA, 2018; pp. 1–11. [CrossRef]
30. Hornik, K. Approximation capabilities of multilayer feedforward networks. *Neural Netw.* **1991**, *4*, 251–257. [CrossRef]
31. Fragkiadaki, K.; Levine, S.; Felsen, P.; Malik, J. Recurrent Network Models for Human Dynamics. In Proceedings of the 2015 IEEE International Conference on Computer Vision (ICCV), Santiago, Chile, 7–13 December 2015; pp. 4346–4354. ISSN: 2380-7504. [CrossRef]
32. Harvey, F.G.; Yurick, M.; Nowrouzezahrai, D.; Pal, C. Robust motion in-betweening. *ACM Trans. Graph.* **2020**, *39*, 60:60:1–60:60:12. [CrossRef]
33. Mall, U.; Lal, G.R.; Chaudhuri, S.; Chaudhuri, P. A Deep Recurrent Framework for Cleaning Motion Capture Data. *arXiv* **2017**, arXiv:1712.03380.
34. Kucherenko, T.; Beskow, J.; Kjellström, H. A Neural Network Approach to Missing Marker Reconstruction in Human Motion Capture. *arXiv* **2018**, arXiv:1803.02665.
35. Holden, D. Robust solving of optical motion capture data by denoising. *ACM Trans. Graph.* **2018**, *37*, 165:1–165:12. [CrossRef]
36. Ji, L.; Liu, R.; Zhou, D.; Zhang, Q.; Wei, X. Missing Data Recovery for Human Mocap Data Based on A-LSTM and LS Constraint. In Proceedings of the 2020 IEEE 5th International Conference on Signal and Image Processing (ICSIP), Nanjing, China, 23–25 October 2020; pp. 729–734. [CrossRef]
37. Kaufmann, M.; Aksan, E.; Song, J.; Pece, F.; Ziegler, R.; Hilliges, O. Convolutional Autoencoders for Human Motion Infilling. *arXiv* **2020**, arXiv:2010.11531.
38. Torres, J.F.; Hadjout, D.; Sebaa, A.; Martínez-Álvarez, F.; Troncoso, A. Deep Learning for Time Series Forecasting: A Survey. *Big Data* **2021**, *9*, 3–21. [CrossRef] [PubMed]

39. Shahid, F.; Zameer, A.; Muneeb, M. Predictions for COVID-19 with deep learning models of LSTM, GRU and Bi-LSTM. *Chaos Solitons Fractals* **2020**, *140*, 110212. [CrossRef] [PubMed]
40. Siami-Namini, S.; Tavakoli, N.; Namin, A.S. The Performance of LSTM and BiLSTM in Forecasting Time Series. In Proceedings of the 2019 IEEE International Conference on Big Data (Big Data), Los Angeles, CA, USA, 9–12 December 2019; pp. 3285–3292. [CrossRef]
41. Czekalski, P.; Łyp, K. Neural network structure optimization in pattern recognition. *Stud. Inform.* **2014**, *35*, 17–32.
42. Srebro, N.; Jaakkola, T. Weighted low-rank approximations. In Proceedings of the Twentieth International Conference on International Conference on Machine Learning, Washington, DC, USA, 21–24 August 2003; AAAI Press: Washington, DC, USA, 2003; pp. 720–727.

sensors

MDPI

Article

Human Action Recognition: A Paradigm of Best Deep Learning Features Selection and Serial Based Extended Fusion

Seemab Khan [1], Muhammad Attique Khan [1,*], Majed Alhaisoni [2], Usman Tariq [3], Hwan-Seung Yong [4], Ammar Armghan [5] and Fayadh Alenezi [5]

1 Department of Computer Science, HITEC University Taxila, Txila 47080, Pakistan; seemab.khan@hitecuni.edu.pk
2 College of Computer Science and Engineering, University of Ha'il, Ha'il 55211, Saudi Arabia; majed.alhaisoni@gmail.com
3 College of Computer Engineering and Science, Prince Sattam Bin Abdulaziz University, Al-Kharaj 11942, Saudi Arabia; u.tariq@psau.edu.sa
4 Department of Computer Science & Engineering, Ewha Womans University, Seoul 120-750, Korea; hsyong@ewha.ac.kr
5 Department of Electrical Engineering, College of Engineering, Jouf University, Sakakah 72311, Saudi Arabia; aarmghan@ju.edu.sa (A.A.); fshenezi@ju.edu.sa (F.A.)
* Correspondence: attique.khan@hitecuni.edu.pk

Abstract: Human action recognition (HAR) has gained significant attention recently as it can be adopted for a smart surveillance system in Multimedia. However, HAR is a challenging task because of the variety of human actions in daily life. Various solutions based on computer vision (CV) have been proposed in the literature which did not prove to be successful due to large video sequences which need to be processed in surveillance systems. The problem exacerbates in the presence of multi-view cameras. Recently, the development of deep learning (DL)-based systems has shown significant success for HAR even for multi-view camera systems. In this research work, a DL-based design is proposed for HAR. The proposed design consists of multiple steps including feature mapping, feature fusion and feature selection. For the initial feature mapping step, two pre-trained models are considered, such as DenseNet201 and InceptionV3. Later, the extracted deep features are fused using the Serial based Extended (SbE) approach. Later on, the best features are selected using Kurtosis-controlled Weighted KNN. The selected features are classified using several supervised learning algorithms. To show the efficacy of the proposed design, we used several datasets, such as KTH, IXMAS, WVU, and Hollywood. Experimental results showed that the proposed design achieved accuracies of 99.3%, 97.4%, 99.8%, and 99.9%, respectively, on these datasets. Furthermore, the feature selection step performed better in terms of computational time compared with the state-of-the-art.

Keywords: human action recognition; deep learning; features fusion; features selection; recognition

Citation: Khan, S.; Khan, M.A.; Alhaisoni, M.; Tariq, U.; Yong, H.-S.; Armghan, A.; Alenezi, F. Human Action Recognition: A Paradigm of Best Deep Learning Features Selection and Serial Based Extended Fusion. *Sensors* **2021**, *21*, 7941. https://doi.org/10.3390/s21237941

Academic Editors: Tomasz Krzeszowski, Adam Świtoński, Michal Kepski and Carlos Tavares Calafate

Received: 4 November 2021
Accepted: 25 November 2021
Published: 28 November 2021

1. Introduction

Human action recognition (HAR) emerged as an active research area in the field of computer vision (CV) in the last decade [1]. HAR has applications in various domains including; surveillance [2], human-computer interaction (HCI) [3], video reclamation, and understanding of visual information [4], etc. The most important application of action recognition is video surveillance [5]. Governments use this application for intelligence gathering, reducing crime rate, for security purposes [6], or even crime investigation [7]. The main motivation of growing research in HAR is due to its use in video surveillance applications [8]. In visual surveillance, HAR plays a key role in recognizing the activities of subjects in public places. Furthermore, these types of systems are also useful in smart cities surveillance [9].

Human actions are of various types. These actions can be categorized into two broad classes, namely voluntary actions and involuntary actions [10]. Manual recognition of these actions in real-time is a tedious and error-prone task; therefore, many CV techniques are introduced in the literature [11,12] to serve this task. Most of the proposed solutions are based on classical techniques such as shape features, texture features, point features, and geometric features [13]. A few techniques are based on the temporal information of the human [14], and a few of them extract human silhouettes before feature extraction [15].

Recently, deep learning has shown promising results in the field of computer vision (CV) [16]. Deep learning makes learning and data representation at multiple levels by mimicking the human brain processing [17] to create models. These models consist of multiple processing layers such as convolutional, ReLu, pooling, fully connected, and Softmax [18]. The functionality of a CNN model is to replicate the working of the human brain as it preserves and makes sense of multidimensional information. There exist multiple methods in deep learning, which include encompassing neural networks, hierarchical probabilistic models, supervised learning, and unsupervised learning models [19].

The HAR process is a challenging task as there are a variety of human actions in daily life. In order to tackle this challenge, deep learning models are utilized. The performance of a deep learning model is always based on the number of training samples [20]. In the action recognition tasks, several datasets are publicly available. These datasets include several actions such as walking, running, leaving a car, waving, kicking, boxing, throwing, falling, bending down, and many more.

Recently proposed systems mainly focus on the hybrid techniques; however, they do not focus on minimizing the computational time [21]. This is an important factor as most time surveillance is performed in real-time. Some of the other key challenges of HAR are as follows: (i) Query video sequences resolution is imperative for the recognition of the focal point in the most recent frame. The background complexity, shadows, lighting conditions, and outfit conditions extract irrelevant information using classical techniques of human action, which later results in inefficient action classification; (ii) with automatic activities recognition under multi-view cameras it is difficult to classify the correct human activities. Change in the motion variation captures the wrong activities under the multi-view cameras; (iii) imbalanced datasets impact the learning of a CNN. A CNN model always needs a massive number of training images for learning; and (iv) features extraction from the entire video sequences includes several irrelevant features, affecting the classification accuracy.

These challenges are considered in this work to propose a fully automated design using deep learning features fusion and best feature selection for HAR under the complex video sequences. The major contributions of this work are summarized as follows:

- Selected two pre-trained deep learning models and removed the last three layers. The new layers are added and trained on the target datasets (action recognition dataset). In the training process, the first 80% of the layers are frozen instead of using all the layers, whereas the training process was conducted using transfer learning.
- Proposed a Serial based Extended (SbE) approach for multiple deep learning features fusion. This approach fused features in two phases for better performance and to reduce redundancy.
- Proposed a feature selection technique named Kurtosis-controlled Weighted KNN (KcWKNN). A threshold function is defined which is further analyzed using a fitness function.
- Performed an ablation study to investigate the performance of each step in terms of advantages and disadvantages.

The rest of the manuscript is organized as follows: Related work is presenting in Section 2. The proposed design for HAR is presented in Section 3, which includes deep learning models, transfer learning, the selection of best features and fusion. Results of the proposed method are presented in Section 4 in terms of tables and confusion matrixes. Finally, Section 5 concludes this work.

2. Related Work

HAR has emerged as an impactful research area in CV from the last decade [22]. It is based on important applications such as visual surveillance [23], robotics, biometrics [24,25], and smart healthcare centers to name a few [26,27]. Several researchers of computer vision developed techniques using machine learning [28] for HAR. Most of these researches focused on deep learning due to its better performance and few of them used barometric sensors for activity recognition [29]. Rasel et al. [30] extracted the spatial features using acidometer sensors and classified using multiclass SVM for final activity recognition. Zhao et al. [31] introduced a combined framework for activity recognition. They combined short-term and long-term features for the final results. Khan et al. [32] combined the attention-based LSTM network with dilated CNN model features for the action recognition. Similarly, a skeleton based attention framework is presented by [33] for action recognition. Maheshkumar et al. [13] presented an HAR framework using both the shape and the OFF features [34]. The presented framework is the combination of Hidden Markov Model (HMM) and SVM. The shape and OFF features are extracted and used for HAR through the HMM classifier. The multi-frame averaging method was adopted for background extraction of the image. A discrete Fourier transform (DFT) was performed to reduce the magnitude on the length feature set from the middle to the body contour. In order to select features, the principal component analysis was implied. The presented framework was tested on videos recorded in real-time settings and achieved maximum accuracy. Weifeng et al. [35] presented a generalized Laplacian Regularized Sparse Coding (LRSC) framework for HAR. It was a nonlinear generalized version of graph Laplacian with a tighter isoperimetric inequality. A fast-iterative shrinkage thresholding algorithm for the optimization of ϱ-LRSC was also presented in this work. The input of the sparse codes learned by the ϱ-LRSC algorithm were placed into the support vector machine (SVM) for final categorization. The datasets used for the experimental process were unstructured social activity attribute (USAA) and HMDB51. The experimental results demonstrated the competence of the presented ϱ-LRSC algorithm. Ahmed et al. [36] presented an HAR model using a depth video analysis. HMM was employed to recognize regular activities of aged people living without any attendant. The first step was to analyze the depth maps through the temporal motion identification method using the segments of human silhouettes in a given scenario. Robust features were selected and fused together to find the gradient orientation change, intensity difference temporal and local movement of the body organs [37]. These fused features were processed via embedded HMM. The experimental process was conducted on three different datasets such as Online Self-Annotated [38], Smart Home, and Three Healthcare, and achieved the accuracies 84.4, 87.3, and 95.97%, respectively. Muhammed et al. [39] presented a smartphone inertial sensors-based framework for human activity recognition. The presented framework was divided into three steps: (i) extract the efficient features; (ii) the features were reduced using the kernel principal component analysis (KPCA) and linear discriminant analysis (LDA) to make them resilient; (iii) resultant features were trained via deep belief neural networks (DBN) to attain improved accuracy. The presented approach was compared with traditional expression recognition approaches such as typical multiclass SVM [40,41] and artificial neural network (ANN) and showed an improved accuracy.

Lei et al. [42] presented a light weight action recognition framework based on DNN using RGB video sequences. The presented framework was constructed using CNNs and LTSM units that was a temporal attention model. The purpose of using CNNs was to segment out the objects from the complex background. LTSM networks were used on spatial feature maps of multiple CNN layers. Three datasets, such as UCF-11, UCF Sports, and UCF-101, were used for experimental processes and achieved 98.45%, 91.89%, and 84.10%, respectively. Abdu et al. [43] presented an HAR framework based on deep learning. They considered the problem of traditional techniques which are not useful for the better accuracy of complex activities. The presented framework used a cross DBNN model that unites the SRUs with GRUs of the neural network. The SRUs were used to execute the sequence multi-modal data input. Then GRUs were used to store and learn the amount of information that can be transferred from past state to future state. Zan et al. [44] presented an action recognition model that served the problem of multi-view HAR. The presented algorithm was based on adaptive fusion and category-level dictionary learning (AFCDL). In order to integrate dictionary learning, query sets were designed, and the regularization scheme was constructed for the adaptive weights assignment. Muhammad et al. [45] presented a new framework of 26-layered CNN for composite action classification. Two layers, the global average pooling layer and fully connected layer (FC) were used for feature extraction. The extracted features are classified using the extreme learning machine (ELM) and Softmax for final action classification. Four datasets named HMDB51, UCF Sports, KTH, and Weizmann were used for the experimentation process and showed better performance. Muhammad et al. [4] presented a new fully automated structure for HAR by fusing DNN and multi-view features. Initially, a pre-trained CNN named VGG19 was implied to take out DNN features. Horizontal and vertical gradients were used to compute multi-view features and vertical directional attributes. Final recognition was performed on the selected features via the Naive Bayes Classifier (NBC). Kun et al. [46] introduced an HAR model based on DNN that combines the convolutional layer with LSTM. The presented model was able to automatically extract the features and perform their classification with the standard parameters.

Recently, the development of deep learning models for HAR using high dimensional datasets has shown immense progress. Classical methods for HAR did not show satisfactory performance, especially for large datasets. In contrast, the modern techniques such as Long Short-Term Memory (LSTM), SV-GCN, and Convolution Neural Networks (CNNs) are showing improved performance and can be considered for further research to obtain an improvement in the accuracy.

3. Proposed Methodology

This section presents the proposed methodology for human action recognition in complex video sequences. The proposed design consists of multiple steps, including feature mapping, feature fusion, and feature selection. Figure 1 represents the proposed design of HAR. In this design, features are extracted from the two pre-trained models such as DenseNet201 and InceptionV3. The extracted deep features are fused using the Serial based Extended (SbE) approach. In the later step, the best features are selected using Kurtosis-controlled Weighted KNN. The selected features are classified using several supervised learning algorithms. Detail of each step is provided below.

3.1. *Convolutional Neural Network (CNN)*

CNN is an innovative technique in deep learning that makes the classification process fast and precise. CNN requires lesser parameters to train compared with the traditional neural networks [47]. A CNN model contains multiple layers where the convolution layer is an integral part. Few other layers contained in the CNN model are pooling layers (min, max, average), the ReLU layer, and some fully connected (FC) layers. The internal structure of a CNN has multiple layers as presented in Figure 2. This figure shows that video sequences are provided as input to this network. In the network, the initially convolutional layer is added to convolve input image features, which are later normalized in pooling and hidden layers. After that, FC layers are added to convert image features into 1D feature vector. The final 1D extracted features are classified in the last layer, which is known as the output layer.

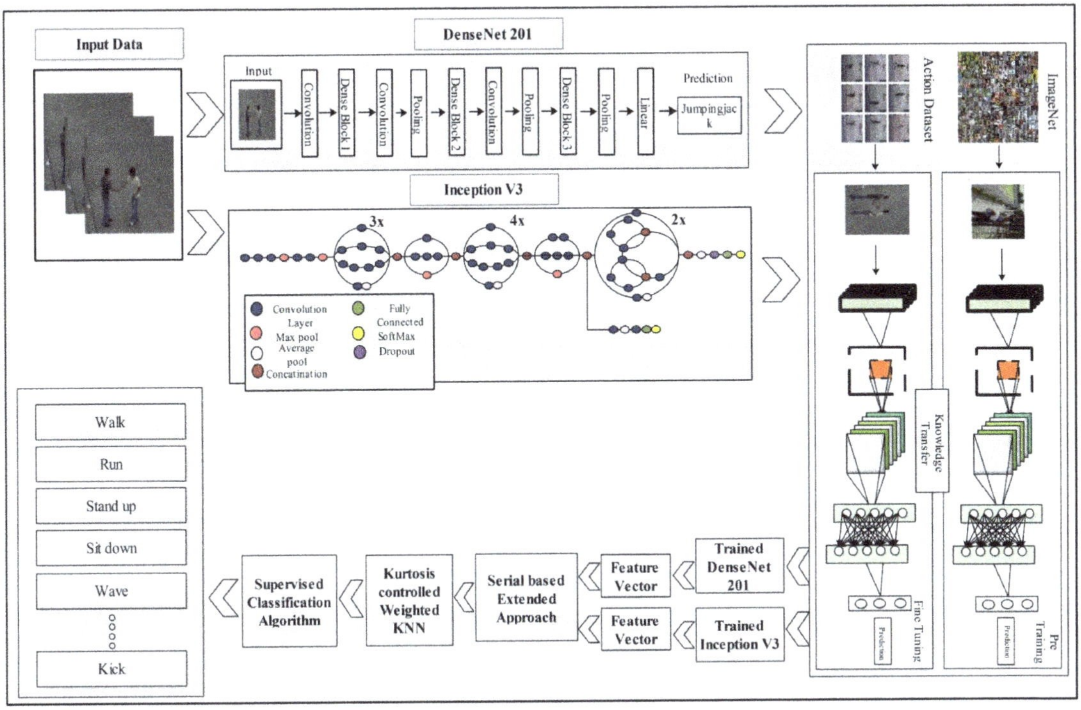

Figure 1. Illustration of a proposed design for HAR using deep learning.

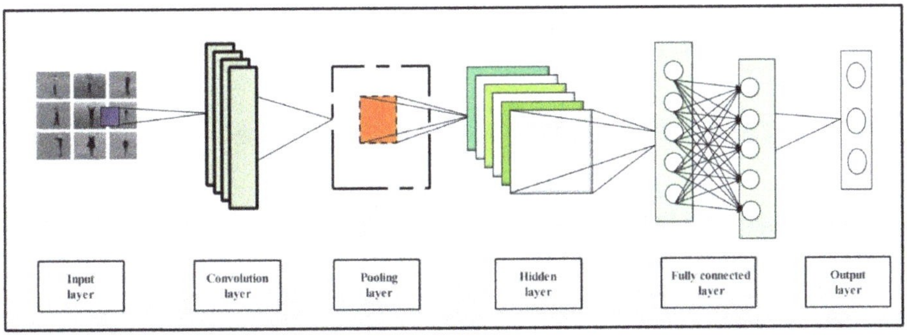

Figure 2. A simple architecture of CNN containing multiple layers for image classification.

3.2. *Densenet201 Pre-Trained Deep Model*

DenseNet is an advanced CNN model where every layer is directly connected with all the layers in subsequent order. These connections help to improve the flow of information in the network, as illustrated in Figure 3. This dense connectivity makes it a dense convolutional network commonly known as DenseNet [48]. Other than the improvement in the information flow, it caters to the vanishing gradient problems as well as it strengthens the feature prorogation process. DenseNet also allows for reusing the features and it reduces required parameters, which eventually reduces the computational complexity of the algorithm. Consider a CNN with ϕ number of layers and ϕ_l layer index has an input stream that starts with x_0. A nonlinear transformation function $F_\phi(.)$ is applied on each layer and it can be a combination of multiple functions such as BN, pooling convolution or

ReLU. In a densely connected network, each layer is connected to its subsequent layers. Output of the ϕth layer is represented by x_{ϕ}.

$$x_{\phi} = F_{\phi}(x_0, \ldots\ldots, x_{\phi-1}) \tag{1}$$

where $(x_0, \ldots\ldots, x_{\phi-1})$ states the concatenation of the feature maps generated in layers $0, \ldots\ldots, \phi - 1$.

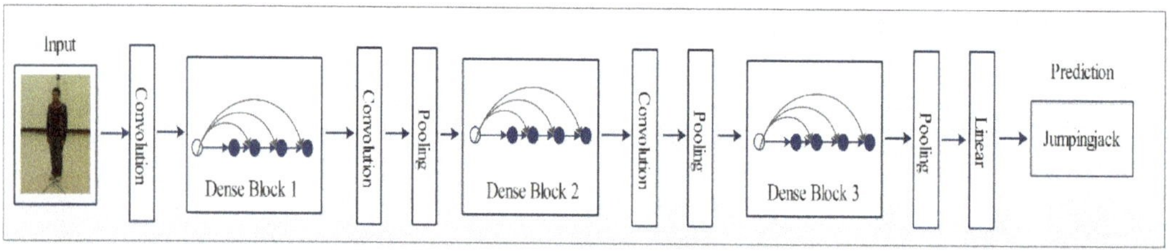

Figure 3. Network architecture of DenseNet201 for action recognition.

3.3. *Inception V3 Pre-Trained Deep Model*

InceptionV3 [49] is an already trained CNN model on the ImageNet dataset. It consists of 316 layers which include convolution layers, pooling layers, fully connected layers, dropout, and Softmax layers. The total number of connections in this model is 350. Unlike a traditional CNN that allows a fixed filter size in a single layer, InceptionV3 has the flexibility to use variable filter sizes and a number of parameters in a single layer which results in better performance. An architecture of InceptionV3 is shown in Figure 4.

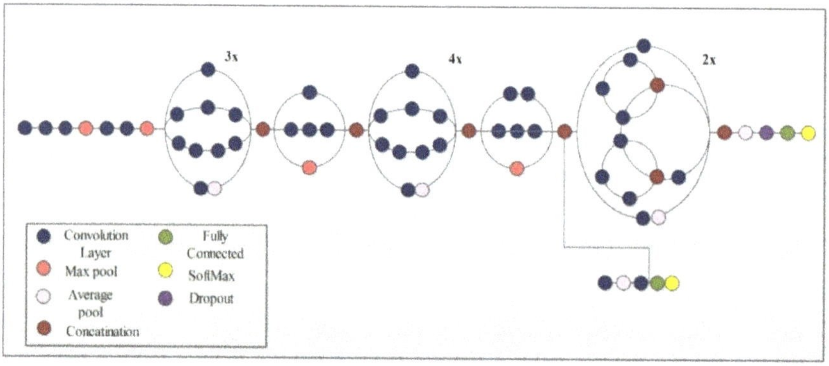

Figure 4. Network architecture of Inceptionv3 model.

3.4. *Transfer Learning Based Learning*

Transfer learning is a well-known technique in the field of deep learning that allows the reusability of a pre-trained model on an advanced research problem [50]. A major advantage of using TL is that it requires less data as input and provides remarkable results. It aims to transfer knowledge from a source domain to a targeted domain, here the source domain refers to a pre-trained model with a very large dataset and the targeted domain is the proposed problem with limited labels [51]. In the source domain, usually a large high-resolution image dataset known as ImageNet is used [52,53]. It contains more than 15 billion labels and 1000 image categories. Image labels in ImageNet are saved according to the wordNet hierarchy, where each node leads to thousands of images belonging to that category. Mathematically, TL is defined as follows:

Given a source domain s_d, defined as:

$$s_d = \left\{\left(x_1^d, y_1^d\right), \ldots, \left(x_i^d, y_i^d\right), \ldots, \left(x_n^d, y_n^d\right)\right\}$$

The learning task is $L_d, L_s, \left(x_m^d, y_m^d\right) \in \varphi$. The target domain is defined as:

$$s_t = \left\{\left(x_1^t, y_1^t\right), \ldots, \left(x_i^t, y_i^t\right), \ldots, \left(x_n^t, y_n^t\right)\right\}$$

The learning task L_t, $\left(x_n^t, y_n^t\right) \in \varphi$, (m, n) will be the size of training data, where $n \ll m$ and y_i^d and y_i^t are the training data labels. Using this definition, both pre-trained models are trained on action datasets. During the training process, the learning rate was 0.01, the mini batch size is 64, the maximum epochs is 100 and the learning method is the stochastic gradient descent. After the fine-tuning process, the output of both models is the number of action classes.

3.5. Features Extraction

Features are extracted from the newly learned models called target models as shown in Figures 5 and 6. Figure 5 represents a DenseNet201 modified model. Using this model, features are extracted using the avg-pool layer. In the output, an $N \times 1920$ dimensional feature vector was obtained, denoted by $\vec{C}$, where N represents number of images in the target dataset.

Using the Inception V3 modified model (depicted in Figure 6), features are extracted from the average pool layer. On this layer, the dimension of the extracted deep feature vector is $N \times 2048$ and it is represented by $\vec{D}$, where N is the number of images in the target dataset.

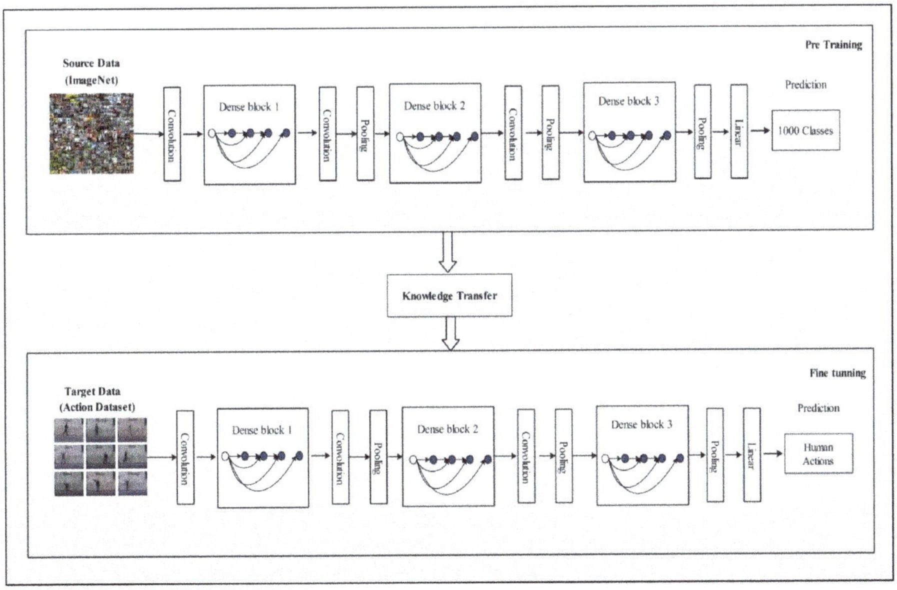

Figure 5. Target model (modified DenseNet201) for feature extraction.

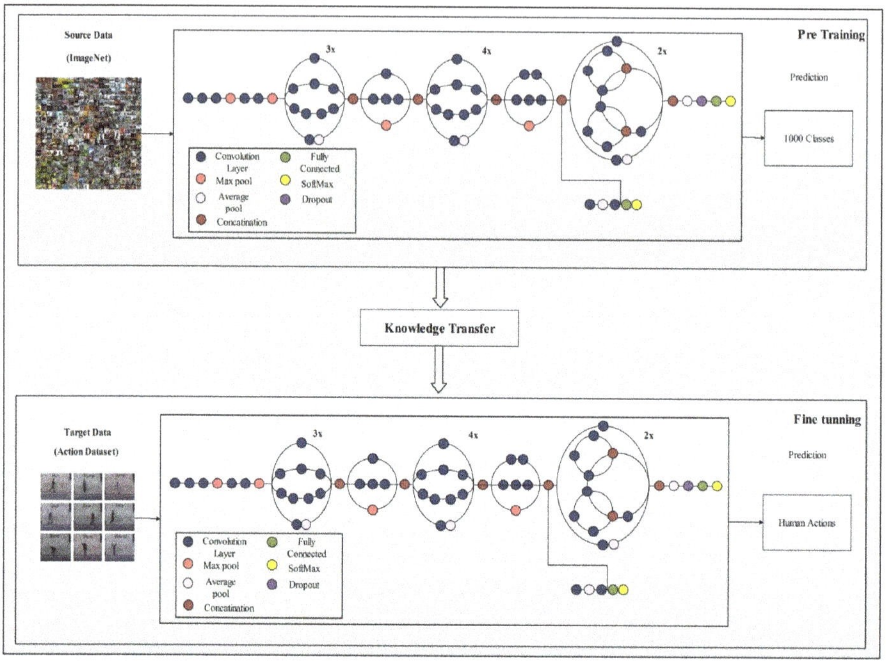

Figure 6. Target model (modified Inception V3) for feature extraction.

3.6. Serial Based Extended Fusion

The fusion of features is becoming a popular technique for improved classification results. The main advantage of this step is to improve the image information in terms of features. The improved feature space increases the classification performance. In the proposed work, a Serial based Extended (SbE) approach is implemented. In this approach, initially features are fused using a serial-based approach. The fused vectors are combined in a single feature vector and to obtain a feature vector of dimension $N \times 3968$ and denoted by β, considering two feature vectors $\vec{C}$ and $\vec{D}$ defined on the outline of sample space $\vec{Z}$. For an arbitrary sample $\eth \in \vec{Z}$, the equivalent two feature vectors are $j \in \vec{C}$ and $k \in \vec{D}$. The serial combined feature of $\eth$ can be defined as $= \begin{pmatrix} j \\ k \end{pmatrix}$. If feature vector $\vec{C}$ has n dimensions and feature vector $\vec{D}$ has m dimensions, then serial fused feature βwill have $(n+m)$ dimensions. After obtaining a βfeature vector, the features are sorted into descending order and the mean value is computed. Based on the mean value, the feature vector is extended in terms of the final fusion.

$$\mu() = \frac{1}{\mathrm{N}} \sum_{i=1}^{\mathrm{N}} (_i) \tag{2}$$

$$Fsn = \begin{cases} Fusion(i) & for\ _i \geq \mu \\ Discard, & ElseWhere \end{cases} \tag{3}$$

Here, $Fusion(i)$ is a final fused feature vector of dimension $N \times K$, where the value of K is always transformed according to the variation in the dataset. Later on, this fusion vector is analyzed using the experimental process and further refined using a feature selection approach.

3.7. Serial Based Extended Fusion

Feature selection is the process of the selection of subset features from the input feature vector [54]. It helps to improve the performance of the algorithm and also reduces the training time. In the proposed design, a new feature selection algorithm is proposed, Kurtosis-controlled Weighted KNN (KcWKNN). The proposed selection method works in the following steps: (i) input fused feature vector; (ii) compute Kurtosis value; (iii) define a threshold function; (iv) calculate fitness, and (v) select the feature vector.

The Kurtosis value is computed as follows:

$$Kr = \frac{\mu_4}{\delta^4} \tag{4}$$

$$\mu_4 = E\left[\left(\widehat{F}_i - E\left[\widehat{F}\right]\right)^n\right], \ \widehat{F}_i \in Fusion(i) \text{ and } n = 4 \tag{5}$$

$$\delta^4 = \sqrt{E\left[\left(\widehat{F}_i - \mu\right)^2\right]} \tag{6}$$

where K is the Kurtosis function, μ_4 is the fourth central moment, and δ is the standard deviation. Kurtosis is a statistical measure that we investigate to find how much the tails of the distribution deviate from the normal. Distributions with higher values are identified in this process. In this work, the main purpose of using Kurtosis is to obtain the higher tail values (outlier features) through the fourth moment that was later employed in the threshold function for the initial feature selection. By using the Kurtosis value, a threshold function is defined as follows:

$$Ts = \begin{cases} FS(i) & for \ \ Fusion(i) \geq Kr \\ Ignore, & Elsewhere \end{cases} \tag{7}$$

The selected feature vector $FS(i)$ is passed into the fitness function WKNN for validation. Mathematically, WKNN is defined as follows:

Consider $\{(x_i, y_i)\}_{i=1}^{N} \in P$ as the training set where x_i is the p-dimensional training vector and y_i is its equivalent class labels set. To determine the label $\overline{y}$ of any $\overline{x}$ from the test set $(\overline{x}, \overline{y})$, the following mathematics takes place.

(a) Compute the Euclidian distance e between $\overline{x}$ and each $(\overline{x}, \overline{y})$, formal given in Equation (8).

$$e(\overline{x}, \ x_i) = \overline{x} - x_{ii_o} \tag{8}$$

(b) Arrange all values in ascending order
(c) Assign a weight $\acute{\omega}_i$ to the ith nearest neighbor using Equation (9).

$$\acute{\omega}_i = \frac{1}{(e(\overline{x}, \ x_i))^2} \tag{9}$$

(d) Assign $\acute{\omega}_i = 1$ for the equally weighted KNN rule,
(e) The class label of $\overline{x}$ is assigned on the basis of majority votes from the neighbors by Equation (10).

$$\overline{y} = \text{argmax} \sum_{(x,y)\in P} \acute{\omega}_i, \ \text{ꟾ}(x = \overline{y}_i) \tag{10}$$

where x is the class label, $\overline{y}_i$ is the class label for ith nearest neighbor and ꟾ(.) is the Dirac-Delta function that takes value = 1 if its argument is true and 0 otherwise.
(f) Compute error.

The error is used as a performance measure, where the number of iterations is initialized as 50. This process is carried out until the error is minimized. Visually, the flow is shown in Figure 7, where it can be seen that the best selected features are finally classified using supervised learning algorithms. Moreover, the complete work of the proposed design is listed in Algorithm 1.

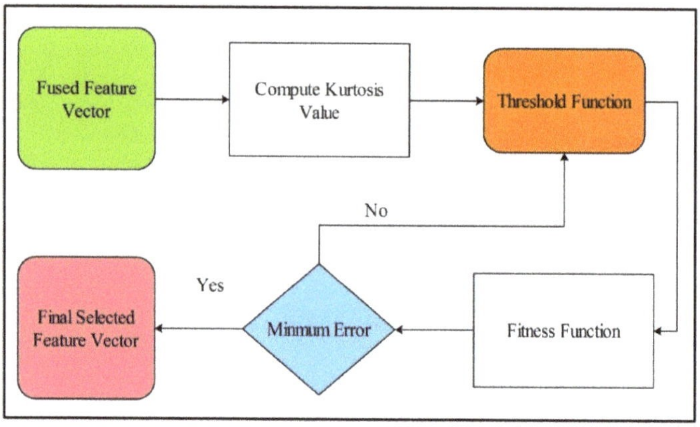

Figure 7. Proposed flow diagram of best feature selection.

Algorithm 1. The complete work of the proposed design.

Input: Action Recognition Datasets
Output: Predicted Action Class
Step 1: Input action datasets
Step 2: Load Pre-trained Deep Models;
- Densenet201
- Inception V3

Step 3: Fine Deep Models
Step 4: Trained Deep Models using TL
Step 5: Feature Extraction from Avg Pooling Layers
Step 6: SbE approach for Features Fusion
Step 7: Best Features Selection using Proposed KcWKNN
Step 8: Predict Action Label

4. Results and Analysis

The experimental process of the proposed method is presented in this section. Four publically available datasets such as KTH [3], Hollywood [38], WVU [39], and IXMAS [40] were used in this work for the experimental process. Each class of these datasets contains 10,000 video frames that are utilized for the experimental process. In the experimental process, 50% of video sequences are used for the training purpose, while the remaining 50% is utilized for the testing purpose. The K-Fold cross validation is adopted, where the value of $K = 10$. Results are computed on several supervised learning algorithms and select the best one is selected based on the accuracy value. All simulations are conducted on MATLAB2020a using a Personal Computer Corei7 with 16 GB of RAM and 8 GB Graphics card.

4.1. Results

A total of four experiments were performed on each dataset to analyze the performance of the middle step. These steps are: (i) performed classification using DenseNet201 deep features; (ii) performed classification using InceptionV3 deep model; (iii) performed classification using the SbE deep features fusion, and (iv) performed classification using KcWKNN-based feature selection.

Experiment 1: Table 1 presents the results of the specific DenseNet201 deep features on selected datasets. In this table, it is noted that the Cubic SVM achieved a better accuracy of 99.3% on the KTH dataset. Other classifiers also achieved a better accuracy of above 94%. For the Hollywood action dataset, the best achieved accuracy is 99.9% for Fine KNN. Similar to the KTH dataset, the rest of the classifiers also performed better on this dataset. The best obtained accuracy for the WVU dataset is 99.8% for Cubic SVM. The rest of the classifiers also performed better and achieved an average accuracy of 97%. The best obtained accuracy of the IXAMAS dataset is 97.3% for Fine KNN.

Table 1. Classification accuracy on specific DenseNet201 deep model. The bold represents the best obtained values.

Classifier	Datasets Accuracy on DenseNet201 Deep Model			
	KTH	Hollywood	WVU	IXAMAS
Linear Discriminant	98.8	99.6	98.3	92.1
Linear SVM	98.0	98.3	97.1	86.6
Quadratic SVM	98.9	99.6	99.7	96.4
Cubic SVM	**99.3**	99.8	**99.8**	95.4
Medium Gaussian SVM	98.6	99.5	97.8	93.1
Fine KNN	98.7	**99.9**	99.3	**97.3**
Medium KNN	96.7	98.8	97.3	88.0
Cosine KNN	96.9	98.8	97.4	88.3
Weighted KNN	97.2	99.7	98.0	92.9
Ensemble Bagged Trees	89.6	98.2	94.5	82.9

Experiment 2: The results of InceptionV3 deep features are provided in Table 2. In this table, it is noted that the best achieved accuracy on the KTH dataset is 98.1%, for the Hollywood dataset it is 99.8%, for the WVU dataset it is 99.1%, and for the IXAMAS dataset it is 96%. From this table, it is observed that the performance of specific DenseNet201 features are better. However, during the computation of results, time significantly increases. Therefore, it is essential to handle this issue with consistent accuracy.

Table 2. Classification accuracy on specific InceptionV3 deep model. The bold represents the best obtained values.

Classifier	Datasets Accuracy on DenseNet201 Deep Model			
	KTH	Hollywood	WVU	IXAMAS
Linear Discriminant	96.6	98.8	96.5	87.3
Linear SVM	95.4	96.3	93.5	81.3
Quadratic SVM	97.6	99.3	99.0	92.1
Cubic SVM	**98.1**	99.5	**99.1**	93.6
Medium Gaussian SVM	97.0	99.3	97.7	91.2
Fine KNN	97.6	**99.8**	98.4	**96.0**
Medium KNN	95.00	98.1	94.8	83.8
Cosine KNN	95.6	98.5	95.1	84.7
Weighted KNN	95.9	99.1	95.8	90.0
Ensemble Bagged Trees	89.0	92.4	90.5	73.3

Experiment 3: After the experiments on specific feature sets, the SbE approach is applied for deep features fusion. The KTH dataset results are provided in Table 3. In this table, The highest performance is recorded for Cubic SVM with an accuracy of 99.3%. Recall

and precision are 99.3% and 99.43% respectively. Moreover, the noted time during the training process is 893.23 s. The second highest accuracy is achieved by a linear discriminant classifier of 99.2%. The rest of the classifiers also performed better. Compared with specific feature vectors, the fusion process results are more consistent. Figure 8 illustrates the true positive rates (TPRs)-based confusion matrix of Cubic SVM that confirms the value of the recall rate. In this figure, the highlighted diagonal values represent the true positive predictions, whereas the values other than the diagonal represent false negative predictions.

Table 3. Achieved results on KTH dataset after fusion of deep features using SbE approach. The bold represents the best obtained values.

Classifier	Recall Rate (%)	Precision Rate (%)	FNR	Time (s)	F1 Score (%)	Accuracy (%)
Linear Discriminant	99.200	99.300	0.80	424.10	99.249	99.2
Linear SVM	98.400	98.616	1.60	487.10	98.508	98.4
Quadratic SVM	99.150	98.283	0.85	706.56	98.714	99.2
Cubic SVM	**99.300**	**99.433**	**0.70**	893.23	**99.366**	**99.3**
Medium Gaussian SVM	98.916	99.083	1.08	1445.8	98.999	98.9
Fine KNN	99.083	99.216	0.91	450.55	99.149	99.1
Medium KNN	96.700	97.233	3.30	447.37	96.965	96.8
Cosine KNN	97.516	97.716	2.48	459.33	97.616	97.5
Weighted KNN	97.483	97.916	2.51	447.59	97.699	97.6
Ensemble Bagged Trees	94.233	94.733	5.76	**192.96**	94.482	94.3

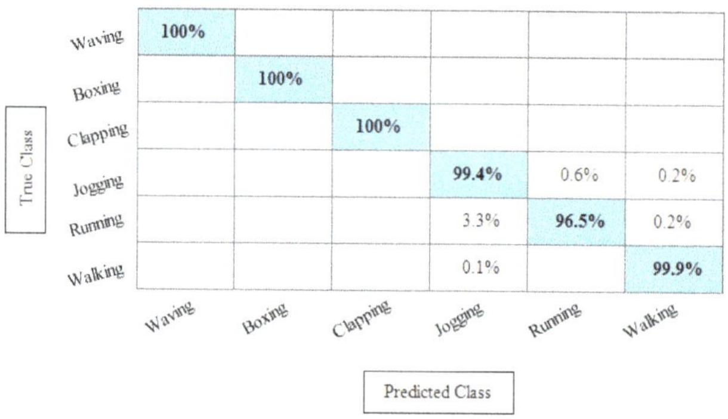

Figure 8. TPR-based confusion matrix of KTH dataset after fusion of deep features using SbE approach.

Table 4 represents the results of the Hollywood action dataset using the SbE approach. In this table, it is noted that the best accuracy is 99.9%, obtained by Fine KNN. Other performance measures such as recall rate, precision rate and F1 score values are 99.1825%, 99.8375%, and 99.5089%, respectively. The rest of the classifiers mentioned in this table performed better and achieved an average accuracy above 98%. Figure 9 illustrates the TPR-based confusion matrix of Fine KNN, where it is clear that each class prediction rate is above 99%. Moreover, compared with the specific deep features experiment on the Hollywood dataset, the fusion process shows more consistent results.

Table 4. Achieved results on Hollywood dataset after fusion of deep features using SbE approach. The bold represents the best obtained values.

Classifier	Recall Rate (%)	Precision Rate (%)	FNR	Time (s)	F1 Score (%)	Accuracy (%)
Linear Discriminant	99.775	99.825	0.22	469.75	99.800	99.9
Linear SVM	99.887	99.25	1.11	734.42	99.567	99.2
Quadratic SVM	99.550	99.725	0.45	1065.4	99.637	99.7
Cubic SVM	99.575	99.775	0.42	1337.4	99.674	99.8
Medium Gaussian SVM	99.287	99.675	0.71	2227.1	99.480	99.7
Fine KNN	**99.182**	**99.837**	**0.18**	**447.76**	**99.508**	**99.9**
Medium KNN	98.500	99.0125	1.50	437.47	98.755	99.1
Cosine KNN	99.037	98.975	0.96	449.13	99.006	99.3
Weighted KNN	99.250	99.45	0.75	439.29	99.349	99.6
Ensemble Bagged Trees	94.425	97.562	5.57	209.63	95.968	96.7

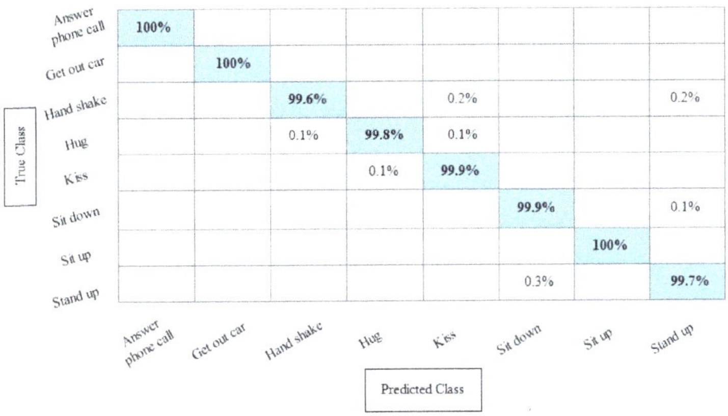

Figure 9. TPR based confusion matrix of Fine KNN using Hollywood dataset after fusion of deep features through SbE approach.

Table 5 presents the results of the WVU dataset using the SbE fusion approach. The highest accuracy is achieved through Linear Discriminant which is 99.8%, where the recall rate, precision rate, and F1 score are 99.79%, 99.78%, and 99.78%, respectively. Quadratic SVM and Cubic SVM performed second best and achieved an accuracy of 99.7% for each. The rest of the classifiers also performed better and gained the average accuracy of above 99%. Figure 10 illustrated the TPR based confusion matrix of the WVU dataset for the Linear Discriminant classifier. This figure showed that the correct prediction rate of each classifier is more than 99%. Compared with this accuracy of WVU on specific features, it is noticed that the fusion process provides consistent accuracy.

Table 5. Achieved results on WVU dataset after fusion of deep features using SbE approach. The bold represents the best obtained values.

Classifier	Recall Rate (%)	Precision Rate (%)	FNR (%)	Time (s)	F1 Score (%)	Accuracy (%)
Linear Discriminant	**99.79**	**99.78**	**0.21**	**2073.1**	**99.785**	**99.8**
Linear SVM	97.74	97.77	2.26	2567.7	97.755	97.7
Quadratic SVM	99.56	99.56	0.44	2824.5	99.560	99.6
Cubic SVM	99.56	99.57	0.44	2267	99.565	99.6
Medium Gaussian SVM	98.56	98.34	1.66	2749	98.449	98.3
Fine KNN	97.0	97.03	3.00	3486	97.015	97.0
Medium KNN	87.15	88.34	12.8	3933.5	87.741	87.2
Cosine KNN	87.98	89.01	12.1	2825.4	88.492	88.0
Weighted KNN	90.89	91.51	9.11	2716.7	91.198	90.9
Ensemble Bagged Trees	94.08	94.12	5.92	965.78	94.100	94.1

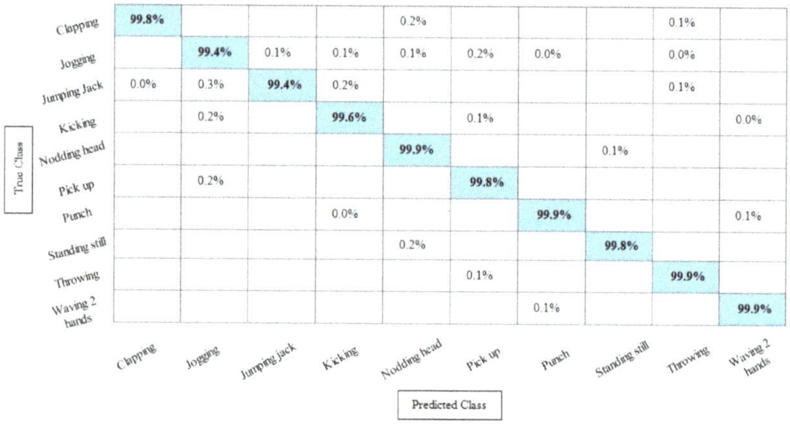

Figure 10. TPR-based confusion matrix of Linear Discriminant classifier after fusion of deep features using SbE approach.

Table 6 presents the results of the IXMAS dataset after SbE features fusion. In this table, it can be seen that the highest accuracy is achieved through Fine KNN of 97.4%, where the recall rate, precision rate, and F1 score are 97.18%, 97.25%, and 97.21%, respectively. Cubic SVM performed second best and achieved an accuracy of 97.3%. The rest of the classifiers also performed better and attained an average accuracy above 93%. Figure 11 illustrates the TPR-based confusion matrix of the Fine KNN for the IXMAS dataset using the SbE approach.

Overall, the results of the SbE approach are improved and are consistent compared with the specific deep features (see results in Tables 1 and 2). However, it is observed that the computational time increases during the fusion process. For a real-time system, this time needs to be minimized. Therefore, a feature selection approach is proposed.

Table 6. Achieved results on IXMAS dataset after fusion of deep features using SbE approach. The bold represents the best obtained values.

Classifier	Recall Rate (%)	Precision Rate (%)	FNR (%)	Time (s)	F1 Score (%)	Accuracy (%)
Linear Discriminant	96.460	96.310	3.54	508.35	96.384	96.5
Linear SVM	91.030	91.230	8.97	1428	91.129	91.3
Quadratic SVM	96.670	96.680	3.33	936.8	96.675	96.7
Cubic SVM	97.216	97.225	2.78	390.9	97.220	97.3
Medium Gaussian SVM	96.016	96.066	3.98	840.3	96.041	96.1
Fine KNN	**97.180**	**97.250**	**2.82**	**570.56**	**97.215**	**97.4**
Medium KNN	88.360	88.890	11.6	560.06	88.624	88.9
Cosine KNN	89.141	89.516	10.8	559.83	89.328	89.7
Weighted KNN	92.475	92.625	7.52	543.5	92.549	92.8
Ensemble Bagged Trees	80.291	81.550	19.7	284.31	80.915	81.4

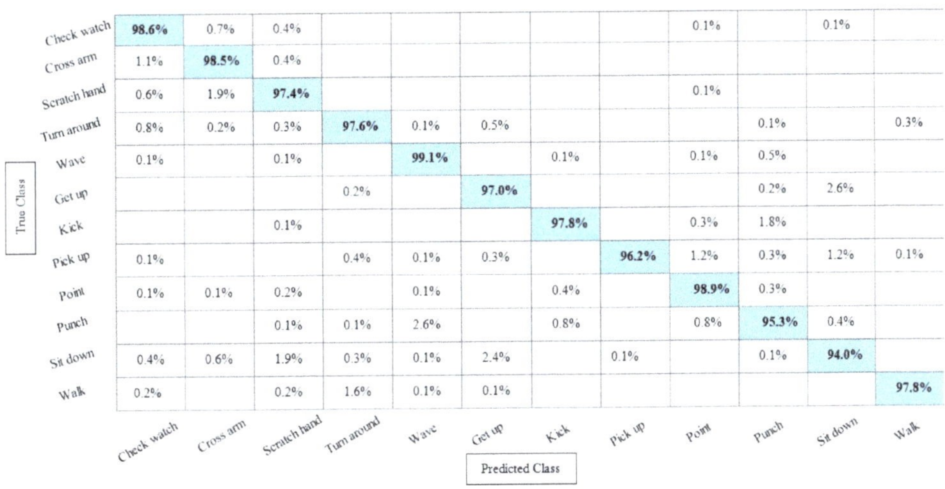

Figure 11. TPR-based confusion matrix of Fine KNN after fusion of deep features using SbE approach.

Experiment 4: In this experiment, the best features are selected using Kurtosis-controlled WKNN and provided to the classifiers. Results are provided in Tables 7–10. Table 7 presents the results of the proposed feature selection algorithm on the KTH dataset. In this table, the highest obtained accuracy is 99%, achieved by Cubic SVM. Other performance measures such as recall, precision and F1 score are 98.1666%, 99.1166% and 99.016%, respectively. Figure 12 illustrates the TPR-based confusion matrix of the Cubic SVM for the best feature selection process. In comparison with Table 3 results, it is noted that the accuracy of Cubic SVM decreases (0.3%), while the computational time expressively declines. The computational time of the Cubic SVM in the fusion process was 893.23 s, which is reduced after the feature selection process to 451.40 s. This shows that the feature selection process not only maintains the recognition accuracy but also minimizes the computational time.

Table 7. Achieved results on KTH dataset after best feature selection using KcWKNN. The bold represents the best obtained values.

Classifier	Recall Rate (%)	Precision Rate (%)	FNR (%)	Time (s)	F1 Score (%)	Accuracy (%)
Linear Discriminant	98.080	98.516	1.92	87.805	98.297	98.1
Linear SVM	97.633	97.933	2.36	255.42	97.783	97.7
Quadratic SVM	98.600	98.866	1.40	360.10	98.733	98.7
Cubic SVM	**98.916**	**99.116**	**1.09**	**451.40**	**99.016**	**99.0**
Medium Gaussian SVM	98.2833	98.483	1.71	687.37	98.383	98.3
Fine KNN	98.616	98.833	1.38	237.93	98.724	98.7
Medium KNN	95.483	96.366	4.51	231.39	95.922	95.7
Cosine KNN	97.016	97.183	2.98	230.18	97.099	97.0
Weighted KNN	96.233	97.000	3.76	222.90	96.615	96.4
Ensemble Bagged Trees	94.150	93.716	5.8	140.57	93.632	94.2

Table 8. Achieved results on Hollywood dataset after best feature selection using KcWKNN. The bold represents the best obtained values.

Classifier	Recall Rate (%)	Precision Rate (%)	FNR (%)	Time (s)	F1 Score (%)	Accuracy (%)
Linear Discriminant	99.087	99.450	0.912	88.375	99.268	99.4
Linear SVM	97.937	98.687	2.062	323.99	98.311	98.6
Quadratic SVM	99.262	99.587	0.737	439.41	99.424	99.5
Cubic SVM	99.387	99.675	0.612	501.67	99.531	99.7
Medium Gaussian SVM	98.587	99.500	1.412	910.78	99.041	99.5
Fine KNN	**99.812**	**99.837**	**0.187**	**213.33**	**99.825**	**99.8**
Medium KNN	97.225	98.550	2.775	224.52	97.883	98.5
Cosine KNN	98.325	98.862	1.675	221.19	98.593	98.9
Weighted KNN	98.575	99.412	1.425	215.89	98.992	99.2
Ensemble Bagged Trees	87.050	94.287	12.95	126.72	90.524	97.7

Table 9. Achieved results on WVU dataset after best feature selection using KcWKNN. The bold represents the best obtained values.

Classifier	Recall Rate (%)	Precision Rate (%)	FNR (%)	Time (s)	F1 Score (%)	Accuracy (%)
Linear Discriminant	98.50	98.53	1.50	241.48	98.515	98.5
Linear SVM	96.51	96.57	3.49	293.2	96.539	96.5
Quadratic SVM	**99.37**	**99.38**	**0.63**	**1064.6**	**99.375**	**99.4**
Cubic SVM	99.43	99.44	0.57	1124.0	99.435	99.4
Medium Gaussian SVM	98.24	98.25	1.76	1363.7	98.245	98.2
Fine KNN	96.55	96.59	3.45	1365.1	96.570	96.5
Medium KNN	86.80	87.98	13.2	1322.0	87.386	86.8
Cosine KNN	87.61	88.73	12.39	1316.2	88.166	87.6
Weighted KNN	90.33	91.07	9.67	1236.8	90.698	90.3
Ensemble Bagged Trees	94.71	94.75	5.29	423.37	94.730	95.7

Table 10. Achieved results on IXMAS dataset after best feature selection using KcWKNN. The bold represents the best obtained values.

Classifier	Recall Rate (%)	Precision Rate (%)	FNR (%)	Time (s)	F1 Score (%)	Accuracy (%)
Linear Discriminant	91.583	91.516	8.41	119.8	91.549	91.7
Linear SVM	88.050	88.400	11.95	714.13	88.224	88.5
Quadratic SVM	95.008	95.083	4.99	634.7	95.045	95.1
Cubic SVM	95.783	95.866	4.21	239.4	95.824	95.9
Medium Gaussian SVM	94.466	94.933	5.53	475.5	94.699	94.6
Fine KNN	**97.075**	**96.991**	**2.92**	**290.69**	**97.033**	**97.1**
Medium KNN	86.383	86.925	13.61	266.24	86.653	86.9
Cosine KNN	88.066	88.233	11.93	270.74	88.149	88.5
Weighted KNN	90.975	91.966	9.02	263.74	91.468	91.2
Ensemble Bagged Trees	83.433	85.108	16.5	175.78	84.261	84.8

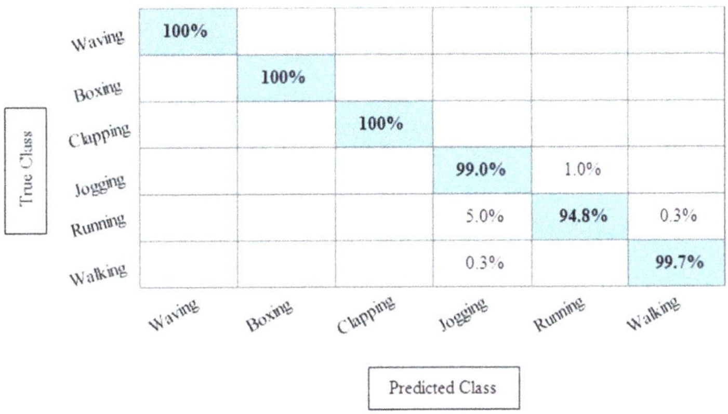

Figure 12. TPR based confusion matrix of Cubic SVM after best feature selection using KcWKNN.

Table 8 presents the best feature selection results on the Hollywood Action dataset and achieved best accuracy by Fine KNN of 99.8%. The other calculated measures such as recall rate, precision rate, and F1 Score are 99.812%, 99.837%, and 99.82%, respectively. For the rest of the classifiers, the average accuracy is above 98% (can be seen in this table). Figure 13 illustrates the TPR-based confusion matrix of Fine KNN for this experiment. The diagonal values in this experiment show the correct predicted values. Comparison with Table 4 shows that the classification accuracy is still consistent, whereas the computational time is significantly reduced. The computational time at the fusion process was 447.76 s, whereas after the selection process, it is reduced to 213.33 s. This shows that the selection of best features using KcWKNN performed significantly better.

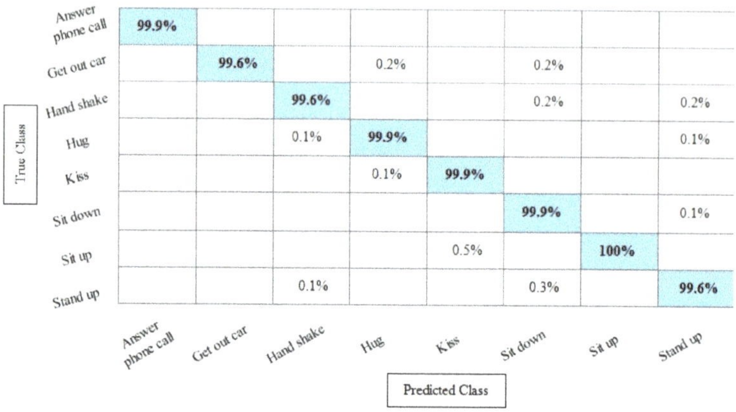

Figure 13. TPR based confusion matrix of Fine KNN after best feature selection using KcWKNN.

Table 9 presents the results of the WVU dataset after the best feature selection using KcWKK. In this table, Quadratic SVM and Cubic SVM performed best with the accuracy of 99.4%, where the recall rate is 99.37% and 99.43%, respectively, and the precision rate is 99.38% and 99.44%, respectively and the F1 score is 99.375%, and 99.43%, respectively. Figure 14 shows the TPR-based confusion matrix of the Cubic SVM for this experiment. This figure shows that the prediction rate of each class is above 99%. Moreover, in comparison with Table 5 (fusion results), the computational time of this experiment on the WVU dataset is almost half and accuracy is still consistent. This shows that the KcWKNN selection approach performed significantly well.

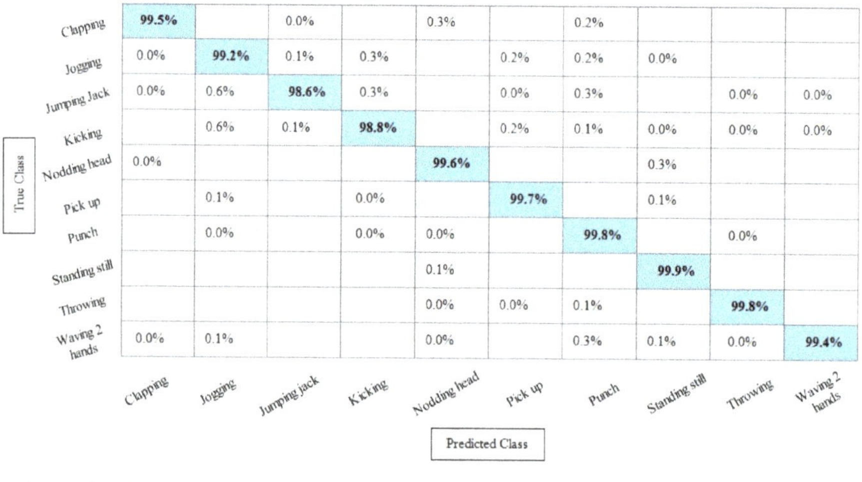

Figure 14. TPR-based confusion matrix of Cubic SVM after best feature selection using KcWKNN.

The results of the KcWKNN-based best features selection on the IXMAS dataset are provided in Table 10. In this table, it is noted that the Fine KNN attained best accuracy of 97.1%, whereas the recall rate, precision rate, and F1 score are 97.075%, 96.9916%, and 97.033%, respectively. Figure 15 illustrates the TPR-based confusion matrix of the Fine KNN for this experiment. The correct prediction value of each class is provided in the diagonal of this figure. Compared with Table 6, this experiment reduces the computational time while maintaining the recognition accuracy.

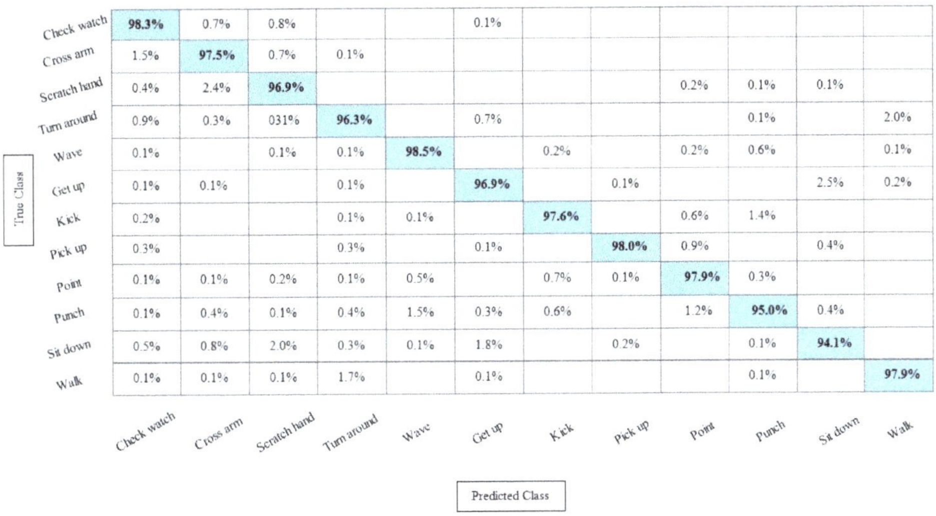

Figure 15. TPR-based confusion matrix of Fine KNN after best feature selection using KcWKNN.

Finally, a detailed analysis was conducted among all experiments in terms of accuracy and time. From Tables 1–10, it is observed that the accuracy value is improved after the proposed fusion process and the time is reduced. However, the noted time was still high and must be reduced further; therefore, a feature selection technique is proposed and time is significantly reduced compared with the original extracted deep features and fusion step (plotted in Figures 16–19). In the selection process, a little change occurred in the accuracy value, but on the other side, a high fall is noted in the computational time.

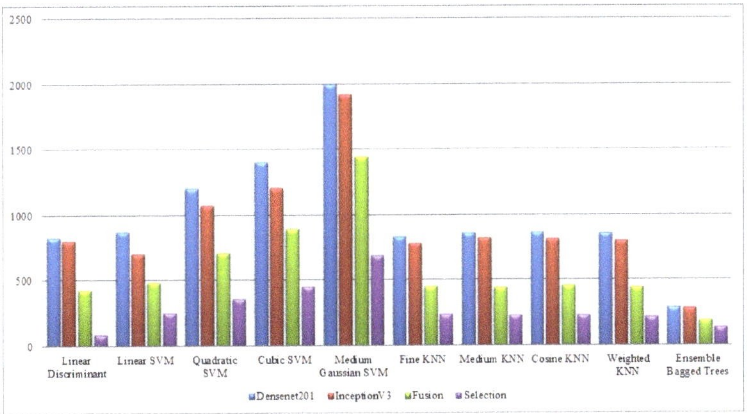

Figure 16. Computational time-based comparison of middle steps on KTH dataset.

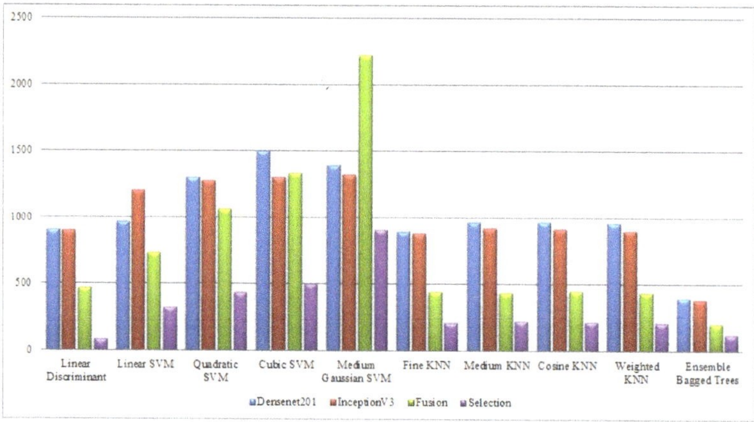

Figure 17. Computational time-based comparison of middle steps on Hollywood dataset.

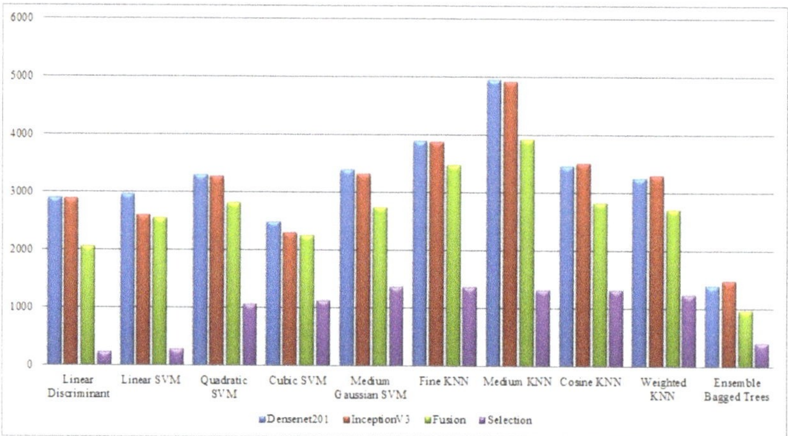

Figure 18. Computational time-based comparison of middle steps on WVU dataset.

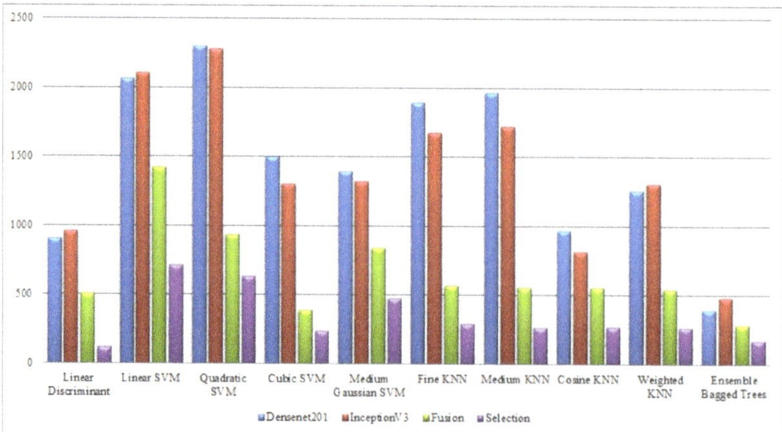

Figure 19. Computational time-based comparison of middle steps on IXMAS dataset.

4.2. Comparison with SOTA

Overall, the feature selection process maintains the classification accuracy while significantly reducing the computational time. A comparison with some recent techniques was also conducted as provided in Table 11. This table shows that the proposed design results are significantly improved. The main strength of the proposed design is the fusion of deep features using the SbE approach and best feature selection using KcWKNN.

Table 11. Comparison of the proposed design with existing techniques in terms of accuracy. The bold represents the best obtained values.

Reference	Dataset	Accuracy (%)
Muhammad et al. [45], 2020	KTH	98.30
Proposed method	**KTH**	**99.00**
Muhammad et al. [4], 2020	IXMAS	95.20
Amir et al. [55], 2021	IXMAS	87.48
Proposed method	**IXMAS**	**97.10**
Muhammad et al. [56], 2020	WVU	99.10
Muhammad et al. [57], 2019	WVU	99.90
Proposed method	**WVU**	**99.40**
Evan et al. [58], 2008	Hollywood	91.80
Proposed method	**Hollywood**	**99.20**

5. Conclusions

HAR has gained a lot of popularity in recent years. Multiple techniques have been used for the accurate recognition of human actions. The problem is to correctly identify the action in real-time and from multiple perspectives. In this work, a design is proposed where the key aim is to improve the accuracy of the HAR process in the complex video sequences using advanced deep learning techniques. The proposed design consists of four steps, namely feature mapping, feature fusion, feature selection, and classification. Two modified deep learning models, DenseNet201 and InceptionV3 were used for feature mapping. Fusion and selection were performed using the serial-based extended approach and Kurtosis-controlled Weighted KNN approach, respectively. The results were obtained after extensive experimentation on state-of-the-art action datasets. Based on the results, it is concluded that the proposed design performed better than the existing techniques in terms of accuracy as well as computational time. Cubic SVM and Fine KNN classifiers were top performers on the proposed HAR method. The key limitation of this work is the computational time that was noted during the original deep extracted features. This step increases the computational time that is not suitable for the real-time applications. As a future study, we intend to test the proposed design on relatively complex action datasets such as HMDB51 and UCF101. Moreover, the recent deep learning models can also be considered for feature extraction and will study the less complexity feature fusion and selection algorithms.

Author Contributions: Conceptualization, S.K., M.A.K. and A.A.; methodology, S.K., M.A.K. and M.A.; software, S.K. and M.A.K.; validation, M.A., U.T. and H.-S.Y.; formal analysis, U.T. and H.-S.Y.; investigation, U.T. and M.A.; resources, M.A.K. and U.T.; data curation, H.-S.Y. and A.A.; writing—original draft preparation, S.K. and M.A.K.; writing—review and editing, M.A., U.T. and F.A.; visualization, A.A. and F.A.; supervision, M.A.K. and H.-S.Y.; project administration, F.A. and A.A.; funding acquisition, H.-S.Y. All authors have read and agreed to the published version of the manuscript.

Funding: This study was partially supported by Ewha Womans University.

Conflicts of Interest: The authors declare no conflict of interest.

References

1. Kim, D.; Lee, I.; Kim, D.; Lee, S. Action Recognition Using Close-Up of Maximum Activation and ETRI-Activity3D LivingLab Dataset. *Sensors* **2021**, *21*, 6774. [CrossRef]
2. Mishra, O.; Kavimandan, P.S.; Tripathi, M.; Kapoor, R.; Yadav, K. Human Action Recognition Using a New Hybrid Descriptor. In *Advances in VLSI, Communication and Signal Processing*; Springer: Singapore, 2021.
3. Chen, X.; Xu, L.; Cao, M.; Zhang, T.; Shang, Z.; Zhang, L. Design and Implementation of Human-Computer Interaction Systems Based on Transfer Support Vector Machine and EEG Signal for Depression Patients' Emotion Recognition. *J. Med. Imaging Health Inform.* **2021**, *11*, 948–954. [CrossRef]
4. Javed, K.; Khan, S.A.; Saba, T.; Habib, U.; Khan, J.A.; Abbasi, A.A. Human action recognition using fusion of multiview and deep features: An application to video surveillance. *Multimed. Tools. Appl.* **2020**, 1–27. [CrossRef]
5. Liu, D.; Xu, H.; Wang, J.; Lu, Y.; Kong, J.; Qi, M. Adaptive Attention Memory Graph Convolutional Networks for Skeleton-Based Action Recognition. *Sensors* **2021**, *21*, 6761. [CrossRef] [PubMed]
6. Ahmed, M.; Ramzan, M.; Khan, H.U.; Iqbal, S.; Choi, J.-I.; Nam, Y.; Kady, S. Real-Time Violent Action Recognition Using Key Frames Extraction and Deep Learning. *Comput. Mater. Continua* **2021**, *69*, 2217–2230. [CrossRef]
7. Wang, J.; Cao, D.; Wang, J.; Liu, C. Action Recognition of Lower Limbs Based on Surface Electromyography Weighted Feature Method. *Sensors* **2021**, *21*, 6147. [CrossRef]
8. Zin, T.T.; Htet, Y.; Akagi, Y.; Tamura, H.; Kondo, K.; Araki, S.; Chosa, E. Real-Time Action Recognition System for Elderly People Using Stereo Depth Camera. *Sensors* **2021**, *21*, 5895. [CrossRef] [PubMed]
9. Farnoosh, A.; Wang, Z.; Zhu, S.; Ostadabbas, S. A Bayesian Dynamical Approach for Human Action Recognition. *Sensors* **2021**, *21*, 5613. [CrossRef] [PubMed]
10. Buehner, M.J. Awareness of voluntary and involuntary causal actions and their outcomes. *Psychol. Conscious. Theory Res. Pract.* **2015**, *2*, 237. [CrossRef]
11. Hassaballah, M.; Hosny, K.M. Studies in Computational Intelligence. In *Recent Advances In Computer Vision*; Hassaballah, M., Hosny, K.M., Eds.; Springer: Berlin/Heidelberg, Germany, 2019.
12. Sharif, M.; Akram, T.; Raza, M.; Saba, T.; Rehman, A. Hand-crafted and deep convolutional neural network features fusion and selection strategy: An application to intelligent human action recognition. *Appl. Soft Comput.* **2020**, *87*, 105986.
13. Kolekar, M.H.; Dash, D.P. Hidden markov model based human activity recognition using shape and optical flow based features. In Proceedings of the 2016 IEEE Region 10 Conference (TENCON), Singapore, 22–25 November 2016.
14. Hermansky, H. TRAP-TANDEM: Data-driven extraction of temporal features from speech. In Proceedings of the 2003 IEEE Workshop on Automatic Speech Recognition and Understanding (IEEE Cat. No. 03EX721), St Thomas, VI, USA, 30 November–4 December 2003.
15. Krzeszowski, T.; Przednowek, K.; Wiktorowicz, K.; Iskra, J. The Application of Multiview Human Body Tracking on the Example of Hurdle Clearance. In *Sport Science Research and Technology Support*; Cabri, J., Pezarat-Correia, P., Vilas-Boas, J., Eds.; Springer: Berlin/Heidelberg, Germany, 2016.
16. Hassaballah, M.; Awad, A.I. *Deep Learning In Computer Vision: Principles and Applications*; CRC Press: Boca Raton, FL, USA, 2020.
17. Voulodimos, A.; Doulamis, N.; Doulamis, A.; Protopapadakis, E. Deep learning for computer vision: A brief review. *Comput. Intell. Neurosci.* **2018**. [CrossRef] [PubMed]
18. Young, T.; Hazarika, D.; Poria, S.; Cambria, E. Recent trends in deep learning based natural language processing. *IEEE Comput. Intell. Mag.* **2018**, *13*, 55–75. [CrossRef]
19. Palacio-Niño, J.-O.; Berzal, F. Evaluation metrics for unsupervised learning algorithms. *arXiv* **2019**, arXiv:1905.05667.
20. Kiran, S.; Khan, M.A.; Javed, M.Y.; Alhaisoni, M.; Tariq, U.; Nam, Y.; Damaševičius, R.; Sharif, M. Multi-Layered Deep Learning Features Fusion for Human Action Recognition. *Comput. Mater. Cont.* **2021**, *69*, 4061–4075. [CrossRef]
21. Khan, M.A.; Alhaisoni, M.; Armghan, A.; Alenezi, F.; Tariq, U.; Nam, Y.; Akram, T. Video Analytics Framework for Human Action Recognition. *Comput. Mater. Cont.* **2021**, *68*, 3841–3859.
22. Sharif, M.; Akram, T.; Yasmin, M.; Nayak, R.S. Stomach deformities recognition using rank-based deep features selection. *J. Med. Econ.* **2019**, *43*, 329.
23. Saleem, F.; Khan, M.A.; Alhaisoni, M.; Tariq, U.; Armghan, A.; Alenezi, F.; Choi, J.; Kadry, S. Human Gait Recognition: A Single Stream Optimal Deep Learning Features Fusion. *Sensors* **2021**, *21*, 7584. [CrossRef] [PubMed]
24. Khan, A.; Javed, M.Y.; Alhaisoni, M.; Tariq, U.; Kadry, S.; Choi, J.; Nam, Y. Human Gait Recognition Using Deep Learning and Improved Ant Colony Optimization. *Comput. Mater. Cont.* **2022**, *70*, 2113–2130. [CrossRef]
25. Mehmood, A.; Tariq, U.; Jeong, C.-W.; Nam, Y.; Mostafa, R.R.; Elaeiny, A. Human Gait Recognition: A Deep Learning and Best Feature Selection Framework. *Comput. Mater. Cont.* **2022**, *70*, 343–360. [CrossRef]
26. Wang, H.; Yu, B.; Xia, K.; Li, J.; Zuo, X. Skeleton Edge Motion Networks for Human Action Recognition. *Neurocomputing* **2021**, *423*, 1–12. [CrossRef]
27. Bi, Z.; Huang, W. Human action identification by a quality-guided fusion of multi-model feature. *Future Gener. Comput. Syst.* **2021**, *116*, 13–21. [CrossRef]
28. Lei, Y.; Yang, B.; Jiang, X.; Jia, F.; Li, N.; Nandi, A.K. Applications of machine learning to machine fault diagnosis: A review and roadmap. *Mech. Syst. Signal Process* **2020**, *138*, 106587. [CrossRef]

29. Manivannan, A.; Chin, W.C.B.; Barrat, A.; Bouffanais, R. On the challenges and potential of using barometric sensors to track human activity. *Sensors* **2020**, *20*, 6786. [CrossRef] [PubMed]
30. Ahmed Bhuiyan, R.; Ahmed, N.; Amiruzzaman, M.; Islam, M.R. A robust feature extraction model for human activity characterization using 3-axis accelerometer and gyroscope data. *Sensors* **2020**, *20*, 6990. [CrossRef]
31. Zhao, B.; Li, S.; Gao, Y.; Li, C.; Li, W. A Framework of Combining Short-Term Spatial/Frequency Feature Extraction and Long-Term IndRNN for Activity Recognition. *Sensors* **2020**, *20*, 6984. [CrossRef] [PubMed]
32. Muhammad, K.; Ullah, A.; Imran, A.S.; Sajjad, M.; Kiran, M.S.; Sannino, G.; Albuquerque, V.H.C. Human action recognition using attention based LSTM network with dilated CNN features. *Future Gener. Comput. Syst.* **2021**, *125*, 820–830. [CrossRef]
33. Li, C.; Xie, C.; Zhang, B.; Han, J.; Zhen, X.; Chen, J. Memory attention networks for skeleton-based action recognition. *IEEE Trans. Neural Netw. Learn. Syst.* **2021**. [CrossRef] [PubMed]
34. Im, W.; Kim, T.-K.; Yoon, S.-E. Unsupervised Learning of Optical Flow with Deep Feature Similarity. In *Computer Vision—ECCV 2020. ECCV 2020*; Lecture Notes in Computer Science; Vedaldi, A., Bischof, H., Brox, T., Frahm, J.M., Eds.; Springer: Cham, Switzerland, 2020; Volume 12369.
35. Liu, W.; Zha, Z.-J.; Wang, Y.; Lu, K.; Tao, D. p-Laplacian regularized sparse coding for human activity recognition. *IEEE Trans. Ind. Electron.* **2016**, *63*, 5120–5129. [CrossRef]
36. Jalal, A.; Kamal, S.; Kim, D. A Depth Video-based Human Detection and Activity Recognition using Multi-features and Embedded Hidden Markov Models for Health Care Monitoring Systems. *Int. J. Interact. Multimed. Artif. Intell.* **2017**, *4*, 54. [CrossRef]
37. Effrosynidis, D.; Arampatzis, A. An evaluation of feature selection methods for environmental data. *Ecol Inform.* **2021**, *61*, 101224. [CrossRef]
38. Melhart, D.; Liapis, A.; Yannakakis, G.N. The Affect Game AnnotatIoN (AGAIN) Dataset. *arXiv* **2021**, arXiv:2104.02643.
39. Hassan, M.M.; Uddin, M.Z.; Mohamed, A.; Almogren, A. A robust human activity recognition system using smartphone sensors and deep learning. *Future Gener. Comput. Syst.* **2018**, *81*, 307–313. [CrossRef]
40. Joshi, A.B.; Kumar, D.; Gaffar, A.; Mishra, D. Triple color image encryption based on 2D multiple parameter fractional discrete Fourier transform and 3D Arnold transform. *Opt. Lasers. Eng.* **2020**, *133*, 106139. [CrossRef]
41. Cervantes, J.; Garcia-Lamont, F.; Rodríguez-Mazahua, L.; Lopez, A. A comprehensive survey on support vector machine classification: Applications, challenges and trends. *Neurocomputing* **2020**, *408*, 189–215. [CrossRef]
42. Wang, L.; Xu, Y.; Cheng, J.; Xia, H.; Yin, J.; Wu, J. Human action recognition by learning spatio-temporal features with deep neural networks. *IEEE Access* **2018**, *6*, 17913–17922. [CrossRef]
43. Gumaei, A.; Hassan, M.M.; Alelaiwi, A.; Alsalman, H. A hybrid deep learning model for human activity recognition using multimodal body sensing data. *IEEE Access* **2019**, *7*, 99152–99160. [CrossRef]
44. Gao, Z.; Xuan, H.-Z.; Zhang, H.; Wan, S.; Choo, K.-K.R. Adaptive fusion and category-level dictionary learning model for multiview human action recognition. *IEEE Internet Things J.* **2019**, *6*, 9280–9293. [CrossRef]
45. Khan, M.A.; Zhang, Y.-D.; Khan, S.A.; Attique, M.; Rehman, A.; Seo, S. A resource conscious human action recognition framework using 26-layered deep convolutional neural network. *Multimed. Tools. Appl.* **2020**. [CrossRef]
46. Xia, K.; Huang, J.; Wang, H. LSTM-CNN architecture for human activity recognition. *IEEE Access* **2020**, *8*, 56855–56866. [CrossRef]
47. Rashid, M.; Sharif, M.; Raza, M.; Sarfraz, M.M.; Afza, F. Object detection and classification: A joint selection and fusion strategy of deep convolutional neural network and SIFT point features. *Multimed. Tools. Appl.* **2019**, *78*, 15751–15777. [CrossRef]
48. Huang, G.; Liu, Z.; Van Der Maaten, L.; Weinberger, K.Q. Densely connected convolutional networks. In Proceedings of the IEEE Conference on Computer Vision and Pattern Recognition, Honolulu, HI, USA, 21–26 July 2017; IEEE Press: Piscataway, NJ, USA.
49. Hussain, N.; Sharif, M.; Khan, S.A.; Albesher, A.A.; Saba, T.; Armaghan, A. A deep neural network and classical features based scheme for objects recognition: An application for machine inspection. *Multimed. Tools. Appl.* **2020**, 1–23. [CrossRef]
50. Akram, T.; Zhang, Y.-D.; Sharif, M. Attributes based skin lesion detection and recognition: A mask RCNN and transfer learning-based deep learning framework. *Pattern Recognit. Lett.* **2021**, *143*, 58–66.
51. Oquab, M.; Bottou, L.; Laptev, I.; Sivic, J. Learning and transferring mid-level image representations using convolutional neural networks. In Proceedings of the IEEE conference on computer vision and pattern recognition, Columbus, OH, USA, 23–28 June 2014.
52. Deng, J.; Dong, W.; Socher, R.; Li, L.-J.; Li, K.; Li, F.-F. Imagenet: A large-scale hierarchical image databas e. In Proceedings of the 2009 IEEE conference on computer vision and pattern recognition, Miami, FL, USA, 20–25 June 2009.
53. Krizhevsky, A.; Sutskever, I.; Hinton, G.E. Imagenet classification with deep convolutional neural networks. *NIPS* **2012**, *25*, 1097–1105. [CrossRef]
54. Naheed, N.; Shaheen, M.; Khan, S.A.; Alawairdhi, M.; Khan, M.A. Importance of features selection, attributes selection, challenges and future directions for medical imaging data: A review. *Comput. Sci. Eng.* **2020**, *125*, 314–344. [CrossRef]
55. Nadeem, A.; Jalal, A.; Kim, K. Automatic human posture estimation for sport activity recognition with robust body parts detection and entropy markov model. *Multimed. Tools. Appl.* **2021**, *22*, 1–34. [CrossRef]
56. Sharif, M.; Zahid, F.; Shah, J.H.; Akram, T. Human action recognition: A framework of statistical weighted segmentation and rank correlation-based selection. *Pattern Anal. Appl.* **2020**, *23*, 281–294. [CrossRef]

57. Akram, T.; Sharif, M.; Javed, M.Y.; Muhammad, N.; Yasmin, M. An implementation of optimized framework for action classification using multilayers neural network on selected fused features. *Pattern Anal. Appl.* **2019**, *22*, 1377–1397.
58. Laptev, I.; Marszalek, M.; Schmid, C.; Rozenfeld, B. Learning realistic human actions from movies. In Proceedings of the 2008 IEEE Conference on Computer Vision and Pattern Recognition, Anchorage, AK, USA, 23–28 June 2008.

sensors

MDPI

Article

Automatic and Efficient Fall Risk Assessment Based on Machine Learning

Nadav Eichler [1,*], Shmuel Raz [2], Adi Toledano-Shubi [3], Daphna Livne [3], Ilan Shimshoni [2] and Hagit Hel-Or [1]

[1] Department of Computer Science, University of Haifa, Haifa 3498838, Israel; hagit@cs.haifa.ac.il
[2] Department of Information Systems, University of Haifa, Haifa 3498838, Israel; razshmu@gmail.com (S.R.); ishimshoni@is.haifa.ac.il (I.S.)
[3] Physiotherapy Institute, Galilee Medical Center, Chicago, IL 60639, USA; adit@gmc.gov.il (A.T.-S.); daphnaniv@gmail.com (D.L.)
* Correspondence: eichler@outlook.com

Abstract: Automating fall risk assessment, in an efficient, non-invasive manner, specifically in the elderly population, serves as an efficient means for implementing wide screening of individuals for fall risk and determining their need for participation in fall prevention programs. We present an automated and efficient system for fall risk assessment based on a multi-depth camera human motion tracking system, which captures patients performing the well-known and validated Berg Balance Scale (BBS). Trained machine learning classifiers predict the patient's 14 scores of the BBS by extracting spatio-temporal features from the captured human motion records. Additionally, we used machine learning tools to develop fall risk predictors that enable reducing the number of BBS tasks required to assess fall risk, from 14 to 4–6 tasks, without compromising the quality and accuracy of the BBS assessment. The reduced battery, termed Efficient-BBS (E-BBS), can be performed by physiotherapists in a traditional setting or deployed using our automated system, allowing an efficient and effective BBS evaluation. We report on a pilot study, run in a major hospital, including accuracy and statistical evaluations. We show the accuracy and confidence levels of the E-BBS, as well as the average number of BBS tasks required to reach the accuracy thresholds. The trained E-BBS system was shown to reduce the number of tasks in the BBS test by approximately 50% while maintaining 97% accuracy. The presented approach enables a wide screening of individuals for fall risk in a manner that does not require significant time or resources from the medical community. Furthermore, the technology and machine learning algorithms can be implemented on other batteries of medical tests and evaluations.

Keywords: fall risk detection; balance; Berg Balance Scale; human tracking; elderly; telemedicine; diagnosis

Citation: Eichler, N.; Raz, S.; Toledano-Shubi, A.; Livne, D.; Shimshoni, I.; Hel-Or, H. Automatic and Efficient Fall Risk Assessment Based on Machine Learning. *Sensors* **2022**, *22*, 1557. https://doi.org/10.3390/s22041557

Academic Editors: Carlos Tavares Calafate, Tomasz Krzeszowski, Adam Świtoński and Michal Kepski

Received: 5 January 2022
Accepted: 14 February 2022
Published: 17 February 2022

1. Introduction

Accidental falls are a major concern in the elderly population, often requiring hospitalization, and may lead to death [1,2]. Falls are one of the main causes of disability, loss of independence, and reduced quality of life. This incurs high expenses on the individuals, their families, and the public health system [3,4]. It has been shown, however, that individuals can significantly reduce the risk of fall by participating in fall prevention programs [5,6]. Thus, there is great importance in performing a wide screening of the elderly population for the risk of fall and, consequently, initiating appropriate intervention programs.

Assessing the risk of fall is typically performed by physiotherapists and other types of medical professionals using various standardized and validated balance tests. One such test is the Berg Balance Scale (BBS) [7,8], a rigorous and time-consuming examination, since it requires the patient to perform 14 different tests. Due to its demand on the medical professional resources, these tests are not widely performed on the general public and are typically administered in the context of rehabilitation. Thus, more efficient testing methods for the risk of fall are crucial for implementing community-wide screening to identify high-risk individuals [5,6].

In this paper, we present a method to alleviate the workload in fall risk assessment. We developed our methods for the Berg Balance Scale (BBS); however, the approach is applicable to any time-consuming battery of tests. We developed an automated system for assessing the risk of fall using the BBS test, which is non-invasive and easy to use. It uses a novel self-calibrating multi-depth camera human motion tracking system previously developed by the authors. Using the data extracted from the cameras, machine learning classifiers were developed to evaluate the performance of the tasks by the patient. Thus, a medical professional is no longer needed to monitor and assess the performance of the test by the patient.

Still, performing 14 tasks is time consuming. Thus, in this paper, we present a machine-learning-based method to predict the fall risk, which enables reducing the number of BBS tasks required to assess fall risk from 14 to 4–6 tasks while maintaining the quality and accuracy of the BBS assessment (at 96%). We term the reordered and reduced BBS battery Efficient-BBS (E-BBS), as it reduces the the number of tasks to be performed and consequently reduces the time required to complete the BBS test. We present the E-BBS task ordering methods, which proceed either in a predefined order of tasks or on a per-patient adaptive task sequencing. The E-BBS can be performed by physiotherapists in a traditional setting or deployed using the automated system, allowing an efficient and effective BBS evaluation.

The automated system was tested in a major hospital, under the guidelines of the Declaration of Helsinki. The results showed high accuracy rates in predicting fall risk and showed a correlation with the BBS scores on individual BBS motion tasks as assessed by medical professionals. The E-BBS was developed by training machine learning algorithms on the data collected at the hospital. The trained E-BBS system was shown to reduce the number of tasks in the BBS test by approximately 50% while maintaining 97% accuracy.

The main scientific contribution of the paper is the novel approach to shortening and creating an adaptive sequence of testing from any given battery of tests (medical or other). The paper implemented the approach on the BBS test, but it can be exploited to reduce any battery of tasks that provides a final score or outcome to a shorter test while maintaining accuracy. The outcome of the study will also hopefully contribute to the medical community, allowing more efficient testing of the risk of fall that can be deployed in medical centers, community centers, as well as in private homes. It will allow a wider reach to the aging community and, as such, help to improve this population's welfare and quality of life together with reducing the burden on families, communities, and at the national level as well.

In the following sections, we review the automated system and introduce the E-BBS. We present a description of the full system including a review of the previously presented study in [9] with additional statistical analysis. We show the results of a pilot study, run on 130 patients in a major hospital, including the accuracy and statistical evaluations. We then present the E-BBS system and show its accuracy and its confidence levels, as well as the average number of BBS tasks that are required to reach the accuracy thresholds.

There is a plethora of balance and risk of fall tests that have been validated and are used in the medical community (see [10] for a review). Most tests involve motor tasks that are scored by a physiotherapist or medical professional. The motor tasks are mostly related to daily actions and movements that are typically performed by humans such as walking, rising from a chair, transitioning between sitting and standing positions, reaching, and more. Some tests are short and easy to administer; others are longer and include a battery of tasks, but are more comprehensive and systematic.

Short tests that focus on walking assess the time or distance required to complete the task and include the 2 m walk [11], 10 m walk [12], and 6 min walk [13]. A more comprehensive gait test is the Dynamic Gait Index [14] with several motor tasks of increasing difficulty.

Tests relying on transitioning into and out of a chair are also very common as this action is important in daily life. They assess lower body strength [15], which is related to the risk of fall [16]. The single task tests in this class include the 30 s chair stand [15], 5X-Sit-to-Stand [17], and 10X-Sit-to-Stand, requiring subjects to rise and lower themselves into a chair as fast as possible. The number of repetitions performed or the time to perform a set number of repetitions serves as the score in these tests.

A very popular balance test, combining both walking and transitioning from a chair, is the Timed Up and Go Test (TuG) [18,19]. It measures the time required to rise from a chair, walk 3 m, turn, return to the chair, and sit back down [20]. This test is popular as it is short and easy to administer, though for reliability, it is often repeated several times [21].

Another type of balance test is those based on static pose including the Unipedal Stance Test [22], Unilateral Forefoot Balance Test [23], and the Romberg Test [24]. These tests have subjects stand on one or both feet in different positions (aligned, tandem, or toe to heel) and with eyes opened or closed. Combining several of these poses in increasing difficulty is used in the 4-Stage Balance Test [25].

Finally, balance tests based on in-place stepping include the Step Test [26], where one foot is repeatedly placed on and off a step, the Four Square Step Test [27,28], where a sequence of steps is performed over objects in a square path, and the Y Balance Test [29], where subjects perform lunging steps in three direction.

The above-described tests rely on a single task or on very few tasks. Though requiring little time to administer, they are not, in general, comprehensive and rigorous. For diagnosis and referral to treatment, medical professionals typically use a more comprehensive testing scheme that includes a larger number of tasks. Though more informative, these tasks are, unfortunately, more time consuming. Common balance tests in this class include the Berg Balance Scale (BBS) [7], the Tinetti Assessment Tool (TAT) [30], the Short Physical Performance Battery (SPPB) [31], and the Balance Evaluation Systems Test (BESTest) [32]. These tests each include a battery of tasks involving holding a pose, walking, sit-to-stand transitions, and more.

As a compromise between comprehensive testing and test administration time, two approaches have been taken. For several of the lengthy tests, shorter versions have been introduced and validated such as the MiniBest [33] and the Short-BBS (SFBBS) [34] (see below). The second approach attempts to incorporate technology and advanced algorithms to assist or replace the balance test. Various sensors have been used to track individuals in their natural environment and assess their balance and risk of fall. Examples include wearable sensors [35], inertial sensors [36], and visual sensors [37,38]. Unfortunately, these intrusive methods are often uncomfortable and expensive and typically do not provide a comprehensive analysis of the patient's balance (e.g., type of imbalance and physiological source of the imbalance). Cameras and other non-contact sensors are advantageous, in being non-intrusive and being capable of collecting a wide range of data per patient. These non-intrusive sensors are desirable for hospitals, old age homes, and home care systems [39,40]. However, video cameras do not capture depth information, which, in assessing balance, may lead to erroneous outcomes and incorrect assessment of the risk of fall [41]. Depth-sensing cameras (such as the Microsoft Kinect [42] and others) can be used to capture depth in the scenes using technologies such as stereo imaging, structured light, and time-of-flight technologies [43]. Indeed, depth sensors have been used on single-task balance tests including the Get-Up-and-Go [44], 10-meter walk test [45], Single-Legged Stance Test [46], and on gait assessment [47]. However, many of the multi-task balance tests require pose and motions that give rise to self-occlusion (for example, the 360° turn in the BBS assessment), in which case multiple cameras are required. However, using more than a single camera requires calibration and synchronization [38], which is inappropriate for an easy-to-use balance assessment system. In our system, we used two depth-sensing cameras in a novel multi-depth camera tracking system, which performs synchronization and calibration automatically and requires no manual intervention [48]. Using this non-invasive technology together with Machine-Learning (ML)-based algorithms, balance and the risk of fall can be successfully and efficiently assessed.

1.1. The Berg Balance Scale

In this study, we implemented our approach on the Berg Balance Scale (BBS) [7,8], a standard and validated measure commonly used by medical professionals to assess the risk of fall.

The BBS is comprehensive and includes 14 motor tasks of varying difficulty, with tasks involving sitting and rising from a chair, holding a pose, turning, stepping, and more. Each task is scored on a five-level scale ranging from zero (unable) to four (independent). The final BBS score is obtained by summing the 14 individual task scores [8]. A BBS score of 36 or less implies a near 100% chance of fall within 6 mo [14]. Scores from 0–20 are considered high fall risk, from 21–40 medium fall risk and scores from 41–56 as low fall risk [7,8].

The BBS measure has been well studied. It has been shown to be valid and to have high sensitivity [14,33,49]. Test–retest reliability has been shown to be very good when tested on elderly individuals [50,51], stroke patients [52,53], and Parkinson's patients [54,55]. The inter- and intra-rater reliability of the BBS was also shown to be good when tested on elderly individuals [7,8,33,56,57], Parkinson's patients [55,58], stroke patients [59,60], and patients following spinal cord injury [61].

The BBS, though comprehensive, is time consuming. To compensate for the lengthy testing, a short form of the BBS was proposed (SFBBS) [34]. This test includes seven of the fourteen BBS tasks, and the rating is on a three-point scale (vs. the five-point scale of the BBS). The SFBBS was shown to have good validity, internal consistency, and reliability on stroke patients [34,62] and on the elderly [63,64] and has been shown to compare well with the standard BBS [62,64,65]. In this paper, we present the Efficient-BBS (E-BBS), an adaptive BBS testing scheme based on machine learning, and show that it significantly improves performance over the SFBBS.

2. Automated Fall Risk Assessment System

The BBS balance assessment task is highly time consuming and thus requires significant resources of the medical professional and of the medical organization as a whole. Currently, this test is most often administered to patients who have already undergone a fall or a medical procedure (stroke, hip/knee replacement, etc.) in order to assess the severity of their condition or assess their rehabilitation. Although it has been shown that timely intervention can reduce the risk of fall, detecting those individuals from the general population that are at risk and would benefit from this intervention is not easily possible, given the expense of balance assessment.

Thus, we propose to develop an automated fall risk assessment system, which is able to administer the BBS procedure and, using machine learning (ML) methods, to automatically predict the risk of fall of the subject. This can be performed without the intervention of a medical specialist and thus can be used for mass screening. Furthermore, since running the complete battery of 14 BBS tasks is time consuming, we propose a method for using a minimal number of BBS tasks that will maintain the accuracy of the standard BBS assessment while significantly reducing the test time.

To be widely used, outside medical centers, the system must be non-intrusive, portable, and easy to use, while still maintaining reliable and consistent BBS score predictions. The proposed system consists of three major components (see Figure 1):

1. Motion tracking system, including 3D cameras;
2. Automatic BBS score prediction algorithms;
3. Final fall risk assessment using machine learning.

The first two components compute the 14 BBS scores by tracking the subject's motion and using machine learning to predict the scores. This work, which was presented in [9], is reviewed in Sections 3 and 4. Section 4 also reviews the machine learning model used to predict the level of risk from the 14 previously predicted BBS scores either as a final score (from 0–56) or as one of three levels of risk (high, medium, or low risk of fall). Finally, in Section 5, we describe our novel machine-learning-based approach for predicting the final BBS score, the E-BBS, which uses an adaptively chosen subset of BBS tasks per subject, based on the subject's scores on these tasks. This approach reduces the number of tasks required to 4–6 tasks per subject.

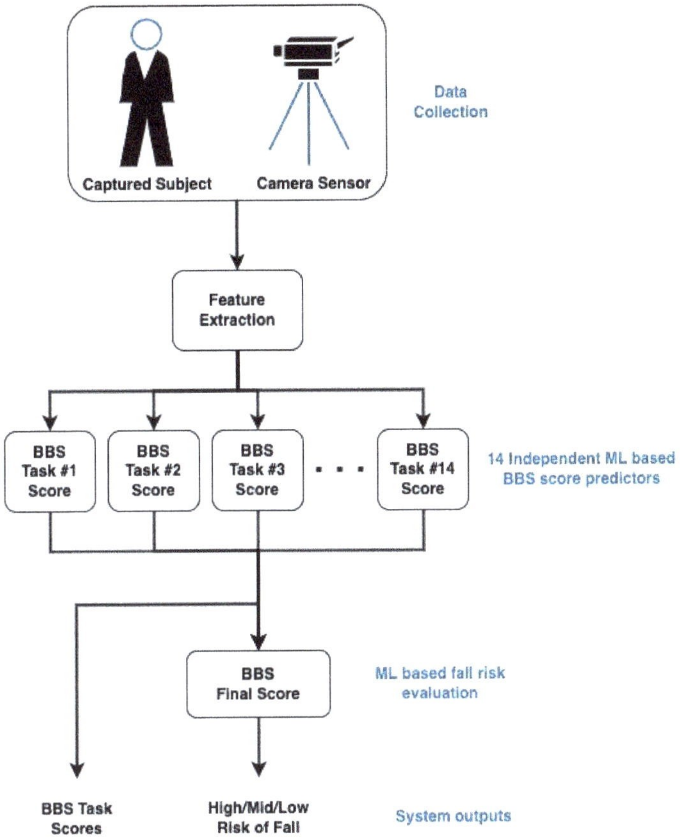

Figure 1. Schematic diagram of the BBS score and fall risk prediction system.

3. Motion Capture and Tracking

To track subjects performing the BBS tasks, we used the Microsoft Kinect [42], a depth sensor camera based on time-of-flight technology [66]. It provides depth information, i.e., the distance from the camera, for every point in the scene for each video frame. When filming human subjects, a skeletal body representation composed of 3D joints and connecting bones (Figure 2) is extracted from the captured depth information using machine learning algorithms [67–69]. For the purpose of tracking and estimating BBS task performance, we also collected the 3D data points in the patient's immediate surroundings, floor position, and orientation, as well as the 3D points of objects in the scene relevant to the BBS task.

Due to the possibility of the self-occlusion of the body during some of the BBS tasks and to ensure full coverage of the subject, we used a two-camera setup where two cameras were placed 3 m from the subject, about 2 m apart and at 45° angles. This ensured full coverage, as well as merging of the data to reduce noise and uncertainty in the skeletal structure.

A major drawback of any multi-camera system is the necessity of performing synchronization and calibration between the cameras. This process typically requires a specialized calibration session with specific calibration tools, a process that is impractical and infeasible for systems such as ours that are targeted for use in the community.

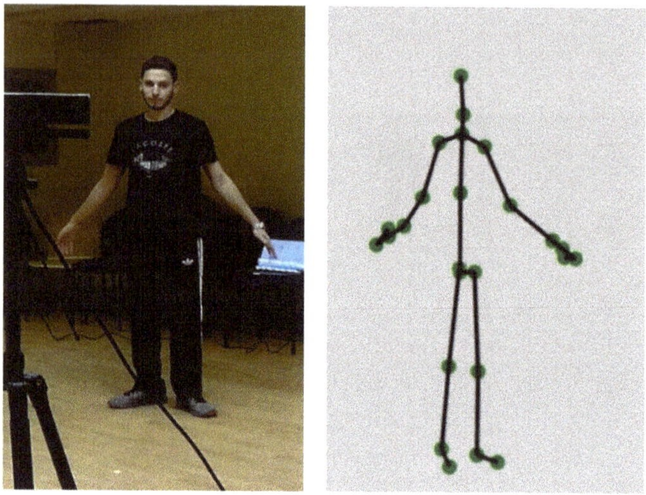

Figure 2. The 3D sensor (**left**) measures the distances of points in the scene from which a skeleton representation of the body pose is produced (**right**).

Thus, we used a novel multi-camera tracking system developed by our team [48,70] in which synchronization and calibration are performed automatically and on the fly by exploiting the patient's motion. The skeletal data acquired by the two calibrated cameras can then be easily integrated. Using this multi-depth-sensing camera tracking system allows motion and pose tracking of subjects to be non-intrusive, portable, and inexpensive.

Kinect allows motion and pose tracking of subjects to be non-intrusive, portable and inexpensive, motion capture system Motion tracking is thus performed non-intrusively.

4. Predicting BBS Scores Using Machine Learning

In this section, we review the system we developed based on computer vision tools and machine learning to predict the BBS scores of a patient on each of the 14 BBS tasks, as well as to predict the final risk of fall. The predicted scores were shown to correlate well with the scores assessed by the physiotherapists. More details can be found in [9].

Following the data collection, spatio-temporal features were extracted from the collected skeletal data and used to train a machine learning model to predict each of the 14 BBS task scores. Given the 14 predicted scores an additional model was trained to predict the final risk of fall of the patients. Figure 1 shows a diagram of the automated system.

4.1. Data Collection

Data for this project were collected in the Physiotherapy Unit at a major public hospital under the guidelines of the Declaration of Helsinki (ID: 0194-15-NHR, Galilee Medical Center). A total of 130 subjects were recruited, 100 of whom were hospital in-patients. Thirty of the subjects were visitors or care givers of patients and were recruited as subjects of low fall risk. All subjects (in-patients and controls) were aged 65 or older. Seventy-six of the subjects were female, and fifty-three were male subjects. All subjects took the BBS test in the hospital's physiotherapy room. The multi-camera tracking system (Section 3) recorded the subjects performing the 14 BBS tasks. Two physiotherapists administered and scored the patient on each of the 14 tasks. The double scoring by the physiotherapists' served to validate the scores. Due to the high BBS inter-rater reliability [7,8,33], only seldomly were the scores of the two therapists inconsistent; in these cases, the more conservative score was used. The physiotherapists' BBS scores for each patient served as the ground truth labels for training the learning models.

4.2. Feature Extraction

To train the BBS score prediction models, sets of features were defined for each of the 14 BBS tasks. The skeletal sequence acquired for each subject per each BBS task (Figure 3) served as the basis for the features. Feature extraction was performed in two steps. First, features were extracted from the skeletal structure of each frame in the sequence. These included: relative positions of skeleton joints, angles between connecting bones, distances between body parts, heights of joints from the ground, and more (Figure 4). Most of the extracted features were independent of the location of the subject and invariant to body size.

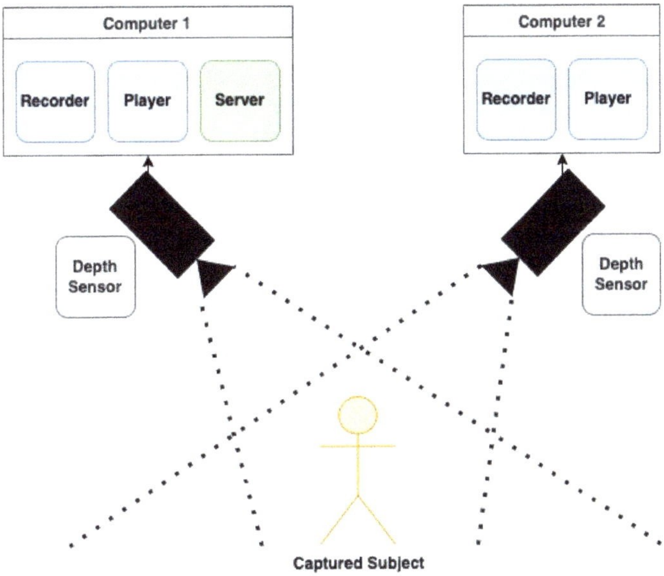

Figure 3. The multi-camera tracking system setup includes two depth sensors allowing the capture of the full range of patient motion, as well as enabling data merging to reduce noise and skeleton errors.

In the second step, spatio-temporal features were calculated from these per-frame features including: maximal/minimal/mean values of the per-frame features across all frames in the sequence, average speed and acceleration of joints across the sequence, motion-paths of the joints, and more. This set of spatio-temporal features served to represent the motion action of a subject performing a single BBS task and were used to train the machine learning algorithms.

To improve model training, the number of features was reduced by selecting the most informative features per BBS task, as computationally derived from the trained models. Feature selection was also guided by recommendations from the physiotherapists as to the most predictive parts of the body and its features. Feature selection resulted in different features per each BBS task, ranging from 100–200 features (for examples, see [9,71]).

4.3. Training

Training and testing were performed using the data collected at the hospital of patients performing the 14 tasks of the BBS test. Each task was recorded as a skeletal sequence, represented using the features described above, and labeled with the BBS score assigned by the physiotherapists. Separate models were trained to predict the BBS score for each of the 14 BBS tasks. An additional model was trained on the BBS scores to predict the final BBS fall risk assessment.

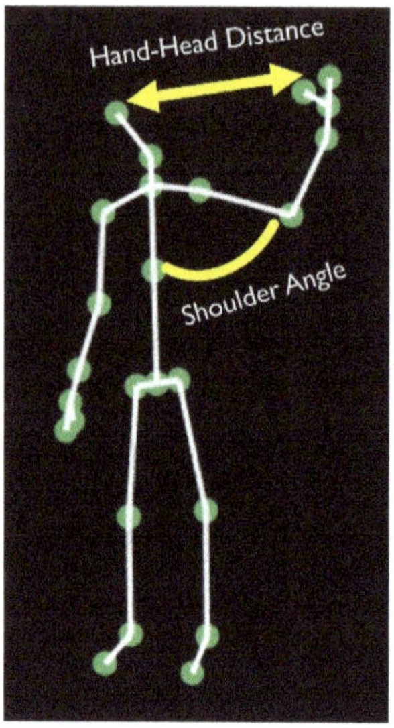

Figure 4. Spatio-temporal features are computed from the skeleton data in each recorded video frame.

For each of the 14 tasks, a random forest classifier [72,73] was trained using leave-one-out cross-validation [74]. The model hyper-parameters were sought using grid search [75]. The number of trees was set to 100 and the depth to 10. The random forest classifier was chosen as its use of bootstrapping enables these models to work well on small datasets. Furthermore, the random forest classifier allows feature ranking [76,77] in which the predictive power of features can be assessed. This in turn assists in feature selection to assist in further reducing over-fitting.

An additional ML-based classifier was trained on the 14 scores predicted by the random forests, to predict the final risk of fall (Figure 1). The risk of fall is defined as one of three categories based on the sum of BBS task scores: high risk (between 0–20), medium risk (21–40), and low risk of fall (41–56) (see Section 1.1). The risk of fall category, calculated from the physiotherapist scores on the subjects, served as the labels of the training data. An SVM classifier [78] was trained for predicting the fall risk category. The Radial Basis Function (RBF) [79] was used as the SVM kernel, with $\gamma = 1/n_f$, where n_f is the number of features, and the regularization parameter $C = 3$. Leave-one-out cross-validation [74] was used to evaluate the model's performance.

4.4. Automatic BBS Score Prediction Results

We tested the performance of the random forest models in predicting each of the BBS task scores and the SVM classifier in predicting the final fall risk category from the 14 task scores.

Table 1 shows the accuracy of the random forest score predictors for each of the 14 BBS tasks. BBS task scores are in 0–4. The number of samples (N) differed between tasks due to some patients' inability to perform tasks or due to technical difficulties in recording (such as occlusion of the subject by the physiotherapist when protecting the patient from falling). Additionally, the distribution of samples across the possible scores was not even since some tasks were very easy (e.g., sitting in a chair in Task 3) and always scored high grades. As seen in the table, the Mean-Squared Error (MSE) of the classifications was very low across tasks, implying that when the classification was incorrect, it was at most one score unit in error. In addition, we also calculated the weighted precision, recall, and F1-score.

It can be seen that the accuracy varied across the different BBS tasks with some tasks showing low performance. However, considering the end goal of assessing the final fall risk, we show that the predicting model compensated for these inaccurate task score predictions and correctly assessed the final risk with high accuracy.

Figure 5a shows the accuracy results in predicting the final risk of fall in one of three categories (high, medium, and low risk of fall). Results are shown as a 3 × 3 confusion matrix comparing the true risk of fall class as determined by the physiotherapists (the sum of the BBS scores assigned by the physiotherapists) with the predicted risk of fall. The overall accuracy was 75.5% correct with an MSE of 0.25. A concern in assessing the risk of fall is the false negative rate (e.g., nine subjects at high risk were classified as medium risk). The ML algorithm allows reducing the false negative rate by adjusting the thresholds. Figure 5b shows the confusion matrix obtained when reducing the false negatives to four subjects. This, however, incurred an increase in false positives and in the MSE (to 0.29).

Table 1. Automatic prediction of BBS scores per task.

BBS Task	Task Description	N	Samples per Class <0,1,2,3,4>	Accuracy	MSE	Recall	Precision	F1
1	Sitting to Standing	102	0,0,0,66,36	87%	0.18	0.87	0.88	0.87
2	Standing Unsupported	111	0,0,15,24,72	73%	0.36	0.73	0.71	0.71
3	Sitting with Back Unsupported	112	0,0,0,0,0,112	100%	0.0	1	1	1
4	Standing to Sitting	105	0,0,0,53,52	88%	0.15	0.88	0.88	0.88
5	Transfers	96	0,0,22,39,35	72%	0.36	0.72	0.72	0.72
6	Standing Unsupported, Eyes Closed	101	0,0,0,49,52	71%	0.32	0.71	0.72	0.71
7	Standing Unsupported, Feet Together	106	13,13,0,33,47	72%	0.37	0.72	0.72	0.72
8	Reaching Forward	75	0,17,0,24,34	73%	0.51	0.73	0.72	0.72
9	Pick up Object from the Floor	99	7,0,0,39,53	72%	0.31	0.72	0.74	0.70
10	Look Behind Shoulders	102	7,9,8,32,46	52%	1.25	0.52	0.50	0.51
11	Turn 360°	100	14,26,20,7,33	66%	0.60	0.66	0.62	0.64
12	Alternate Feet on Step	93	39,11,12,0,31	74%	0.34	0.74	0.69	0.71
13	Standing Unsupported, One Foot in Front	93	30,14,30,0,19	68%	0.54	0.68	0.64	0.64
14	Standing on One Leg	109	39,40,8,0,22	66%	0.80	0.66	0.64	0.65

Finally, feature ranking was performed on the final fall risk prediction model. Features were ranked according to their F-statistic [80]. The most predictive features were found to be:

- Turn 360° (Task #11);
- Alternate feet on step (Task #12);
- Transfers between chairs (Task #5);
- Reaching forward with outstretched arm (Task #8).

Indeed, the first two are considered in practice to be highly informative (as confirmed by the physiotherapists who co-authored this paper).

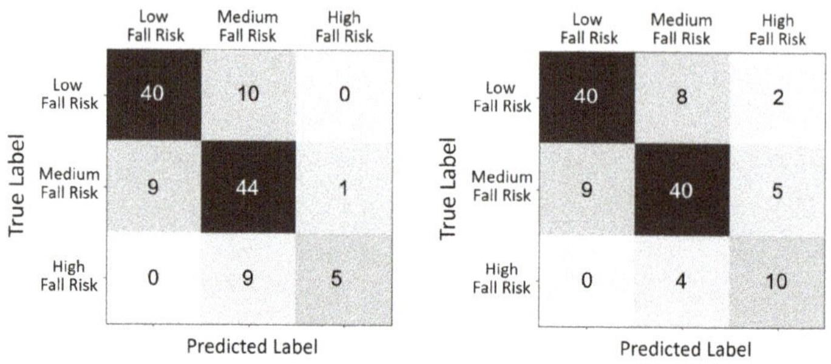

Figure 5. Confusion matrix between the true risk of fall as determined by the physiotherapists and the predicted risk of fall (**left**). False negatives can be reduced by manipulating the thresholds (**right**). The MSE values are 0.25 and 0.29, respectively.

4.5. Statistical Analysis

Statistical analysis was performed to evaluate the correlations between the physiotherapist scores of the BBS and the predicted scores produced by our automated system (termed ML predictions). Two physiotherapists scored each of the patients performing the 14 BBS tasks. For each patient, an ML prediction was calculated for each BBS task. The overall level of risk was categorized into three risk levels: high, medium, and low risk of fall. The overall level of the risk of fall was determined by the sum of the 14 scores: 0–20: high fall risk; 21–40: medium fall risk; 41–56: low fall risk.

An intraclass correlation (two-way mixed-model, single measure) [81] was used for measuring inter-rater reliability of the BBS final score between the two physiotherapists and the ML prediction. Included in the analysis also was the minimal score between the two physiotherapists (MIN(A,D)), calculated on each sample independently. This is in accord with a conservative scoring that tends toward fewer false alarms (see Section 4.4). AN Intraclass Correlation Coefficient (ICC) above 0.8 reflects high reliability, 0.6–0.79 moderate reliability, and less than 0.6 low reliability. Table 2 shows the ICC results. The ICC measure of the raters' consistency in measuring final BBS scores was higher between the physiotherapists than between the physiotherapists and the ML prediction (Table 2). Saying that, the correlation between the prediction results and the physiotherapists' measures was high (>0.83) both between the two physiotherapists and between each physiotherapist and the ML prediction.

Table 2. The intra-class correlation coefficient between physicians and ML of the BBS scores. All p-values < 0.001.

	D	Min(A,D)	ML Prediction
A	0.981	0.989	0.839
D		0.992	0.834
Min(A,D)			0.824

5. Efficient Fall Risk Evaluation Algorithm

The automated system for BBS assessment presented above is an effective method for reducing physiotherapist resources and allowing a wider screening of the elderly community for the risk of fall. In this section, we introduce an additional enhancement in which machine learning was used to reduce the number of BBS tasks required to be performed. This approach can reduce the number of tasks from 14 to an average of 4–6 tasks per subject, thus reducing the amount of time spent by the patient and the medical staff member (physiotherapist or the person supervising the automatic process) required for assessing fall risk. The approach can be applied both to the physical BBS and to the automatic system and in essence can be exploited for any other battery of tests.

The standard BBS assessment carried out either by a physiotherapist or performed using the automated method described above includes 14 BBS tasks that are performed by the subject in a predefined sequential order. The subject is scored on each of the tasks. The scores are then either summed (if collected by the physiotherapist) or run through our automated ML algorithm (Section 4) in order to assess the final fall risk of the subject into one of three classes (high, medium, or low fall risk).

Considering the BBS assessment as an iterative process (where one task is performed per iteration), every iteration can be considered as a "partial predictor" of the final fall risk assessment category. As additional tests are performed and task scores are accumulated, the prediction becomes more accurate. Thus, we used ML to develop a method in which the BBS tasks were ordered in a manner that optimized for accuracy of the final fall risk prediction and allowed for the testing to terminate early when the prediction reached a high confidence level. The BBS tasks may be administered in a predetermined optimal order constant across all subjects or may be adaptively determined per subject. Either way, the number (and consequently, the time required to perform the BBS assessment) was significantly reduced, making the whole process more efficient.

5.1. Preprocessing: Building a Dataset of Fall Risk Predictors

The goal of the adaptive fall risk evaluation algorithm was to find the minimal subset of BBS tasks that would ensure the highest classification (prediction) accuracy for the risk of fall. To this end, we built a dataset of ML-based fall risk predictors. We considered all subsets of the 14 BBS tasks ($2^{14} - 1$ subsets) and, for each subset, trained a machine learning classifier to predict the final fall risk assessment using as the input only the scores associated with the tasks in the subset. Together with the prediction, each classifier also output a measure of confidence in the prediction.

The fall risk predictors were trained using the patient data collected for the automated BBS system as described in Section 4. We created two different datasets of predictors. One dataset consisted of predictors trained on the physiotherapists' BBS scores with the ground truth risk category determined by the sum of these scores. The second dataset consisted of predictors trained on the BBS scores computed by our automated BBS assessment system described in Section 4. The fall risk category determined by the physiotherapists served as the ground truth in this case as well. Three types of machine learning algorithms were tested as predictors: SVM [78], decision trees [82] and random forest [72]. Each of these algorithms outputs the predicted risk class, as well as the confidence in the prediction. The random forest models produced the most accurate predictors, both in terms of accuracy and in terms of the average confidence level. Thus, we considered only the random forest models in this study. The random forests were trained with 100 trees.

For each dataset, the trained predictors were ranked according to the accuracy in prediction (proportion of correct fall risk predictions), as well as the average confidence of the predictions over the training set.

5.2. Efficient Re-Ordering of the BBS Tasks

The enhancement of the BBS testing that we propose involved re-ordering the BBS tasks and interactively predicting the risk of fall after each task is performed and scored. Together with the fall risk prediction, the confidence in the prediction is given after each task as well. Given a confidence threshold CT, the BBS testing terminates when the confidence exceeds the threshold. A schematic diagram of the system is shown in Figure 6. We term the new ordering and shortened sequence of BBS tasks Efficient-BBS (E-BBS), where the process is efficient in the number of tasks the patient has to perform.

The algorithm for determining the E-BBS task order requires: (a) the first BBS task (or a subset of initial tasks) and (b) a method to determine the next BBS task to perform. Let x_i be the BBS scores of the ith subject in the training set and be y_i the risk class (high, medium, low) associated with x_i (assume there are N such pairs (x_i, y_i)). Recall that the preprocessing step (Section 5.1) created a dataset of ML-based predictors for every subset of the BBS tasks. We define $Pred(SS, x_i)$ as the fall risk class prediction for x_i according to the trained predictor associated with the BBS task subset SS. The function $Conf(SS, x_i)$ returns the confidence associated with the prediction. E-BBS is an iterative process with a single BBS task performed at each iteration. Let CSS be the current subset of BBS tasks (tasks that have been performed and scored), and denote by NT the next task to be determined from among the unused set of tasks UT.

We developed and tested four different selector methods (see Figure 6) for choosing the next BBS task to be performed:

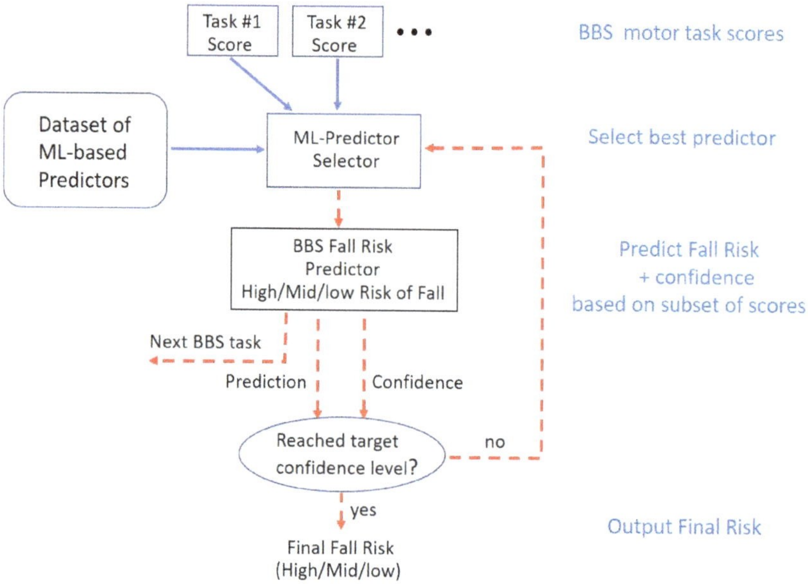

Figure 6. Schematic diagram of the E-BBS fall risk prediction system with efficient and adaptive ordering of the BBS tasks.

- Method 1. The next task NT is selected as that which when augmented to CSS creates a subset whose predictor has the highest accuracy over the complete training set.

$$NT = \arg\max_{T \in UT} \sum_{i=1}^{N} \mathbb{I}(Pred(\{CSS, T\}, x_i) = y_i),$$

  where $\mathbb{I}$ is the indicator function;
- Method 2. NT is determined as above, but with the accuracy score of the augmented subset predictor calculated only on the training examples x_i for which the CSS predictor gives a confidence below the confidence threshold CT, i.e., the x_i's for which the classifier did not yet make a decision.

$$\begin{aligned} NT &= \arg\max_{T \in UT} \sum_{i=1}^{N} \mathbb{I}(Pred(\{CSS, T\}, x_i) = y_i) \\ &\times \mathbb{I}(Conf(CSS, x_i) < CT); \end{aligned}$$

- Method 3. The third method is an adaptive method that depends on the scores x_p of the patient being tested for BBS. NT is determined as above, but the ith training example's contribution to the sum is weighted by its similarity to the scores x_p of the patient. The greater the similarity, the higher the weight is.

$$\begin{aligned} NT &= \arg\max_{T \in UT} \sum_{i=1}^{N} \mathbb{I}(Pred(\{CSS, T\}, x_i) = y_i) \\ &\times \mathbb{I}(Conf(CSS, x_i) < CT) \\ &\times d(CSS(x_i), CSS(x_p)), \end{aligned}$$

where $CSS(x_p)$ and $CSS(x_i)$ are the BBS scores of the patient and of the ith training sample restricted to the tasks in CSS. As a similarity measure, we used $d(x_i, x_j) = \exp(-||x_i - x_j||^2/\sigma^2)$, where the parameter σ^2 controls the contribution of the point as a function of the distance;

- Method 4. The fourth method extends the third method by considering only the examples in the training set for which the algorithm correctly classified the example.

$$\begin{aligned} NT &= \arg\max_{T \in UT} \sum_{i=1}^{N} \mathbb{I}(Pred(\{CSS, T\}, x_i) = y_i) \\ &\times \mathbb{I}(Conf(CSS, x_i) < CT) \\ &\times d(CSS(x_i), CSS(x_p)) \\ &\times \mathbb{I}(y_i = \hat{y}_i), \end{aligned}$$

where $\hat{y}_i$ is the final prediction of the algorithm, i.e., $Pred(AT, x_i) = \hat{y}_i$, where AT is the set of all tasks.

It can be seen that the first two selector methods produced a task sequence that was independent of the patient input. Thus, these selector methods produced a constant order of BBS tasks that was later used on all patient data when testing. Selector Methods 3 and 4 are adaptive, as the NT task is chosen based on training data, which are dependent on the data of the patient being tested. Thus, for each patient, a different BBS sequence of tasks is produced. However, we show later in Section 5.3 that, in fact, all E-BBS sequences shared the same initial portion of the task sequence.

5.3. Results: Efficient BBS

Given a starting subset of BBS tasks, a confidence threshold, and a training set, each of the four selector methods produces a different optimal ordering of BBS tasks. To evaluate the performance of each such ordering, we used five-fold cross-validation on the training set. For consistency, we also compared the results with the standard ordering of BBS tasks [7,8], as well as the Short-Form BBS (SFBBS), which selected a subset of seven tasks to be performed [34] (see Section 1.1).

The quality of the performance of a specific ordering of tasks was evaluated using two measures: the accuracy of predicting the fall risk category and the average number of BBS tasks required to complete the prediction process. Since the Efficient-BBS assessment terminates the testing when the confidence of the prediction reaches the desired threshold, the number of required BBS tasks was significantly lower than the number of BBS tasks in the standard BBS test (14).

We compared the performance of the adaptive ordering across selector methods, using confidence thresholds of 90, 92, 94, 96, 98, and 100. The initial subset of BBS tasks considered were of size 1, 2, and 3 (a discussion on the significance of starting with an initial subset of tasks is given in the Discussion Section 6). Finally, we compared the results across the two types of datasets: based on the physiotherapist scoring and based on the automatic BBS scoring.

Figure 7 plots the accuracy and the average number of BBS tasks required for the E-BBS ordering produced by the four selector methods trained on the physiotherapists scoring, as well as the standard BBS ordering. For each method, the plot shows values for the six different confidence thresholds. Naturally, the higher the confidence threshold, the longer the length of the sequence is. The initial test set was selected as the optimal set of three tasks, as discussed below, and included the three BBS tasks numbered {8,9,11} (see [7]). As can be seen, all orderings of BBS tasks reached an accuracy of around 97% correct risk of fall predictions. However, the different selector methods showed a significant reduction in average BBS tasks compared to the standard BBS, requiring from 4–6 tasks on average compared to the 14 tasks of the standard BBS. Additionally, we plot the performance of the SF-BBS [34] with seven BBS tasks at an accuracy rate of 87% on our patient data, showing that the E-BBS significantly outperforms the SF-BBS (the SF-BBS uses a three-unit scoring scale, whereas we relied on a five-unit scale used in the standard BBS testing). The four selector methods showed comparable performance with a slight advantage for Method 3.

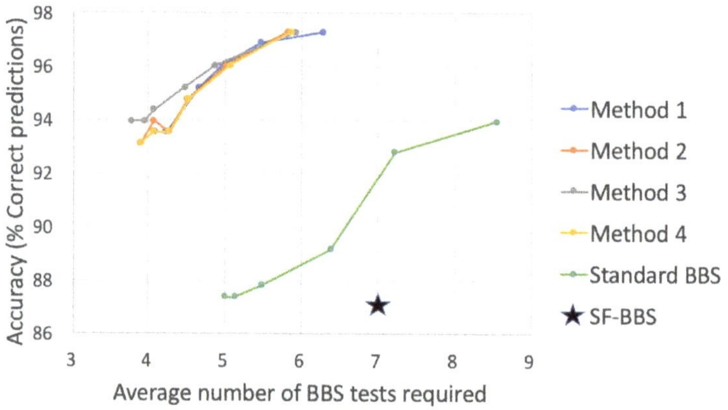

Figure 7. Accuracy vs. average number of BBS tests for different selector methods (Section 5.2) trained on the physiotherapist scoring. For each method, the plot shows values for 6 different confidence thresholds (90, 92, 94, 96, 98, and 100).

Figure 8 displays the same results as Figure 7 when training was performed on the scores predicted by the automatic BBS system. One can observe a lower rate of performance, but, as before, the standard BBS was strongly outperformed by the four selector methods, with Method 4 showing the best performance. However, in this case, all methods reached an accuracy of 76–77% correct risk of fall classification. Furthermore, it can be observed that there was a drop in accuracy when the confidence threshold reached 100. This was due to the fact that the automatic BBS score assessment was inconsistent in its performance with some of the BBS tasks showing low prediction accuracy, as shown in Table 1. The trained predictors selected the high-accuracy tasks first in the E-BBS ordering, leaving those with low accuracy to later in the ordering. When the confidence threshold was low, the BBS assessment of a subject was able to predict confidently without relying on those BBS tasks with low accuracy. However, when the confidence threshold approached 100, those tasks must be recruited, and their inaccuracy led to incorrect predictions of the overall fall risk. Albeit that there was this fault, the average number of required BBS tasks was still significantly lower than 14. We note that when continuing up to the fourteenth task, the four selector methods did not improve in accuracy beyond that shown in the plot, which is consistent with the non-adaptive results shown in Figure 5.

We now question the initial BBS tasks used by the E-BBS test. The reason for allowing a definition of an initial set of BBS tasks is that the iterative method of BBS testing and the design of the selector methods inherently imply that the optimal ordering was determined following a greedy algorithm. As such, a local minimum may be reached in the optimization. To mitigate this effect, we allowed a global optimal subset to be chosen as the initial set of tasks in the ordering.

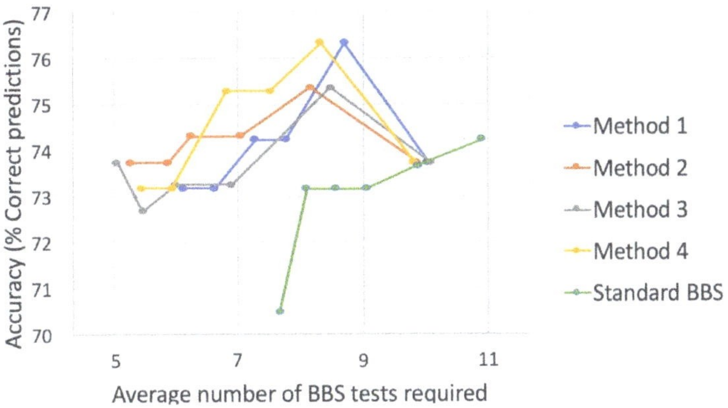

Figure 8. Accuracy vs. average number of BBS tests for different selector methods (Section 5.2) trained on the automatic BBS scoring. For each method, the plot shows values for 6 different confidence thresholds (90, 92, 94, 96, 98, and 100).

Without any external constraints on the initial task set, we chose the set to be that which performed optimally. Since the predictors trained in the preprocessing stage (Section 5.1) were each ranked by their prediction accuracy, we chose a subset of a predefined size whose predictor showed the best accuracy. We considered subsets of size 1, 2, and 3. Table 3 shows the accuracy of the predictors associated with subsets of size 1 when trained on the patient data with physiotherapists' scoring. The results in the table can be interpreted as the predictive quality of each individual task of the BBS. It can be seen that Task #9, as a single task, was the best predictor of fall risk on our test set with 85.5% accuracy. Similarly, for subsets of size 2 and 3, we found that the optimal initial task sets were {9,11} and {8,9,11}, respectively.

Table 3. Singe BBS tasks—predictor accuracy.

BBS Task	Accuracy (%)
9	85.5
7	81.4
6	81.2
11	80.8
8	80.0
4	77.8
5	77.4
12	76.2
1	74.2
10	72.6
2	70.7
13	67.5
14	67.3
3	50.8

Figure 9 shows the accuracy vs. the average number of BBS tasks required, when using different initial subsets of BBS tasks. For comparison, also shown are the results for Subset {1} and for the standard BBS test sequence. Results are shown for Selector Method 3. As can be seen, all E-BBS orderings were significantly better than the standard BBS and also better than the Subset {1} case. The accuracy was highest for the subset of size 3, reaching 97% accuracy at a confidence level of 100. All orderings required only 3–6 BBS tasks on average. Using BBS Task 1 as the initial task, as is used in the standard BBS test, showed the least accurate results of the E-BBS orderings. This is indicative of the structure of the standard BBS test where "easier" tasks are performed at the beginning of the testing sequence. These, however, are less informative and have a lower predictive quality (see Table 3). In the optimal ordering, these would appear later in the ordering, with the more informative tasks appearing first.

Finally, we studied the new order of BBS tasks as expressed in the E-BBS. We first considered the physiotherapist training set and, for simplicity, focused on the initial task subset with the single BBS Task #9, which was determined as the optimal starting task, and we set the confidence threshold to 100. We considered the four task selector methods (Section 5.2) and considered the E-BBS task sequence they produced over a test set of patients. To present the results, we used occurrence matrices, as shown in Figures 10 and 11. Columns of the matrix indicate the order in the E-BBS sequence. Each row indicates a standard BBS task enumerated 1–14. The value in each matrix entry (i,j) indicates the proportion of times that BBS task i appeared in an E-BBS sequence in position j across all E-BBS sequences produced over the test set.

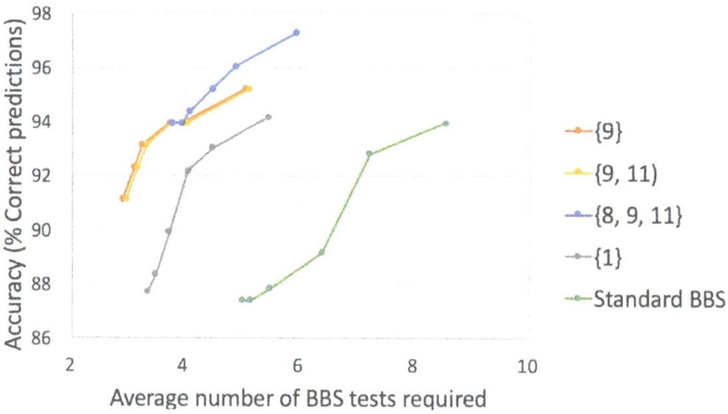

Figure 9. Accuracy vs. average number of BBS tests for the different initial subset of tasks. Results are shown for Selector Method 3 and training on the physiotherapists' data. For each initial subset of the tasks, the plot shows values for 6 different confidence thresholds (90, 92, 94, 96, 98, and 100).

Figure 10 displays four occurrence matrices trained and tested on the physiotherapist data. Matrices (a) to (d) show results for Selector Methods 1 to 4, respectively. It can be seen that the number of BBS tasks used in the E-BBS sequences decreased along the order. This was due to the fact that for most patients, the number of tasks required to reach the confidence threshold was much lower than 14, and the E-BBS evaluation was terminated before all 14 tasks were performed.

As expected, Selector Methods 1 and 2, which are not-adaptive, produced a constant sequence of the E-BBS, which is a permutation of the standard BBS. Selector Methods 3 and 4 are adaptive and thus produced a different E-BBS sequence for each subject. However, it can be seen that the first two tasks in the sequence were always the same—Tasks #9 and #11 (followed by #8 with high probability)—and then showed variability in the subsequent tasks, with Selector Method 3 showing a wider variability than Method 4. More interesting is the fact that the initial part of the E-BBS sequence was similar across all four selector methods (all four matrices showed initial BBS Tasks 9, 11, 8, and even 7 with high values). This indicates that regardless of whether the adaptive or constant E-BBS is used, the same BBS tasks will be invoked initially, implying that these tasks are predictive of the final assessment of the risk of fall.

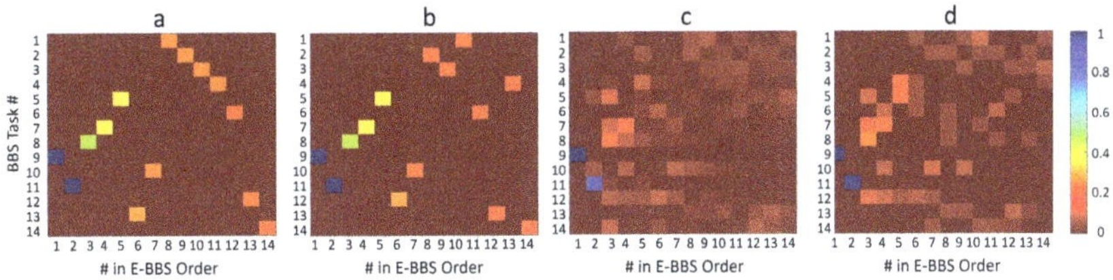

Figure 10. Occurrence matrices depicting the ordering of BBS tasks in the E-BBS. Columns indicate the order in the E-BBS sequence. Each row indicates a standard BBS task as defined in [7]. The matrix entry value indicates the proportion of times a BBS task was used in a certain E-BBS sequence position across the test set. (**a–d**) Occurrence matrices of E-BBS sequences as trained on the physiotherapist data and using the 4 Selector Methods 1 to 4, respectively.

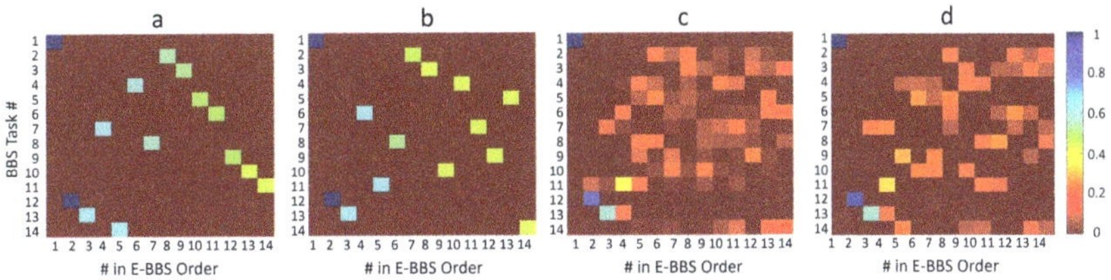

Figure 11. (**a–d**) same as Figure 10, but trained on the automatic BBS scoring data.

Figure 11 displays similar occurrence matrices trained on the automatic scoring of BBS patients. Here too we see similar characteristics, albeit noisier. The common initial tasks in the E-BBS sequence on these data were BBS Tasks 1, 12, and 13. The distinction between this sequence and that obtained for the physiotherapist data was due to the fact that the automatic system introduces errors in the BBS scoring itself. Thus, the tasks appearing early in the E-BBS are those that are predictive of fall risk, as well as reliable in terms of automatic BBS scoring.

The outcome of this analysis implies that the E-BBS order of BBS tasks can be set as constant for the first three tasks (namely, Tasks 9, 11, and 8), followed by either the constant sequence determined by Selector Methods 1 and 2 or performed adaptively per patient using Selector Methods 3 or 4. Considering that most patient testing terminated early due to reaching the desired confidence level, the E-BBS sequence beyond the first 3-6 tasks was rare.

We summarize the orderings of tasks in the E-BBS testing in Tables 4 and 5. In Table 4, sequences are shown for Selector Methods 1 and 2 and for the physiotherapist data and the automatic scoring data. As described above, the first three tasks are common to all E-BBS options, diverging only later. Regarding Methods 3 or 4, tasks were selected adaptively for each subject according to the BBS scores achieved until this step. Table 5 shows an example of a single subject for both methods. Note that this task sequence terminated at different points for each subject dependent on the subject's scores and the configured confidence threshold CT.

Table 4. E-BBS order of tasks using Methods 1 and 2. Task numbers are the standard BBS task numbers [7].

Data	Method #	T 1	T 2	T 3	T 4	T 5	T 6	T 7	T 8	T 9	T 10	T 11	T 12	T 13	T 14
Physiotherapist	1	9	11	8	7	5	13	10	1	2	3	4	6	12	14
	2	9	11	8	7	5	12	10	2	3	1	6	13	4	14
Automatic	1	1	12	13	7	14	4	8	2	3	5	6	9	10	11
	2	1	12	13	6	11	8	2	3	10	4	7	9	5	14

Table 5. E-BBS order of tasks using Methods 3 and 4. Task numbers are the standard BBS task numbers [7].

Data	Method #	T 1	T 2	T 3	T 4	T 5	T 6	T 7	T 8	T 9	T 10	T 11	T 12	T 13	T 14
Physiotherapist	3	9	11	8	7	4	5	10	1	3	14	2	13	12	6
	4	9	11	8	7	2	5	10	1	4	3	6	13	14	12
Automatic	3	1	12	13	11	5	4	9	2	10	14	8	7	6	3
	4	1	12	13	11	4	5	7	10	14	8	2	9	3	6

6. Discussion and Conclusions

We presented an approach to automating the BBS fall risk assessment test. The approach involves two main parts. First, a computer vision and ML-based system tracks the motion and pose of human subjects performing the BBS tasks, and then, a two-level ML model first predicts the BBS score for each of the fourteen tasks, the output of which is fed into another ML model, which then predicts the final fall risk category. In addition, we presented an ML-based method that determines an Efficient-BBS (E-BBS) battery of tests, requiring the patient to perform only a subset of the original BBS tests, while achieving the same quality of prediction as the full BBS test in a significantly shorter time. We emphasize that the E-BBS can be implemented on the outputs predicted by the automated BBS score predictor or directly on the scores supplied by the physiotherapists.

The approaches presented in this paper were tested on data collected at a major hospital where physiotherapists provided BBS scores and the level of fall risk for hospital patients and healthy subjects. The system showed high accuracy rates on assessing fall risk and good correlation with ground truth scores on the individual BBS tasks. In our experiments, we used real test results, where the tests were performed in the standard order, but we simulated the order of the tests for the E-BBS evaluation. In a real setting, the physiotherapists (our co-authors) stated that the order of tests has some importance and starting first with easier tests might produce better scores by the patients. Thus, additional considerations could be added into the subset selection process, possibly incurring a slight decrease in performance. This is a topic of future research.

The complete system is non-invasive and easy to use in a set-up-and-go form, well suited to be used by non-technically-savvy individuals. Furthermore, the E-BBS allows the testing to be significantly more time efficient. Thus, the system is well suited for expanding testing beyond the confines of hospitals, medical centers, and doctors' offices. It allows implementing a wide-scale screening of the elderly population for a high risk of fall. The system can efficiently determine those at low risk and, more importantly, direct those found to be at high risk to further medical assessment and preventive treatment.

Finally, we note that this study focused on evaluating the risk of fall and the BBS scores. However, the motion analysis, as well as the efficient sequencing approach can be applied to any other sequence of assessment tests.

Author Contributions: Conceptualization: All authors; data collection—test administrations: A.T.-S. and D.L.; data collection filming: N.E. and S.R.; software: N.E.; analysis—feature extraction: All authors; machine learning: N.E., I.S. and H.H.-O.; statistics: S.R.; writing: N.E., I.S. and H.H.-O.; intro, background, and literature survey: N.E., A.T.-S. and D.L.; article review and editing: All authors; graphics: N.E., I.S. and H.H.-O.; supervision: I.S. and H.H.-O.; project administration: S.R. All authors have read and agreed to the published version of the manuscript.

Funding: This research was funded by a grant from the Israel Innovation Authority (Dockets 63436 and 67323) and from the Israeli Science Foundation Grant No. 1455/16.

Institutional Review Board Statement: The study was conducted according to the guidelines of the Declaration of Helsinki and approved by the Institutional Review Board Galilee Medical Center, Israel. Approval Number 0115-18-NHR.

Informed Consent Statement: Informed consent was obtained from all subjects involved in the study.

Acknowledgments: The authors would like to thank Said Touré for assistance in editing and labeling the videos for analysis.

Conflicts of Interest: The authors declare no conflict of interest.

References

1. McCarthy, M. Falls are leading cause of injury deaths among older people, US study finds. *BMJ* **2016**, *354*, i5190. [CrossRef]
2. Bergen, G. Falls and fall injuries among adults aged $\geq$ 65 years—United States, 2014. *Morb. Mortal. Wkly. Rep.* **2016**, *65*, 993–998. [CrossRef] [PubMed]
3. Florence, C.S.; Bergen, G.; Atherly, A.; Burns, E.; Stevens, J.; Drake, C. Medical costs of fatal and nonfatal falls in older adults. *J. Am. Geriatr. Soc.* **2018**, *66*, 693–698. [CrossRef] [PubMed]
4. Czerwiński, E.; Białoszewski, D.; Borowy, P.; Kumorek, A.; Białoszewski, A. Epidemiology, clinical significance, costs and fall prevention in elderly people. *Ortop. Traumatol. Rehabil.* **2008**, *10*, 419–428.
5. Gillespie, L.D.; Robertson, M.C.; Gillespie, W.J.; Sherrington, C.; Gates, S.; Clemson, L.M.; Lamb, S.E. Interventions for preventing falls in older people living in the community. *Cochrane Database Syst. Rev.* **2012**, *9*, 2. [CrossRef]
6. Stevens, J.A.; Lee, R. The potential to reduce falls and avert costs by clinically managing fall risk. *Am. J. Prev. Med.* **2018**, *55*, 290–297. [CrossRef]
7. Berg, K.O.; Wood-Dauphine, S.; Williams, I.J.; Gayton, D. Measuring balance in the elderly: Preliminary development of an instrument. *Physiother. Can.* **1989**, *41*, 304–311. [CrossRef]
8. Berg, K.O.; Wood-Dauphinee, S.L.; Williams, I.J.; Maki, B. Measuring balance in the elderly: Validation of an instrument. *Can. J. Public Health* **1992**, *83*, S7–S11.
9. Masalha, A.; Eichler, N.; Raz, S.; Toledano-Shubi, A.; Niv, D.; Shimshoni, I.; Hel-Or, H. Predicting Fall Probability Based on a Validated Balance Scale. In Proceedings of the Computer Vision and Pattern Recognition (CVPR) CVPM Workshop, Seattle, WA, USA, 14–19 June 2020.
10. Soubra, R.; Chkeir, A.; Novella, J.L. A Systematic Review of Thirty-One Assessment Tests to Evaluate Mobility in Older Adults. *BioMed Res. Int.* **2019**, *2019*, 1354362 . [CrossRef]
11. Butland, R.J.; Pang, J.; Gross, E.R.; Woodcock, A.A.; Geddes, D.M. Two-, six-, and 12-minute walking tests in respiratory disease. *Br. Med. J. (Clin. Res. Ed.)* **1982**, *284*, 1607. [CrossRef]
12. Flansbjer, U.B.; Holmbäck, A.M.; Downham, D.; Patten, C.; Lexell, J. Reliability of gait performance tests in men and women with hemiparesis after stroke. *J. Rehabil. Med.* **2005**, *37*, 75–82. [PubMed]
13. ATS Committee on Proficiency Standards for Clinical Pulmonary Function Laboratories. ATS statement: Guidelines for the six-minute walk test. *Am. J. Respir. Crit. Care Med.* **2002**, *166*, 111–117. [CrossRef] [PubMed]
14. Shumway-Cook, A.; Baldwin, M.; Polissar, N.L.; Gruber, W. Predicting the probability for falls in community-dwelling older adults. *Phys. Ther.* **1997**, *77*, 812–819. [CrossRef] [PubMed]
15. Jones, C.; Rikli, R.; Beam, W. A 30-s chair-stand test as a measure of lower body strength in community-residing older adults. *Res. Q. Exerc. Sport* **1999**, *70*, 113–119. [CrossRef] [PubMed]
16. Cheng, Y.Y.; Wei, S.H.; Chen, P.Y.; Tsai, M.W.; Cheng, I.C.; Liu, D.H.; Kao, C.L. Can sit-to-stand lower limb muscle power predict fall status? *Gait Posture* **2014**, *40*, 403–407. [CrossRef]
17. Buatois, S.; Miljkovic, D.; Manckoundia, P.; Gueguen, R.; Miget, P.; Vançon, G.; Perrin, P.; Benetos, A. Five times sit to stand test is a predictor of recurrent falls in healthy community-living subjects aged 65 and older. *J. Am. Geriatr. Soc.* **2008**, *56*, 1575–1577. [CrossRef]
18. Mathias, S.; Nayak, U.; Isaacs, B. Balance in elderly patients: The "get-up and go" test. *Arch. Phys. Med. Rehabil.* **1986**, *67*, 387–389.
19. Podsiadlo, D.; Richardson, S. The timed "Up & Go": A test of basic functional mobility for frail elderly persons. *J. Am. Geriatr. Soc.* **1991**, *39*, 142–148.
20. Shumway-Cook, A.; Brauer, S.; Woollacott, M. Predicting the probability for falls in community-dwelling older adults using the Timed Up & Go Test. *Phys. Ther.* **2000**, *80*, 896–903.
21. Bloch, M.L.; Jønsson, L.R.; Kristensen, M.T. Introducing a third timed up & go test trial improves performances of hospitalized and community-dwelling older individuals. *J. Geriatr. Phys. Ther.* **2017**, *40*, 121.
22. Fregly, A.R.; Graybiel, A. An ataxia test battery not requiring rails. *Aerosp. Med.* **1968**, *39*, 277–282. [PubMed]
23. Clark, M.S. The Unilateral Forefoot Balance Test: Reliability and validity for measuring balance in late midlife women. *N. Z. J. Physiother.* **2007**, *35*, 110.
24. Rogers, J. Romberg and his test. *J. Laryngol. Otol.* **1980**, *94*, 1401–1404. [CrossRef]
25. Rossiter-Fornoff, J.E.; Wolf, S.L.; Wolfson, L.I.; Buchner, D.M.; Group, F. A cross-sectional validation study of the FICSIT common data base static balance measures. *J. Gerontol. Ser. Biol. Sci. Med. Sci.* **1995**, *50*, M291–M297. [CrossRef] [PubMed]
26. Hill, K.D.; Bernhardt, J.; McGann, A.M.; Maltese, D.; Berkovits, D. A new test of dynamic standing balance for stroke patients: reliability, validity and comparison with healthy elderly. *Physiother. Can.* **1996**, *48*, 257–262. [CrossRef]
27. Dite, W.; Temple, V.A. Four Square Step Test (FSST). *Arch. Phys. Med. Rehabil.* **2002**, *83*, 1566–1571. [CrossRef]
28. Moore, M.; Barker, K. The validity and reliability of the four square step test in different adult populations: A systematic review. *Syst. Rev.* **2017**, *6*, 187. [CrossRef]
29. Neves, L. The Y Balance Test—How and Why to Do it? *Int. Phys. Med. Rehabil. J.* **2017**, *2*, 48.

30. Tinetti, M.E.; Williams, T.F.; Mayewski, R. Fall risk index for elderly patients based on number of chronic disabilities. *Am. J. Med.* **1986**, *80*, 429–434. [CrossRef]
31. Guralnik, J.M.; Ferrucci, L.; Pieper, C.F.; Leveille, S.G.; Markides, K.S.; Ostir, G.V.; Studenski, S.; Berkman, L.F.; Wallace, R.B. Lower extremity function and subsequent disability: Consistency across studies, predictive models, and value of gait speed alone compared with the short physical performance battery. *J. Gerontol. Ser. Biol. Sci. Med. Sci.* **2000**, *55*, M221–M231. [CrossRef]
32. Horak, F.B.; Wrisley, D.M.; Frank, J. The balance evaluation systems test (BESTest) to differentiate balance deficits. *Phys. Ther.* **2009**, *89*, 484–498. [CrossRef] [PubMed]
33. Viveiro, L.A.P.; Gomes, G.C.V.; Bacha, J.M.R.; Junior, N.C.; Kallas, M.E.; Reis, M.; Jacob Filho, W.; Pompeu, J.E. Reliability, Validity, and Ability to Identity Fall Status of the Berg Balance Scale, Balance Evaluation Systems Test (BESTest), Mini-BESTest, and Brief-BESTest in Older Adults Who Live in Nursing Homes. *J. Geriatr. Phys. Ther.* **2019**, *42*, E45–E54. [CrossRef] [PubMed]
34. Chou, C.Y.; Chien, C.W.; Hsueh, I.P.; Sheu, C.F.; Wang, C.H.; Hsieh, C.L. Developing a short form of the Berg Balance Scale for people with stroke. *Phys. Ther.* **2006**, *86*, 195–204. [CrossRef] [PubMed]
35. Sun, R.; Sosnoff, J.J. Novel sensing technology in fall risk assessment in older adults: A systematic review. *BMC Geriatr.* **2018**, *18*, 14. [CrossRef]
36. Howcroft, J.; Kofman, J.; Lemaire, E.D. Review of fall risk assessment in geriatric populations using inertial sensors. *J. Neuroeng. Rehabil.* **2013**, *10*, 91. [CrossRef]
37. Luque, R.; Casilari, E.; Morón, M.J.; Redondo, G. Comparison and characterization of android-based fall detection systems. *Sensors* **2014**, *14*, 18543–18574. [CrossRef]
38. Yang, L.; Ren, Y.; Hu, H.; Tian, B. New fast fall detection method based on spatio-temporal context tracking of head by using depth images. *Sensors* **2015**, *15*, 23004–23019. [CrossRef]
39. Aslan, M.; Sengur, A.; Xiao, Y.; Wang, H.; Ince, M.C.; Ma, X. Shape feature encoding via fisher vector for efficient fall detection in depth-videos. *Appl. Soft Comput.* **2015**, *37*, 1023–1028. [CrossRef]
40. Vallabh, P.; Malekian, R. Fall detection monitoring systems: A comprehensive review. *J. Ambient. Intell. Humaniz. Comput.* **2018**, *9*, 1809–1833. [CrossRef]
41. Kwolek, B.; Kepski, M. Human fall detection on embedded platform using depth maps and wireless accelerometer. *Comput. Methods Programs Biomed.* **2014**, *117*, 489–501. [CrossRef]
42. Microsoft. Kinect V2 RGB-D Sensor Website. Available online: https://developer.microsoft.com/en-us/windows/kinect (accessed on 2 January 2022).
43. Sarbolandi, H.; Lefloch, D.; Kolb, A. Kinect range sensing: Structured-light versus Time-of-Flight Kinect. *Comput. Vis. Image Underst.* **2015**, *139*, 1–20. [CrossRef]
44. Kargar, A.B.; Mollahosseini, A.; Struemph, T.; Pace, W.; Nielsen, R.D.; Mahoor, M.H. Automatic measurement of physical mobility in Get-Up-and-Go Test using kinect sensor. In Proceedings of the International Conference, IEEE Engineering in Medicine and Biology Society, Chicago, IL, USA, 26–30 August 2014; pp. 3492–3495.
45. Geerse, D.J.; Coolen, B.H.; Roerdink, M. Kinematic validation of a multi-Kinect v2 instrumented 10-meter walkway for quantitative gait assessments. *PLoS ONE* **2015**, *10*, e0139913. [CrossRef]
46. Eltoukhy, M.; Kuenze, C.; Oh, J.; Signorile, J. Balance Assessment using Microsoft Xbox Kinect: 1136 Board number 315. *Med. Sci. Sport. Exerc.* **2017**, *49*, 315. [CrossRef]
47. Clark, R.; Vernon, S.; Mentiplay, B.; Miller, K.; Mcginley, J.; Pua, Y.; Paterson, K.; Bower, K. Instrumenting gait assessment using the Kinect in people with stroke: Reliability and association with balance tests. *J. Neuroeng. Rehabil.* **2015**, *12*, 15. [CrossRef] [PubMed]
48. Eichler, N.; Hel-Or, H.; Shimshoni, I.; Itah, D.; Gross, B.; Raz, S. 3D motion capture system for assessing patient motion during Fugl-Meyer stroke rehabilitation testing. *IET Comput. Vis.* **2018**, *12*, 963–975. [CrossRef]
49. Bogle Thorbahn, L.D.; Newton, R.A. Use of the Berg Balance Test to predict falls in elderly persons. *Phys. Ther.* **1996**, *76*, 576–583. [CrossRef]
50. Newstead, A.H.; Hinman, M.R.; Tomberlin, J.A. Reliability of the Berg Balance Scale and balance master limits of stability tests for individuals with brain injury. *J. Neurol. Phys. Ther.* **2005**, *29*, 18–23. [CrossRef]
51. Donoghue, D.; Stokes, E.K. How much change is true change? The minimum detectable change of the Berg Balance Scale in elderly people. *J. Rehabil. Med.* **2009**, *41*, 343–346. [CrossRef]
52. Hiengkaew, V.; Jitaree, K.; Chaiyawat, P. Minimal detectable changes of the Berg Balance Scale, Fugl-Meyer Assessment Scale, Timed "Up & Go" Test, gait speeds, and 2-minute walk test in individuals with chronic stroke with different degrees of ankle plantarflexor tone. *Arch. Phys. Med. Rehabil.* **2012**, *93*, 1201–1208.
53. Flansbjer, U.B.; Blom, J.; Brogårdh, C. The reproducibility of Berg Balance Scale and the Single-leg Stance in chronic stroke and the relationship between the 2 tests. *PM&R* **2012**, *4*, 165–170.
54. Steffen, T.; Seney, M. Test-retest reliability and minimal detectable change on balance and ambulation tests, the 36-item short-form health survey, and the unified Parkinson disease rating scale in people with parkinsonism. *Phys. Ther.* **2008**, *88*, 733–746. [CrossRef] [PubMed]
55. Leddy, A.L.; Crowner, B.E.; Earhart, G.M. Functional gait assessment and balance evaluation system test: Reliability, validity, sensitivity, and specificity for identifying individuals with Parkinson disease who fall. *Phys. Ther.* **2011**, *91*, 102–113. [CrossRef]

56. Conradsson, M.; Lundin-Olsson, L.; Lindelöf, N.; Littbrand, H.; Malmqvist, L.; Gustafson, Y.; Rosendahl, E. Berg balance scale: Intrarater test-retest reliability among older people dependent in activities of daily living and living in residential care facilities. *Phys. Ther.* **2007**, *87*, 1155–1163. [CrossRef] [PubMed]
57. Holbein-Jenny, M.A.; Billek-Sawhney, B.; Beckman, E.; Smith, T. Balance in personal care home residents: A comparison of the Berg Balance Scale, the Multi-Directional Reach Test, and the Activities-specific Balance Confidence Scale. *J. Geriatr. Phys. Ther.* **2005**, *28*, 48–53. [CrossRef] [PubMed]
58. Scalzo, P.L.; Nova, I.C.; Perracini, M.R.; Sacramento, D.R.; Cardoso, F.; Ferraz, H.B.; Teixeira, A.L. Validation of the Brazilian version of the Berg balance scale for patients with Parkinson's disease. *Arq.-Neuro-Psiquiatr.* **2009**, *67*, 831–835. [CrossRef] [PubMed]
59. Mao, H.F.; Hsueh, I.P.; Tang, P.F.; Sheu, C.F.; Hsieh, C.L. Analysis and comparison of the psychometric properties of three balance measures for stroke patients. *Stroke* **2002**, *33*, 1022–1027. [CrossRef]
60. Berg, K.; Wood-Dauphinee, S.; Williams, J. The Balance Scale: Reliability assessment with elderly residents and patients with an acute stroke. *Scand. J. Rehabil. Med.* **1995**, *27*, 27–36.
61. Wirz, M.; Müller, R.; Bastiaenen, C. Falls in persons with spinal cord injury: Validity and reliability of the Berg Balance Scale. *Neurorehabilit. Neural Repair* **2010**, *24*, 70–77. [CrossRef]
62. Liaw, L.J.; Hsieh, C.L.; Hsu, M.J.; Chen, H.M.; Lin, J.H.; Lo, S.K. Test–retest reproducibility of two short-form balance measures used in individuals with stroke. *Int. J. Rehabil. Res.* **2012**, *35*, 256–262. [CrossRef]
63. Kim, S.G.; Kim, M.K. The intra-and inter-rater reliabilities of the Short Form Berg Balance Scale in institutionalized elderly people. *J. Phys. Ther. Sci.* **2015**, *27*, 2733–2734. [CrossRef]
64. Karthikeyan, G.; Sheikh, S.G.; Chippala, P. Test-retest reliability of short form of berg balance scale in elderly people. *Glo Adv. Res. J. Med. Med. Sci.* **2012**, *1*, 139–144.
65. Jogi, P.; Spaulding, S.J.; Zecevic, A.A.; Overend, T.J.; Kramer, J.F. Comparison of the original and reduced versions of the Berg Balance Scale and the Western Ontario and McMaster Universities Osteoarthritis Index in patients following hip or knee arthroplasty. *Physiother. Can.* **2011**, *63*, 107–114. [CrossRef] [PubMed]
66. Hansard, M.; Lee, S.; Choi, O.; Horaud, R.P. *Time-of-Flight Cameras: Principles, Methods and Applications*; Springer Science & Business Media: Berlin/Heidelberg, Germany, 2012.
67. Shotton, J.; Sharp, T.; Kipman, A.; Fitzgibbon, A.; Finocchio, M.; Blake, A.; Cook, M.; Moore, R. Real-time human pose recognition in parts from single depth images. *Commun. ACM* **2013**, *56*, 116–124. [CrossRef]
68. Han, J.; Shao, L.; Xu, D.; Shotton, J. Enhanced Computer Vision With Microsoft Kinect Sensor: A Review. *IEEE Trans. Cybern.* **2013**, *43*, 1318–1334.
69. Wang, Q.; Kurillo, G.; Ofli, F.; Bajcsy, R. Evaluation of Pose Tracking Accuracy in the First and Second Generations of Microsoft Kinect. In Proceedings of the International Conference on Healthcare Informatics (ICHI), Dallas, TX, USA, 21–23 October 2015; pp. 380–389.
70. Eichler, N.; Hel-Or, H.; Shmishoni, I.; Itah, D.; Gross, B.; Raz, S. Non-invasive motion analysis for stroke rehabilitation using off the shelf 3d sensors. In Proceedings of the International Joint Conference on Neural Networks (IJCNN), Rio de Janeiro, Brazil, 8–13 July 2018; pp. 1–8.
71. Masalha, A. Predicting Fall Probability Based on a Validated Balance Scale. Master's Thesis, University of Haifa, Haifa, Israel, 2020.
72. Ho, T.K. The random subspace method for constructing decision forests. *IEEE Trans. PAMI* **1998**, *20*, 832–844.
73. Breiman, L. Random forests. *Mach. Learn.* **2001**, *45*, 5–32. [CrossRef]
74. Molinaro, A.M.; Simon, R.; Pfeiffer, R.M. Prediction error estimation: A comparison of resampling methods. *Bioinformatics* **2005**, *21*, 3301–3307. [CrossRef]
75. Hsu, C.W.; Chang, C.C.; Lin, C.J. A Practical Guide to Support Vector Classification. 2003. Available online: https://www.csie.ntu.edu.tw/~cjlin/papers/guide/guide.pdf (accessed on 2 January 2022)
76. Chizi, B.; Maimon, O. Dimension Reduction and Feature Selection. In *Data Mining and Knowledge Discovery Handbook*; Springer: Berlin/Heidelberg, Germany, 2009; pp. 83–100.
77. Chizi, B.; Rokach, L.; Maimon, O. A survey of feature selection techniques. In *Encyclopedia of Data Warehousing and Miningn*, 2nd ed.; IGI Global: Hershey, PA, USA, 2009; pp. 1888–1895.
78. Cortes, C.; Vapnik, V. Support-vector networks. *Mach. Learn.* **1995**, *20*, 273–297. [CrossRef]
79. Chang, Y.W.; Hsieh, C.J.; Chang, K.W.; Ringgaard, M.; Lin, C.J. Training and testing low-degree polynomial data mappings via linear SVM. *J. Mach. Learn. Res.* **2010**, *11*, 1471–1490.
80. Hahs-Vaughn, D.L.; Lomax, R.G. *Statistical Concepts-A Second Course: A Second Course*; Routledge: London, UK, 2013.
81. Altman, D.G. *Practical Statistics for Medical Research*; CRC Press: Boca Raton, FL, USA, 1990.
82. Quinlan, J.R. Induction of decision trees. *Mach. Learn.* **1986**, *1*, 81–106. [CrossRef]

sensors

MDPI

Article

Application of Fuzzy and Rough Logic to Posture Recognition in Fall Detection System

Barbara Pękala [1,2,*], Teresa Mroczek [2], Dorota Gil [2] and Michal Kepski [1]

[1] Institute of Computer Science, University of Rzeszów, 35-310 Rzeszów, Poland; mkepski@ur.edu.pl
[2] Department of Artificial Intelligence, University of Information Technology and Management, 35-225 Rzeszów, Poland; tmroczek@wsiz.edu.pl (T.M.); dgil@wsiz.edu.pl (D.G.)
* Correspondence: bpekala@ur.edu.pl

Abstract: Considering that the population is aging rapidly, the demand for technology for aging-at-home, which can provide reliable, unobtrusive monitoring of human activity, is expected to expand. This research focuses on improving the solution of the posture detection problem, which is a part of fall detection system. Fall detection, using depth maps obtained by the Microsoft Kinect sensor, is a two-stage method. We concentrate on the first stage of the system, that is, pose recognition from a depth map. For lying pose detection, a new hybrid FRSystem is proposed. In the system, two rule sets are investigated, the first one created based on a domain knowledge and the second induced based on the rough set theory. Additionally, two inference aggregation approaches are considered with and without the knowledge measure. The results indicate that the new axiomatic definition of knowledge measures, which we propose has a positive impact on the effectiveness of inference and the rule induction method reducing the number of rules in a set maintains it.

Keywords: precedence indicator; knowledge measure; fuzzy inference; rule induction; posture detection; aggregation function

Citation: Pękala, B.; Mroczek, T.; Gil, D.; Kepski, M. Application of Fuzzy and Rough Logic to Posture Recognition in Fall Detection System. *Sensors* **2022**, *22*, 1602. https://doi.org/10.3390/s22041602

Academic Editor: Gwanggil Jeon

Received: 4 January 2022
Accepted: 16 February 2022
Published: 18 February 2022

1. Introduction and Description Problem

Fuzzy [1] and rough [2] sets provide tools for the analysis of significant imperfections of data and knowledge. The former allows classification of objects as belonging to a given degree to a set or relation. The latter provides approximations in cases where the information is incomplete. In this paper, we demonstrate how the mentioned theories can be merged into a hybrid system to improve the solution of the posture detection problem, which is a part of a fall detection system.

Considering that the population is aging rapidly, the demand for assistive technology for aging at home which can provide reliable, unobtrusive monitoring of human activity is expected to expand. One important aim of assistive technology is to provide prolonged independent living in a safe, home like environment without changing everyday lifestyle. Falls are a severe problem within the growing aging population. Many efforts have been undertaken to develop reliable methods of fall detection. The increasing number of studies in this area have allowed us to identify the major challenges and issues for fall detection technology, especially: performance, usability, and acceptance by the elderly. Fall detection systems need to be as accurate and reliable as possible both in terms of high sensitivity and specificity. In practice, this means that fall detectors must reliably distinguish between falls and activities of daily living (ADL) robustly, sustaining at low false alarm ratio. The method should not limit the placement of the sensors, or be sensitive to volatile environmental conditions. Such detection systems fall into two major categories, that is, wearable sensors and context-aware systems [3]. The main advantages of wearable sensors are size, usability, power consumption, and costs of use. The availability of cheap, embedded inertial sensors used in smartphones and smartwatches has contributed to the growth in their popularity in recent years. Usually, such approaches use threshold-based techniques to check if a person's movement exceeds a predetermined threshold [4]. Some of the methods incorporate gyroscopes to obtain the person's orientation [5]. Unfortunately, none of the above-mentioned methods provides satisfactory accuracy. Moreover, body-worn devices cannot be worn during certain activities, such as sleeping, changing clothes, and washing, moreover elderly people may forget to wear such devices. Context-aware systems are based on different kinds of sensors located in the user's environment: cameras, microphones, pressure sensors, Doppler radar, and so forth. The main benefit of using context-aware systems is that no sensors need to be attached to the body of the monitored person, hence the reliability does not depend on the user's willingness to wear the device. On the other hand, this form of activity monitoring is more expensive, invasive, and sometimes requires time to install and calibrate. Camera-based systems, which are one type of context-aware detectors, offer a promising way to detect falls and have been a subject of extensive research. Numerous attempts have been made to detect falls based on a single CCD camera, multiple cameras, stereo-pair cameras, and omnidirectional ones. Although CCD cameras offer several advantages, like the possibility to recognize various daily activities, the lack of ability to work in nightlight conditions and preserve privacy well may be considered serious drawbacks. Compared with the above-mentioned solutions, depth maps are insensitive to lightning conditions and provide 3D information that may substantially contribute towards the robust analysis of human activity.

This paper is focused on human pose recognition which is one part of the hierarchical system proposed in [6]. The mentioned system consists of two input fuzzy-reasoning engines (analyzing pose and movement separately) and a triggering alert Sugeno engine. The fuzzy reasoning on disjoint subsets of the linguistic variables performed by the engines leads to the reduction of the number of fuzzy rules needed for input-output mapping. Analyses of fuzzy and rough inference algorithms for posture detection, which are a part of the fall detection system, require methods that take into account uncertainty, for example, fuzzy set theory and rough set theory. These two theories model different types of uncertainty. The rough set theory takes into consideration the indiscernibility between objects. The second, that is, fuzzy set theory deals with the ill-definition of the boundary of a class through a continuous generalization of set characteristic functions. Given that these approaches pursue different goals, it is more natural to combine the two models of uncertainty than to force them to compete on the same problems. Thus, both approaches will be used in the proposed decision-making system.

The main objective of our research is to improve the solution to the posture detection problem. Therefore, a new hybrid system, based on fuzzy and rough sets, has been developed; the concept of the fuzzy information measure has been investigated and a new axiomatic definition of the knowledge measure has been introduced. In the system, two rule sets are investigated, the first one created based on a domain knowledge and the second induced based on the rough set theory, and two inference aggregation approaches are considered with and without knowledge measure. These measures together with various aggregation methods are used to evaluate the accuracy of the classification of rule sets in the decision-making process (the aim is also to indicate individual operators and fuzzification methods included in the tested system that meet the adopted assumptions, that is, to take into account the uncertainty represented by approximated values). The efficiency of the system is compared to [6]. The knowledge measure can be considered as a dual measure of fuzzy entropy or uncertainty. An entropy measure cannot capture all uncertainties in FSs. Knowledge measure has been studied in fuzzy environments, for example, in [7,8] and in intuitionistic fuzzy environments [9,10], which introduced knowledge measures in an IFS theory as a dual axiom system of intuitionistic fuzzy entropy. In this paper, the new knowledge measure is used to solve the problems of fuzzy inference (in a posture detection system) and tested using different aggregations in the process of aggregating premises. Its effectiveness is then compared using other measures known from the literature.

The following points summarize the main contribution of this study:

(i) New measures:
- A new subsethood measure for fuzzy values is proposed and its validity is proved with the help of the example of use;
- A new knowledge measure for FSs is introduced and its significance is proved with the help of the example of use;

(ii) A new hybrid system is proposed and used in a real decision making problem, i.e., a fall detection system for the elderly, in particular in a posture detection system:
- The proposed knowledge measure is applied to fuzzy inference problems;
- A rule induction method is applied to reduce the number of rules in a set while maintaining the effectiveness of the inference process and significantly improve the performance of a approximate reasoning.

The paper is organized as follows. In Section 2 related works are presented. In Section 3 methodology and data descriptions are proposed. In addition, elements of the fuzzy and rough sets theory as well as new measures of precedence and knowledge based on precedence indicators with their applications to fuzzy inference are presented. Finally, the experimental results of simulations of a hybrid approach to the fall detection problems are described in Section 4.

2. Related Work

Recently, depth cameras have been used in fall detection [11,12]. Ref. [13] applied the skeletal model obtained from Kinect SDK to fall detection. Ref. [14] proposed employing 3D joint tracking information to estimate the walking speed and to extract features describing the movements of a person going down the stairs. However, a person can be in one of many poses before a fall, so the skeleton extraction model may fail, or be unreliable during fall motion [15,16]. In [16] a two-stage fall detection method is proposed. Temporal segmentation of the vertical state time series of a person tracked in 3D is used in the first stage to identify on-ground events. In the second stage the confidence that the event was preceded by a fall is calculated, using a set of decision trees and features extracted from ground-based events. The improvement of fall detection reliability by combining depth and inertial sensors was proposed in [17]. Recent work demonstrates that merging the depth with accelerometer signal improves human activity recognition [18]. A more detailed overview of recent fall detection methodology using depth sensors is provided in [19]. Other approaches are based, for example, on convolutional neural networks (CNNs). However, due to the limited amount of data, their performance is limited. In [20] the authors used transfer learning where pre-traning on the ImageNet dataset AlexNet architecture was applied to accelerometric data, achieving an accuracy of 96.4%. Additionally, the authors of [21] also used depth data, however extracted from videos and thus applied to 3D-CNN. The detection of falls base on videos relies on multiple frames and uses more complex models, thus it can be considerably slower. By using data augmentation, they increased the model accuracy from 69.6% to 92.4% [22]. In this work we perform detection and classification of body contour on depth images. This approach ensures the privacy of the monitored person and is very effective in terms of processing speed. Our method involves merging the techniques mentioned above, fuzzy sets theory and rough sets theory. Despite the popularity of machine learning approaches, issues may arise with the use of simulated human fall event data. Firstly, the small number of actors, may not be sufficient to represent the entire population in terms of variability in human properties (i.e., height) or human biomechanics [23]. Scarcity of data may be problematic (especially for deep learning) so approaches other than traditional supervised classification are being investigated [24]. Another solution to address the lack of data is a customization of the parameters of the decision system to a person's physical characteristics [25]. Our approach leverages the ease of customization and explainability of a fuzzy inference system by reducing the number of rules, allowing to build a linguistically understandable classifier maintaining high detection accuracy.

3. Methodology, Data, Theory and Tools Descriptions

For the purpose of to this article, we propose a new hybrid diagnostic system based on fuzzy and rough sets theory. To be specific, two rule sets are investigated, the first one created based on a domain knowledge and the second constructed by the rough set theory along with the main area of research which is concentrated on the concept of fuzzy information measure, and therefore the knowledge measure. These measures together with various aggregation methods are used to evaluate the accuracy of the classification of rule sets in the decision-making process.

3.1. Methodology and Data

The main goal of this research was to compare two approaches to posture recognition in fall detection: **I. Knowledge Approach** and **II. Rough Set Approach**. In the first approach a method based on a domain knowledge was used to generate a set of rules, the cardinality of which results from the combinatorial characteristic of this method. In turn, in the second approach induction method based on rough sets (described in Section 3.3) was used to reduce a set of rules. Next, both sets of rules were used in the fuzzy inference and evaluation process separately. Additionally, expert knowledge was used for modeling the selection of the parameters for the fuzzification function (described in Section 4). This combination of fuzzy and rough solutions is a novelty to the systems studied in the literature on fall detection problems. The concept of a hybrid approach (that we call a FuzzyRoughSystem, or FRSystem), presented in Figure 1, was based on three processes: *Data Acquisition Process*, *Fuzzy Inference Process* and *Evaluation Process*.

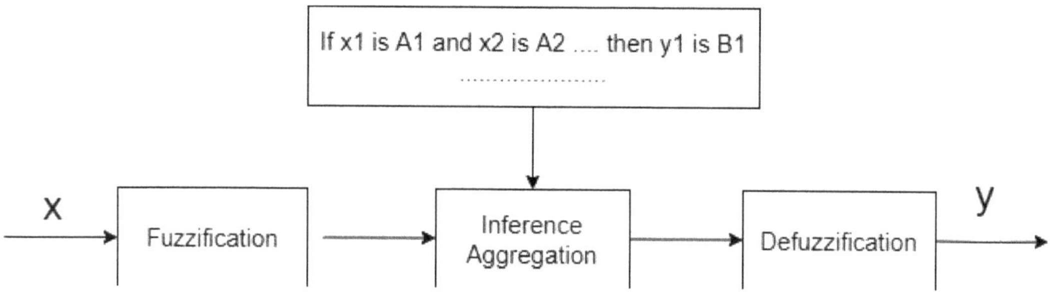

Figure 1. The FRSystem flowchart.

In the Data Acquisition Process, Kinect v1 cameras and an inertial motion sensor were used. The inertial sensors: PS Move and x-IMU collected data at 60 Hz and 256 Hz rates, respectively. The cameras were placed in different locations (one the front of the room parallel to the floor and the second one on the ceiling, facing down), in each case, the camera could be static or mounted on an active head. To preserve the user's privacy, only the depth maps were analyzed. Depth maps were acquired using USB protocol, while accelerometric data were streamed wirelessly from the accelerometer using the Bluetooth protocol. For data acquisition, the OpenNI library was used, while the IMU sensor's software was prepared based on the source codes provided by the manufacturer.

As a result, 5990 depth maps were collected in the UR Fall Detection Dataset. These depth maps were acquired using two Microsoft Kinect cameras from two different viewpoints. Each of the 30 distinct falls had about 150 labelled frames. The depth maps were stored as PNG16 images with 640×480 resolution.

The fall detection system, based on the images, was carried out in two stages: detection of a lying pose based on a single depth map and character movement analysis using dynamic transitions, however, in this work, we focused on the first stage of the system. Features describing the silhouette of a person at a given moment were determined as a result of the clustering of 600 images depicting characters in various poses, including during a fall and while performing ADL actions were analyzed. Ultimately, the following descriptors were selected from the set of features:

- H/W—the ratio of the height of the person's bounding box to its width in the segmented point cloud.
- H/H_{max}—the ratio of the height of the person's surrounding box in the current frame to the physical height of the person.
- $max(\sigma_x, \sigma_z)$—the maximum standard deviation of the values of points belonging to the character from its center of gravity along the axes of the Kinect camera coordinate system.
- P_{40}—the ratio of the number of points, lying no more than 40 cm above the floor, to the number of all points (belonging to the character point cloud).

Before we present and discuss the implementation of the new system (Section 4), we will recall some facts and introduce new elements in the fuzzy sets theory or rough sets theory.

3.2. Fuzzy Set Theory

Firstly, we recall the concept of a fuzzy set (relation) (cf. [26]). We consider fuzzy sets in a set $P \neq \emptyset$.

Definition 1 ([1]). *An arbitrary operation $R : p \to [0,1]$ is a fuzzy set on P.*

All fuzzy sets on P will be denoted per $FS(P)$ and the membership function describing the degree of belonging of $p \in P$ to R is $\mu_R(P)$.

3.2.1. Basic Operations

In this chapter, we will focus on the elementary operations (fuzzy negations and implication functions built on $[0,1]$) used in fuzzy reasoning, which is the basis of our novel system and which will also be recalled in Section 3.2.3.

Definition 2 (cf. [27]). *A non-increasing operation $N : [0,1] \to [0,1]$ which satisfies $N(0) = 1$ and $N(1) = 0$ is called a fuzzy negation N, which is strong if $N(N(p)) = p$, $p \in [0,1]$.*

Example 1 (cf. [28]). *Examples of fuzzy negations N are:*
- $N_k(p) = 1 - p$ *(strong negation called classical/standard negation);*
- $N_w(p) = (1 - p^w)^{\frac{1}{w}}$, $w > 0$;
- $N(p) = 1 - p^2$, *which is strict but not strong;*
- $N_S^\lambda(p) = \frac{1-p}{1+\lambda p}$, *the Sugeno family of fuzzy (strong) negations, where $\lambda \in (-1, \infty)$ and for $\lambda = 0$ we get the classical fuzzy negation.*

Definition 3 ([29]). *An operation $I : [0,1]^2 \to [0,1]$ which is a decreasing in the first component and increasing in the second component also fulfilling $I(1,0) = 0$, $I(0,1) = I(0,0) = I(1,1) = 1$ is called a fuzzy implication.*

Examples of fuzzy implications I are:

- Łukasiewicz implication—$I_{LK}(p,q) = \begin{cases} 1, & \text{if } p \leq q \\ 1 - p + q, & \text{otherwise;} \end{cases}$
- Fodor implication—$I_{FD}(p,q) = \begin{cases} 1, & \text{if } p \leq q \\ \max(1 - p, q), & \text{otherwise;} \end{cases}$

- Rescher implication—$I_{RS}(p,q) = \begin{cases} 1, & \text{if } p \leq q \\ 0, & \text{otherwise;} \end{cases}$
- Reichenbach implication—$I_{RC}(p,q) = 1 - p + pq$;
- Kleene-Dienes implication—$I_{KD}(p,q) = \max(1-p,q)$.

Now, we recall the basic and the most important operation on fuzzy sets, i.e., an aggregation function.

Definition 4 (cf. [30]). *An operation $A : [0,1]^n \to [0,1]$, $n \geq 2$ which is increasing and fulfils boundary conditions $A(0,\ldots,0) = 0$, $A(1,\ldots,1) = 1$ is called an aggregation function.*

Example 2. *Examples of aggregation functions are:*
• *lattice: $T_M(p,q) = \min(p,q)$, $S_M(p,q) = \max(p,q)$;*
• *algebraic: $T_P(p,q) = pq$, $S_P(p,q) = p + q - pq$;*
• *Łukasiewicz: $T_L(p,q) = \max(0, p+q-1)$,*
$S_L(p,q) = \min(1, p+q)$;

Arithmetic mean

$$A_{mean}(p_1,\ldots,p_n) = \frac{1}{n}(p_1 + \ldots + p_n); \tag{1}$$

Geometric mean

$$A_{gmean}(p_1,\ldots,p_n) = \sqrt[n]{p_1 \cdots p_n}; \tag{2}$$

Square mean

$$A_{2mean}(p_1,\ldots,p_n) = \sqrt{\frac{{p_1}^2 + \ldots + {p_n}^2}{n}}; \tag{3}$$

The OWA operator (ordered weighted averaging) $OWA : [0,1]^n \to [0,1]$

$$OWA(p_1,\ldots,p_n) = \sum_{i=1}^{n} w_i p_{(i)}, \tag{4}$$

(i) means a permutation of $\{1,\ldots,n\}$ such that $p_{(1)} \geq p_{(2)} \geq \ldots \geq p_{(n)}$ and $w = (w_1,\ldots,w_n) \in [0,1]^n$ is a vector of weights (i.e., $w_i \in [0,1]$ and $\sum_{i=1}^{n} w_i = 1$) for $p_1,\ldots p_n \in [0,1]$, $n \in N$.

We will also employ the concept of pre-aggregation function [31], which satisfies the same boundary conditions as an aggregation function, but, in return to requiring monotonicity, directional monotonicity is needed, that is:

Definition 5. *An operation $F : [0,1]^n \to [0,1]$ is a pre-aggregation function if it fulfils*
(1) There exists $\vec{r} \in [0,1]^n$ ($\vec{r} \neq \vec{0}$) a real vector which F is $\vec{r}$-increasing, that is, for all points $(p_1,\ldots,p_n) \in [0,1]^n$ and for all $c > 0$ such that $(p_1 + cr_1,\ldots,p_n + cr_n) \in [0,1]^n$, holds $F(p_1 + cr_1,\ldots,p_n + cr_n) \geq F(p_1,\ldots,p_n)$.
(2) F fulfils the boundary conditions: $F(0,\ldots,0) = 0$ and $F(1,\ldots,1) = 1$.

Example 3 ([31]). *Examples of pre-aggregation functions:*
1. $F(p,q) = p - (\max(0,p-q))^2$ is $(0,1)$—increasing (not an aggregation function).
2. $L_\lambda(p,q) = \frac{\lambda p^2 + (1-\lambda)q^2}{\lambda p + (1-\lambda)q}$ (with convention $0/0 = 0$) is $(1-\lambda,\lambda)$—increasing, for $\lambda \in [0,1]$ (the weighted Lehmer mean).

3.2.2. Knowledge Measure

We will focus on an important measure, that is, the measure of fuzzification, that is, the knowledge measure. We propose to use this measure in the process of fuzzy inference when drawing conclusions from premises (in aggregating premises). Before we move on to a new idea of measuring knowledge in the fuzzy set environment/theory, we need to present a certain tool useful for the operation of fuzzy values, that is, a measure of inclusion of fuzzy values called a precedence indicator.

Precedence Indicator

Research on fuzzy sets began with the concept of Zadeh (1965), where $K \leq L$ iff $\forall_{p \in P} K(x) \leq L(x)$, but Bandler and Kohout (1980) proposed a new measure subsethood grade/precedence indicator of a fuzzy set in another fuzzy set which is based on a considering the infimum of an appropriate aggregation of implication operators. This idea of Bandler and Kohout inspired many authors to study fuzzy subsethood measures as the type of function $\sigma : FS(P) \times FS(P) \to [0,1]$ with the different axiomatizations that have been proposed are not equal and they hinge on the examined applications. Based on this fact, and drawing inspiration from the works [32–35] in this paper we propose a new list of axiomatization for fuzzy precedence measure $\text{Prec} : [0,1] \times [0,1] \to [0,1]$ as the class of implication operators which allows us to:

1. Construct a new precedence indicator inspired by the axiomatic definition of the fuzzy subsethood measures;
2. Construct new knowledge measures using a new precedence indicator;
3. Apply new knowledge measures in fuzzy inference, as an illustrative example of the effectiveness of the proposed new measures.

Definition 6. *An operation* $\text{Prec} : ([0,1])^2 \to [0,1]$ *is called a* ***precedence indicator*** *if it fulfils:*

P1 $\text{Prec}(p,q) = 0$ *iff* $p = 1$ *and* $q = 0$*;*

P2 $\text{Prec}(p,q) = 1$ *iff* $p \leq q$ *for any* $p,q \in [0,1]$*;*

P3 *If* $p \leq q \leq r$*, then* $\text{Prec}(r,p) \leq \text{Prec}(q,p)$ *and* $\text{Prec}(r,p) \leq \text{Prec}(r,q)$ *for any* $p,q,r \in [0,1]$*.*

Now we propose the constructive method of the precedence indicator based on an aggregation and negation functions.

Proposition 1. *Let N denote a fuzzy negation (i.e., an antytonic operation that fulfils* $N(0) = 1$*,* $N(1) = 0$*) and A is the aggregation* $A \leq \max$*. Then*

$$\text{Prec}_A(p,q) = \begin{cases} 1, & \text{if } p \leq q, \\ A(N(p),q), & \text{otherwise} \end{cases} \tag{5}$$

is the precedence indicator.

Here are some examples of the precedence indicators that satisfy Proposition 1.

Example 4. *For* $A = A_{mean}$ *and standard negation N we have*

1.
$$\text{Prec}_A(p,q) = \begin{cases} 1, & \text{if } p \leq q, \\ \frac{1-p+q}{2}, & \text{otherwise} \end{cases} \tag{6}$$

or for Sugeno negation with $\lambda = 1$ *we have*

2.
$$\text{Prec}_A(p,q) = \begin{cases} 1, & \text{if } p \leq q, \\ \frac{1}{2}\frac{1-p}{1+p} + \frac{q}{2}, & \text{otherwise} \end{cases} \tag{7}$$

for $p,q \in [0,1]$*.*

We pay attention to the fact that precedence indicators create a subclass of fuzzy implication functions as we observe in the following example.

Example 5. *The following operations are implication function but not precedence indicators:*

$$I(p,q) = \begin{cases} 1, & \text{if } p \le q, \\ 0, & \text{if } p = 1, q \neq 1, \\ \frac{1}{2}, & \text{otherwise,} \end{cases} \tag{8}$$

$$I(p,q) = \begin{cases} |p-q|, & \text{if } p < q, \\ 1-|p-q|, & \text{if } p = q, \\ A(N(p),q), & \text{otherwise} \end{cases} \tag{9}$$

for $p, q \in [0,1]$.

Knowledge Measure

In this part of the work, we consider the crucial concept of information in the setting of uncertainty, that is, the idea of the knowledge measure of a fuzzy set, and suggest a new construction process for it by use of a precedence indicator. Cognitively, the knowledge measure is dual to the entropy measure of the arbitrary fuzzy set which gives the average values/height of fuzziness/ambiguity existing in the fuzzy set. Similarly, we can wonder about the average amount of knowledge present in the fuzzy set. Thus, a knowledge measure of a fuzzy set needs to satisfy the following axiomatic postulates. We propose some generalisation (in the fourth axiom) of the axiomatic definition of knowledge measure presented in [7,8].

Definition 7. *For* $R \in FS(P)$ *a knowledge measure would satisfy the following properties:*

K1 $K(R)$ *has maximum value iff* R *is a crisp set, i.e.,* $R(p_i) = 0$ *or* 1 *for all* $p_i \in P$,

K2 $K(R)$ *has minimum value iff* R *is the most fuzzy set, i.e.,* $R(p_i) = 0.5$ *for all* $p_i \in P$,

K3 $K(R^*) \ge K(R)$, *where* R^* *is a crisped version (sharpened) of* R,

K4 $K(R) = K(R^N)$, *where* R^N *is the duality (complement) of set* R *for strong fuzzy negation* N, *i.e.,* $R^N(p) = N(R(p)), p \in P$ *(for classic negation* N *we obtain a complement relation of* R*).*

We suggest the following construction method of the knowledge measure.

Proposition 2. *Let* Prec *be a precedence indicator that satisfies Proposition 1, where aggregation* A *is symmetric and* N *is the strong negation with an equilibrium point* 0.5 *(i.e.,* $N(0.5) = 0.5$*) for* $R \in FS(P)$, $card(P) = n$, $n \in N$, *then*

$$K(R) = \frac{1}{n}\sum_{i=1}^{n} \frac{|\mathrm{Prec}(1, R(p_i)) - \mathrm{Prec}(R(p_i), 0)|}{1 - \min(\mathrm{Prec}(1, R(p_i)), \mathrm{Prec}(R(p_i), 0))} \tag{10}$$

is a knowledge measure.

Proof. Let $i = 1, \ldots, n$. At the beginning let us note that $0 \le K(R) \le 1$.

(K1) is obvious with the assumption about R, Prec, and their properties. Because for a crisp relation of R we have:

1. for $R(p_i) = 1$ $\mathrm{Prec}(1,1) = 1$ and $\mathrm{Prec}(1,0) = 0$ or
2. for $R(p_i) = 0$ $\mathrm{Prec}(1,0) = 0$ and $\mathrm{Prec}(0,0) = 1$

and as consequence we obtain $K(R) = 1$.

Conversely, suppose $K(R) = 1$, this is possible for $|\text{Prec}(1, R(p_i)) - \text{Prec}(R(p_i), 0)| = 1$ for all i, which implies

$$(\text{Prec}(1, R(p_i)) = 1 \; and \; \text{Prec}(R(p_i), 0) = 0) \; or \; (\text{Prec}(1, R(p_i)) = 0 \; and \; \text{Prec}(R(p_i), 0) = 1),$$

so from P1 and P2 we obtain $R(p) \in \{0, 1\}, p \in P$, that is, R is crisp relation.

(K2) By Proposition 1 and $R(p_i) = 0.5$ for all i and from the symmetry property of A and for the equilibrum point 0.5 of N we observe $\text{Prec}(1, 0.5) = A(N(1), 0.5) = A(0, 0.5) = A(0.5, 0) = \text{Prec}(0.5, 0)$, i.e., $K(R) = 0$. Conversely, by assumption $K(R) = 0$ we obtain $|\text{Prec}(1, R(p_i)) - \text{Prec}(R(p_i), 0)| = 0$ for all i, thus $\text{Prec}(1, R(p_i)) = \text{Prec}(R(p_i), 0)$, which implies $R(p_i) = 0.5$ for all i.

(K3) If R^* is crisper than R, that is,

1. $R^*(p_i) \geq R(p_i)$ for $R(p_i) \geq 0.5$,
2. $R^*(p_i) \leq R(p_i)$ for $R(p_i) < 0.5$.

Based on Proposition 1 and for

$$\text{Prec}(R^*(p_i), 0) \leq \text{Prec}(R(p_i), 0), \; \text{Prec}(1, R(p_i)) \leq \text{Prec}(1, R^*(p_i))$$

and

$$\text{Prec}(1, R(p_i)) \geq \text{Prec}(R(p_i), 0) \; for \; R(p_i) \geq 0.5.$$

Thus

$$|\text{Prec}(1, R^*(p_i)) - \text{Prec}(R^*(p_i), 0)| \geq |\text{Prec}(1, R(p_i)) - \text{Prec}(R(p_i), 0)|,$$

that is, $K(R^*) \geq K(R)$. In a similar way we consider the case $R(p_i) < 0.5$.

(K4) Based on Proposition 1 we observe for the symmetric aggregation A:

$$|A(0, R^N(p_i)) - A(R(p_i), 0)| = |A(0, R(p_i)) - A(R^N(p_i), 0)| \; for \; all \; i,$$

as a consequence we have $K(R^N) = K(R)$, which completes the proof. □

Example 6. *If in Proposition 2 we used precedence indicators satisfying Proposition 1 with* $A \in \{A_{mean}, A_{gmean}, A_{2mean}, \min, \max\}$ *and N is standard (classical) negation, then we obtain knowledge measure* $K(R)$ *for* $R \in FS(P)$.

3.2.3. Knowledge Measure and Fuzzy Inference (Mamdani)

The known and popular area of fuzzy logic and its extensions application is approximate reasoning, where from uncertainty/imprecise inputs/fuzzy premises or rules we often obtain uncertainty/imprecise inferences. Approximate reasoning has been used in many fields, for example, medical diagnosis, expert systems and control systems.

The main goal of this part of the paper is to explore the more general algorithm of approximate reasoning by using the general modus ponens property with the arbitrary aggregation functions next to the new knowledge measure. In the beginning, an algorithm for multi conditional approximate reasoning based on the new aggregation-based composition rules is proposed. The use of knowledge measure in fuzzy reasoning is a new accent in the classical model of inference. Thus we obtain a modification of the standard fuzzy reasoning method.

Approximate reasoning is the procedure where a possible uncertainty/imprecise conclusion is implied from a collection of uncertainty/imprecise premises. The classical modus ponens schema, was extended by Zadeh [36] to fuzzy reasoning in the following way and we obtained the GMP, that is, Generalized Modus Ponens:

```
Proposition:  If  p  is  D  then  q  is  E
Fact:             p  is  D'
-----------------------------------------
q  is  E',
```

where E' is the fuzzy set in the universe Q. The main plus of the GMP is that we can obtain new information even if D' and D are different. Usually, in the GMP the fuzzy rule is represented using a fuzzy relation R on the referential set $P \times Q$. Existing different methods to build R can be used [37]. The most promising:

$R(p,q) = I(D(p), E(q))$, where I is an implication function. We may build the implication function from the aggregation function: $I(p,q) = A(1-p,q)$ with $A(1,0) = A(0,1) = 1$. Thus we can also create the relation R using the aggregation function by specific assumptions.

The fuzzy inference process is as follows

$$E'(q) = A_{p \in P}\, B(D'(p), R(p,q));\ i.e.\ E' = D' \circ R, \tag{11}$$

where A, B are aggregation functions on $[0,1]$. The basic inference process has the form presented in Figure 1.

Our novelty in the fuzzy inference in the process of aggregating premises is the proposal to use the combination of aggregation and knowledge measure as the following new operator:

$$O_R = B(A_{i=1}^{n}(p_i), K(R)), \tag{12}$$

where R is a fuzzy set on P, where $cardP = n$. Thus premises data in the given rule and K knowledge measure created by Proposition 2 and A, B are aggregation functions.

3.3. Rough Set Theory

The rough set theory use the indiscernibility relation to discover information about objects in an information system.

Definition 8 ([38])**.** *An information system (IS) is an ordered quadruple (U,AT,V,f) where U is a finite nonempty set of objects, AT is a finite nonempty set of attributes,* $V = \bigcup_{a \in AT} V_a$*; is a nonempty finite set of values of attributes, where* V_a *is the domain of attribute a, and* $f : U \times AT \to V$ *is an information function such that* $f(x,a) \in V_a$ *for all* $x \in U$ *and* $a \in AT$*.*

A *decision table* is a type of information system. In the decision table the set $AT = A \cup D$; A is a set of attributes, and D is set of decisions, $D \cap A = \emptyset$. Whereas, a *concept* is the set of all cases with the same decision value [39].

Definition 9 ([2])**.** *For each subset of attributes* $A \subseteq AT$ *a binary indiscernibility relation IND(A) on U can be determined as follows:*

$$IND(A) = \{(x,y) \in U \times U | \forall a \in A, f(x,a) = f(y,a)\}.$$

Let $a \in A$, $v \in V$, and p = (a, v) be an attribute-value pair. The set [p] of all cases from U for which attribute a has value v is called a block of attribute-value pairs [40]. The rule induction Algorithm 1 LEM2 [39], in order to find a local covering of an input set, explores the space of attribute-value pairs.

Let X be a subset of U and P be a nonempty collection of nonempty sets of attribute-value pairs. The set P is a *minimal complex* of X if and only if X depends on P and no proper subset P' of P exists such that X depends on P' [39]. ρ is a *local covering* of X if and only if the following conditions are satisfied:

1. each member P of ρ is a minimal complex of X,
2. $\bigcup_{p \in \rho}[P] = X$
3. ρ is minimal [39].

Algorithm 1 LEM2

Input: a set X
Output: a single local covering ρ of set X

```
X := G;
ρ := ∅;
while G ≠ ∅ do
    P := ∅
    P_G = {p|[p] ∩ G ≠ ∅}
    while P = ∅ or [P] ⊈ X do
        select a pair p ∈ P_G such that |[p] ∩ G| is maximum;
        if a tie occurs, select a pair p ∈ P_G with the smallest cardinality of [p];
        if another tie occurs, select first pair;
        P := P ∪ {p}
        G := [p] ∩ G
        P_G := {p|[p] ∩ G ≠ ∅} − P
    end while
    for each p ∈ P do
        if [P − {p}] ⊆ X then P := P − {p};
        end if
        ρ := ρ ∪ {P};
        G := X − ⋃_{P∈ρ}[P];
    end for
end while
for each ρ ∈ P do
    if ⋃_{P'∈ρ−{P}}[P'] = X then ρ := ρ − P;
    end if
end for
```

The LEM2 algorithm has been used successfully in many areas, recently in [41–45].

4. Implementation and Results

We implemented the inference system of FRSystem (Figure 1) in the following way: for the values of each input, that is, H/W, $H/_{max}$, $max(\sigma_x, \sigma_z)$, P_{40} we generated the fuzzy sets by using the adequate membership function needed for suitable rules, so for Lo (low value of the feature), Me (average value of the feature), Hi (high value of the feature) and the value of *isLy (lying position)*, *myLy (maybe lying position)* and *notLy (not lying position)* we use function type Z, Gaussian and type S, respectively (the Gaussian function is uniquely built by two different Gaussian functions). For the above functions we propose the following parameters:

1. H/W:
   $\mu^{Lo}_{H/W}(p, 0.5, 1.25, 2)$, $\mu^{Me}_{H/W}(p, 2, 0.5, 2, 0, 4)$, $\mu^{Hi}_{H/W}(p, 2, 2.6, 3.2)$;
2. H/max:
   $\mu^{Lo}_{H/max}(p, 0.25, 0.4, 0.6)$, $\mu^{Me}_{H/max}(p, 0.6, 0.1, 0.6, 0.2)$, $\mu^{Hi}_{H/max}(p, 0.6, 0.8, 1)$;
3. $max(\sigma_x, \sigma_z)$:
   $\mu^{Lo}_{max(\sigma_x,\sigma_z)}(p, 260, 285, 310)$, $\mu^{Me}_{max(\sigma_x,\sigma_z)}(p, 310, 17, 310, 33)$,
   $\mu^{Hi}_{max(\sigma_x,\sigma_z)}(p, 310, 360, 410)$;
4. P_{40}:
   $\mu^{Lo}_{P_{40})}(p, 0.18, 0.3, 0.42)$, $\mu^{Me}_{P_{40}}(p, 0.42, 0.08, 0.42, 0.09)$, $\mu^{Hi}_{P_{40}}(p, 0.42, 0.55, 0.68)$;
5. $Pose$:
   $\mu^{isLy}_{Pose}(p, 0.22, 0.36, 0.5)$, $\mu^{mayLy}_{Pose}(p, 0.5, 0.09, 0.5, 0.09)$, $\mu^{notLy}_{Pose}(p, 0.5, 0.63, 0.77)$.

Based on the collected data, two rule sets were generated independently. The first one, a result of the Rough Set Approach, contained 44 rules: 10 rules for the pose notLy, 34 rules for the pose mayLy and 10 rules for the pose isLy. The second one, a result of the Knowledge

Approach (FRSystem, Figure 1), contained 81 rules $((3\ cases(functions))^{4features}$, [46]): 13 for the pose notLy, 52 rules for the pose mayLy and 16 rules for the pose isLy.

Next, in the *Fuzzy Inference Process*, a modified version of the basic Mamdani model was applied to obtain a posture decision (lying or not). Namely, in fuzzy inference, in the process of aggregating premises, a combination of aggregation and knowledge measure was used (new aspect by applying the operator O_R, see Section 3.2.3) constructed using a new precedence indicator. The effectiveness of the new measure was compared with the classic model without using the knowledge measure (the Sections 3 and 4 in the FRSystem (Figure 2)) and also the effectiveness of applying different aggregations in the fuzzy inference process was analyzed.

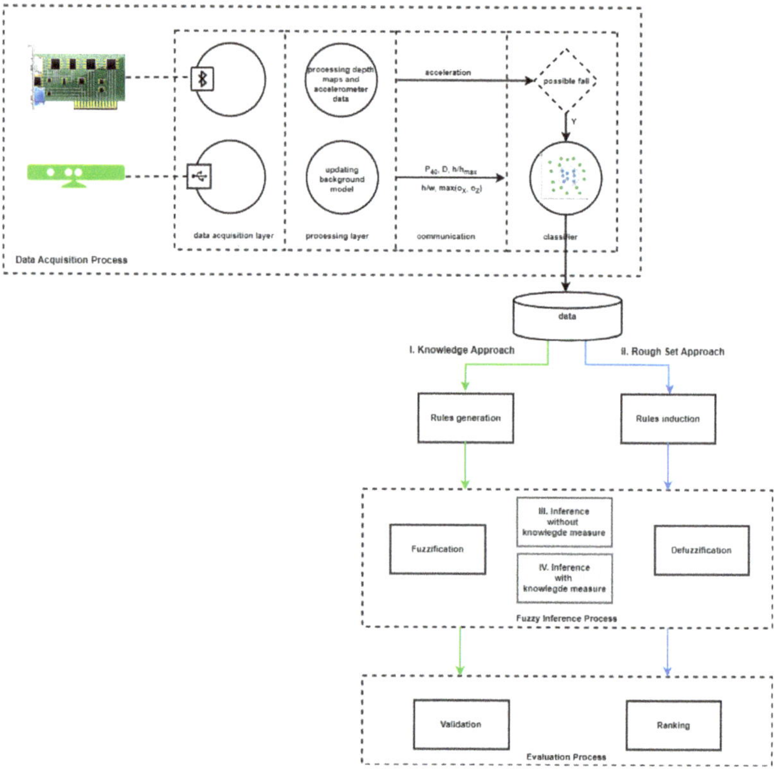

Figure 2. The Scheme of the fuzzy inference process.

To demonstrate the effectiveness of the proposed hybrid approach the following characteristics were used:

- accuracy

$$ACC = \frac{TP + TN}{TP + TN + FP + FN}, \tag{13}$$

where TP is the number of correct isLy classifications, TN is the number of correct notLy classifications, FP is the number of notLy classifications as isLy and FN the number of isLy classifications as notLy

- specificity

$$SPE = \frac{TN}{TN + FP}, \tag{14}$$

- precision

$$PRE = \frac{TP}{TP + FP}, \tag{15}$$

- sensitivity

$$REC = \frac{TP}{TP + FN} \tag{16}$$

in the *Evaluation Process*. Note that accuracy means how close a measurement is to the actual or expected value. The precision says how close the sets of measurements are to each other. The recall is characterized as the percentage of relevant results that are correctly classified by the used model, and specificity means the percentage of true negative results.

Finally, the rules used in inference (I. Rough Set Approach and II. Knowledge Approach) were assessed based on: the number of correct classifications of the rule, the effectiveness of the rule in the set and the effectiveness of the rule within the decision class.

We assumed that the effectiveness of the rule in the set can be expressed as follows:

$$\frac{the\ number\ of\ correct\ classifications\ of\ the\ rule}{the\ number\ of\ objects\ in\ the\ set}. \tag{17}$$

In turn the effectiveness of the rule within the decision class can be determined as follows:

$$\frac{the\ number\ of\ correct\ classifications\ of\ the\ rule\ in\ the\ decision\ class}{the\ number\ of\ objects\ in\ the\ decision\ class}. \tag{18}$$

Based on the above-mentioned measures, a rule ranking was created. First, the strongest rules from the set classification point of view were identified. Then, among the strongest rules, the ones which turned out to be the most effective within the decision class were selected. In this way, the rules that were critical to pose detection were indicated. The rules that were critical to pose detection were indicated. Finally, we use the center of gravity method for the defuzzification process.

To measure the effectiveness of our approach, the above-mentioned characteristics: accuracy (ACC), specificity (SPE), precision (PRE), and recall (REC) (sensitivity) were used. We studies the following cases:

- Determination of the effectiveness of classic fuzzy inference (without the knowledge measure and without the rule reduction) in fall detection problems, Table 1;
- Assessment of the impact of different aggregation functions and different knowledge measures, i.e., precedence indicators, on the effectiveness of classification of the reduced and nonreduced rules, using the FRSystem, Table 2;
- Verification of the effectiveness of the different knowledge measure construction methods in the FRSystem, proposed by us and others known from the literature, Table 3.
- Estimation of the effectiveness of each rule in the whole set and within the decision class.

Tables 1–3 show the experimental results obtained during the given dataset analysis. Presented outcomes in Table 1 maintain a high level of classification comparable to [6]. However, the next studies show that we observe progress in our classification results if we use the FRSystem (as can be seen in the result in Tables 1 and 2) where the results are grouped for the original set of rules and after their selection by the rough method. Moreover, we compare the effectiveness of different aggregation functions used in the fuzzy inference, in the process of aggregating premises. We present the best results obtained for knowledge measures that satisfy Proposition 2 and are used in FRSystem. In particular, in K_1 we use in operator O_R aggregation functions A_{2mean} and $B = F$ (A_{2mean}

used in the precedence indicator used in the Knowledge measure K) (which we denote as $K_1(A_{2mean}, B_F, A_{2mean})$. Similarly, we created $K_2(A_{2mean}, B_F, \max)$, $K_3(\min, B_F, A_{2mean})$, $K_4(A_{mean}, B_F, A_{mean})$, $K_5(A_{mean}, B_{min}, \max)$. In the presented results, we assume the results of each class we aggregate by the maximum.

Table 1. Confusion and classification evaluation metrics by the standard fuzzy inference system with aggregations from examples 2 and 3.

	T_m	T_p	A_{mean}	OWA	F
TP	7303	7303	7420	7375	7420
TN	1968	1968	2069	2056	1969
FP	405	405	304	317	404
FN	149	149	32	77	32
ACC	0.944	0.944	**0.966**	0.960	0.956
PRE	0.947	0.947	**0.961**	0.959	0.948
REC	0.980	0.980	0.996	0.990	**0.996**
SPE	0.829	0.829	**0.872**	0.866	0.830

Table 2. Confusion and classification evaluation metrics with the operator K used in the FRSystem, where All and Red. means test on full and on reduced set of rules. respectively.

	K_1		K_2		K_3		K_4		K_5	
	All	Red.	All	Red.	All	Red.	All	Red.	All	Red.
TP	7443	7446	7442	7446	7430	7445	7440	7446	7440	7445
TN	2066	1938	2076	1956	2083	1939	2063	1961	2064	1985
FP	307	435	297	417	290	434	310	412	309	388
FN	9	6	10	6	22	7	12	6	12	7
ACC	0.968	0.956	**0.969**	0.957	0.968	0.956	0.967	0.957	0.967	**0.96**
PRE	0.960	0.945	**0.962**	0.947	0.962	0.945	0.96	**0.948**	0.96	0.95
REC	**0.999**	**0.999**	**0.999**	**0.999**	0.997	**0.999**	**0.999**	**0.999**	**0.999**	**0.999**
SPE	0.871	0.817	0.875	0.824	**0.878**	0.818	0.869	0.826	0.87	**0.84**

Table 3. Confusion and classification evaluation metrics with different knowledge measures used in the FRSystem.

	K	K^{SLS}	K^{AK}
TP	7442	7434	7444
TN	2076	2026	2064
FP	297	347	309
FN	10	18	8
ACC	**0.969**	0.963	0.968
PRE	**0.962**	0.956	0.960
REC	**0.999**	0.998	**0.999**
SPE	**0.875**	0.854	0.870

The best results we obtained are marked in bold. As can be seen, the best performance is obtained for K_2 used in the FRSystem, with the following measures: ACC (96.9%), PRE (96.2%), SPE (87.8%) and REC (99.9%). What is more, we may say that the application of a reduced set of rules retained the classification level, that is, we obtained results with an acceptable difference of error, in a limit of the error at the level of about 0.01 (see Table 2). Thus, paths I and II in the FRSystem are comparable in the effectiveness aspect, but reducing the number of rules also has another important and positive effect on our model because we do not have to take into account all the attribute-value relationships. Only the most important relationships are selected in the induction process. A smaller and at the same time, optimal set of rules is easier for experts to evaluate.

Moreover, in Table 3 we compare our best results (we denote by K the knowledge measure built-in to the proposed method and used in the FRSystem) with other methods to build knowledge measures known in the literature (unlike our approach, the dependence (precedence indicator) of a given fuzzy value on the extreme (certain value) is not taken into account), such as: $K^{SLS}(F) = \frac{1}{n}\sum_{i=1}^{n} 2[F^2(p_i) + (1 - F(p_i))^2] - 1$ [8], $K^{AK}(F) = log_2[\frac{2}{n}\sum_{i=1}^{n}(F^2(p_i) + (1 - F(p_i))^2)]$ [7].

There, the fuzzy and dual values are taken into account while in our approach the given fuzzy value is compared by subsethood measure with the extremes (the largest and the smallest certain value), which gives a more complete picture of the uncertainty contained in the measurements. We observe the higher effectiveness of the proposed new knowledge measure (see Table 3). For comparison we take K from case K_2 from the result presented in Table 2:

$$K(F) = \frac{1}{n}\sum_{i=1}^{n} \frac{|\mathrm{Prec}_{\max}(1, F(p_i)) - \mathrm{Prec}_{\max}(F(p_i), 0)|}{1 - \min(\mathrm{Prec}_{\max}(1, F(p_i)), \mathrm{Prec}_{\max}(F(p_i), 0))}, \tag{19}$$

where for $p, q \in [0, 1]$ we have

$$\mathrm{Prec}_{\max}(p, q) = \begin{cases} 1, & \text{if } p \leq q, \\ \max(N(p), q), & \text{otherwise.} \end{cases}$$

In order to identify the most relevant attribute values (from a classification view point) for each decision class the rules were assessed first on the whole set, and then on the concepts. As a result, the values of the attributes clearly defining the detection of a lying or non-lying position are indicated and presented in Table 4. It should be noted that, the H/W attribute did not occur in the reduced set of rules, among the conditions of the most efficient rules for the notLy decision class. The absence of this attribute did not affect the quality of classification within this class in relation to the non-reduced set of rules. The remaining conjunctions of conditions for the most effective reduced and non-reduced rules were identical.

Table 4. Specification of the most relevant attribute values for decision classes, where Lo, Me and Hi means low, average and high value of the feature, respectively and Ly means lying position.

H/W	H/H_{max}	$max(\sigma_x, \sigma_z)$	P_{40}	Concept
Hi	Hi	Lo	Lo	notLy
Me	Hi ∨ Me	Lo	Lo	
Lo	Lo	Lo ∨ Me	Hi	∼notLy

5. Conclusions

In this paper, we have provided the initial results of a very interesting new approach to the selection of appropriate aggregation functions and a set of rules for fuzzy inference in the problem of fall detection, especially posture detection. Moreover, the main research was concentrated on investigating the concept of a fuzzy information measure, presenting a new axiomatic definition for the knowledge measure, and using theirs in the proposed hybrid system. The results obtained for the mentioned aspects indicate the positive results of the new approach. Out of 81 rules (see [46]), by applying the LEM2 algorithm we indicate 44 rules (see [47]) which allow us to significantly reduce the dimensionality of the studied problem and facilitate its analysis while maintaining a high level of classification comparable to [6].

Our goal for future work is to develop this research on both theoretical and practical grounds. For example, we would like, in cooperation with an Elderly care home in Rzeszow, to expand the data set and develop some new methods to represent data, for example, a hybrid method that uses fuzzy and rough sets concerning uncertainty, so we will use interval-valued fuzzy set theory. In addition, the developed hybrid inference method seems to be very promising for use with different input data sets in the future. In particular, new measures of information may prove useful for the issues or methodologies observed in the works [7,8], where the proposed knowledge measure is utilized to calculate the weights vector, when weights are partially known and other when weights are completely unknown in economic terms, in multiple attribute decision-making methods or in image thresholding based on a fuzzy accuracy measure.

Author Contributions: Conceptualization and Methodology, B.P., T.M. and M.K.; Software, M.K., D.G.; Validation and Investigation, B.P., T.M. and M.K. and D.G.; Data Curation, D.G., M.K.; Writing—Original Draft, B.P., T.M. and D.G.; Writing—Review, Editing, and visualization, B.P., T.M., D.G. and M.K.; supervision, B.P. All authors have read and agreed to the published version of the manuscript.

Funding: This research receives funding from the University of Information Technology and Management, Rzeszow, Poland.

Institutional Review Board Statement: Not applicable (Data were collected as simulations under laboratory conditions).

Informed Consent Statement: Not applicable (Data were collected as simulations under laboratory conditions).

Data Availability Statement: The dataset used in this research work is on the website http://fenix.univ.rzeszow.pl/~mkepski/.

Conflicts of Interest: The authors declare no conflict of interest.

References

1. Zadeh, L.A. Fuzzy sets. *Inf. Contr.* **1965**, *8*, 338–353. [CrossRef]
2. Pawlak, Z. Rough sets. *Int. J. Comput. Inf. Sci.* **1982**, *11*, 341–356. [CrossRef]
3. Igual, R.; Medrano, C.; Plaza, I. Challenges, issues and trends in fall detection systems. *BioMed. Eng. Online* **2013**, *12*, 1–24. [CrossRef] [PubMed]
4. Bourke, A.K.; O'Brien, J.V.; Lyons, G.M. Evaluation of a threshold-based tri-axial accelerometer fall detection algorithm. *Gait Posture* **2007**, *26*, 194–199. [CrossRef] [PubMed]
5. Li, Q.; Stankovic, J.A.; Hanson, M.A.; Barth, A.T.; Lach, J.; Zhou, G. Accurate, Fast Fall Detection Using Gyroscopes and Accelerometer-Derived Posture Information. In Proceedings of the IEEE International Workshop on Wearable and Implantable Body Sensor Networks, Berkeley, CA, USA, 3–5 June 2009; pp. 138–143.
6. Kwolek, B.; Kepski, M. Fuzzy inference-based fall detection using kinect and body-worn accelerometer. *Appl. Soft Comput.* **2016**, *40*, 305–318. [CrossRef]
7. Arya, V.; Kumar, S. Knowledge measure and entropy: A complementary concept in fuzzy theory. *Granul. Comput.* **2021**, *6*, 631–643. [CrossRef]
8. Singh, S.; Lalotra, S.; Sharma, S. Dual concepts in fuzzy theory: Entropy and knowledge measure. *Int. J. Intell. Syst.* **2019**, *34*, 1034–1059. [CrossRef]
9. Szmidt, E.; Kacprzyk, J.; Bujnowski, P. How to measure amount of knowledge conveyed by Atanassov's intuitionistic fuzzy sets. *Inf. Sci.* **2014**, *257*, 276–285. [CrossRef]
10. Wang, G.; Zhang, J.; Song, Y.; Li, Q. An entropy-based knowledge measure for Atanassov's intuitionistic fuzzy sets and its application to multiple attribute decision making. *Entropy* **2018**, *20*, 981. [CrossRef]
11. Mastorakis, G.; Makris, D. Fall detection system using Kinect's infrared sensor. *J. Real-Time Image Process.* **2014**, *9*, 635–646. [CrossRef]
12. Rougier, C.; Meunier, J.; St-Arnaud, A.; Rousseau, J. Robust Video Surveillance for Fall Detection Based on Human Shape Deformation. *IEEE Trans. Circuits Syst. Video Technol.* **2011**, *21*, 611–622. [CrossRef]
13. Planinc, R.; Kampel, M. Introducing the use of depth data for fall detection. *Pers. Ubiquitous Comput.* **2012**, *17*, 1063–1072. [CrossRef]
14. Parra-Dominguez, G.S.; Taati, B.; Mihailidis, A. 3D Human Motion Analysis to Detect Abnormal Events on Stairs. In Proceedings of the IEEE International Conference on 3D Imaging, Modeling, Processing, Visualization and Transmission, Zurich, Switzerland, 13–15 October 2012; pp. 97–103.
15. Rojas, I.; Joya, G.; Cabestany, J. Indoor Activity Recognition by Combining One-vs.-All Neural Network Classifiers Exploiting Wearable and Depth Sensors. In *Advances in Computational Intelligence*; Lecture Notes in Computer Science; Springer: Berlin/Heidelberg, Germany, 2013.

16. Stone, E.E.; Skubic, M. Fall Detection in Homes of Older Adults Using the Microsoft Kinect. *IEEE J. Biomed. Health Inform.* **2015**, *19*, 290–301. [CrossRef] [PubMed]
17. Kwolek, B.; Kepski, M. Human fall detection on embedded platform using depth maps and wireless accelerometer. *Comput. Methods Programs Biomed.* **2014**, *117*, 489–501. [CrossRef] [PubMed]
18. Chen, J.; Kwong, K.; Chang, D.; Luk, J.; Bajcsy, R. Wearable sensors for reliable fall detection. In Proceedings of the IEEE Annual International Conference of the IEEE Engineering in Medicine and Biology Society, Shanghai, China, 17–18 January 2005; pp. 3551–3554.
19. Webster, D.; Celik, O. Systematic review of Kinect applications in elderly care and stroke rehabilitation. *J. Neuroeng. Rehabil.* **2014**, *11*, 1–24. [CrossRef]
20. Yhdego, H.; Li, J.; Morrison, S.; Audette, M.; Paolini, C.; Sarkar, M.; Okhravi, H. Towards musculoskeletal simulation-aware fall injury mitigation: Transfer learning with deep cnn for fall detection. In Proceedings of the IEEE Spring Simulation Conference (SpringSim), Tucson, AZ, USA, 29 April–2 May 2019; pp. 1–12.
21. Li, H.; Cryer, S.; Acharya, L.; Raymond, J. Video and image classification using atomisation spray image patterns and deep learning. *Biosyst. Eng.* **2020**, *200*, 13–22. [CrossRef]
22. Hwang, S.; Ahn, D.; Park, H.; Park, T. Maximizing accuracy of fall detection and alert systems based on 3d convolutional neural network. In Proceedings of the IEEE/ACM Second International Conference on Internet-of-Things Design and Implementation (IoTDI), Pittsburgh, PA, USA, 18–21 April 2017; pp. 343–344.
23. Santoyo-Ramón, J.A.; Casilari-Pérez, E.; Cano-García, J.M. A study on the impact of the users' characteristics on the performance of wearable fall detection systems. *Sci. Rep.* **2021**, *11*, 23011. [CrossRef]
24. Khan, S.S.; Hoey, J. Review of fall detection techniques: A data availability perspective. *Med. Eng. Phys.* **2017**, *39*, 12–22. [CrossRef]
25. Ren, L.; Shi, W. Chameleon: Personalised and adaptive fall detection of elderly people in home-based environments. *Int. J. Sens. Netw.* **2016**, *20*, 163–176. [CrossRef]
26. Zadeh, L.A. Similarity relations and fuzzy orderings. *Inf. Sci.* **1971**, *3*, 177–200. [CrossRef]
27. Klement, E.P.; Mesiar, R.; Pap, E. *Triangular Norms*; Kluwer Academic Publications: Dordrecht, The Netherlands, 2000.
28. Pradera, A.; Beliakov, G.; Bustince, H.; de Baets, B. A review of the relationships between implication, negation and aggregation functions from the point of view of material implication. *Inf. Sci.* **2016**, *329*, 357–380. [CrossRef]
29. Baczyński, M.; Jayaram, B. *Fuzzy Implications*; Studies in Fuzziness and Soft Computing; Springer: Berlin/Heidelberg, Germany, 2008; Volume 231.
30. Calvo, T.; Kolesárová, A.; Komorniková, M.; Mesiar, R. Aggregation operators: Properties, classes and construction methods. In *Aggregation Operators*; Calvo, T., et al., Eds.; Physica-Verlag: Heidelberg, Germany, 2002; pp. 3–104.
31. Lucca, G.; Sanz, J.A.; Dimuro, G.P.; Bedregal, B.; Mesiar, R.; Kolesárová, A.; Bustince, H. Preaggregation Functions: Construction and an Application. *IEEE Trans. Fuzzy Syst.* **2016**, *24*, 260–272. [CrossRef]
32. Bandler, W.; Kohout, L. Fuzzy power sets and fuzzy implication operators. *Fuzzy Sets Syst.* **1980**, *4*, 13–30. [CrossRef]
33. Bustince, H.; Mohedano, V.; Barrenechea, E.; Pagola, M. Definition and construction of fuzzy DI-subsethood measures. *Inf. Sci.* **2006**, *176*, 3190–3231. [CrossRef]
34. Santos, H.; Couso, I.; Bedregal, B.; Takáč, Z.; Minárová, M.; Asiain, A.; Barrenechea, E.; Bustince, H. Similarity measures, penalty functions, and fuzzy entropy from new fuzzy subsethood measures. *Int. J. Intell. Syst.* **2019**, *34*, 1281–1302. [CrossRef]
35. Sinha, D.; Dougherty, E.R. Fuzzification of set inclusion: Theory and applications. *Fuzzy Sets Syst.* **1993**, *55*, 15–42. [CrossRef]
36. Zadeh, L.A. A theory of approximate reasoning. In *Machine Intelligence*; Hayes, J.E., Michie, D., Mikulich, L.I., Eds.; Elsevier: New York, NY, USA, 1979 ; Volume 9, pp. 149–194.
37. Klir, G.J.; Yuan, B. *Fuzzy Sets and Fuzzy Logic: Theory and Applications*; Prentice-Hall: Hoboken, NJ, USA, 1995.
38. Pawlak, Z. *Rough Sets. Theoretical Aspects of Reasoning about Data*; Kluwer Academic Publishers: Dordrecht, The Netherlands; Boston, MA, USA; London, UK, 1991.
39. Grzymala-Busse, J.W. Rule Induction. In *Data Mining and Knowledge Discovery Handbook*; Maimon, O., Rokach, L., Eds.; Springer: Boston, MA, USA, 2009.
40. Grzymala-Busse, J.W. Rough Set Theory with Applications to Data Mining. In *Real World Applications of Computational Intelligence*; Studies in Fuzziness and Soft Computing; Gh. Negoita, M., Reusch, B., Eds.; Springer: Berlin/Heidelberg, Germany, 2005; Volume 179.
41. Charchalis, A.; Pawletko, R. The Use of Expert System for Marine Diesel Engine Diagnosis. *Sci. Pap. Pol. Nav. Acad.* **2012**, *1*, 49–55. [CrossRef]
42. Chien-Chung, C.; Sengottiyan, S. BLEM2: Learning Bayes' rules from examples using rough sets. In Proceedings of the 22nd International Conference of the North American Fuzzy Information Processing Society NAFIPS, Chicago, IL, USA, 24–26 July 2003.
43. Inuiguchi, M.; Tsuji, M.; Kusunoki, Y.; Tsurumi, M. LEM2-Based Rule Induction from Data Tables with Imprecise Evaluations. In Proceedings of the International Conference on Rough Sets and Knowledge Technology, RSKT 2011 , Banff, AB, Canada, 9–12 October 2011.
44. Inuiguchi, M.; Tsurumi, M.; Fukuda, D.; Yamanaka, K. LEM2-based rule induction via clustering decision classes. In Proceedings of the IEEE International Conference on Systems, Man and Cybernetics, Waikoloa, HI, USA, 12 October 2005.

45. Narsale, N.; Agarwal, V. Implementation of LEM2 algorithm On FPGA. In Proceedings of the 2019 3rd International conference on Electronics, Communication and Aerospace Technology (ICECA), Coimbatore, India, 12–14 June 2019; pp. 1–5.
46. Kepski, M. Fall Detection and Selected Action Recognition Using Image Sequences. Ph.D. Thesis, AGH University of Science and Technology, Kraków, Poland, 2016. (in Polish)
47. The Rule Set Produced by LEM2 Algorithm. Available online: https://tiny.pl/rnk88 (accessed on 16 March 2021).

sensors

MDPI

Article

Markerless vs. Marker-Based Gait Analysis: A Proof of Concept Study

Matteo Moro [1,2,3,*], Giorgia Marchesi [1,3], Filip Hesse [1], Francesca Odone [1,2] and Maura Casadio [1,3]

1 Department of Informatics, Bioengineering, Robotics and Systems Engineering (DIBRIS), University of Genova, 16145 Genova, Italy; giorgia.marchesi@edu.unige.it (G.M.); filip_hesse@yahoo.de (F.H.); francesca.odone@unige.it (F.O.); maura.casadio@unige.it (M.C.)
2 Machine Learning Genoa (MaLGa) Center, 16146 Genova, Italy
3 Spinal Cord Italian Laboratory (S.C.I.L.), 17027 Pietra Ligure, Italy
* Correspondence: matteo.moro@edu.unige.it

Abstract: The analysis of human gait is an important tool in medicine and rehabilitation to evaluate the effects and the progression of neurological diseases resulting in neuromotor disorders. In these fields, the gold standard techniques adopted to perform gait analysis rely on motion capture systems and markers. However, these systems present drawbacks: they are expensive, time consuming and they can affect the naturalness of the motion. For these reasons, in the last few years, considerable effort has been spent to study and implement markerless systems based on videography for gait analysis. Unfortunately, only few studies quantitatively compare the differences between markerless and marker-based systems in 3D settings. This work presented a new RGB video-based markerless system leveraging computer vision and deep learning to perform 3D gait analysis. These results were compared with those obtained by a marker-based motion capture system. To this end, we acquired simultaneously with the two systems a multimodal dataset of 16 people repeatedly walking in an indoor environment. With the two methods we obtained similar spatio-temporal parameters. The joint angles were comparable, except for a slight underestimation of the maximum flexion for ankle and knee angles. Taking together these results highlighted the possibility to adopt markerless technique for gait analysis.

Keywords: markerless; human motion analysis; gait analysis; computer vision; deep learning

Citation: Moro, M.; Marchesi, G.; Hesse, F.; Odone, F.; Casadio, M. Markerless vs. Marker-Based Gait Analysis: A Proof of Concept Study. *Sensors* **2022**, *22*, 2011. https://doi.org/10.3390/s22052011

Academic Editors: Carlos Tavares Calafate, Tomasz Krzeszowski, Adam Świtoński and Michal Kepski

Received: 31 December 2021
Accepted: 2 March 2022
Published: 4 March 2022

Publisher's Note: MDPI stays neutral with regard to jurisdictional claims in published maps and institutional affiliations.

1. Introduction

Gait analysis is a fundamental tool in medicine and rehabilitation [1]. It helps expert physicians to characterize and monitor motion patterns after orthopedic injuries and in people with neurological diseases, e.g., stroke, spinal cord injury, multiple sclerosis, or Parkinson [2]. Furthermore, gait analysis can be used to tailor appropriate and specific rehabilitation treatments. Quantitative assessments ensure repeatability and objectivity of the analysis, compared to visual observations only [3]. This kinematic quantification has been a major technical challenge for many years in the mid 90s [4].

Infrared marker-based motion capture systems (MoCap) have been developed to track continuous motion in 3D space [5]. Due to their high level of precision, infrared marker-based systems are considered the gold standard in modern gait analysis [6] and, in general, in accurate tracking of human motion. However, these approaches have limitations. First of all, they require many markers to be attached firmly to the body of the person. This process is time consuming and results in a cumbersome setup that can influence the naturalness of the motion [7]. These systems are also expensive and require skilled personnel to apply the markers correctly and to post-process the recorded data, making the overall analysis operator dependent [7]. Thus, the entire process requires many resources in terms of time and personnel. For these reasons, recently, many efforts have been made to study cheaper, faster, and simpler approaches to characterize human motion and, consequently, gait analysis [8]. Among the possible alternatives, systems based on wearable sensors (such as Inertial Measurements Units (IMU)) are less expensive, but suffer from the same issues of marker-based approaches.

In the last decades, deep learning algorithms have moved forward in solving computer vision problems [9]. In particular, recent advances on markerless pose estimation algorithms, based on computer vision and deep neural networks, are opening the possibility of adopting efficient methods for extracting motion information starting from common red-green-blue (RGB) video data [10]. This leads to the question of whether deep learning-based approaches can be adopted to analyze human motion in different domains and, specifically, if they can be adopted to perform accurate gait analysis for clinical applications [8]. Video-based techniques present many advantages with respect to marker-based systems. First of all, markerless video-based approaches are less expensive; they are not cumbersome and do not affect the naturalness of the motion, thus, they can be adopted to study human motion in an unconstrained environment. Lastly, they can be fully automatic and, hence, not operator dependent [6]. However, there are few studies that quantitatively compare the information extracted with video-based markerless techniques with those retrieved with gold standard marker-based systems [11–13]. As reported in the following section, most of them focus on 2D analysis, while for 3D analysis, to the best of our knowledge, there is still a lack of evidence related to the differences between video-based markerless and standard marker-based systems when describing meaningful kinematic variables and spatio-temporal parameters of human gait. In this work we aim at filling this gap by comparing both the spatio-temporal parameters and the joint angles changes during the gait cycle, computed from the keypoints extracted with these two techniques in 3D space.

Indeed, in this work, we defined an algorithm that, taking as inputs three RGB videos (acquired from 3 different viewpoints) and the calibration parameters, computes 3D keypoints positions. More precisely, our algorithm is composed by the following steps:

1. Keypoints detection in the image planes with a state-of-the-art Convolutional Neural Network (CNN): Pose ResNet-152 [14].
2. Keypoints refinement of the 2D detections, adopting Adafuse [15], that leverages epipolar geometry. In this step, also the weights of Pose ResNet-152 [14] are refined.
3. Keypoints' trajectories temporal filtering to increase the spatio-temporal consistency.
4. 3D reconstruction: Combining the detected keypoints from the different viewpoints, we reconstructed the 3D positions of each keypoint following a geometric approach [16].

First, we trained our algorithm on the Human3.6M dataset [17]. Then, we used the trained model to extract the 3D keypoints positions from our acquired data. Starting from the 3D keypoints coordinates, we computed spatio-temporal and kinematic gait parameters. Then, we compared our method with the gold standard marker-based method. Figure 1 summarizes the main steps addressed in this work.

In this context, the main contributions of this work can be summarized as follows:

- Implementation of a video-based markerless pipeline for gait analysis. The pipeline takes as input RGB videos (multiple viewpoints of the same scene) and camera calibration parameters, computes the 3D keypoints following the algorithm summarized above, and gives as outputs the kinematic parameters usually computed in gait analysis.
- Comparison between marker and markerless systems. We tested the reliability and the stability of the implemented pipeline. To do that, we acquired the gait of 16 healthy subjects with both a marker-based system (*Optitrack*) and a multi-view RGB camera system. Then, by using a biomechanical model (OpenSim Software [18]), we computed the spatio-temporal and kinematic gait parameters [4] with data from both the gold standard motion capture system and our implemented markerless pipeline. Then, we compared the results from the two systems. Experimental results obtained in a preliminary study focusing on 2D data (single viewpoint) [19] provide initial evidence of the comparability of the two approaches.

The paper is organized as follow: In Section 2, related works that drives this research are presented; in Section 3 we present our sample and how we collected data; in Section 4, the 3D extraction's procedure for both marker and markerless data are presented (Section 4.1 and Section 4.2, respectively); in Section 5, we presented the data filtering and the computation of spatio-temporal and kinematic parameters; in Section 6, the statistical tests used to compare the two approaches are presented; and in Sections 7 and 8, we present our results and its related discussion.

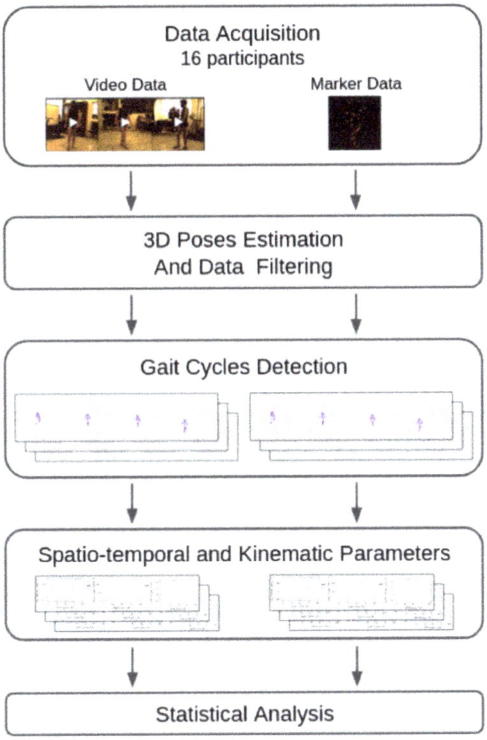

Figure 1. Summary of the workflow.

2. Related Works

Many efforts have been spent in the last few years to implement and test video-based systems that are able to characterize human gait without using cumbersome and intrusive markers placed on the body skin. In this section, we present works that addressed this problem by following approaches that differ for: the dimensionality of the considered space (2D or 3D analysis), type of cameras, e.g., depth cameras (RGBD) or RGB cameras, and type of algorithms (deep learning or classical approaches).

Rodrigues et al. [20] developed a markerless multimodal motion capture system using multiple RGBD cameras and IMUs mounted to the lower limbs of the participants to estimate spatio-temporal parameters and joint angles. Corazza et al. [21] extracted the walking people's silhouettes from 16 RGB camera views. These 2D silhouettes extracted from images recorded from different perspectives allowed the researchers to reconstruct the visual hull of the subject as a 3D model. By post-processing this model, the relevant joint angles could be determined. The authors could achieve a good performance determining the angles on the sagittal plane, however with larger errors on smaller angles, such as the knee adduction angle. Examples of similar approaches that used one or more RGB cameras and extracted silhouettes or used RGBD cameras can be found in [11,22–25].

Recently, due to the continuous progress in terms of accuracy and computational costs of pose estimation algorithms based on deep learning architecture, there is an increasing interest in the study of video-based systems for gait analysis. Kidzinski et al. [26] performed 2D gait analysis starting from the detection of keypoints in the image plane and, then, analyzing their trajectories extracting the joint angles and their changes on the gait cycle. They analyzed data from 1792 videos of 1026 patients with cerebral palsy. This approach has the potential to assess early symptoms of neurological disorders by using not expensive and commonly used technology. We followed a similar approach in Moro et al. [19] to investigate gait patterns in 10 stroke survivors. These works succeeded in performing a quantitative movement analysis using single camera videos in a stable way with results comparable to standard marker-based methods. Unfortunately, the 2D nature of the images limited the analysis to elevation angles [27] and to a subset of spatio-temporal parameters.

Vafadar et al. [28] performed markerless gait analysis by first reconstructing an accurate human pose in 3D from multiple camera views. They collected a gait-specific dataset composed by 31 participants, 22 with normal gait and 9 with pathological gait. The researchers recorded the gait of the participants with a standard marker-based system and with 4 RGB cameras. For 3D pose estimation, they relied on the approach proposed by [29]. They were successfully able to reconstruct the human pose while walking in 3D. However, they did not include in the detection keypoints on the feet and, consequently, they were not able to extract significant spatio-temporal parameters as the stride width and the ankle joint motion.

3. Dataset

A total of 16 unimpaired participants (6 females, mean age ± standard deviation: 27 ± 2 years old) without a known history of orthopaedic injuries or neurological diseases walked naturally in straight lines from one side of a room to the opposite. The path was 6 m long. Each participant performed 20 trials, 10 for each direction.

The setup for data acquisition (see Figure 2) included (i) a calibrated multi-view camera system consisting of 3 RGB Mako G125 GigE cameras with Sony ICX445 CCD sensor, resolution 1292 × 964, 30 frames per second (fps) for markerless analysis and (ii) a calibrated motion capture system, the Optitrack Flex 13 Motion Capture system, 1.3 MP, 56° Horizontal FOV, 46° Vertical FOV, 28 LEDs, 8.33 ms latency, with 8 cameras acquiring at 100 Hz. With the motion capture system, we acquired the 3D position of 22 infrared passive markers placed on the body of the participants following the Davis protocol [30]. RGB cameras calibration was performed according to Zhang's method [31]. As a calibration pattern, we used a checkerboard with squares 40 × 40 mm. The calibration covered a volume of 6.5 × 2.5 × 2 m.

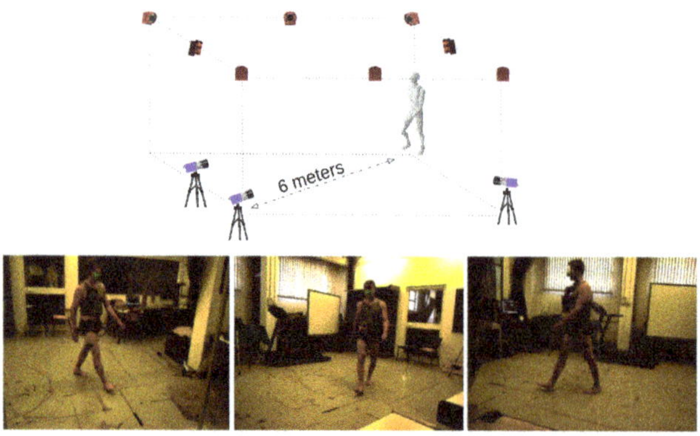

Figure 2. Setup adopted for data acquisition. The upper panel shows the sketch of the setup with the position of the 8 infrared (red) and 3 RGB (blue) cameras. The lower panel shows the three view points of the RGB cameras.

The study was conducted according to the guidelines of the Declaration of Helsinki, and approved by the Institutional Review Board of the Department of Informatics, Bioengineering, Robotics and Systems Engineering (DIBRIS), University of Genoa, Genova, Italy (protocol code CE DIBRIS-008/2020 approved on 18/05/2020). All the participants involved in the study signed an informed consent form.

4. 3D Keypoints

In this section, we present the processing for obtaining the 3D positions of meaningful keypoints. These steps are different for the marker-based and markerless approaches. More precisely, in the marker-based approach, we used the software Motive [32] to extract the 3D trajectories of the markers. In the markerless approach, we adapted the algorithm Adafuse [15] to detect and refine keypoints from the RGB videos. The two procedures are described in detail below.

4.1. Marker Data

The motion capture system reconstructed the trajectories of the markers in the 3D reference system, starting from 8 infrared cameras. To perform the motion analysis, we needed to add a feature matching and tracking step. The process of *sorting and tracking* the markers is a standard procedure performed after data acquisition with a motion capture system. The software Motive [32] provided with the Optitrack motion capture system automatically performed this procedure by applying a model of the human body, indicating the position of the markers (Figure 3A), defined by the user. However, in cases of markers occlusions or presence of disturbances as reflexes, this procedure required the manual intervention of the operator, resulting in a time consuming procedure. This workload emphasizes one drawback of the marker-based motion capture system. At the end of this process, we obtained 16 matrices $Pmarker^j$ with $j = 1, \ldots, 16$ indicating the index for each participant, of shape $22 \times 3 \times M_j$ (22 representing the number of markers, 3 the $(X, Y, Z)_m$ markers' coordinates in the 3D space in the markers reference system ($_m$) and M_j for the number of samples for the acquisition of the j-th participant).

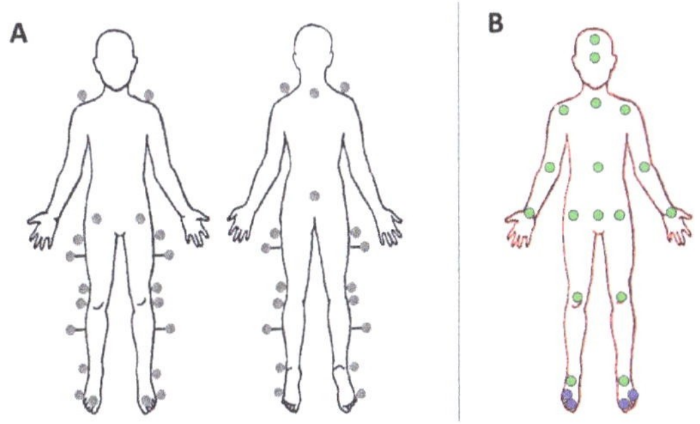

Figure 3. (**A**) Frontal and back views of the positions of the 22 markers positioned in this study according to the Davis protocol [30]. Specifically they were placed on the spinal process of C7 and on the spinal process of the sacrum (both visible in the back view) and bilaterally on: the acromion, the Anterior Superior Iliac Spine (ASIS), the greater trochanter, the middle between the greater trochanter and the lateral epicondyle of the femur (with bars 5 cm long), the lateral epicondyle of the femur, the fibula head, the middle between the fibula head and the lateral malleolus (with bars 5 cm long), the lateral malleolus, the first metatarsal phalangeal joint, and the fifth metatarsal phalangeal joint on the lateral aspect of the foot. (**B**) 2D keypoints (green and blue dots) considered in this work from the Human3.6 dataset. The two blue keypoints in each foot are highlighted because they are not included in [15] and we added them in our training.

4.2. Video Data

The RGB cameras produced video streams acquired from three views. To obtain the 3D points, we needed to detect semantic features in 2D and then triangulate them in 3D. The resulting 3D points were easily tracked, since each one of them was associated with a semantic meaning. Thus, the aim of this step was the detection of the 3D positions of keypoints that represent the analogous of markers and that can be adopted to perform gait analysis. To perform this step, it is possible to proceed in two different ways: (i) rely on a 2D pose estimator to detect the positions of the keypoints in the image planes of each viewpoint and then reconstruct the positions of each keypoint in the 3D space with a 3D reconstruction algorithm (e.g., [16])or (ii) rely directly on an end-to-end 3D pose estimator (see the review [8] for examples). We opted for the first option in order to have higher control in the number and in the position of the body keypoints detected in the image planes.

For this task we relied on AdaFuse [15]: A deep learning-based algorithm that allows one to accurately detect the positions of specific keypoints in the image plane and leverages classical stereo vision algorithms [16] to reconstruct the 3D positions of the detected keypoints. We selected Adafuse as it is one of the most recent (2021) and most precise [15] algorithms for 3D pose estimation. Its precision is due to the refinements in the image planes (2D) of the detected keypoints: It leverages epipolar geometry and on stereo vision algorithms to refine 2D detection. In this way, the 3D keypoints estimates are also more precise. In addition, the CNN (Convolutional Neural Network) for the 2D keypoints detection (2D backbone in the following sections) can be accurately selected based on the specific goal. In Section 4.2.1, we present and justify our choices.

Adafuse is mainly divided into the three following parts:

- A 2D pose estimator backbone (*Pose ResNet* [14]).
- A fusing deep learning architecture that refines the probability maps of each view generated in the first step. To accomplish this, the algorithm takes into account the information from neighboring views and it leverages epipolar geometry [16]. In this way it, is possible to enrich the information of each probability map at any point x by adding the information of the probability maps of its neighbor viewpoints.
- A geometric 3D reconstruction part that leverages the intrinsic and extrinsic camera parameters obtained during calibration.

4.2.1. Adafuse Training

The pretrained 2D backbone models provided by AdaFuse authors [15] do not consider keypoints on the feet. Since these keypoints are necessary for gait analysis to compute the kinematic parameters related with the ankle joint (i.e., ankle dorsi-/plantar-flexion), we had to train the model with new data that also included keypoints on the feet. Moreover, to effectively train our model, we also needed a dataset with the 3D ground truth positions of each keypoint. The direct outputs of the AdaFuse algorithm are 2D probability maps ($U_t^{j,i,l}$) of each keypoint l for each input frame (I_t^j, at t-th time instant and for the j-th participant) for each viewpoint i ($i = \{1,2,3\}$). The final 3D pose could be computed by geometric triangulation. This is true if the 2D ground truth positions of each keypoint are consistent between the different viewpoints. Unfortunately, this is not the case for most of the available datasets.

Among the public available datasets (well summarized in [11]), we relied on the Human3.6m dataset [17] because it included almost all the characteristics required by our analysis and described below. The Human3.6m dataset contains recordings of 11 professional actors (6 male, 5 female), performing in 17 different scenarios. Those scenarios are, for example discussion, smoking, taking photos, or walking. The actors wear natural clothes while having markers attached to their clothes (or skin, if the skin is visible). In total, the dataset includes over 3.6 million images with human poses. Each scene only shows one actor at a time, so this dataset is only suitable for single human pose estimation. The dataset includes both a multi-view RGB camera system (with 4 cameras) and a motion capture system with infrared cameras and 32 markers (see [17] for further details). Leveraging *Vicon's* skeleton fitting procedure [33] and by applying forward kinematics, the Human3.6m dataset [17] provides both the 3D ground truth (i.e., the positions of the keypoints in the 3D space), and the 2D ground truth (i.e., the positions of the keypoints projected into the 2D image planes) of the different viewpoints (see Figure 3B). The reader is referred to [17] for more details on how 3D and 2D ground truth were provided. Human3.6m was our best option, even if it presented drawbacks for our main goal. For example, the feet sometimes get rather blurry, mainly in the swing phase where one foot moves quickly. Additionally, the background carpet, under the lighting condition during the recordings, has color similar to the skin, so contrast decreases to a low level, where even for human observers, it would be hard to detect the keypoints precisely.

We fine tuned the Adafuse architecture in two steps:

1. **2D backbone**. We first focused on the 2D backbone network creating independent probability maps of the keypoints in Figure 3B for each separate input image and we fine tuned the Pose ResNet-152 [14] pretrained on the COCO dataset [34]. We did not train the network directly from scratch to reduce time and the amount of computational resources needed. We fine tuned the network by adopting a subset of the Human3.6m training images, i.e., we considered one image every 20 frames (for a total of 180,000 training images). This allowed us to have a training set with a reasonable number of frames sufficiently different from one another.
2. **Full architecture**. Then we focused on the fusing network which refines the maps with the help of the neighboring views. This second part of the AdaFuse architecture should not be trained separately (as mentioned in [15]), but jointly with the 2D backbone. Thus, we initialized the first part (2D backbone) with the weights obtained with the fine tuning described above and the fusion network with random weights. In this case, the inputs of the process are not just single images (as for the previous step), but a group of images representing the same time instant but coming from different viewpoints. Additionally, we input the calibration information for the group of images containing intrinsic and extrinsic parameters. These parameters are not used by the neural network itself, but in an immediate post-processing step which computes the 3D poses at the end. The target and output for the neural network is a group of probability maps corresponding to the input images. It is worth remembering here that the outputs of the full Adafuse process are just probability maps and not 3D points, however they are more precise than those from the 2D backbone because additional information from neighboring views is fused with the backbone prediction. The 3D pose is then computed via triangulation.

4.2.2. Inference

We applied the model trained as described in Section 4.2.1 to our dataset for retrieving the 3D positions of the Human3.6m keypoints highlighted in Figure 3B. Since Pose ResNet-152 requires as input a bounding box also localizing the person in the image plane for each frame composing the videos, we relied on CenterNet [35], which is a state-of-the-art object detector, to create these bounding boxes for our dataset. Thus, we input to the model the 3 images coming from the 3 different viewpoints at the same time instant t, the bounding boxes, and the intrinsic and extrinsic parameters retrieved with cameras calibration. Firstly, we obtained the probability maps for different keypoints at the same time instant (Figure 4A for some examples) and then the final 2D locations of each keypoint (Figure 4B). At the end, the final output is a vector of shape 21 × 3 (21 keypoints with the corresponding $(X, Y, Z)_v$ coordinates in the 3D space in the camera reference system $_v$) with $j = 1, \ldots, 16$ representing the number of videos (i.e., the number of participants) and $t = 1, \ldots, N_j$, which is the index for the number of frames composing the j-th video (N_j is the total number of frame for the video of the j-th participant). At the end of this step, 16 matrices were left $Pmarkerless^j$ with a shape of $21 \times 3 \times N_j$ (see Figure 4C for examples of 3D poses).

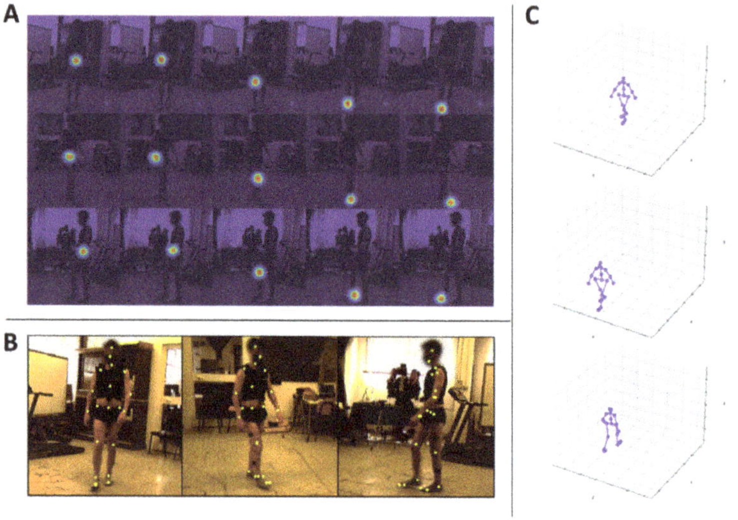

Figure 4. (**A**) Examples of the detected probability maps ($U_t^{j,i,l}$) for the j-th participant at a specific time instant t. The rows represent the 3 different viewpoints i. Each column represents a different keypoint l detected on the right leg (from left to right: hip, knee, heel, toe). (**B**) Examples of the detected keypoints (yellow dots) on the three views composing our dataset. (**C**) Examples of the final 3D skeleton of the video pre-processing.

4.3. Keypoints Detection Evaluation Metrics

To evaluate the accuracy of our trained model, we relied on two metrics usually adopted to evaluate the accuracy of 2D and 3D pose estimation algorithms.

For the evaluation of the 2D backbone, we relied on the Percentage of Correct Keypoints (PCK) [36]. Given the ground truth (as defined in [17]) and the estimate position detected by our model of a certain keypoint l at the time instant t (x_t^l and $\tilde{x}_t^l$, respectively), the PCK defines how close the estimate $\tilde{x}_t^l$ is with respect to the ground truth position x_t^l. In particular, $\tilde{x}_t^l$ is considered correctly detected if:

$$|\tilde{x}_t^l - x_t^l| < r_{thr} \quad (1)$$

where $|\tilde{x}_t^l - x_t^l|$ represents the Euclidean distance between the estimate and the ground truth position of the keypoint l. This means that to be considered correctly detected, the estimate $\tilde{x}_t^l$ should fall inside a circle centered in the ground truth x_t^l and with radius r_{thr}. In many works regarding 2D pose estimation algorithm [10], the PCK is computed considering r_{thr} as a percentage of: (i) the torso diameter (usually the 20%); (ii) the head bone link (usually the 50%, $PCKh@0.5$ with h indicating the head bone link and @0.5 indicating a 50% threshold). In this work, we adopted $PCKh@\tau$ considering different thresholds τ, e.g., $PCKh@0.5$, $PCKh@0.75$, $PCKh@1$, corresponding 0.5, 0.75, and 1 time to the length of the head bone link.

On the other side, for the evaluation of the accuracy of the full process ending with the 3D reconstruction, we relied on the Mean Per Joint Position Error (MPJPE). The MPJPE is the most common metric to evaluate 3D estimates and it is defined for each keypoint as the mean euclidean distance in the 3D space between the estimated keypoint ($\tilde{x}_t^l$) and the correspondent ground truth (x_t^l).

5. 3D Keypoints Trajectories Processing

The 3D trajectories of the keypoints extracted with marker-based ($Pmarker^j$) and markerless ($Pmarkerless^j$) systems were processed in the same way to extract quantitative parameters describing the gait of each participant.

5.1. *Gait Cycle Detection*

One gait cycle is defined as the period that starts with the heel strike (first instant when the heel hits the ground) of one foot and ends with the following heel strike of the same foot. A typical approach for automatic gait cycle detection in the absence of force platforms is to analyze the speed of the feet keypoints [37]. The cycle starts when the heel hits the ground; in this time instant, the speed of the heel is close to zero. It remains close to zero for the entire stance phase (the phase starting with the heel strike and ending when the foot leaves the ground) and it goes up in the swing phase (complementary to the stance phase). Then, the swing phase ends and the heel speed goes close to zero again. This first time instant where the speed is close to zero is the one representing the end of the current gait cycle and also the start of the following one.

For both the marker and the markerless approaches, we detected the start and the end of the gait cycle by following this procedure and considering the vertical component of the heel keypoint, low pass filtered with a Butterworth filter (4-th order, 3 Hz cut off frequency). We computed the derivative of the filtered vertical (Y) heel coordinates, obtaining the velocity profiles. Then, we computed the speed absolute value by combining the coordinates and we automatically detected the gait cycles following the considerations mentioned before. It is worth mentioning here that the 3 Hz cut off frequency filter was only used for gait cycle detection. To process the keypoints' signals in later steps, we filtered the original raw signals as described in the following sections.

5.2. *Spatio-Temporal Parameters and Joint Angles*

The 3D coordinates trajectories of each keypoint during the gait cycles were low pass filtered (Butterworth, 4-th order, $12Hz$ cut off frequency) [4].

Starting from the heels' markers trajectories, we extracted the spatio-temporal parameters that characterize the human gait. In particular, we computed the following parameters: (i) Stride length: the distance (in meters) walked during a gait cycle; (ii) Stride time: the time (in seconds) necessary to walk one gait cycle; (iii) Stance phase: percentage of the gait cycle during which the foot of interest is touching the ground; (iv) Swing phase: percentage of the cycle complementary to the stance phase, when the foot of interest is not touching the ground; (v) Stride width: the distance (in meters) between the right and the left foot across the cycle; and (vi) Speed: mean speed of the center of mass of the body during the cycle.

To estimate the joint angles during the gait cycle, we relied on the open source software Opensim [18]. Opensim is commonly adopted to estimate joint angles during gait analysis because it allows associating the detected keypoints/markers to human biomechanical skeleton models and analyze the kinematics and the relative muscular activation. In this work, we adopted the Rajagopal Model [38], a full body musculoskeletal model for dynamic simulations of human movements, widely used in gait analysis applications. In Opensim, two tools are specifically designed to solve our problem, *Scaling* and *Inverse Kinematics*. The first was adopted to scale a generic skeleton model to fit the input markers/keypoints data. The latter was used to simulate the motion of the skeleton and to estimate the joint angles for each gait cycle for each subject. Following the steps explained above, we extracted the joint angles for the central gait cycle of each trial (for a total of 20 gait cycle) for each participant involved in the study both with marker-based and markerless systems.

6. Statistical Analysis

To compare the time profile of the joint angles during the gait cycle obtained with the markerless and the marker-based gait analysis we used the statistical parametric mapping method, which is specifically designed for continuous field analysis [39] and is already used in similar applications in gait analysis [19]. In this study we applied this method to the 1D spatio-temporal variables describing the variations of the joint angles during the gait cycle by using the open source software spm1d [39]. Specifically, we performed a one dimensional paired t-test. We tested the following null hypothesis: "There are no statistically significant differences between the gait angles obtained with our markerless approach and the gait angles obtained with the gold standard marker-based system". The alpha level indicating the probability of incorrectly rejecting the null hypothesis was set at 0.05. Small values of p allow for the rejection of the null hypothesis. Indeed, if we obtain $p > 0.05$, we can conclude that our statistical tests did not find significant differences between the gait angles obtained with our markerless approach and those obtained with the gold standard marker-based system. To follow a conservative approach, i.e., to maximize the possibility to find statistically significant differences between the results obtained with the two methods, we did not apply Bonferroni corrections. Notice that the application of corrections for multiple comparisons would decrease the probability to find significant differences between the single point curves. Furthermore, we compared the spatio-temporal parameters obtained with the two methods with a paired t-test. Again, statistical significance was set for all statistics at the family-wise error rate of $\alpha = 0.05$.

7. Results

7.1. Architecture Evaluation

To evaluate the accuracy of our trained 2D backbone, we computed the *PCKh* for each keypoint (see Figure 5 for a qualitative result). As threshold value r_{thr}, we selected a percentage of the head bone link for each participant (indicated by the *h* in *PCKh*). The following multiplication factors were chosen: 1 (*PCKh*@1), 0.75 (*PCKh*@0.75), and 0.5 (*PCKh*@0.5). Table 1 summarizes the obtained results.

Table 1. Accuracy (%) of the 2D backbone, i.e., the percentage of corrected keypoints (PCKh) considering different threshold values: 1, 0.75, and 0.5 times the head bone link (*PCKh*@1, *PCKh*@0.75, and *PCKh*@0.5, respectively).

Keypoints	PCKh@1	PCKh@0.75	PCKh@0.5
head	96.3	95.8	95.2
root	96.6	95.6	94.8
nose	96.1	94.3	87.2
neck	96.1	89.3	77.2
right shoulder	93.4	87.4	66.7
right elbow	89.1	79.8	70.7
right wrist	85.5	78.6	67.8
left shoulder	95.2	88.9	72.7
left elbow	90.6	82.2	77.1
left wrist	85.0	78.7	70.0
belly	94.2	80.7	72.0
right hip	96.0	87.6	73.2
right knee	93.4	85.5	76.2
right foot1	91.6	79.7	61.4
right foot2	92.3	84.5	68.6
right foot3	89.2	77.3	63.0
left hip	95.8	85.1	72.1
left knee	92.4	79.9	66.7
left foot1	90.3	75.9	52.8
left foot2	91.7	83.4	67.7
left foot3	88.7	78.4	64.4

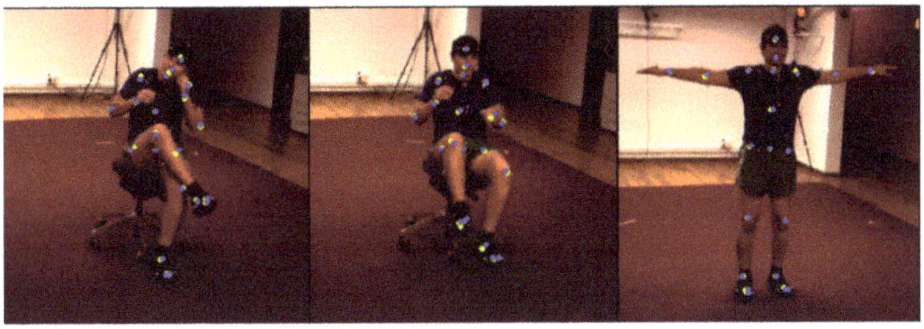

Figure 5. Examples of the keypoints detected with our model (yellow dots) with respect to the ground truth (blue dots).

The neural network indeed learned to detect also the new keypoints (toes and heels) with a high accuracy. The *PCKh* for these keypoints is comparable to the one of the others, and also to the results presented in other works (see for instance [15]).

To evaluate the accuracy of the full architecture, we computed the MPJPE across all the detected keypoints and obtained an error of 23.65 millimeters, again comparable to the one obtained in [15] (e.g., 19.5 millimeters on the same dataset, however with fewer keypoints–the feet were excluded) and also comparable with the error obtained in the best performing recent works about 3D pose estimation (between 19 and 30 millimeters) [40–43].

7.2. *Joint Angles and Spatio-Temporal Parameters*

We computed the spatio-temporal parameters described in the previous section for each gait cycle for every participant and compared the results obtained with the two different techniques. In Table 2, we report the mean and standard deviation across all the subjects. Note that parameters obtained with our markerless pipeline are similar to the ones extracted with the gold standard marker-based technique, as highlighted by the statistical comparisons: All p-values > 0.050, see Table 2 for more details.

Table 2. Spatio-temporal parameters computed with marker-based and markerless systems, and statistical results of the comparison between the two methods (last row). We report the mean ± the standard deviation of each parameter. The stance and swing phases are reported in % with respect to the whole gait cycle; stride length and step width and expressed in meters (m); stride time in seconds (s); and the speed in meters per second (m/s).

	Stance Phase (%)	**Swing Phase (%)**	**Stride Length (m)**	**Step Width (m)**	**Stride Time (s)**	**Speed (m/s)**
Marker	59.2 ± 2.6	40.8 ± 2.6	1.35 ± 0.11	0.10 ± 0.02	1.13 ± 0.02	1.31 ± 0.10
Markerless	59.6 ± 3.1	40.4 ± 3.1	1.40 ± 0.21	0.12 ± 0.02	1.11 ± 0.04	1.35 ± 0.16
p-values	0.644	0.644	0.474	0.132	0.291	0.341

We compared the joint angles obtained by our markerless approach to those obtained with the marker-based method. We selected the following meaningful angles: hip flexion/extension, knee flexion/extension, ankle dorsi-/planta-flexion, hip ab-/ad-duction, and pelvis tilt. Figure 6 shows the mean and standard deviation of the angles previously mentioned across all the participants (black: marker-based, red: markerless) and the results of the paired t-test. No statistical differences were found between the two techniques with the exception of a slight underestimation of the knee flexion and the ankle dorsiflexion angle between 70% and 80% of the gait cycle (during the swing phase, see gray areas in the paired t-tests in the right column of Figure 6 in correspondence of these two angles). Note that those statistical differences are not robust to multiple comparison, i.e., applying a Bonferroni correction the differences are not below the threshold for significance.

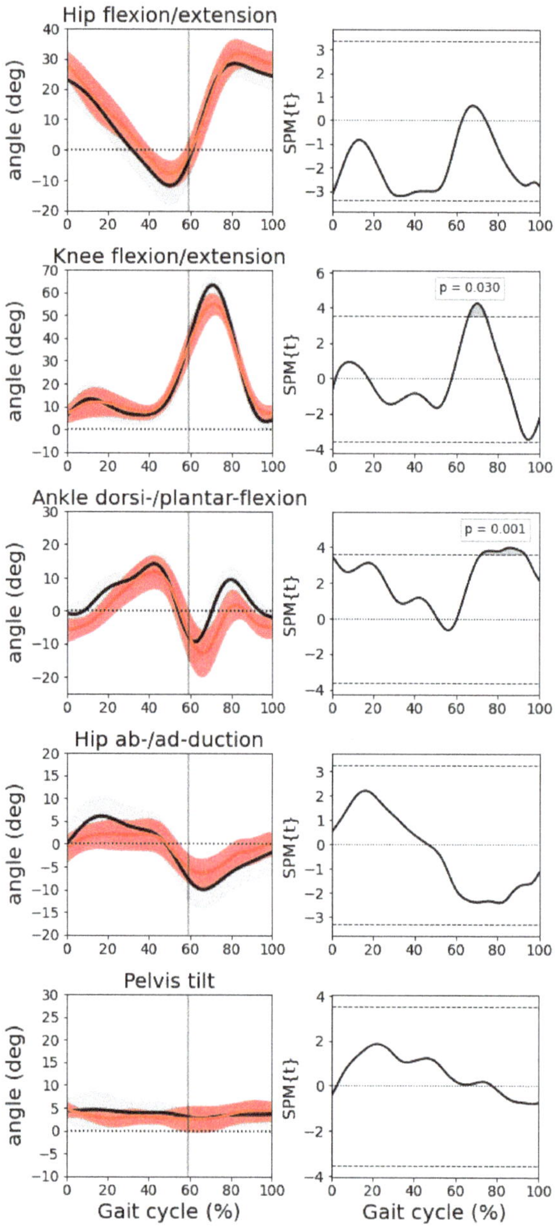

Figure 6. Left column: joint angles (mean and std). From top to bottom: hip flexion/extension, knee flexion/extension, ankle dorsi-/planta-flexion, hip ab-/ad-duction, and pelvis tilt. In black shows the results obtained with the marker-based system and in red shows the results with the markerless pipeline. Right column: results of the correspondent paired t-tests.

8. Discussion and Conclusions

This paper presents an approach for markerless gait analysis relying only on RGB video acquisition and leveraging computer vision and deep learning algorithms. Our approach presents the following advantages with respect to the gold standard marker-based methods:

1. It requires less expertise and has no bias introduced by any operators. In fact, while the operator during marker-based data acquisition needs to place markers carefully on the subjects skin in order to avoid biased results, our pipeline works fully automatically, and it is independent of any human performance;
2. It does not affect the naturalness of gait in any ways since it does not require cumbersome markers and sensors. Furthermore, it makes the data acquisition easier and faster because it is not necessary to place markers on the body skin;
3. It is less expensive and with a simpler setup and is easier to use outside laboratory environments, since it requires only RGB cameras.

Conversely, the results obtained with our markerless system present differences with respect to the ones obtained with the gold standard, especially during the swing phase in the maximum flexion of the knee and the ankle joint angles. These differences are statistically significant, however they appear to be small. Nonetheless, this limitation should be accounted and further investigated when adopting this markerless pipeline to detect and monitor abnormal motion patterns in people with orthopaedic injury or neurological diseases. If we focus on the errors related to the knee and the ankle joint angles during the swing phase, we can observe that they are mainly due to small errors in the detection of the feet keypoints. In fact, during the swing phase, the foot moves quickly and the image tends to become blurry, making it is difficult also for human beings to detect keypoints with high confidence. The immediate way to reduce the motion blur is to adopt RGB cameras with a higher temporal resolution, meaning a higher acquisition rate (fps). In this way, the motion blur will be reduced and, consequently, the detection error will also be lower.

Apart from inputting higher quality data to our pipeline, we can also improve the 2D backbone itself. In fact, the one adopted in this work and in AdaFuse [15] (Simple Baselines) is not the best performer according to multiple benchmarks. For example, the neural network HRNet [44] had been proven to provide better results on the Human3.6m dataset. Improving the accuracy of the detection will reduce the errors highlighted before.

In conclusion, the results suggest that the proposed markerless pipeline is a promising alternative to compute the marker-based system to most spatio-temporal and kinematic parameters. We highlighted also the limits of our pipeline and we presented possible solutions to overcome them in our future works.

Author Contributions: Conceptualization, M.C. and F.O.; methodology, M.C. and F.O.; software, F.H., M.M. and G.M.; validation, M.M. and G.M.; formal analysis, F.H., M.M. and G.M.; investigation, M.C. and F.O.; resources, M.C. and F.O.; data curation, M.M., G.M. and F.H.; writing—original draft preparation, M.M., G.M. and F.H.; writing—review and editing, M.C. and F.O.; visualization, M.M., G.M. and F.H.; supervision, M.C. and F.O.; project administration, M.C. and F.O.; funding acquisition, M.C. and F.O. All authors have read and agreed to the published version of the manuscript.

Funding: This research was funded by Fondazione Italiana Sclerosi Multipla (FISM–2019/PR-single050). G.M. was supported by Regione Liguria.

Institutional Review Board Statement: The study was conducted according to the guidelines of the Declaration of Helsinki, and approved by the Institutional Review Board of the Department of Informatics, Bioengineering, Robotics and Systems Engineering (DIBRIS), University of Genoa, Genova, Italy (protocol code CE DIBRIS-008/2020 approved on 18 May 2020).

Informed Consent Statement: Informed consent was obtained from all subjects involved in the study.

Data Availability Statement: In our study, we relied on the public available dataset Human3.6m that you can find at http://vision.imar.ro/human3.6m/description.php (accessed on 30 December 2021).

Acknowledgments: The authors thank Issa Mouawad for his support in the work.

Conflicts of Interest: The authors declare no conflict of interest. The funders had no role in the design of the study; in the collection, analyses, or interpretation of data; in the writing of the manuscript, or in the decision to publish the results.

References

1. Fritz, N.E.; Marasigan, R.E.R.; Calabresi, P.A.; Newsome, S.D.; Zackowski, K.M. The impact of dynamic balance measures on walking performance in multiple sclerosis. *Neurorehabilit. Neural Repair* **2015**, *29*, 62–69. [CrossRef]
2. di Biase, L.; Di Santo, A.; Caminiti, M.L.; De Liso, A.; Shah, S.A.; Ricci, L.; Di Lazzaro, V. Gait analysis in Parkinson's disease: An overview of the most accurate markers for diagnosis and symptoms monitoring. *Sensors* **2020**, *20*, 3529. [CrossRef]
3. Wren, T.A.; Tucker, C.A.; Rethlefsen, S.A.; Gorton, G.E., III; Õunpuu, S. Clinical efficacy of instrumented gait analysis: Systematic review 2020 update. *Gait Posture* **2020**, *80*, 274–279. [CrossRef]
4. Whittle, M.W. *Gait Analysis: An Introduction*; Butterworth-Heinemann: Oxford, UK, 2014.
5. Cloete, T.; Scheffer, C. Benchmarking of a full-body inertial motion capture system for clinical gait analysis. In Proceedings of the 2008 30th Annual International Conference of the IEEE Engineering in Medicine and Biology Society, Vancouver, BC, Canada, 20–25 August 2008; pp. 4579–4582.
6. Colyer, S.L.; Evans, M.; Cosker, D.P.; Salo, A.I. A review of the evolution of vision-based motion analysis and the integration of advanced computer vision methods towards developing a markerless system. *Sport. Med.-Open* **2018**, *4*, 1–15. [CrossRef]
7. Carse, B.; Meadows, B.; Bowers, R.; Rowe, P. Affordable clinical gait analysis: An assessment of the marker tracking accuracy of a new low-cost optical 3D motion analysis system. *Physiotherapy* **2013**, *99*, 347–351. [CrossRef]
8. Desmarais, Y.; Mottet, D.; Slangen, P.; Montesinos, P. A review of 3D human pose estimation algorithms for markerless motion capture. *Comput. Vis. Image Underst.* **2021**, *212*, 103275. [CrossRef]
9. Voulodimos, A.; Doulamis, N.; Doulamis, A.; Protopapadakis, E. Deep learning for computer vision: A brief review. *Comput. Intell. Neurosci.* **2018**, *2018*. [CrossRef]
10. Zheng, C.; Wu, W.; Yang, T.; Zhu, S.; Chen, C.; Liu, R.; Shen, J.; Kehtarnavaz, N.; Shah, M. Deep learning-based human pose estimation: A survey. *arXiv* **2020**, arXiv:2012.13392.
11. Kwolek, B.; Michalczuk, A.; Krzeszowski, T.; Switonski, A.; Josinski, H.; Wojciechowski, K. Calibrated and synchronized multi-view video and motion capture dataset for evaluation of gait recognition. *Multimed. Tools Appl.* **2019**, *78*, 32437–32465. [CrossRef]
12. Moro, M.; Casadio, M.; Mrotek, L.A.; Ranganathan, R.; Scheidt, R.; Odone, F. On The Precision Of Markerless 3d Semantic Features: An Experimental Study On Violin Playing. In Proceedings of the 2021 IEEE International Conference on Image Processing (ICIP), Anchorage, AK, USA, 19–22 September 2021; pp. 2733–2737.
13. Needham, L.; Evans, M.; Cosker, D.P.; Wade, L.; McGuigan, P.M.; Bilzon, J.L.; Colyer, S.L. The accuracy of several pose estimation methods for 3D joint centre localisation. *Sci. Rep.* **2021**, *11*, 20673. [CrossRef]
14. Xiao, B.; Wu, H.; Wei, Y. Simple baselines for human pose estimation and tracking. In Proceedings of the European Conference on Computer Vision (ECCV), Munich, Germany, 8–14 September 2018; pp. 466–481.
15. Zhang, Z.; Wang, C.; Qiu, W.; Qin, W.; Zeng, W. AdaFuse: Adaptive Multiview Fusion for Accurate Human Pose Estimation in the Wild. *Int. J. Comput. Vis.* **2021**, *129*, 703–718. [CrossRef]
16. Hartley, R.; Zisserman, A. *Multiple View Geometry in Computer Vision*; Cambridge University Press: Cambridge, UK, 2003.
17. Ionescu, C.; Papava, D.; Olaru, V.; Sminchisescu, C. Human3. 6m: Large scale datasets and predictive methods for 3d human sensing in natural environments. *IEEE Trans. Pattern Anal. Mach. Intell.* **2013**, *36*, 1325–1339. [CrossRef] [PubMed]
18. Delp, S.L.; Anderson, F.C.; Arnold, A.S.; Loan, P.; Habib, A.; John, C.T.; Guendelman, E.; Thelen, D.G. OpenSim: Open-source software to create and analyze dynamic simulations of movement. *IEEE Trans. Biomed. Eng.* **2007**, *54*, 1940–1950. [CrossRef] [PubMed]
19. Moro, M.; Marchesi, G.; Odone, F.; Casadio, M. Markerless gait analysis in stroke survivors based on computer vision and deep learning: A pilot study. In Proceedings of the 35th Annual ACM Symposium on Applied Computing, Brno, Czech Republic, 30 March–3 April 2020; pp. 2097–2104.
20. Rodrigues, T.B.; Salgado, D.P.; Catháin, C.Ó.; O'Connor, N.; Murray, N. Human gait assessment using a 3D marker-less multimodal motion capture system. *Multimed. Tools Appl.* **2020**, *79*, 2629–2651. [CrossRef]
21. Corazza, S.; Muendermann, L.; Chaudhari, A.; Demattio, T.; Cobelli, C.; Andriacchi, T.P. A markerless motion capture system to study musculoskeletal biomechanics: Visual hull and simulated annealing approach. *Ann. Biomed. Eng.* **2006**, *34*, 1019–1029. [CrossRef]
22. Castelli, A.; Paolini, G.; Cereatti, A.; Della Croce, U. A 2D markerless gait analysis methodology: Validation on healthy subjects. *Comput. Math. Methods Med.* **2015**, *2015*. [CrossRef]
23. Clark, R.A.; Bower, K.J.; Mentiplay, B.F.; Paterson, K.; Pua, Y.H. Concurrent validity of the Microsoft Kinect for assessment of spatiotemporal gait variables. *J. Biomech.* **2013**, *46*, 2722–2725. [CrossRef]
24. Gabel, M.; Gilad-Bachrach, R.; Renshaw, E.; Schuster, A. Full body gait analysis with Kinect. In Proceedings of the 2012 Annual International Conference of the IEEE Engineering in Medicine and Biology Society, San Diego, CA, USA, 28 August–1 September 2012; pp. 1964–1967.

25. Saboune, J.; Charpillet, F. Markerless human motion tracking from a single camera using interval particle filtering. *Int. J. Artif. Intell. Tools* **2007**, *16*, 593–609. [CrossRef]
26. Kidziński, Ł.; Yang, B.; Hicks, J.L.; Rajagopal, A.; Delp, S.L.; Schwartz, M.H. Deep neural networks enable quantitative movement analysis using single-camera videos. *Nat. Commun.* **2020**, *11*, 4054. [CrossRef]
27. Borghese, N.A.; Bianchi, L.; Lacquaniti, F. Kinematic determinants of human locomotion. *J. Physiol.* **1996**, *494*, 863–879. [CrossRef]
28. Vafadar, S.; Skalli, W.; Bonnet-Lebrun, A.; Khalifé, M.; Renaudin, M.; Hamza, A.; Gajny, L. A novel dataset and deep learning-based approach for marker-less motion capture during gait. *Gait Posture* **2021**, *86*, 70–76. [CrossRef] [PubMed]
29. Iskakov, K.; Burkov, E.; Lempitsky, V.; Malkov, Y. Learnable triangulation of human pose. In Proceedings of the IEEE/CVF International Conference on Computer Vision, Seoul, Korea, 27–28 October 2019; pp. 7718–7727.
30. Davis III, R.B.; Ounpuu, S.; Tyburski, D.; Gage, J.R. A gait analysis data collection and reduction technique. *Hum. Mov. Sci.* **1991**, *10*, 575–587. [CrossRef]
31. Zhang, Z. A flexible new technique for camera calibration. *IEEE Trans. Pattern Anal. Mach. Intell.* **2000**, *22*, 1330–1334. [CrossRef]
32. Motive: Optical Motion Capture Software. Available online: https://www.vicon.com/ (accessed on 1 November 2021).
33. Vicon. Available online: https://optitrack.com/software/motive/ (accessed on 1 November 2021).
34. Lin, T.Y.; Maire, M.; Belongie, S.; Hays, J.; Perona, P.; Ramanan, D.; Dollár, P.; Zitnick, C.L. Microsoft coco: Common objects in context. In Proceedings of the European Conference on Computer Vision, Zurich, Switzerland, 6–12 September 2014; pp. 740–755.
35. Zhou, X.; Wang, D.; Krähenbühl, P. Objects as points. *arXiv* **2019**, arXiv:1904.07850.
36. Yang, Y.; Ramanan, D. Articulated human detection with flexible mixtures of parts. *IEEE Trans. Pattern Anal. Mach. Intell.* **2012**, *35*, 2878–2890. [CrossRef]
37. O'Connor, C.M.; Thorpe, S.K.; O'Malley, M.J.; Vaughan, C.L. Automatic detection of gait events using kinematic data. *Gait Posture* **2007**, *25*, 469–474. [CrossRef]
38. Rajagopal, A.; Dembia, C.L.; DeMers, M.S.; Delp, D.D.; Hicks, J.L.; Delp, S.L. Full-body musculoskeletal model for muscle-driven simulation of human gait. *IEEE Trans. Biomed. Eng.* **2016**, *63*, 2068–2079. [CrossRef]
39. Pataky, T.C.; Vanrenterghem, J.; Robinson, M.A. Zero-vs. one-dimensional, parametric vs. non-parametric, and confidence interval vs. hypothesis testing procedures in one-dimensional biomechanical trajectory analysis. *J. Biomech.* **2015**, *48*, 1277–1285. [CrossRef]
40. Reddy, N.D.; Guigues, L.; Pishchulin, L.; Eledath, J.; Narasimhan, S.G. TesseTrack: End-to-End Learnable Multi-Person Articulated 3D Pose Tracking. In Proceedings of the IEEE/CVF Conference on Computer Vision and Pattern Recognition, Nashville, TN, USA, 20–25 June 2021; pp. 15190–15200.
41. He, Y.; Yan, R.; Fragkiadaki, K.; Yu, S.I. Epipolar transformers. In Proceedings of the IEEE/CVF Conference on Computer Vision and Pattern Recognition, Seattle, WA, USA, 14–19 June 2020; pp. 7779–7788.
42. Li, W.; Liu, H.; Ding, R.; Liu, M.; Wang, P.; Yang, W. Exploiting Temporal Contexts with Strided Transformer for 3D Human Pose Estimation. *arXiv* **2021**, arXiv:2103.14304.
43. Shan, W.; Lu, H.; Wang, S.; Zhang, X.; Gao, W. Improving Robustness and Accuracy via Relative Information Encoding in 3D Human Pose Estimation. In Proceedings of the 29th ACM International Conference on Multimedia, Chengdu, China, 20 October 2021; pp. 3446–3454.
44. Sun, K.; Xiao, B.; Liu, D.; Wang, J. Deep high-resolution representation learning for human pose estimation. In Proceedings of the IEEE/CVF Conference on Computer Vision and Pattern Recognition, Salt Lake City, UT, USA, 18–23 June 2019; pp. 5693–5703.

sensors

MDPI

Article

Augmentation of Human Action Datasets with Suboptimal Warping and Representative Data Samples

Dawid Warchoł * and Mariusz Oszust

Department of Computer and Control Engineering, Faculty of Electrical and Computer Engineering, Rzeszów University of Technology, W. Pola 2, 35-959 Rzeszów, Poland; marosz@kia.prz.edu.pl
* Correspondence: dawwar@prz.edu.pl; Tel.: +48-17-865-1614

Abstract: The popularity of action recognition (AR) approaches and the need for improvement of their effectiveness require the generation of artificial samples addressing the nonlinearity of the time-space, scarcity of data points, or their variability. Therefore, in this paper, a novel approach to time series augmentation is proposed. The method improves the suboptimal warped time series generator algorithm (SPAWNER), introducing constraints based on identified AR-related problems with generated data points. Specifically, the proposed ARSPAWNER removes potential new time series that do not offer additional knowledge to the examples of a class or are created far from the occupied area. The constraints are based on statistics of time series of AR classes and their representative examples inferred with dynamic time warping barycentric averaging technique (DBA). The extensive experiments performed on eight AR datasets using three popular time series classifiers reveal the superiority of the introduced method over related approaches.

Keywords: data augmentation; skeletal data; human action recognition; time series classification

Citation: Warchoł, D.; Oszust, M. Augmentation of Human Action Datasets with Suboptimal Warping and Representative Data Samples. *Sensors* **2022**, *22*, 2947. https://doi.org/10.3390/s22082947

Academic Editor: Subhas Mukhopadhyay

Received: 16 February 2022
Accepted: 9 April 2022
Published: 12 April 2022

1. Introduction

The automatic interpretation of actions performed by the human body is both challenging and desired. Well-designed action recognition (AR) algorithms could be put into practice in the detection of aggressive behavior, video surveillance, interaction with humans and robots, or advanced control over virtual reality avatars. In recent years, many methods for human action recognition have been developed [1]. However, similarly to other subfields of pattern recognition, they suffer from overfitting or inability to create more robust machine learning models due to lack of diverse training samples. Therefore, the data augmentation techniques designed to enrich AR databases are desired. Furthermore, their usability in practice is also supported by the difficulty of creating AR databases with various samples covering feature space well enough to train a classifier. Consequently, the data augmentation methods used for multidimensional data samples (e.g., synthetic minority over-sampling technique (SMOTE) [2]) cannot be directly used for augmenting time series of AR classes since they take into account a relationship between consecutive measurements or often non-linear distortions affecting the duration variability of registered time series of a class [3]. Additionally, such time-related feature space prevents a simple addition of new data points (i.e., entire time series) between existing samples. Considering the challenges of the time series data augmentation techniques, in the literature, several approaches have been proposed. They perform operations that stretch, cut, shrink, or perturb input time series [4,5]. In more advanced solutions, new time-series are generated using deep network-based models [6], the weighting of aligned averages [7,8], or concatenating parts of two perturbed time series by the dynamic time warping (DTW) technique [9]. However, those methods are considering time series classification problems without addressing issues related to the AR time series domain, in which data samples often belong to a relatively large number of similar classes with irregular, partially-overlapping boundaries.

The literature review reveals the scarcity of time series augmentation approaches devoted to AR problems. Additionally, the existing solutions are often associated with mandatory data processing steps damaging important temporal information or architectures that require large-scale datasets and dedicated hardware for efficient training. Therefore, in this paper, a novel method for time series data augmentation is introduced. It uses SPAWNER (**S**ubo**P**tim**A**l **W**arped time series ge**NE**rato**R**) algorithm [9] to generate new data samples and incorporates a set of constraints to provide time series suitable for AR datasets. The constraints are defined to reject samples that do not introduce new knowledge to the dataset and samples likely to be generated in a solution space occupied by a neighboring class. To achieve such a goal, new time series is compared with one of its input samples and a representative solution created for a class using Dynamic Time Warping Barycenter Averaging (DBA) [7]. In the proposed **A**ction **R**ecognition SPAWNER (ARSPAWNER), the comparison is performed taking into account statistics of samples within a considered class.

The contributions of this study are as follows.

1. A novel method for AR time series augmentation with small amount of data;
2. A novel and efficient method for determining constraints on generated data samples using statistics for a class and its representatives along with their incorporation into the data augmentation approach to address AR-related characteristics;
3. Comparative evaluation of the method with related approaches on eight AR datasets using popular classifiers.

The paper is arranged as follows. Section 2 reviews previous work on human action recognition and time-series data augmentation. Section 3 introduces the proposed approach. In Section 4, feature extraction techniques used to process AR time series benchmarks employed in experiments are described. Section 5 presents comparative evaluation of the method with related approaches. Finally, Section 6 concludes the paper and indicates possible directions of future work.

2. Related Work

The classification results of a machine learning method depend on the availability of learning data samples. Hence, they should cover enough feature space to provide the classifier with information that allows for unequivocal determination of class labels of unknown samples. With only a few learning examples, the classifier in most cases would not be able to correctly infer differences between classes, identify class boundaries, or address the variability of samples within a class. Similarly, the imbalanced distribution of data samples per class or the occupation by the most samples of a small area may lead to a drop in the classification performance. Therefore, many approaches to enrich class diversity or determine artificial samples close to class boundaries are proposed based on linear data transformation [2,10]. However, such approaches cannot be used with time series as most of them are nonlinearly transformed in the time scale, which causes variations in their lengths, even for the same class. Hence, simpler approaches to time series augmentation consider removal of a part of a time series, adding data points between existing values (i.e., warping), or introducing noise, rotation, and scaling [4]. In a more developed solution proposed by Forestier et al. [8], DBA, averages of multiply aligned data samples are iteratively weighted. As a result, for a set of time series, a new example is generated that can be seen as their representative. However, its usage for time series of large dimensionality and length, aiming at generating more samples from selected subsets from the input dataset, is challenging due to its computation demands [7,9]. In the previous authors' work, SPAWNER, time series are generated in the warped space between two data samples using their suboptimal alignment [9]. Specifically, the method uses DTW [11] for the alignment of perturbed parts of two input time series and concatenates aligned parts. The suboptimality arises from the usage of two randomly selected parts of each sample and the concatenation of their result instead of the DTW-based optimal alignment of the entire (i.e., non-divided) sequences. As those approaches are devoted to augmenting time series databases from many domains, there are works devoted to data generation techniques devoted to enriching time series from a single domain, addressing its characteristics. For example, Haradal et al. [6] introduced a method for augmentation of electrocardiogram (ECG) and electroencephalogram (EEG) datasets using generative adversarial networks (GAN) for the generation and discrimination of synthetic biosignals. In the work of Ramponi et al. [12], similar signals are generated with conditional GAN. The electrocardiograms are generated by Cao et al. [13] using samples of different classes and by Delaney et al. [14] using a variety of GAN architectures. Electroencephalographic data are augmented by Krall et al. [15] introducing distortions that consider temporal, spatial, or rotational changes. The data augmentation technique introduced by Le Guennec et al. [4] adds noise and magnitude changes to the input time series. Additionally, it warps them and removes some of their fragments (the cropping operation).

Some works address the augmentation of human action recognition datasets. For example, Shen et al. [16] proposed the Imaginative GAN (IGAN) and assessed it from a perspective of diversity and affinity of resulting samples. IGAN is a modification of the conventional GAN using unsupervised learning. The method approximates the distribution of the input data and samples new data. Additionally, it learns the latent behavioral (speed of actions) and physical (sizes of body parts) attributes. Ramachandra et al. [17] proposed an approach in which human activities measured by inertial sensors are recognized using data augmented by the proposed transformer GAN. Song et al. [18] specified an Interactive Action Translation (IAT) task that, taking into account rules of interaction, learns a model to generate a response for a given stimulation during inference. The method uses the Pair Embedding (PE) that utilizes Gaussian distributions of paired relationships to cluster individual actions in an embedding space and generate new pairs in their respective neighborhood. Here, encoders in a Paired Variational Auto-Encoders (PVAEs) and PCA-based linear dimension reductions are employed. Hoelzemann et al. [19] proposed human action data augmentation using a recurrent GAN based on a set of long short-term memory (LSTM) cells of four trained DeepConvLSTM models.

Despite promising performance of recent GAN-based data augmentation approaches, the GAN solutions require large-scale data to obtain stable models [16,18] or can be sensitive to outlying data samples [17]. Additionally, they may require data prepossessing in which human actions are unified to the same length due to architecture constraints. Consequently, the unification, or interpolation, negatively affects the input data and limits the variability of obtained samples. Furthermore, GAN, as other deep learning techniques, require demanding hyperparameter tuning [17], time-consuming training, and are associated with additional input data modifications to avoid overfitting.

Since, in this work, the augmentation of time series representing human actions is considered, main methods for their recognition are briefly introduced. They can be divided into deep learning and handcrafted approaches, where the techniques that belong to the first category extract suitable features and train a classifier in an end-to-end manner, while handcrafted approaches have separate feature extraction and classification steps. Furthermore, some of the deep learning methods are based on feature vectors but require a large amount of training data to provide acceptable models.

Among recently introduced AR methods, the approach by Sidor and Wysocki [20] uses a handcrafted Viewpoint Feature Histogram (VFH) point cloud description method [21] to calculate features for cells dividing point clouds of registered human actions. The cells represent different parts of the human body, and, therefore, such calculated features are more distinctive than those extracted for the whole cloud. Additionally, the method fuses two classifiers to improve its effectiveness. In the works of Pazhoumand-Dar et al. and Lillo et al. [22,23], the recognition is based on skeletal joint locations, angles between them, and more complex relationships between body parts. Skeletal data combined with local features extracted from depth images in the area around the projected joints can be found in the works of Raman and Maybank and Shahroudy et al. [24,25]. In these solutions, a two-level hierarchical Hidden Markov Model (HMM) [24] or hierarchical mixed norm with three levels of regularization over learning weights [25] are employed. One of the latest and most effective approaches to applying deep learning techniques to AR is presented by Farnoosh et al. [26]. In that work, a low-dimensional deep generative latent model encoding highly correlated skeletal data into a few sets of switching autoregressive temporal processes is introduced. The model decodes from the low-dimensional representations to the skeletal data and associated labels. Wang et al. [27] proposed the Skeleton Edge Motion Networks (SEMM) with spatio-temporal building blocks consisting of the concatenated spatial branch and temporal branch. It is observed that the spatial branch is effective when human actions do not have rich temporal information, while the temporal branch performs well with actions having a lot of movement of specific body parts. To boost the performance of SEMM, a progressive ranking loss that facilitates maintaining temporal order information in a self-supervised manner is employed. The spatial–temporal transformer network (ST-TR) is introduced by Plizzari et al. [28]. It models dependencies between skeletal joints using the transformer self-attention operator. Additionally, a spatial self-attention module (SSA) and a temporal self-attention module (TSA) are applied to understand intra-frame interactions between particular body parts and model inter-frame correlations. Then, the SSA and TSA are combined in a two-stream network. Donahue et al. [29] proposed an approach to human activity recognition based on video recordings using the long-term recurrent convolutional network (LRCN) with jointly trained convolutional (spatial) and recursive (temporal) parts.

In this study, to better highlight the capabilities of data augmentation techniques and offer results that can be easily replicated without additional hardware needed by recent deep learning models, handcrafted features, and popular classifiers are taken into account. Consequently, the relationship between generated samples of AR datasets that contain effective handcrafted features and the performance of several classifiers is investigated.

3. Proposed Method

In ARSPAWNER, two input time series of a given class are divided into two parts for a separate alignment using DTW and, after their concatenation, a new time series example is formed. This part of the time series processing is performed by the SPAWNER technique. Then, the resulted time series is rejected if it does not satisfy a set of constraints based on the AR time series characteristics.

In the approach, M-dimensional time series $X = [x^1, x^2, \ldots, x^L]$ of the length L is processed. Specifically, each $x^l \in \mathbb{R}^M$, $l = 1, 2, \ldots, L$, and $X \in \mathbb{R}^{L \times M}$. Then, a dataset of N samples, $L_n, n = 1, 2, \ldots, N$, $X_n \in \mathbb{R}^{L_n \times M}$, L_n is length of n-th sample, forms a collection $U = \{(X_1, C_1), (X_2, C_2), \ldots, (X_N, C_N)\}$, where $C \in \{1, K\}$ are class labels (K). Consequently, a classifier trained on U assigns a label C to test time series $Y \in \mathbb{R}^{L \times M}$.

To generate new time series based on a combination of two input samples X_1 and X_2 of the same class, the method employs DTW. In DTW, for $X_1 = [x_1^1, x_1^2, \ldots, x_1^i, \ldots, x_1^{L_1}]$ and $X_2 = [x_2^1, x_2^2, \ldots, x_2^j, \ldots, x_2^{L_2}]$, so-called *warping path* is determined which indicates optimal sequence $W = [w_1, w_2, \ldots, w_P]$, where P is the length of the path, p-th element $w_p = (i, j)$, and $max(L_1, L_2) \le P < L_1 + L_2$. Therefore, a $L_1 \times L_2$ matrix D is calculated. For all (i, j), it contains distances between time series $[x_1^1, \ldots, x_1^i]$ and $[x_2^1, \ldots, x_1^j]$. To select the optimal alignment between X_1 and X_2, the path W^* minimizing the total cumulative distance is found by calculating $D(i, j) = (x_1^i - x_2^j)^2 + min(D(i-1, j), D(i, j-1), D(i-1, j-1))$. The warping path satisfies three conditions: (1) The boundary condition which forces the path to start at the beginning of the time series, $w_1 = (1, 1)$, and finish at their ends, $w_P = (L_1, L_2)$; (2) The monotonicity condition according to which the time series indices in the path are monotonically increasing: $(i_1 \le i_2 \le \ldots \le L_1, j_1 \le j_2 \le \ldots \le L_2)$; (3) The continuity condition which limits the acceptable path steps to adjacent matrix elements. It can be written as $w_{p+1} - w_p \in \{(1, 0, (0, 1), (1, 1)\} \forall_{p \in \{1, 2, \ldots, P-1\}}$. The warping window ξ limits the elements of X_1 and X_2 that can be aligned, i.e., $\forall_{(i,j) \in w_p} ||i - j|| \le \xi$. DTW is used to calculate the distance $d = D(L_1, L_2)$ between time series.

To generate new examples in a suboptimal manner, an additional fourth constraint on the warping path is considered that forces it to contain the element $w_p = (R_1, R_2)$, where $R_1 = \lceil rL_1 \rceil, R_2 = \lceil rL_2 \rceil$, r is a single, uniformly distributed, randomly generated number in the interval $(0, 1)$. Here, $\lceil \cdot \rceil$ denotes ceiling operation. To prevent the calculation of $L_1 \times L_2$ matrix D and reducing the computational cost, two matrices $R_1 \times R_2$ D_1 and $|(L_1 - R_1)| \times |(L_2 - R_2)|$ D_2 are used. Then, $[x_1^1, x_1^2, \ldots, x_1^{R_1}]$ is aligned with $[x_2^1, x_2^2, \ldots, x_2^{R_2}]$ and $[x_1^{R_1+1}, x_1^{R_1+2}, \ldots, x_1^{L_1}]$ is aligned with $[x_2^{R_2+1}, x_2^{R_2+2}, \ldots, x_2^{L_2}]$. The resulting warping paths W_1^* and W_2^* are optimal due to the fourth constraint and the separate usage of D_1 and D_2. However, after their concatenation the obtained path is suboptimal. Moreover, ξ_1 and ξ_2 used to determine W_1^* and W_2^* are taken from $\lceil 0.1 \cdot max(R_1, R_2) \rceil$ and $\lceil 0.1 \cdot max(|L_1 - R_1|, |L_2 - R_2|) \rceil$, respectively. They reduce the flexibility of the path from the perspective of the matrix D, as well as the concatenated paths W_1^* and W_2^*. After the paths are concatenated to $W_{1,2}^*$, the algorithm aligns X_1 to X_2 generating sequences $X_1^\star$ and $X_2^\star$ with the length of $W_{1,2}^*$. To produce a new time series, $X^\star$, $X_1^\star$ and $X_2^\star$ are merged, where $x^\star \in X^\star$, is a random number chosen from a normal distribution with a small σ, $x^\star \sim \mathcal{N}(\mu, \sigma^2)$, $\mu = 0.5(x_1^\star + x_2^\star)$, $\sigma = 0.05|x_1^\star - x_2^\star|$.

To improve the quality of a AR dataset involving artificial example, $X^{\star,C}$, generated by the method, additional constraints limiting the possibility of its acceptance are introduced. At first, average $\tilde{d}_k$ and standard deviation $\hat{d}_k$ of the DTW distances is computed between all samples that belong to each k-th ($k = 1, 2, \ldots, K$) class. Then, the DBA approach is employed to provide representative sample for the class $\acute{X}_k = DBA(X_1^C, X_2^C, \ldots, X_N^C), C = k$, where X_N^C is the number of samples that belong to the $C = k$ class [8]. Specifically, it is computed as

$$argmin \acute{X}_k \in E \sum_{i=1}^{N^C} DTW^2(\acute{X}_k, X_i^C), \tag{1}$$

where E is a space induced by DTW and the optimization problem is solved using an expectation-maximization scheme and iterative refining of the $\acute{X}_k$ [8]. Finally, the $X^{\star,C}$ is introduced to the dataset if the following conditions are met (Equations (2) and (3)):

$$d_1 > r_1 \tilde{d}_k \wedge d_2 > r_1 \tilde{d}_k, \tag{2}$$

$$d_1 < T \wedge d_2 < T, \tag{3}$$

where $d_1 = DTW(X_1, X^{\star,C})$, $d_2 = DTW(\acute{X}_k, X^{\star,C})$, $T = r_1 \tilde{d}_k + \hat{d}_k(r_2 + \hat{d}_k / \tilde{d}_k)$, and (r_1, r_2) are parameters.

The proposed condition accepts only such time series which introduce new knowledge to the dataset, assuming that close proximity of the already present examples makes new examples redundant. The upper limit prevents the emergence of new examples in areas occupied by other classes.

To highlight the differences between SPAWNER and ARSPAWNER, 2D Multi-Dimensional Scaling (MDS) [30] embeddings of DTW dissimilarities for the exemplary time series from the MSRA I dataset are presented in Figure 1. The figure contains class boundaries of similar or overlapping classes to better indicate areas in which the methods created new examples. Input data samples are denoted by filled triangles. To facilitate the analysis, the same examples are connected by arcs. As shown, the SPAWNER produces examples that are filtered out by ARSPAWNER. For example, two newly created members of the "orange" class by SPAWNER are rejected by ARSPAWNER due to their close proximity to the input data samples. Consequently, one member of the "green" class and three members of the "purple" class were rejected by ARSPAWNER. Interestingly, the scattered input examples of the "blue" class resulted in the emergence of two samples produced by SPAWNER that are too far from them. Hence, ARSPAWNER removed them, significantly altering the class boundary. It is worth noticing that the MDS embeddings strongly depend on the examples that are considered while it is calculated. Overall, the class boundaries with examples introduced by ARSPAWNER are compact, without time series that could negatively impact the recognition of samples from other classes.

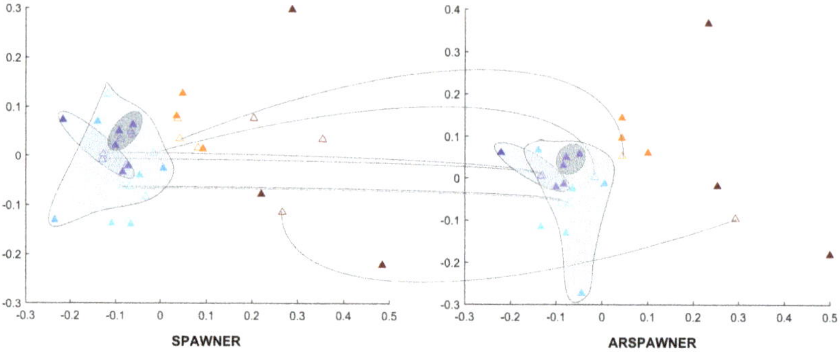

Figure 1. Class boundaries in the 2D MDS embeddings of DTW dissimilarities for the exemplary time series from the MSRA I dataset generated by SPAWNER and ARSPAWNER. Boundaries of neighboring classes are highlighted.

4. Action Recognition Descriptors and Features

The action recognition features employed to show the effectiveness of the proposed data augmentation approach are using successful Bone Pair Descriptor (BPD) [31] and Distance Descriptor (DD) [32].

4.1. Distance Descriptor

The Distance Descriptor represents relationships among pairwise joint distances in skeletal data. DD can be calculated based on 3D joint coordinates, without using vector data. The descriptor features are obtained for N joints as follows.

1. For each joint P_i, $1 \leq i \leq N$ do:
    (a) Calculate distances between the other joints P_j, $j \neq i$;
    (b) Sort joints P_j by the calculated distances in ascending order;
    (c) Assign consecutive integers a_{ij} to the ordered joints P_j, starting from 1.
2. Create a feature vector consisting of integer values assigned to the joints P_j in step 1(c) in the following order: $[a_{12}, a_{13}, a_{14}, a_{15}, a_{21}, \ldots, a_{NN-1}]$;
3. Reduce the feature vector by adding together integers a corresponding to the same pair of indices i, j: $[a_{12} + a_{21}, a_{13} + a_{31}, \ldots, a_{N-1N} + a_{NN-1}]$.

Finally, each feature value is divided by $2(N-1)$ to normalize them to the interval [0–1]. Note that an input set of joints should be selected from the whole skeleton before the calculation of DD to reduce the computation time and increase the classification accuracy. DD is calculated using the Euclidean distance.

4.2. Bone Pair Descriptor

The Bone Pair Descriptor encodes the angular relations between particular pairs of bones. The descriptor is calculated as follows. Let P_c be the skeleton central joint, b_c the central vector associated with the joint P_c, P_i the i-th non-central joint, and b_i the vector associated with that joint (Figure 2). Vectors b_c and b_i coincide with a bone or a part of the spine.

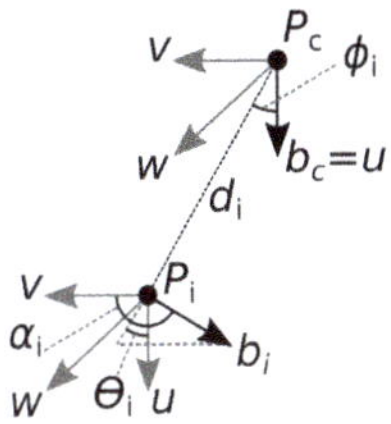

Figure 2. Calculation of Bone Pair Descriptor.

The relative position of vectors b_c and b_i is described by α, ϕ, and Θ according to Equations (4)–(6) [33]:

$$\alpha_i = a\cos(v_i \cdot b_i) \tag{4}$$

$$\phi_i = a\cos\left(u \cdot \frac{d_i}{|d_i|}\right) \tag{5}$$

$$\Theta_i = a\tan\left(\frac{w_i \cdot b_i}{u \cdot b_i}\right) \tag{6}$$

where the vectors u, v_i, and w_i define the Darboux frame [34]:

$$u = b_c \tag{7}$$

$$v_i = \frac{d_i}{|d_i|} \times u \tag{8}$$

$$w_i = u \times v_i \tag{9}$$

with $\cdot$ and $\times$ representing the scalar product and the vector product, respectively. Let N be the number of non-central joints. The BPD has $3N$ features calculated for each non-central joint using Equations (4)–(6):

$$V = [\alpha_1, \phi_1, \Theta_1, \alpha_2, \phi_2, \Theta_2, \ldots, \alpha_N, \phi_N, \Theta_N] \tag{10}$$

Finally, the features are normalized to the interval [0–1], dividing them by the maximum of π for α or ϕ, and 2π for Θ. BPD requires the selection of central joint P_c, non-central joints P_i, and joints determining vectors, b_c b_i, from the whole skeleton.

In the experiments, only α and ϕ features were used since Θ proved ineffective and its calculation is time-consuming [31].

5. Experiments and Discussion

5.1. Datasets

For the evaluation of the approach, six human action datasets with skeletal data were used: MSR Action3D (MSRA) [35], UTD Multimodal Human Action Dataset (UTD-MHAD) [36], UTKinect-Action3D (UTK) [37], Florence 3D Action Dataset (FLORENCE) [38], SYSU 3D Human–Object Interaction Set (SYSU) [39], and Kinect Activity Recognition Dataset (KARD) [40]. The MSRA dataset is split into three separate subsets, MSRA I, MSRA II, MSRA III, as suggested by its authors [35]. Each subset contains different action classes, although some of them appear in two subsets. That makes a total of eight datasets used in experiments. Detailed information about the datasets, including the length variability of time series, the number of input examples, and the number of augmented examples produced by each approach, is presented in Table 1.

Table 1. Characteristics of datasets used in experiments.

Name	Classes	Subjects	Sequences (Actions)	Time Series Length	Input Examples	Augmented Examples	Validation Protocol
MSRA I	8	10	224	13–76	118	611	50-50 validation
MSRA II	8	10	207	15–100	118	573	50-50 validation
MSRA III	8	10	225	13–71	113	438	50-50 validation
UTD-MHAD	27	8	861	41–125	431	1163	50-50 validation
UTK	10	10	199	5–110	179	744	10-fold cross-validation
FLORENCE	9	10	215	8–35	194	1109	10-fold cross-validation
SYSU	12	40	480	58–638	240	1087	50-50 validation
KARD	18	10	540	42–310	270	685	50-50 validation

According to the original paper introducing the MSRA dataset, there are seven subjects performing actions. However, the larger version, consisting of 10 subjects, is publicly available and can be downloaded from the authors' website [41]. This version was used in the experiments.

In this study, two types of validation were performed. For MSRA, SYSU, UTD-MHAD, and KARD, 50-50 validation tests were used, in which the training and testing sets were split in half based on the subjects performing actions. The protocol for UTD-MHAD and FLORENCE is 10-fold cross-validation. For each dataset, the validation protocols proposed by the authors were used. In the case of KARD, 50-50 validation was used instead of the 10-fold cross-validation due to excessive computation time. All performed tests were subject independent, which means that in each test, the training set contains actions performed by subjects not present in the testing set. Such tests simulate the behavior of a recognition application in practice, where people performing actions do not participate in the creation of the training data.

Actions from all datasets were recorded using a Microsoft Kinect sensor. In this work, only skeletal joints were used to characterize human actions. The skeletons for actions present in all datasets except FLORENCE and KARD consist of 20 joints, while the skeletons used to capture actions in the FLORENCE and KARD datasets consist of 15 joints (see Figure 3).

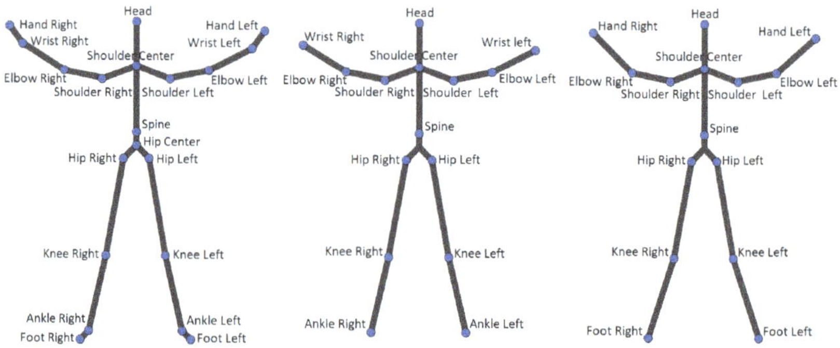

Figure 3. Three skeletons available in datasets: (**left**) MSRA, UTD-MHAD, UTK, and SYSU; (**middle**) FLORENCE; (**right**) KARD.

The same subsets of joints and bones cannot be used for 20-joint datasets and 15-joint datasets. Furthermore, FLORENCE and KARD do not have identical joint sets despite having the same number of joints. Therefore, for the experiments, three groups of joint subsets and bone subsets were selected separately for the Distance Descriptor and the Bone Pair Descriptor. They are listed in Tables 2 and 3 .

Table 2. Subsets of joints used for the Distance Descriptor. "L." and "R." denote Left and Right, respectively.

MSRA, UTD-MHAD, UTK, SYSU	FLORENCE	KARD
Hand L.	Wrist L.	Hand L.
Hand R.	Wrist R.	Hand R.
Shoulder L.	Shoulder L.	Shoulder L.
Shoulder R.	Shoulder R.	Shoulder R.
Head	Head	Head
Spine	Spine	Spine
Hip L.	Hip L.	Hip L.
Hip R.	Hip R.	Hip R.
Ankle L.	Ankle L.	Foot L.
Ankle R.	Ankle R.	Foot R.

Table 3. Subsets of bones used for the Bone Pair Descriptor. "L." and "R." denote Left and Right, respectively.

MSRA, UTD-MHAD, UTK, SYSU	FLORENCE	KARD
Spine–Head (central)	Spine–Head (central)	Spine–Head (central)
Elbow R.–Wrist R.	Elbow R.–Wrist R.	Elbow R.–Wrist R.
Wrist R.–Hand R.	Shoulder R.–Elbow R.	Shoulder R.–Elbow R.
Shoulder R.–Elbow R.	Elbow L.–Wrist L.	Elbow L.–Wrist L.
Elbow L.–Wrist L.	Shoulder L.–Elbow L.	Shoulder L.–Elbow L.
Wrist L.–Hand L.	Hip R.–Knee R.	Hip R.–Knee R.
Shoulder L.–Elbow L.	Knee R.–Ankle R.	Knee R.–Foot R.
Hip R.–Knee R.	Hip L.–Knee L.	Hip L.–Knee L.
Knee R.–Ankle R.	Knee L.–Ankle L.	Knee L.–Foot L.
Ankle R.–Foot R.		
Hip L.–Knee L.		
Knee L.–Ankle L.		
Ankle L.–Foot L.		

The subsets of joints and bones were selected experimentally as a part of the previous work on the subject of human action recognition [31]. Different configurations were also tested, however, the chosen subsets yielded the best results in terms of recognition rate and computation time.

5.2. Visual Examples of Augmented Time Series

To show exemplary time series, in Figure 4, two actions from the MSRA II dataset [35] (i.e., "draw circle" and "high arm wave") are presented along with the additional time series generated by ARSPAWNER. The curves of the first action represent the first DD feature related to Hand Left and Hand Right joints, and the curves of "high arm wave" action represent ϕ feature of BPD, for which the non-central vector is determined by Wrist Right and Hand Right joints.

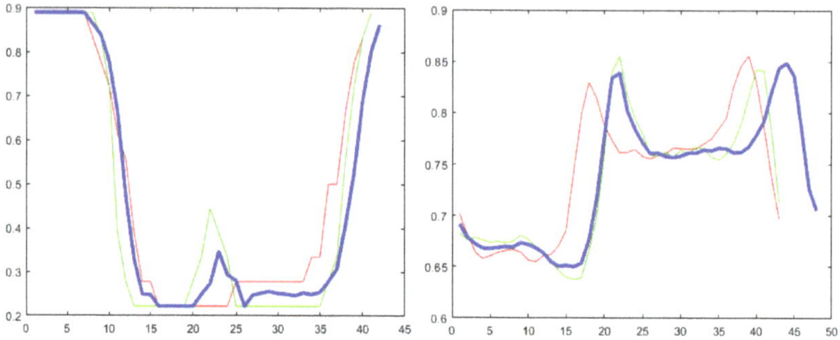

Figure 4. Time series generated by ARSPAWNER (blue curve) based on two exemplary timeseries (red and green curves). The left plot represents "draw circle" action and the right plot represents "high arm wave" action from MSRA II dataset.

5.3. Classifiers

Among classification methods, the classical Dynamic Time Warping (DTW) and two recent methods were used: LogDet Divergence-based Metric Learning with Triplet constraints (LDMLT) [42] and Time series Cluster Kernel (TCK) [43]. LDMLT is a classifier based on Mahalanobis distance and the so-called triplet constraints used for its learning [42]. TCK is a method that calculates similarities between time series using Gaussian Mixture Models (GMM) augmented with informative prior distributions. It can handle missing data without the usage of imputation methods [43]. The output of DTW and LDMLT is the distance between two given sequences, i.e., each testing sequence is compared to each training sequence. Therefore, there is a need to apply the nearest neighbor classifier to determine the class represented by the closest sequence.

In Table 4, the configuration of parameters for each classifier is presented. The parameter values were set experimentally in the spirit of fairness, i.e., by changing them and checking whether the recognition rate is improved.

Table 4. Parameters of the classifiers.

Classifier	Parameter Name	Parameter Value
DTW	Window size	5
	Metric	Euclidean
LDMLT	Triplets factor	20
	Alpha factor	5
	Number of iterations	15
TCK	Maximum number of mixture components	5
	Number of randomizations	50
	Number of iterations	20

5.4. Results

The feature vectors used in the experiments are concatenations of the DD and BPD features without Θ, which makes a total of 69 features (45 for DD and 24 for BPD).

ARSPAWNER generates new data based on a pair of input time series, and therefore, the number of generated examples by other methods is aligned with the number of returned samples. This ensures a fair comparison of algorithms.

In this study, four augmentation methods are compared using the classification accuracy obtained for each dataset and classifier. Due to the randomness of the augmentation algorithms and TCK classifier, each accuracy is calculated for 10 runs and averaged. Then, the following values are calculated: average accuracy, average rank, geometric average rank, and a number determining how many times a method achieved the best accuracy (count best). These values are considered as the comparative criteria. To compare the methods, ranks from 1 to 5 are used, where a lower rank means a method has greater accuracy. The compared approaches are: SPAWNER, ARSPAWNER, Window Slicing [4], and Window Warping [4]. The results for each method, and for the case in which the augmentation is not performed (non-augmentation case), are presented in Table 5.

The experimental results reveal that the proposed ARSPAWNER is the most effective augmentation method and outperforms the non-augmentation case according to all comparative criteria. The method shows the greatest advantage over the others when used with DTW. However, in the case of the other two classifiers, ARSPAWNER and SPAWNER have close average effectiveness. They both significantly outperform the other methods, as well as the results of the non-augmentation case.

The LDMLT classifier yielded better results than the other two methods for all datasets, and its suitability for the action recognition problems was proven in the previous study [31]. The study of Kamycki et al. [9], in which SPAWNER was introduced, does not address action recognition problems considering time series from different domains. Interestingly, in that study, the LDMT classifier showed inferior performance. From Table 5, it can also be seen that the TCK classifier obtained the worst results among all three methods for all datasets, except UTK, for which it outperformed DTW.

Overall, it can be seen that the introduced ARSPAWNER outperforms the remaining data augmentation methods on action recognition datasets, since it considers the specificity of such data collections, with many similar and overlapping action classes.

Table 5. Experimental comparison of augmentation methods for three classifiers in terms of classification accuracy. The two best results for each classifier and dataset are written in bold.

Dataset/Aug. Method	None	WW	WS	SPAWNER	ARSPAWNER
		DTW			
MSRA I	71.7	70.6	74.3	**74.4**	**76.1**
MSRA II	69.0	69.7	**73.1**	69.3	**71.7**
MSRA III	83.9	84.2	84.0	**86.5**	**86.5**
UTD-MHAD	86.3	86.3	83.9	**86.5**	**86.7**
UTK	81.9	80.7	**86.4**	85.4	**86.4**
FLORENCE	78.6	78.4	**81.7**	81.5	**81.8**
SYSU	69.2	67.2	70.8	**71.2**	**72.5**
KARD	89.6	**90.9**	**91.6**	88.0	89.7
		LDMLT			
MSRA I	75.5	80.6	82.6	**86.2**	**86.5**
MSRA II	78.8	77.3	73.2	**80.6**	**83.2**
MSRA III	**90.2**	88.8	88.6	89.4	**89.6**
UTD-MHAD	**92.1**	90.4	84.4	**92.4**	89.2
UTK	91.5	92	91.9	**95.4**	**95.7**
FLORENCE	86.0	84.7	84.7	**88.5**	**87.4**
SYSU	68.8	61.4	64.4	**70.9**	**70.5**
KARD	95.9	96.4	94.0	**97.0**	**97.6**
		TCK			
MSRA I	55.8	62.8	62.1	**65.7**	**66.5**
MSRA II	54.9	58.0	**58.5**	54.9	**58.1**
MSRA III	75.7	79.3	77.1	**81.7**	**81.4**
UTD-MHAD	**62.0**	56.6	57.7	**61.5**	60.3
UTK	92.6	**93.3**	**93.7**	93.2	**93.3**
FLORENCE	78.0	79.7	79.4	**81.6**	**80.4**
SYSU	62.7	62.8	62.3	**66.5**	**66.2**
KARD	85.5	88.0	**88.3**	**88.9**	85.2
		Overall results			
Average rank	3.88	3.65	3.38	2.21	1.90
Geometric average rank	3.6	3.53	2.95	1.93	1.68
Count best	2	0	5	8	11
Average accuracy	78.2	78.3	78.7	80.7	80.9

5.5. Visual Comparison

To show the areas in which new samples are generated by the augmentation methods from the MSRA I dataset, Kruskal's nonmetric MDS [30] is employed. To facilitate the analysis, the first 60 actions are considered. MDS reduces the dimensionality of data samples and can be used with time series of different lengths by the usage of the DTW dissimilarity matrix. The matrix contains pairwise DTW distances between examples. The MDS representations of exemplary time series are shown in Figure 5. Input time series are filled while the colors indicate their classes. The proximity of samples from different classes or existing overlapped class boundaries illustrate the recognition problems. However, the introduction of new data samples in most cases improved the recognition accuracy of classifiers, it can be assumed that methods generating time series in areas within class boundaries are likely to lead to a higher recognition rate. As presented, ARSPAWNER generates fewer examples in areas occupied by representatives of other classes than in the case of the remaining augmentation approaches.

The recognition problems can also be highlighted by showing testing examples together with training data and augmented data. Therefore, in Figure 6, solid triangles represent 2D MDS embeddings of testing samples from the entire MSRA I dataset, and empty triangles denote training data (Figure 6a) and augmented data (Figure 6b), respectively. The placement of testing samples in the feature space indicates recognition problems as the class boundaries are difficult to determine due to the presence of clusters of similar examples from different classes in close proximity. Even classes that seem to be easily distinguished, represented here by yellow and bright green triangles are close to each other while training examples of the bright green class are far from that boundary (Figure 6a). This means that training examples do not carry enough information to be able to successfully recognize examples from these two classes. The emergence of augmented samples (Figure 6b) cannot solve this problem, since such knowledge cannot be obtained, but adds more examples in vital areas, shrinking overlapped class areas. Similar observations can be made for other datasets. It is worth noticing that the reported results strongly depend on the capabilities of used classifiers. Some of them may not be suitable to recognize human actions as can be seen in the TCK case.

5.6. Comparison with CGAN

Since there are approaches based on GAN architecture to augment time series in different domains, the performance of ARSPAWNER is compared with those of Conditional GAN on three MSRA datasets. Due to the lack of Matlab implementations of GAN-based approaches designed to augment action recognition time series in the literature, the available Matlab CGAN example designed to generate synthetic time series was adapted (MathWorks, https://www.mathworks.com (accessed on 13 March 2022)) [44]. The employed CGAN uses 1-D convolutional networks and is designed to perform the two-class augmentation. The generator network projects and reshapes the $1 \times 1 \times 100$ noise arrays to $4 \times 1 \times 1024$ arrays. It converts data labels to embedding $4\times 1 \times 1$ vectors. Then, it concatenates the outputs of the two inputs and upsamples them to $1201 \times 1 \times 1$ arrays with 1-D transposed convolution layers and ReLU layers. The dimensionality of the arrays is determined by the application of the origin of the adapted example. The discriminator network takes two inputs and classifies original and synthesized $1201 \times 1 \times 1$ signals. It reshapes and concatenates them. Then, after downsampling, a series of 1-D convolution layers with leaky ReLU (a scale of 0.2) are employed.

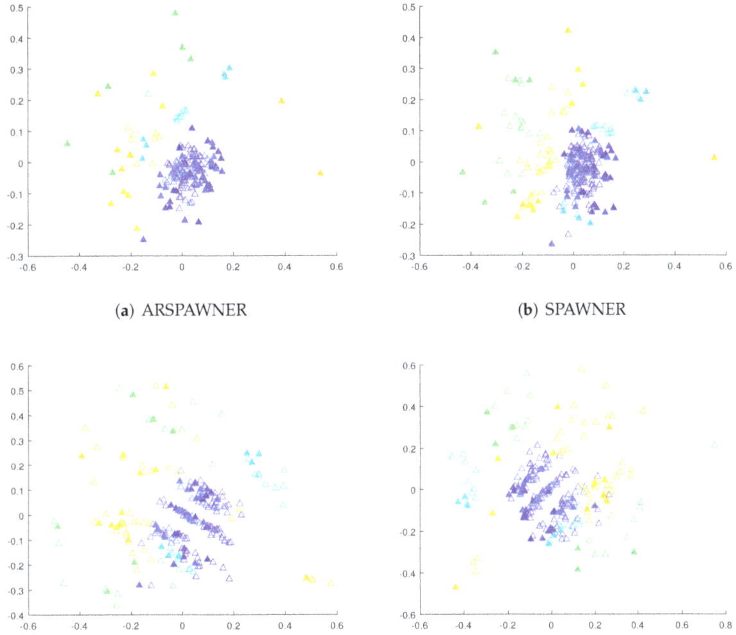

(**a**) ARSPAWNER (**b**) SPAWNER

(**c**) Cropping (**d**) Warping

Figure 5. The 2D MDS embeddings of DTW dissimilarities between training and augmented sequences from the MSRA I dataset for the compared augmentation methods. Colors are used to differentiate the classes, the filled triangles denote input examples.

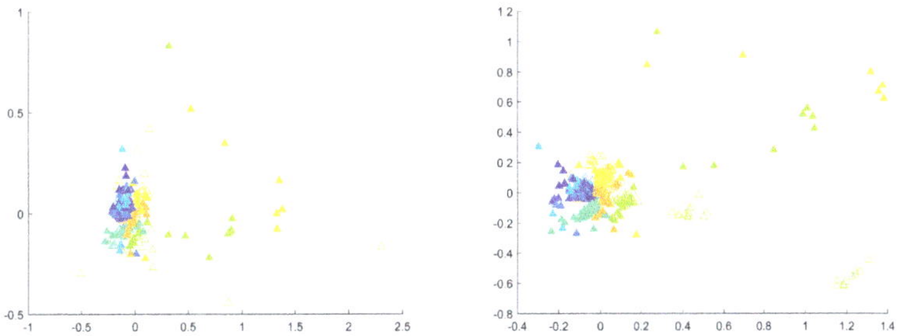

(**a**) Training and testing data (**b**) Augmented and testing data

Figure 6. The 2D MDS embeddings of DTW dissimilarities between testing and training or testing and augmented sequences from the MSRA I dataset. Colors differentiate the classes, the filled triangles denote testing examples.

The network was adapted to perform the augmentation of action recognition MSRA datasets that contain time series of different lengths, belonging to 8 classes and composed of 69 features. Specifically, due to the ability to generate two class time series, it was run eight times with input samples divided into two classes (i.e., the class considered in a given run and the rest). Additionally, since it is not designed to process multivariate time series and to avoid time-consuming computations, the PCA technique was applied to reduce the feature dimensionality from 69 to 5 and CGAN was run for each new feature independently with the concatenation of data to form synthesized five-dimensional vectors. Furthermore, since time series in MSRA datasets are of different lengths, they were interpolated to the same length, imposed by the network architecture. The finally obtained augmented examples were added to the original samples and employed by the nearest neighbor classifier with the DTW distance. The parameters of the network were set as recommended by the network designers, with the reduced number of iterations since the model converged earlier, allowing for the reduction in the training time. Important parameters of CGAN: number of iterations = 1000, learning rate = 0.0002, the Adam optimizer, batch size = 256, latent dimension = 100, and embedding dimension = 100. In experiments with CGAN, a PC with Nvidia Quadro RTX 4000 MAX-Q GPU, i9-10885H CPU, and 64 GB RAM was used. To ensure a fair comparison, ARSPAWNER was also run on the same five-dimensional feature vectors resulting from PCA.

The accuracy of the classifier for three augmented MSRA datasets after PCA feature reduction is presented in Table 6. It can be seen that the classifier equipped with data generated by ARSPAWNER improves its accuracy by a large margin. The improvement can also be visible for CGAN-created data in the case of MSRA I. However, for the MSRA II and III datasets, creating synthetic samples led to a significant drop in the performance of the classifier. The problems with the generation of suitable data examples of CGAN are possibly caused by the lack of a sufficient number of learning data examples, challenging data examples in the dataset after reduction by PCA, and inefficiency of the employed network architecture. To better highlight encountered problems with CGAN architecture, the 2D MDS embeddings were created for the entire MSRA I dataset (Figure 7). As shown, input data samples are close to each other due to the usage of PCA reducing the dimensionality of the time series in the dataset. However, ARSPAWNER was able to create samples in large clusters (Figure 7d) in their proximity (Figure 7c). CGAN, in turn, created many samples across the feature space, with their representatives also located in places that belong to the neighboring classes (Figure 7a,b).

Table 6. Comparison of CGAN and ARSPAWNER on the MSRA datasets. The best result for each dataset is written in bold.

Dataset	None	CGAN	ARSPAWNER
MSRA I	0.7075	0.7453	**0.8118**
MSRA II	0.6283	0.5487	**0.6994**
MSRA III	0.8125	0.6964	**0.8393**

5.7. Impact of Parameters

The next experiment concerns the impact of the ARSPAWNER parameters r_1 and r_2 on the classification accuracy. Figure 8 shows 3D surface plots calculated for each classifier and MSRA II dataset. The values of r_1 and r_2 are within the range [0.1–1.0] with step 0.1. The classification accuracy for DTW ranges from 63.2% to 72.6%, for LDMLT the range is [78.8–85.5%], and for TCK the range is [53.3–62.4%]. For each classifier, the difference between the lowest and the highest result is greater than 5 percentage points and smaller than 10 percentage points. Therefore, it can be concluded that the parameters r_1 and r_2 have a moderate impact on the classification accuracy.

The r_1 and r_2 parameters govern two constraints on the generated time series. Hence, a more detailed experiment, involving all three MSRA datasets, shows the impact of lower and upper constraints on the performance of ARSPAWNER with the nearest neighbor classifier with the DTW distance. Additionally, it allows for assessing the importance of the class representatives used in the conditions. The results presented in Table 7 indicate that both conditions should be present to obtain the best recognition rate for the MSRA datasets. However, the condition that rejects examples created near to a given input sample or a representative sample of a class (Equation (2)) is more influential than the upper limit (Equation (3)), responsible for acceptance of candidates closer to the class borders. Since both conditions are based on two distances to a considered input sample (d_1) and the DBA representative (d_2), their calculation reveals that they both should be used. It is justified by a larger drop in the performance of the ARSPAWNER in the case in which distance to the input sample is not employed. This confirms the usability of the introduced usage of the representative time series for each class.

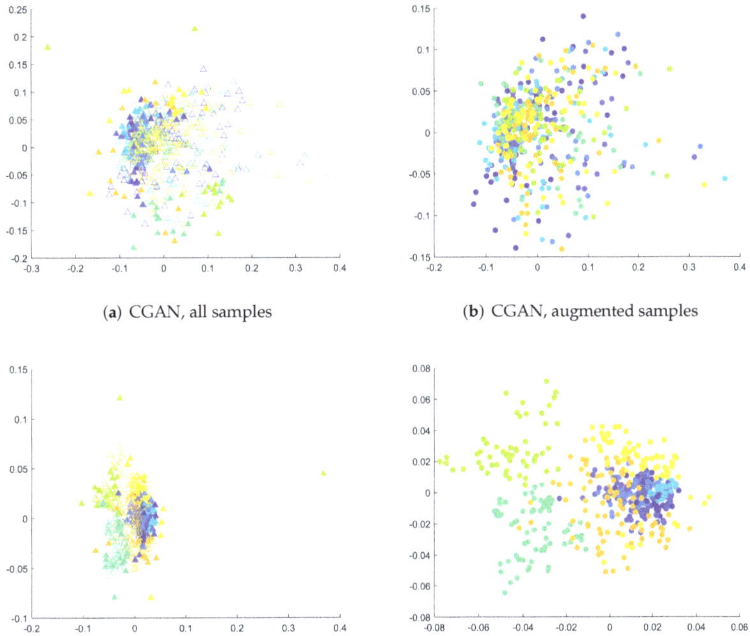

(**a**) CGAN, all samples (**b**) CGAN, augmented samples

(**c**) ARSPAWNER, all samples (**d**) ARSPAWNER, augmented samples

Figure 7. The 2D MDS embeddings of DTW dissimilarities between sequences of reduced dimensionality from the MSRA I dataset for CGAN and ARSPAWNER. Colors are used to differentiate the classes, the filled triangles denote input examples (**a**,**c**), while filled circles denote augmented samples (**b**,**d**).

Table 7. Performance of ARSPAWNER with active conditions.

Active Condition	MSRA I	MSRA II	MSRA III
Equations (2)–(3)	76.1	71.7	86.5
Equation (2)	76.1	70.4	86.5
Equation (3)	76.0	70.1	84.9
Lack of d_1 in Equations (2)–(3)	76.5	70.1	85.6
Lack of d_2 in Equations (2)–(3)	75.6	70.8	85.6

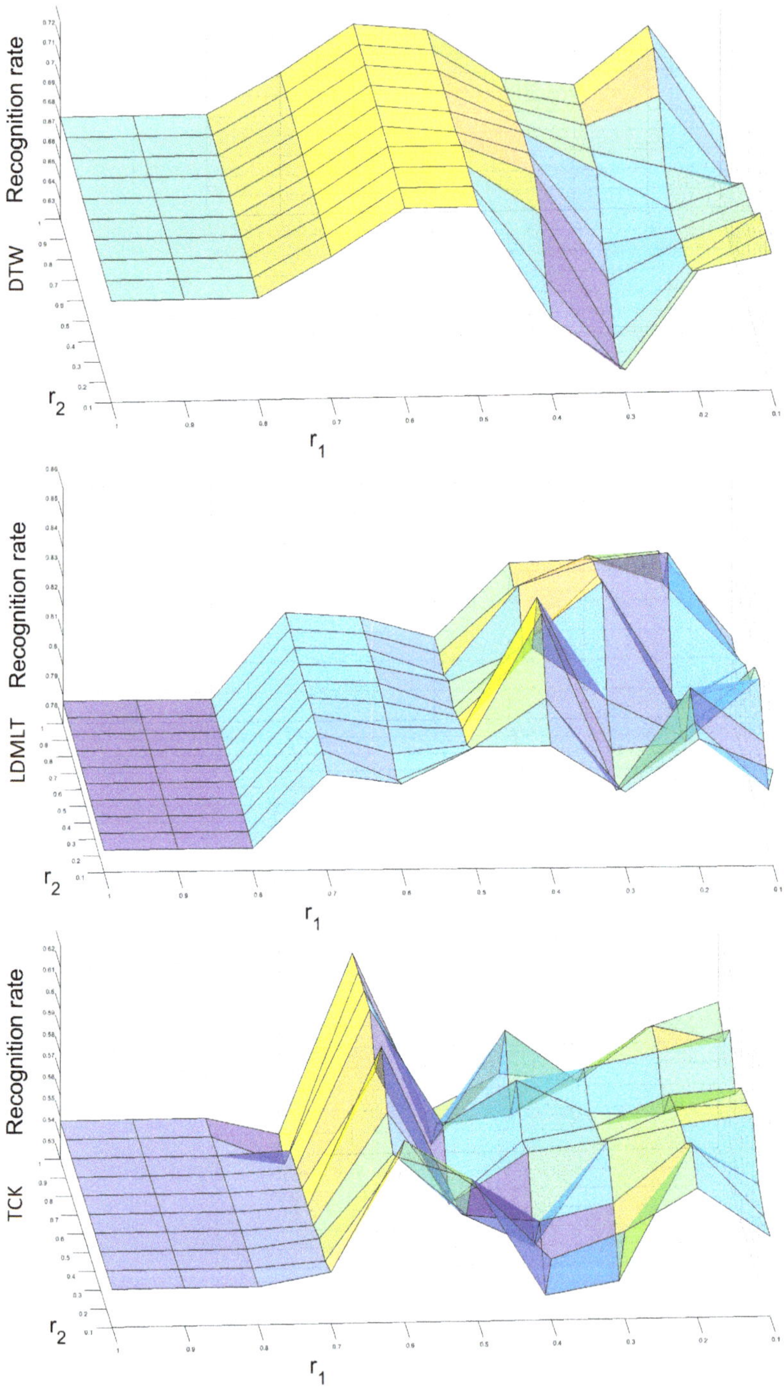

Figure 8. Three-dimensional surface plots presenting the impact of ARSPAWNER parameters r_1 and r_2 on classification accuracy with MSRA II dataset. The upper, middle, and lower plots represent the results of DTW, LDMLT, and TCK, respectively.

5.8. Performance with Small Number of Training Examples

To determine the capability of the introduced ARSPAWNER to augment small datasets, composed of a small number of training examples per class, it was tested using the MSRA I-III datasets varying the number of input time series. This experiment also indicates problems with small benchmark datasets in which class boundaries cannot be easily established due to an insufficient amount of available data and a relatively large number of classes (i.e., there are eight classes in the MSRA datasets). In the experiment, 3 to 15 input examples per class were randomly selected and used by ARSPAWNER to generate synthetic data. Then, the average accuracy of the nearest neighbor classifier with the DTW distance based on ten draws is reported in Figure 9. Overall, as reported, ARSPAWNER can improve the results of the classifier based only on a few available training samples. Depending on the dataset and the way testing examples are scattered in the feature space, the positive effect of the augmentation is visible even for five input examples.

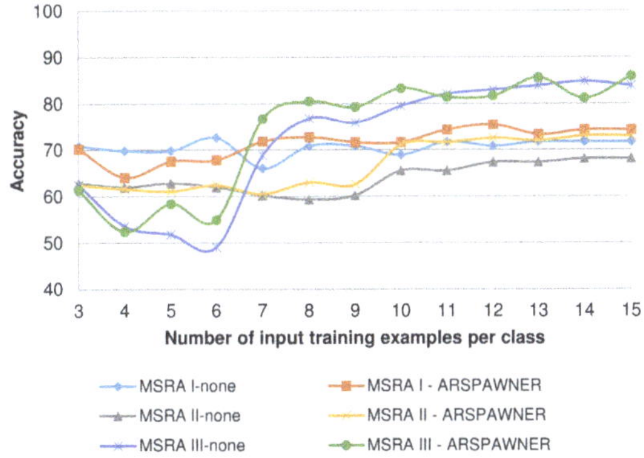

Figure 9. Average accuracy of the nearest neighbor classifier with the DTW distance based on a small number of augmented training examples per class.

6. Conclusions

In this paper, a novel method for the augmentation of datasets with time series representing human actions has been presented. The introduced ARSPAWNER improves the original SPAWNER by introducing action recognition-related constraints addressing problems present in this domain. The approach identifies data samples, i.e., time series, that are far enough from input samples and still do not cross the boundaries of other classes. Additionally, data samples that are in the proximity of the input time series, and consequently do not introduce new knowledge, are rejected. The constraints are based on distances between a new sample and an input sample and a sample generated as a representative time series characterizing a class. It has been shown that the introduced constraints provide to the augmentation leading to the improved performance of classifiers. The method has been experimentally compared with related approaches using three classifiers on eight action recognition datasets.

Future work will involve an application of optimization techniques to select a suitable set of generated time series based on data clustering quality indices. Such an approach can be seen as an extension of the study presented in this paper since constraints that remove augmented samples may be replaced with a step in which their suitability is assessed based on the quality criteria describing clusters of generated samples. Another interesting research direction is to employ augmentation methods like ARSPAWNER to augment small datasets and train time-consuming deep learning classifiers.

To facilitate the reproducibility of the approach, the Matlab implementation of the introduced ARSPAWNER is available at www.marosz.kia.prz.edu.pl/ARSPAWNER.html

(accessed on 13 March 2022). The scripts for Distance Descriptor and Bone Pair Descriptor are also publicly available and can be downloaded [45].

Author Contributions: Conceptualization, M.O.; methodology, M.O. and D.W.; software, M.O. and D.W.; validation, D.W.; data curation, D.W.; writing—original draft preparation, M.O. and D.W.; writing—review and editing, M.O. and D.W.; visualization, M.O and D.W.; supervision, M.O. and D.W. All authors have read and agreed to the published version of the manuscript.

Funding: This project is financed by the Minister of Education and Science of the Republic of Poland within the "Regional Initiative of Excellence" program for years 2019–2022. Project number: 027/RID/2018/19, amount granted: 11 999 900 PLN.

Data Availability Statement: In this study, publicly available datasets were analyzed. They can be found here: 1. MSRA (I, II, III)—https://sites.google.com/view/wanqingli/data-sets/msr-action3d; 2. UTD—https://personal.utdallas.edu/~kehtar/UTD-MHAD.html; 3. UTK—http://cvrc.ece.utexas.edu/KinectDatasets/HOJ3D.html; 4. FLORENCE—https://www.micc.unifi.it/resources/datasets/florence-3d-actions-dataset; 5. SYSU—http://www.isee-ai.cn/~hujianfang/ProjectJOULE.html; 6. KARD—https://data.mendeley.com/datasets/k28dtm7tr6/1; The source codes of our methods can be found here: http://vision.kia.prz.edu.pl (accessed on 13 March 2022).

Conflicts of Interest: The authors declare no conflict of interest.

References

1. Zhang, H.B.; Zhang, Y.X.; Zhong, B.; Lei, Q.; Yang, L.; Du, J.X.; Chen, D.S. A comprehensive survey of vision-based human action recognition methods. *Sensors* **2019**, *19*, 1005. [CrossRef] [PubMed]
2. Chawla, N.V.; Bowyer, K.W.; Hall, L.O.; Kegelmeyer, W.P. SMOTE: Synthetic minority over-sampling technique. *J. Artif. Intell. Res.* **2002**, *16*, 321–357. [CrossRef]
3. Handhika, T.; Murni; Lestari, D.P.; Sari, I. Multivariate time series classification analysis: State-of-the-art and future challenges. *IOP Conf. Ser. Mater. Sci. Eng.* **2019**, *536*, 012003. [CrossRef]
4. Le Guennec, A.; Malinowski, S.; Tavenard, R. Data Augmentation for Time Series Classification using Convolutional Neural Networks. In Proceedings of the AALTD 2016: Second ECML/PKDD International Workshop on Advanced Analytics and Learning on Temporal Data, Riva del Garda, Italy, 19–23 September 2016; p. 11.
5. Um, T.T.; Pfister, F.M.J.; Pichler, D.; Endo, S.; Lang, M.; Hirche, S.; Fietzek, U.; Kulić, D. Data Augmentation of Wearable Sensor Data for Parkinson's Disease Monitoring using Convolutional Neural Networks. In Proceedings of the ACM International Conference on Multimodal Interaction, Glasgow, UK, 13–17 November 2017. [CrossRef]
6. Haradal, S.; Hayashi, H.; Uchida, S. Biosignal Data Augmentation Based on Generative Adversarial Networks. In Proceedings of the 2018 40th Annual International Conference of the IEEE Engineering in Medicine and Biology Society (EMBC), Honolulu, HI, USA, 17–21 July 2018; pp. 368–371. [CrossRef]
7. Fawaz, H.I.; Forestier, G.; Weber, J.; Idoumghar, L.; Muller, P.A. Data Augmentation Using Synthetic Data for Time Series Classification with Deep Residual Networks. *arXiv* **2018**, arXiv:1808.02455.
8. Forestier, G.; Petitjean, F.; Dau, H.A.; Webb, G.I.; Keogh, E. Generating Synthetic Time Series to Augment Sparse Datasets. In Proceedings of the 2017 IEEE International Conference on Data Mining (ICDM), New Orleans, LA, USA, 18–21 November 2017; pp. 865–870. [CrossRef]
9. Kamycki, K.; Kapuściński, T.; Oszust, M. Data Augmentation with Suboptimal Warping for Time-Series Classification. *Sensors* **2020**, *20*, 98. [CrossRef]
10. Douzas, G.; Bacao, F. Geometric SMOTE a geometrically enhanced drop-in replacement for SMOTE. *Inf. Sci.* **2019**, *501*, 118–135. [CrossRef]
11. Sakoe, H.; Chiba, S. Dynamic programming algorithm optimization for spoken word recognition. *IEEE Trans. Acoust. Speech* **1978**, *26*, 43–49. [CrossRef]
12. Ramponi, G.; Protopapas, P.; Brambilla, M.; Janssen, R. T-CGAN: Conditional Generative Adversarial Network for Data Augmentation in Noisy Time Series with Irregular Sampling. *arXiv* **2018**, arXiv:1811.08295.
13. Cao, P.; Li, X.; Mao, K.; Lu, F.; Ning, G.; Fang, L.; Pan, Q. A novel data augmentation method to enhance deep neural networks for detection of atrial fibrillation. *Biomed. Signal Process. Control* **2020**, *56*, 101675. [CrossRef]
14. Delaney, A.M.; Brophy, E.; Ward, T.E. Synthesis of Realistic ECG using Generative Adversarial Networks. *arXiv* **2019**, arXiv:1909.09150.
15. Krell, M.M.; Seeland, A.; Kim, S.K. Data Augmentation for Brain-Computer Interfaces: Analysis on Event-Related Potentials Data. *arXiv* **2018**, arXiv:1801.02730.
16. Shen, J.; Dudley, J.J.; Kristensson, P.O. The Imaginative Generative Adversarial Network: Automatic Data Augmentation for Dynamic Skeleton-Based Hand Gesture and Human Action Recognition. In Proceedings of the 2021 16th IEEE International Conference on Automatic Face and Gesture Recognition (FG 2021), Jodhpur, India, 15–18 December 2021.

17. Ramachandra, S.; Hölzemann, A.; Laerhoven, K.V. Transformer Networks for Data Augmentation of Human Physical Activity Recognition. *arXiv* **2021**, arXiv:2109.01081.
18. Song, Z.; Yuan, Z.; Zhang, C.; Chi, W.; Ling, Y.; Zhang, S. Learning End-to-End Action Interaction by Paired-Embedding Data Augmentation. In *Computer Vision–ACCV 2020*; Ishikawa, H.; Liu, C.L.; Pajdla, T.; Shi, J., Eds.; Springer International Publishing: Cham, Switzerland, 2021; pp. 191–206.
19. Hoelzemann, A.; Sorathiya, N.; Van Laerhoven, K. Data Augmentation Strategies for Human Activity Data Using Generative Adversarial Neural Networks. In Proceedings of the 2021 IEEE International Conference on Pervasive Computing and Communications Workshops and other Affiliated Events (PerCom Workshops), Kassel, Germany, 22–26 March 2021; pp. 8–13. [CrossRef]
20. Sidor, K.; Wysocki, M. Recognition of Human Activities Using Depth Maps and the Viewpoint Feature Histogram Descriptor. *Sensors* **2020**, *20*, 2940. [CrossRef] [PubMed]
21. Rusu, R.B.; Bradski, G.; Thibaux, R.; Hsu, J. Fast 3D recognition and pose using the Viewpoint Feature Histogram. In Proceedings of the 2010 IEEE/RSJ International Conference on Intelligent Robots and Systems, Taipei, Taiwan, 18–22 October 2010; pp. 2155–2162.
22. Pazhoumand-Dar, H.; Lam, C.P.; Masek, M. Joint movement similarities for robust 3D action recognition using skeletal data. *J. Vis. Commun. Image Represent.* **2015**, *30*, 10–21. [CrossRef]
23. Lillo, I.; Niebles, J.C.; Soto, A. Sparse composition of body poses and atomic actions for human activity recognition in RGB-D videos. *Image Vis. Comput.* **2017**, *59*, 63–75. [CrossRef]
24. Shahroudy, A.; Ng, T.T.; Yang, Q.; Wang, G. Multimodal Multipart Learning for Action Recognition in Depth Videos. *IEEE Trans. Pattern Anal. Mach. Intell.* **2016**, *38*, 2123–2129. [CrossRef]
25. Raman, N.; Maybank, S. Activity Recognition using a supervised non-parametric Hierarchical HMM. *Neurocomputing* **2016**, *199*, 163–177. [CrossRef]
26. Farnoosh, A.; Wang, Z.; Zhu, S.; Ostadabbas, S. A Bayesian Dynamical Approach for Human Action Recognition. *Sensors* **2021**, *21*, 5613. [CrossRef]
27. Wang, H.; Yu, B.; Xia, K.; Li, J.; Zuo, X. Skeleton edge motion networks for human action recognition. *Neurocomputing* **2021**, *423*, 1–12. [CrossRef]
28. Plizzari, C.; Cannici, M.; Matteucci, M. Skeleton-based action recognition via spatial and temporal transformer networks. *Comput. Vis. Image Underst.* **2021**, *208–209*, 103219. [CrossRef]
29. Donahue, J.; Hendricks, L.A.; Rohrbach, M.; Venugopalan, S.; Guadarrama, S.; Saenko, K.; Darrell, T. Long-Term Recurrent Convolutional Networks for Visual Recognition and Description. *IEEE Trans. Pattern Anal. Mach. Intell.* **2017**, *39*, 677–691. [CrossRef] [PubMed]
30. Kruskal, J.B. Multidimensional scaling by optimizing goodness of fit to a nonmetric hypothesis. *Psychometrika* **1964**, *29*, 1–27. [CrossRef]
31. Warchoł, D.; Kapuściński, T. Human Action Recognition Using Bone Pair Descriptor and Distance Descriptor. *Symmetry* **2020**, *12*, 1580. [CrossRef]
32. Kapuściński, T.; Warchoł, D. Hand Posture Recognition Using Skeletal Data and Distance Descriptor. *Appl. Sci.* **2020**, *10*, 2132. [CrossRef]
33. Rusu, R.B.; Marton, Z.C.; Blodow, N.; Beetz, M. Learning informative point classes for the acquisition of object model maps. In Proceedings of the 2008 10th International Conference on Control, Automation, Robotics and Vision, Hanoi, Vietnam, 2–5 December 2018; pp. 643–650.
34. Spivak, M. *A Comprehensive Introduction to Differential Geometry*, 3rd ed.; Publish or Perish: Houston, TX, USA, 1999; Volume 3.
35. Li, W.; Zhang, Z.; Liu, Z. Action recognition based on a bag of 3D points. In Proceedings of the 2010 IEEE Computer Society Conference on Computer Vision and Pattern Recognition Workshops, San Francisco, CA, USA, 13–18 June 2010; pp. 9–14.
36. Chen, C.; Jafari, R.; Kehtarnavaz, N. UTD-MHAD: A Multimodal Dataset for Human Action Recognition Utilizing a Depth Camera and a Wearable Inertial Sensor. In Proceedings of the 2015 IEEE International Conference on Image Processing (ICIP), Quebec City, QC, Canada, 27–30 September 2015; pp. 168–172. [CrossRef]
37. Xia, L.; Chen, C.C.; Aggarwal, J.K. View invariant human action recognition using histograms of 3D joints. In Proceedings of the 2012 IEEE Computer Society Conference on Computer Vision and Pattern Recognition Workshops, Providence, RI, USA, 16–21 June 2012; pp. 20–27.
38. Seidenari, L.; Varano, V.; Berretti, S.; Del Bimbo, A.; Pala, P. Recognizing Actions from Depth Cameras as Weakly Aligned Multi-part Bag-of-Poses. In Proceedings of the 2013 IEEE Conference on Computer Vision and Pattern Recognition Workshops, Portland, OR, USA, 23–28 June 2013; pp. 479–485.
39. Hu, J.F.; Zheng, W.S.; Lai, J.; Zhang, J. Jointly Learning Heterogeneous Features for RGB-D Activity Recognition. In Proceedings of the IEEE Conference on Computer Vision and Pattern Recognition (CVPR), Boston, MA, USA, 7–12 June 2015.
40. Gaglio, S.; Re, G.L.; Morana, M. Human Activity Recognition Process Using 3-D Posture Data. *IEEE Trans. Hum.-Mach. Syst.* **2015**, *45*, 586–597. [CrossRef]
41. MSRA Dataset. Available online: https://sites.google.com/view/wanqingli/data-sets/msr-action3d (accessed on 11 April 2022).
42. Mei, J.; Liu, M.; Wang, Y.F.; Gao, H. Learning a Mahalanobis Distance-Based Dynamic Time Warping Measure for Multivariate Time Series Classification. *IEEE Trans. Cybern.* **2016**, *46*, 1363–1374. [CrossRef]

43. Øyvind Mikalsen, K.; Bianchi, F.M.; Soguero-Ruiz, C.; Jenssen, R. Time series cluster kernel for learning similarities between multivariate time series with missing data. *Pattern Recognit.* **2018**, *76*, 569–581. [CrossRef]
44. Radford, A.; Metz, L.; Chintala, S. Unsupervised Representation Learning with Deep Convolutional Generative Adversarial Networks. *arXiv* **2016**, arXiv:1511.06434.
45. Matlab Scripts for Distance Descriptor and Bone Pair Descriptor. Available online: http://vision.kia.prz.edu.pl (accessed on 1 January 2022).

sensors

MDPI

Article

Pattern Recognition of EMG Signals by Machine Learning for the Control of a Manipulator Robot

Francisco Pérez-Reynoso, Neín Farrera-Vazquez, César Capetillo, Nestor Méndez-Lozano, Carlos González-Gutiérrez and Emmanuel López-Neri *

Centro de Investigación, Innovación y Desarrollo Tecnológico UVM (CIIDETEC-UVM), Universidad del Valle de Mexico, Querétaro 76230, Mexico; francisco.perez@uvmnet.edu (F.P.-R.); nein.farrera@uvmnet.edu (N.F.-V.); cesar.capetillo@uvmnet.edu (C.C.); nestor.mendez@uvmnet.edu (N.M.-L.); calberto.gonzalez@uvmnet.edu (C.G.-G.)

* Correspondence: emmanuel.lopezne@uvmnet.edu

Abstract: Human Machine Interfaces (HMI) principles are for the development of interfaces for assistance or support systems in physiotherapy or rehabilitation processes. One of the main problems is the degree of customization when applying some rehabilitation therapy or when adapting an assistance system to the individual characteristics of the users. To solve this inconvenience, it is proposed to implement a database of surface Electromyography (sEMG) of a channel in healthy individuals for pattern recognition through Neural Networks of contraction in the muscular region of the biceps brachii. Each movement is labeled using the One-Hot Encoding technique, which activates a state machine to control the position of an anthropomorphic manipulator robot and validate the response time of the designed HMI. Preliminary results show that the learning curve decreases when customizing the interface. The developed system uses muscle contraction to direct the position of the end effector of a virtual robot. The classification of Electromyography (EMG) signals is obtained to generate trajectories in real time by designing a test platform in LabVIEW.

Keywords: EMG; pattern recognition; machine learning; robot; cyber-physical systems

Citation: Pérez-Reynoso, F.; Farrera-Vazquez, N.; Capetillo, C.; Méndez-Lozano, N.; González-Gutiérrez, C.; López-Neri, E. Pattern Recognition of EMG Signals by Machine Learning for the Control of a Manipulator Robot. *Sensors* **2022**, *22*, 3424. https://doi.org/10.3390/s22093424

Academic Editors: Tomasz Krzeszowski, Adam Świtoński, Michal Kepski and Carlos Tavares Calafate

Received: 24 March 2022
Accepted: 25 April 2022
Published: 30 April 2022

1. Introduction

A person with a disability is an individual who has one or more physical or mental deficiencies that prevent their full and effective participation in equal conditions when interacting with different social environments. In recent years, the development of HMI for people with motor disabilities has been oriented towards the use of systems based on Electromyography (EMG). In [1], a review of the state of the art in EMG monitoring is presented in terms of applications in rehabilitation and minimally invasive acquisition devices; among the advantages that it highlights are in the fields of physiotherapy and telemedicine. In [2], through three EMG channels, they control the position of a robot with two degrees of freedom; the processing is done as a function of time through the amplitude of the signal when movements are made with the elbow and the shoulder joint. Four channels of surface electromyography acquisition are proposed in [3], where pairs of electrodes are placed according to the position and orientation of the target muscles. Selecting materials with excellent properties for devices on the skin, the fabricated electrodes achieve low skin electrode impedance and record sEMG signals with a high signal-to-noise ratio. In [4], a review on signal acquisition and pattern recognition through Machine Learning is presented. In [5], a myoelectric pattern recognition-driven hand exoskeleton was designed for stroke rehabilitation. It detects and recognizes the intention of movement based on EMG signals, and then the exoskeleton helps the user to perform six types of hand movements in a real way. One of the main challenges in the design of interfaces based on sEMG is the obtention of a signal function or model that allows for the reliable control of a care system. Due to the non-stationary signal behavior, three methods are generally used for sEMG analysis to extract information, which are in the time [6,7], frequency, and time–frequency domain [8]. There are some practical factors, such as the change in arm position, that prevent robust myoelectric control. In [9], an experiment with 14 subjects is carried out to accurately characterize factors that alter the EMG recording. Using regression algorithms, they obtain real-time feedback on changes in the position of the arm and displacement of the electrodes. Pattern recognition has been studied further to develop control algorithms for electric hand prostheses [10,11]. These works have shown excellent accuracy when classifying different types of hand movement (>95% for 10 classes), [12–14]. Most of the pattern recognition approaches have the limitation that only one of the functions of the prosthetic hand can be controlled, due to its sequential and binary control. Such control strategies make it impossible to perform natural movements of the hand that consist of the simultaneous activation of different degrees of freedom. Some studies have introduced new pattern recognition schemes that classify combined movements [15–19]. The disadvantage of the new approach is the total number of classes, as it increases drastically when new classes are considered. Recently, regression-based approaches have been on the rise, as they provide control information that allows for multi-degree-of-freedom control. In this work, a regression algorithm using neural networks is proposed to obtain a model through multiclass categorization that allows for the control of a robotic system with three degrees of freedom of the anthropomorphic type. The analysis of a single channel of sEMG that classifies signals with different times of muscle contraction is implemented, with the objective that the robot moves accurately according to predetermined positions in a state machine and demonstrates the correct operation of an HMI by reducing the learning curve. In [20], a study of multichannel electromyography signals is carried out, which is one of the methods used in the recognition of human movement patterns. An exoskeleton robot is controlled and EMG signals are measured during dynamic or isometric muscle contractions. As a result, they developed a pattern recognition model of dynamic and isometric muscle contractions using the Short Time Fourier Transform (STFT).

Section 2 presents the fundamentals of the EMG signal. Section 3 presents the design of the HMI from the acquisition of the EMG signal and its analog and digital processing. The multiclass classification model obtained using neural networks is described. The different classes to be detected, the training algorithm and the operation of a state machine that determines the position of the robot according to the result obtained from the model are shown. The classification to reach the desired position is explained, obtaining the dynamics and inverse kinematics of an anthropomorphic robot with three degrees of freedom. A PD+ control is implemented to apply the necessary torque to each joint of the robot and to validate the operation of the HMI. It designs a graphical user interface in LabVIEW software by interacting a virtual robot and the EMG signal. Section 4 describes the results obtained with the classifier, the experimental tests and the response time for each test.

2. HMI Systems Based on EMG

The neuron is the cellular unit of the central nervous system. It has two properties: (1) Sensory, which gives it the ability to respond to physical and chemical agents with the initiation of a nerve impulse; and (2) Conductivity, which gives it the property of transmitting these impulses from one side to another. The dendrites that originate in the cell body are responsible for receiving impulses from other neurons and sending them to the soma of their own neuron. The axon is an extension from the neuronal soma that conducts the impulse to the muscle; it is surrounded by a myelin sheath that allows for better impulse conductivity. The neuron that originates the EMG biopotential is called a motor neuron, which conducts the impulse through the neuromuscular junction to the muscle fiber, as shown in Figure 1 [21].

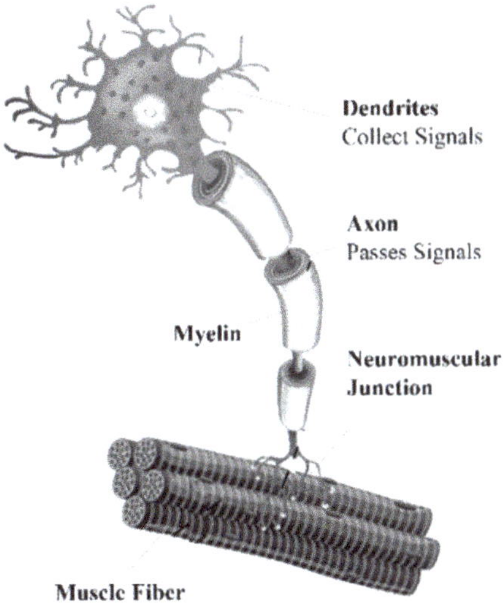

Figure 1. Components of the motor neuron, [21].

EMG is an electrical exploration of the peripheral nerves by the stimulation of the muscles to achieve their contraction. The differential potential in the biceps brachii is measured by placing two silver/silver chloride (Ag/AgCl) electrodes and a reference electrode located at the junction of the forearm and hand, as shown in Figure 2. [20].

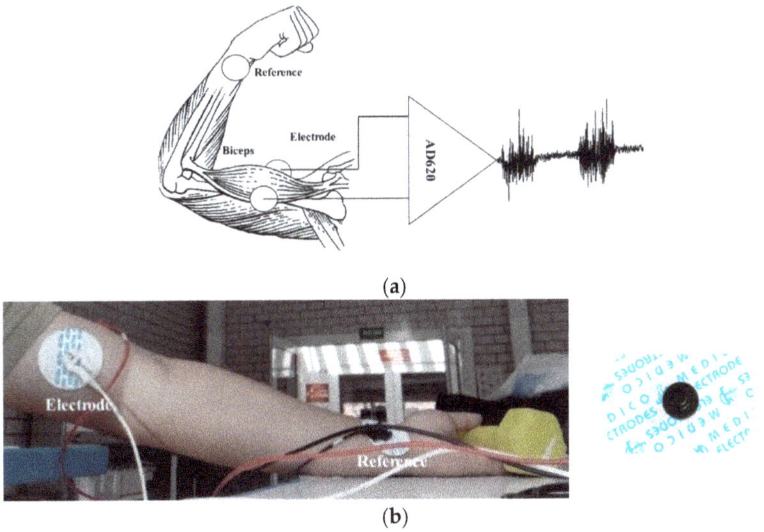

Figure 2. (a) Electrode placement diagram and the AD620 instrumentation amplifier, (b) Physical representation of the EMG signal acquisition protocol and the Silver/Silver Chloride (Ag/AgCl) electrode implemented.

When muscle contraction is performed, there are two types: (1) Isometric contraction, which is a static form of exercise in which a muscle contracts to produce force without an appreciable change in muscle length; and (2) Isotonic contraction, which is without appreciable change in the force of contraction. The distance between the origin of the muscle and its insertion becomes smaller. For EMG acquisition study purposes, in the protocol carried out, isometric contractions are recorded by placing a weight in the user's hand with a value of 5 pounds. This process is carried out in order to avoid the acquisition of noise due to involuntary movements and to keep the arm static while the biceps brachii contraction is performed for short periods of time. This process is carried out in order to avoid the acquisition of noises due to involuntary movements and to keep the arm static while the contraction of the biceps brachii is performed for short periods of recording time, no longer than 45 s, preventing the user from making an unwanted movement due to fatigue. When performing the acquisition, it was observed that the muscle relaxation periods of 5 s made it possible to accurately obtain the muscle contraction times, thus avoiding the introduction of noise due to muscle fatigue. The goal of this work is to demonstrate that, with a correct training of the neural network, adding dynamic muscle contractions due to involuntary movements as an extra class of recognition allows the system to rule out this muscle noise as a motion control command. The EMG signals have amplitudes from 0.1 mV to 5 mV, with a bandwidth of 0 to 5 KHz [21]. With this information, a first acquisition is made using a BIOPAC® commercial system, which allows for the recording of the differential signal taken from two electrodes and a reference, as indicated in Figure 3a. This system records the waveforms of the EMG signal in order to validate the implemented acquisition protocol. Tests were performed for contraction times of 1, 3 and 5 s. In Figure 3b, the response obtained from the EMG signal to a contraction of 5 s with rest pauses also of 5 s is presented. The inconvenience presented in the acquisition protocol when using this system is that the record is stored in a numerical database and cannot be read directly by any other acquisition card. Real-time implementation of the Fast Fourier Transform (FFT) is necessary to verify the spectrum in frequency and obtain the value of the cutoff frequency for the implementation of filters. The next section presents the instrumentation implemented and the digital processing for the acquisition of the database.

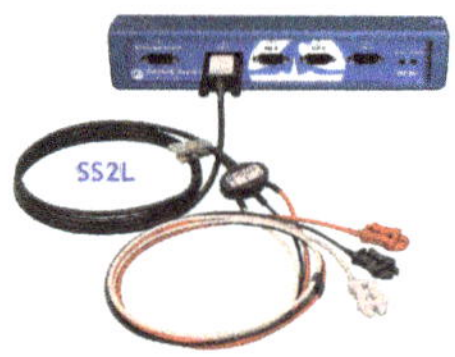

(a)

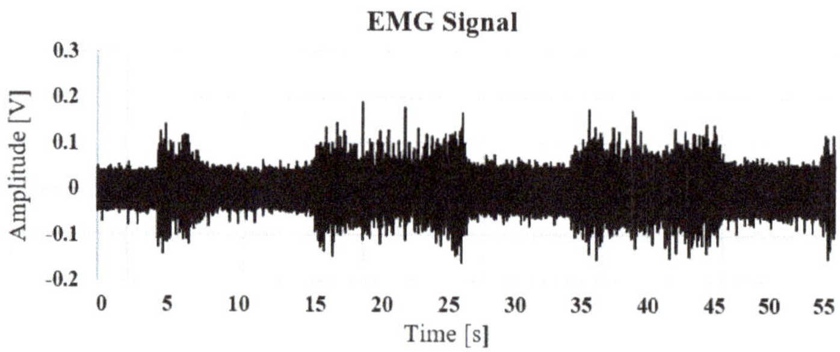

(b)

Figure 3. (**a**) Biopac® System, (**b**) Database obtained from the Biopac of the EMG signal with sustained isometric contraction of 5 s.

3. Materials and Methods

The amplifier used is the IC AD620 due to its characteristic of a common rejection ratio of 100 dB and the gain adjustment with an external resistor. A circuit with a basal corrector and a Common Mode Rejection (CMRR) configuration connected to the junction of the forearm and hand is implemented as a circuit reference. According to the amplitude and frequency characteristics of the EMG signal, the analog processing stage is designed, which includes amplification, isolation and filtering.

A. *Amplification with basal corrector*

An instrumentation amplifier CI AD620 is implemented as a preamplification system to acquire the differential EMG signal with a gain of 500. A basal correction circuit is conditioned to eliminate the level of direct current (DC) caused by involuntary movements of the user or an incorrect connection of the electrodes. The circuit is a IC TL084 operational amplifier in its integrator configuration that is connected in feedback to the Ref and V_{out} outputs of the instrumentation amplifier, as shown in Figure 4, implementing a high pass filter that eliminates the DC bias voltage and preventing op amps from reaching their maximum power limits.

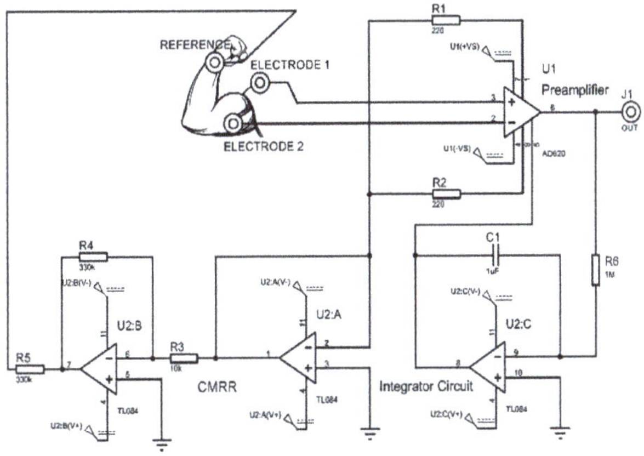

Figure 4. Amplification module and basal corrector.

B. *Analog Filter*

To filter the frequency components that are not within the bandwidth of the EMG signal, a range from 0.5 Hz to 5 KHz, a second order bandpass filter in Butterworth configuration with unity gain is designed, with a ratio of 40 dB per decade using high impedance TL084 operational amplifiers, precision resistors and electrolytic capacitors; see Figure 5.

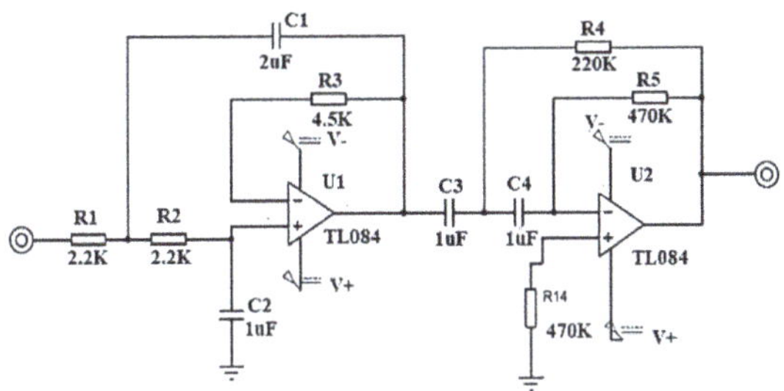

Figure 5. Filter in second order Butterworth configuration at 40 dB/decade.

The output of the analog filter stage is connected to the absolute voltage input of a DAQ6009 acquisition card connected via USB port to a laptop, with a sample rate of 10KHz. An acquisition card with a ground plane is designed to decrease inductive noise, as indicated in Figure 6a. Figure 6b shows the response of the acquisition card in the

Tektronix® oscilloscope. Analog noise is observed, which is subsequently eliminated by means of a digital filter.

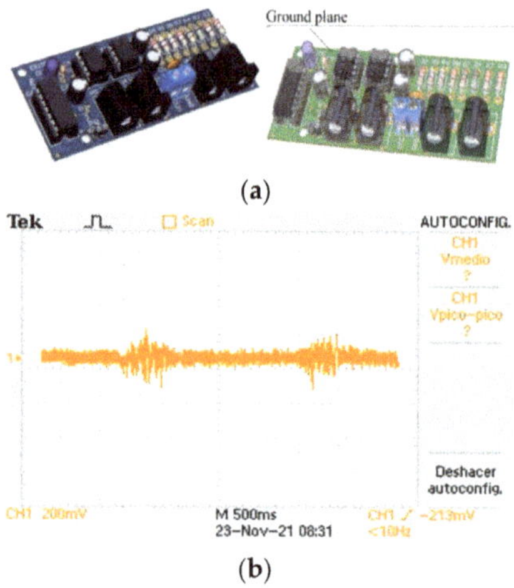

Figure 6. (**a**) EMG signal acquisition cards, (**b**) EMG signal response in the Tektronix oscilloscope.

C. *Digital Filter*

Due to the acquisition system being subject to the interference of electromagnetic noise induced by lamps or some other external device, and in order to digitally tune the response of the filter, the design of a digital low pass filter is implemented. First, the analog/digital conversion is done with the National Instrument DAQ6009 card at an acquisition frequency of 10 KHz at 9600 bauds with 11 bits of resolution. The procedure consists of obtaining samples of the continuous signal at instants of time, defining $v_i[n] = v_n(nT)$, where T is the sampling period.

The response of the digital first order low pass filter is obtained with the aim of reducing the computational cost when applying the filter in real time. The filter configuration is indicated in Figure 7, indicating its response in terms of the complex frequency s. In Equation (1), the filter response is plotted as a function of the complex discrete frequency z.

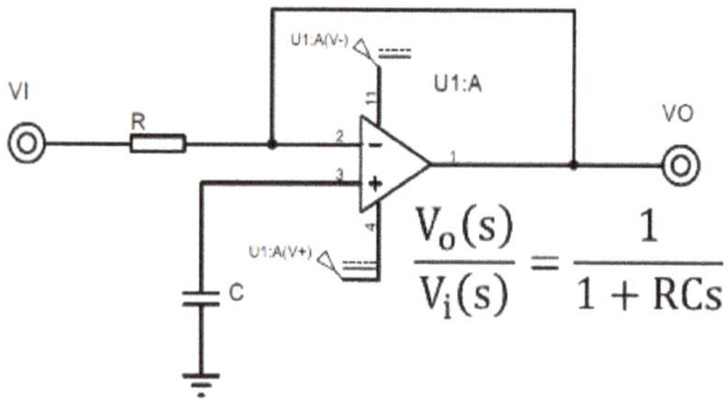

Figure 7. First order low pass filter and its transfer function as a function of the complex variable s.

$$G(z) = \frac{\left(1 - e^{\frac{T}{RC}}\right) z^{-1}}{1 - e^{\frac{T}{RC}} z^{-1}} \tag{1}$$

In Equation (2) the filter equation is indicated as a function of the discrete variable n by means of difference equations when implementing the inverse z-transform of Equation (1).

$$v_o[n] = e^{-2\pi f_c T}\, v_o[n-1] + \left(1 - e^{-2\pi f_c T}\right) v_i[n-1] \quad (2)$$

To obtain the value of the cutoff frequency (f_c) and tune the digital filter, the Discrete Fourier Transform (FFT) is implemented. First, the EMG signal is digitized by means of a convolution with a Dirac delta pulse train as a function of time, where $v_i[n]$ is a signal represented in an exponential Fourier series, as in Equation (3). a_k represents the amplitude of the signal energy.

$$v_i[n] = \sum_{k=N} a_k e^{j\frac{2\pi}{N}kn} = a_0 e^{j\frac{2\pi}{N}0n} + a_1 e^{j\frac{2\pi}{N}1n} \ldots + a_{N-1} e^{j\frac{2\pi}{N}(N-1)n} \quad (3)$$

The frequency spectrum analysis is performed by applying the Fourier Transform on the discrete signal $v_i[n]$, obtaining as a result a train of delta functions in frequency $X(e^{j\varpi})$, as indicated by Equation (4), whose amplitude is determined by the weighting of coefficients a_k, through the results of the spectrum in Frequency. The component that provides more energy to the signal is calculated; thus, the frequency of the induced noise is determined, and the cutoff frequency is obtained with precision (f_c) for the design of the digital filter. Figure 8 is the result of the implementation of the digital filter in the acquisition of the EMG signal.

$$X\left(e^{j\omega}\right) = \sum_{k=-\infty}^{+\infty} a_k 2\pi\delta\left(\omega - \frac{2\pi}{N}k\right) \quad (4)$$

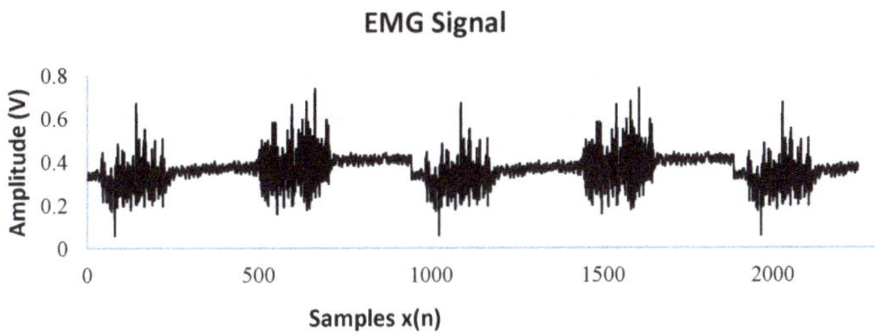

Figure 8. Filtered EMG signal.

D. *Multiclass Classifier: One Hot Encoding*

In this section, the method used is presented so that, in real time, the movements determined through the EMG interface are executed on a manipulator robot. An intelligent system for muscle contraction classification was implemented. Using a Multilayer Neural Network (MNN), a model is obtained that identifies four different classes of muscle contraction. The first class is described as Sharp muscle pulse (SMP), the second class as Smooth muscle pulse 3 s (SMP3), the third class as Smooth Muscle Pulse 5 s (SMP5) and, finally, the fourth class is described as Noise Involuntary Movements (NIM). These signals are classified using the One-Hot Encoding technique that labels the waveform of each signal with an integer. Thus, the digital inputs of a state machine are obtained, which determine the predetermined position of a manipulator robot with three degrees of freedom in the Cartesian plane (x, y, z) inside the robot workspace. In Figure 9, the architecture of the HMI based on EMG is presented.

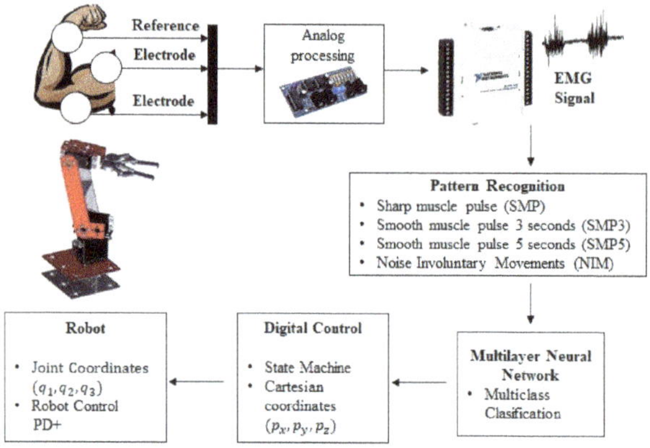

Figure 9. Architecture of the EMG signal classification method for the control of a manipulator robot.

To perform the identification of patterns in the EMG signal of a single channel, they are divided into action potentials with different time intervals. The SMP (Sharp Muscle Pulse) class has an instantaneous contraction interval of 1 s and muscle relaxation intervals of 5 s. The SMP3 (Smooth Muscle Pulse 3 s) class has a contraction interval of 3 s and muscle relaxation intervals of 5 s. The SMP5 (Smooth Muscle Pulse 5 s) class has a muscle contraction interval of 5 s and muscle relaxation intervals of 5 s. The NIM (Noise Involuntary Movements) class is a class that records the resting state of users as well as involuntary arm movements recorded during acquisition. All these samples are stored in a vector called p. Figure 10 indicates the waveform of each class. The SMP, SMP3 and SMP5 classes indicate a position change control order in the manipulator robot, while the NIM class indicates a total stop state, so the MNN has as inputs the different signals identified in classes stored in the vector $p^{1\times n}$. An integer is assigned to each class through supervised training; this labeling is stored in a vector called $T^{1\times n}$, where n is the total number of samples.

Action potential (muscle contraction time intervals)

- Class 1: Sharp Muscle Pulse (SMP) = p_1

Sustained contraction for a time interval of 1 s, rest interval of 5 s.

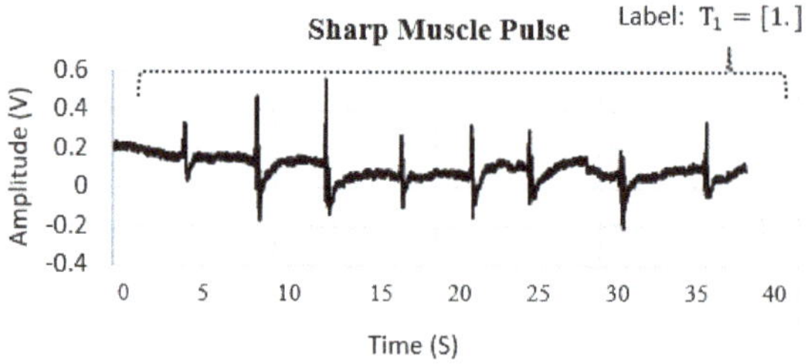

- Class 2: Smoot Muscle Pulse 3 s (SMP3) = p_2

Sustained contraction for a time interval of 3 s, rest interval of 5 s.

Figure 10. *Cont.*

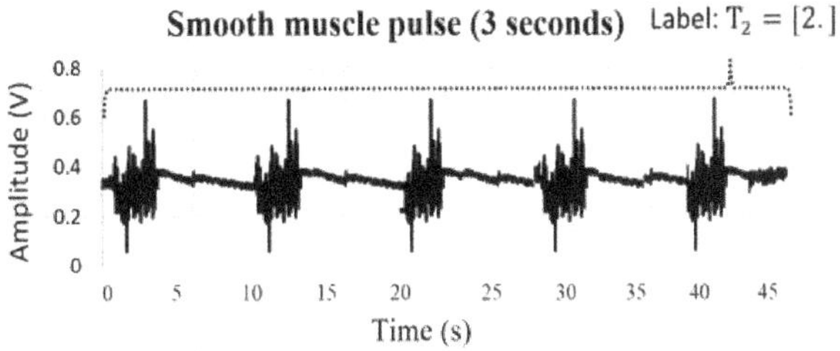

- Class 3: Smoot Muscle Pulse 5 s (SMP5)

Sustained contraction for a time interval of 5 s, rest interval of 5 s.

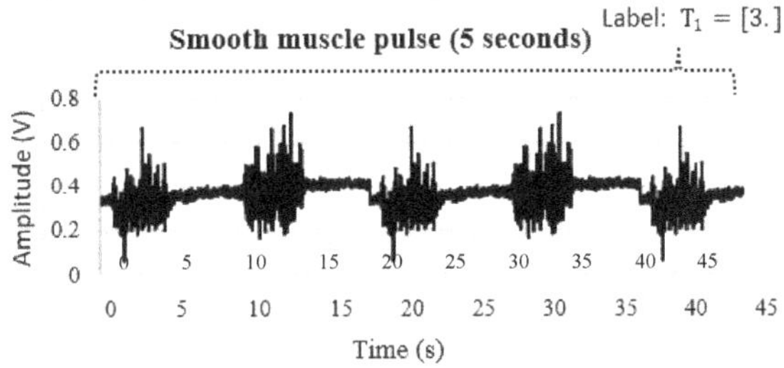

Resting Potential

- Class 4: Noise Involuntary Movements (NIM)

Relaxed biceps muscle and acquisition of involuntary arm movements.

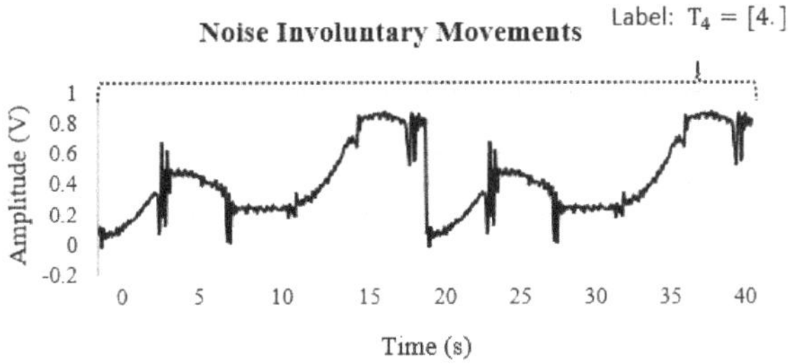

Figure 10. Graphical representation of the data set (input vector $p^{1\times n}$). The output vector $T^{1\times n}$ stores the labels of each class using integer data.

E. *Multiclass Classifier: Multilayer Neural Network*

In this section, the implementation of an intelligent system for the classification of EMG signals is presented. The representation of the multilayer neural network is presented in Figure 11, where $p = [p^T]$ is the vector of the R inputs, $b = [b^T]$ represents the polarization of S neurons, $n = [n^T]$ represents the net inputs of each of the S neurons and $W = [W_{SR}{}^T]$ is the matrix of synaptic weights.

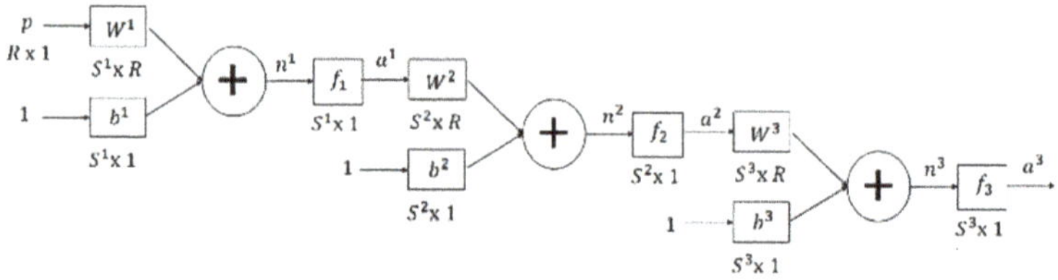

Figure 11. Structure of the Multilayer Neural Network.

The first stage consists of data normalization because the EMG signals have different voltage thresholds. The description of this procedure is presented in Equation (5), where p represents the data set of the EMG signal by means of a vector of an acquisition channel. The mean of the data is subtracted, with a standard deviation equal to 1 to minimize the computational cost when the network performs the learning process.

$$p = \frac{p - p^{mean}}{\sqrt{p^{var}}} = \frac{p - p^{mean}}{p^{std}} \tag{5}$$

Algorithm 1 describes the pseudocode for the implementation of the Neural Network in Python; the training consists of assigning to each sample the value of a constant that is stored in the vector T. This vector is the desired result for each class and has the same dimensions as the input vector p.

In Figure 12, an association between the precision of the neural network with new data (Trian loss) and the value of the loss function (Val loss) after 3000 epochs is presented. Both graphs have a tendency to zero as training progresses, indicating a correct functioning of the optimizer. In [22], the authors designed multiclass classification on two channels of electrooculography signals and controlled an omnidirectional mobile robot in the X, Y plane. In this work, it is shown that, according to the muscle contraction time, the multiclass classification allows for the control of robotic systems that work in space (X, Y, Z) and that are adaptive to the individual characteristics of the user, achieving a personalization of the Interface.

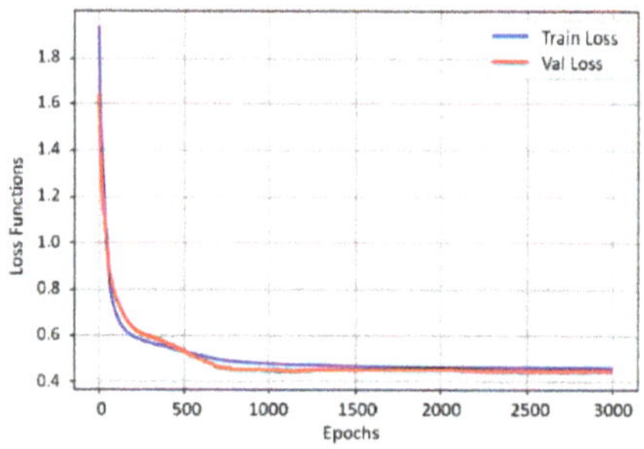

Figure 12. Graph of the accuracy trend of the neural network with new data (Train loss) and the trend of the loss function (Val loss). Neural network accuracy ratio after 3000 epochs.

Algorithm 1: Multilayer Perceptrón algorithm implemented for the EMG

$p \leftarrow$ *Input_vector*
$T \leftarrow$ *Output_vector*

*/*** output vector T where the labeling value is stored by one-hot-encoding of each the classes***/*
$T \leftarrow [\{0,0,0,1\},\{0,0,2,0\},\{0,3,0,0\},\{4,0,0,0\}]$

scaler $\leftarrow$ *StandardScaler( ).fit(P)*

$p \leftarrow$ *scaler.transform(P)*

*/**Divide p into a test* (P_{test}) *and a training set* (P_{train})***/*

one_hot_labels = to_categorial(T, num_classes $\leftarrow$ 4)

P_train, P_test, T_train, T_test $\leftarrow$ *train_test_split(P, one_hot_labels, test_size* $\leftarrow$ 0.20, *random_state* $\leftarrow$ 42)

*/**Random Initialization**/*
$W \leftarrow 2 \times (random - 0.5) \times scale$

epochs $\leftarrow$ 3000
hiddenNodes $\leftarrow$ 4

model $\leftarrow$ *Sequential( )*

model.add(Dense(hiddenNodes, activation $\leftarrow$ *relu, input_*dim $\leftarrow$ 4)
$a[1] \leftarrow max(0, n)$ *//ReLu activation function*

model.add(Dense(4, activation $\leftarrow$ *'softmax'))*
$a[2] \leftarrow e^{n_4} / \sum_1^5 e^{n_4}$ *//Softmax activation function*

model.summary( )

loss $\leftarrow$ *categorical_crossentropy*
*/**Loss function (categorial cross entropy**/*
$L(y, \hat{y}) \leftarrow \frac{1}{N} \sum_{j=1}^{M} \sum_{i=1}^{N} \left(y_{ij} \log \left(\hat{y}_{ij} \right) \right)$

optimizer $\leftarrow$ *tf.keras.optimzers.Adam( )*
$W \leftarrow W - \frac{\alpha m}{\sqrt{v + \epsilon}}$

model.compile(loss $\leftarrow$ *loss, optimizer* $\leftarrow$ *optimizer, metrics* $\leftarrow$ *['accuracy'])*
history $\leftarrow$ *model.fit(P_train, T_train , epochs* $\leftarrow$ *epochs, vebose* $\leftarrow$ 1 *, validation_split* $\leftarrow$ 0.1)

test, test $\leftarrow$ *model.evaluate(P, t, verbose* $\leftarrow$ 1)
weights(model.layers, 3)
scaling(scaler, 3)
layers(model.layers)

The obtained values of the synaptic weights W and the polarization vector b of the two neurons, after 3000 epochs:

$$W_1 = [4][1] = \begin{bmatrix} -0.321 \\ 1.016 \\ 1.322 \\ 1.564 \end{bmatrix}$$

$$W_2 = [4][4] = \begin{bmatrix} -0.363 & 0.232 & 0.222 & -0.123 \\ 0.543 & -0.127 & 0.142 & -0.234 \\ 0.126 & -0.123 & -0.118 & 0.233 \\ -0.217 & 0.147 & 0.156 & -0.126 \end{bmatrix}$$

$$b_1 = \begin{bmatrix} -0.321 & 0.087 & 0.123 & 0.224 \end{bmatrix}$$

$$b_2 = \begin{bmatrix} 0.457 & -0.121 & 0.789 & 0.389 \end{bmatrix}$$

Once the model recognizes each of the classes by means of integers, a comparison system is implemented using the premise, "If the Network output is: (integer) [1–4] then 1 is enabled when the network recognizes the waveform that corresponds to each label, otherwise it is 0". This process allows for a combination of digital pulses for the activation of a state machine.

F. *State Machine*

The combination of digital signals obtained from the pattern recognition of the neural network by means of class classification allows for the transition change of a state machine. A Mealy-type machine is implemented, which generates an output based on its current state and an input. Three finite sets determined by the inputs, outputs and states are defined.

In Figure 13, the transitions of the digital inputs are indicated and the NIM class is represented as the most significant bit. In the next position the SMP3 class is, then the SMP5 class and finally the SMP class, so that there is an input 4 bits for transition change. Each of the states indicates a predetermined position of the manipulator robot with three degrees of freedom in Cartesian coordinates (p_x, p_y, p_z). Subsequently, these coordinates are converted to joint coordinates (q_1, q_2, q_3) using the inverse kinematics of the manipulator robot. There is an input IN9 that, when detecting a status at 1 of noise or involuntary movements, completely deactivates the operation of the robot; this is taken as a security measure to not activate the robot when this class of signals occurs.

Digital inputs of the EMG signal				Input Number
NIM	SMP3	SMP5	SMP	
0	0	0	0	IN1
0	0	0	1	IN2
0	0	1	0	IN3
0	0	1	1	IN4
0	1	0	0	IN5
0	1	0	1	IN6
0	1	1	0	IN7
0	1	1	1	IN8
1	0	0	0	IN9

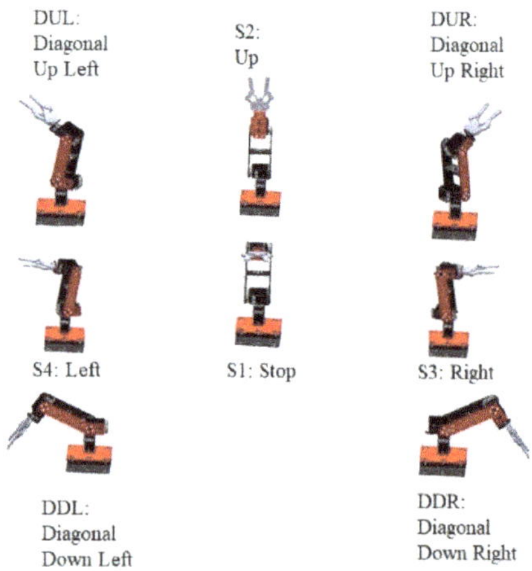

Figure 13. The table presents the inputs of the digital system, and the system outputs are indicated by means of the robot diagram.

In Figure 14, the designed machine has eight possible states for muscle movement, with four digital inputs corresponding to the high and low pulses of the Neural Network recognition. Table 1 describes the position in Cartesian coordinates of each of the robot states.

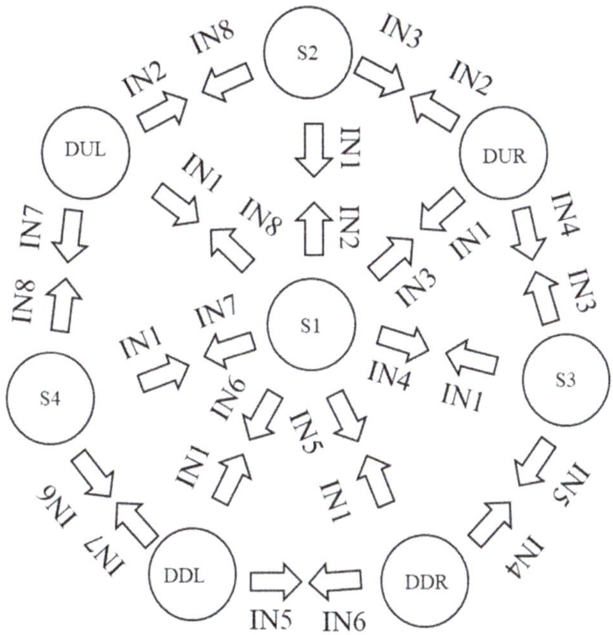

Figure 14. Eight-state Mealy-type machine, transition indicated.

Table 1. Description of each of the desired positions of each state.

State	Input EMG	Desired Value in Meters			Desired Movement
		p_x	p_y	p_z	
S1	IN1	0	−0.34	0.38	*Stop*
S2	IN2	0	−0.11	0.46	*Up*
S3	IN4	0.34	−0.34	0.38	*Right*
S4	IN7	−0.34	−0.34	0.38	*Left*
DUL	IN8	−0.34	−0.11	0.46	*Diagonal Up Left*
DUR	IN3	0.34	−0.11	0.46	*Diagonal Up Right*
DDL	IN6	−0.34	−0.34	0.28	*Diagonal Down Left*
DDR	IN5	0.34	−0.34	0.28	*Diagonal Down Right*

The selected robot is an anthropomorphic robot with three degrees of freedom and rotational joints whose operation is similar to the human arm (Figure 15), where l_1, l_2 and l_3 represent the total length of the links, l_{c1}, l_{c2} and l_{c3} represent the length from the initial end to the center of mass of each of the links that make up the robot, m_1, m_2 and m_3 are the values of the center of mass of each link, $x_{0\ldots3}, y_{0\ldots3}, z_{0\ldots3}$ represent the cartesian axes indicating the orientation of the position and q_1, q_2 and q_3 represent each degree of freedom of each rotational joint of the robot.

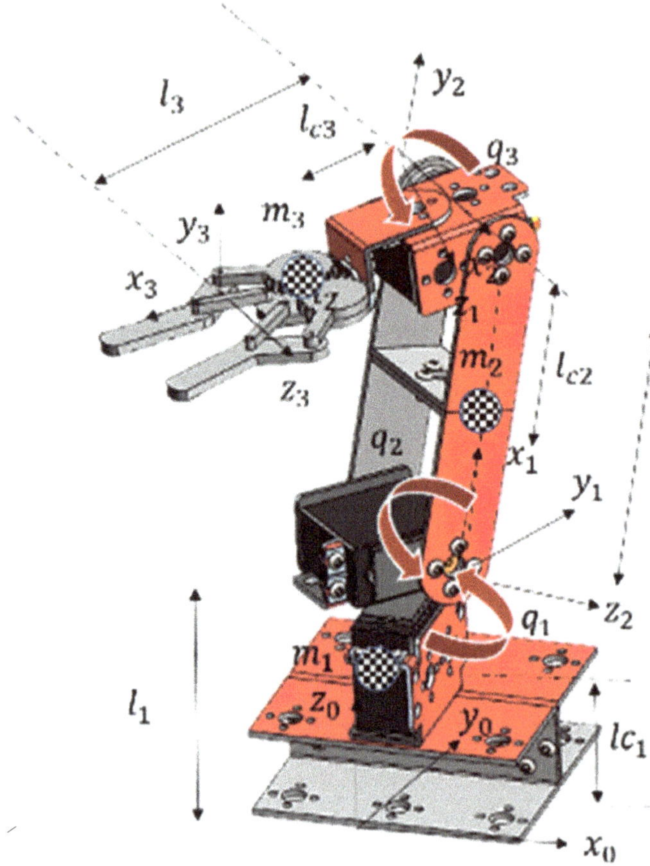

Figure 15. Virtual model of an anthropomorphic robot with three degrees of freedom.

To determine the workspace of the anthropomorphic robot, the calculation of the forward kinematics is performed, which determines the position of the end effector in Cartesian coordinates (p_x, p_y, p_z) based on joint coordinates (q_1, q_2, q_3) indicated in Equation (6). These equations are fundamental for the calculation of the robot dynamics.

$$\begin{aligned} p_x &= \cos(q_1)(l_{c3}\cos(q_2+q_3)+l_{c2}\cos(q_2)) \\ p_y &= \sin(q_1)\,(l_{c3}\cos(q_2+q_3)+l_{c2}\cos(q_2)) \\ p_z &= l_1+l_{c3}\sin(q_2+q_3)+l_{c2}\sin(q_2) \end{aligned} \tag{6}$$

Because the state machine has the coordinates of the end effector position in meters for each of the Cartesian axes x,y y z, the inverse kinematics of the robot defined in Equation (7), these equations determine the value of the position in radians for each of the degrees of freedom (q_1, q_2, q_3).

$$q_1 = \tan^{-1}\left(\frac{py}{px}\right) q_2 = 2\tan^{-1}\left(\frac{b+\sqrt{b^2+a^2-c^2}}{a+c}\right) \tag{7}$$

where:

$$c = p_x^2\cos^2(q_1) + 2\,p_x\,p_y\,\mathrm{sen}(q_1)\cos(q_1) + p_z^2 - 2\,p_z\,l_1 + l_1^2 + l_2^2 - l_3^2$$
$$a = 2\,p_x\,l_2\cos(q_1) + 2\,p_y\,l_2\,\mathrm{sen}(q_1)$$
$$b = 2\,p_z\,l_2 - 2\,l_1\,l_2$$
$$q_3 = \tan^{-1}\left(\frac{p_z\cos(q_2) - l_1\cos(q_2) - p_x\cos(q_1)\sin(q_2) - p_y\sin(q_1)\sin(q_2)}{p_z\cos(q_2) - l_1\cos(q2) - p_x\cos(q_1)\sin(q_2) - p_y\sin(q_1)\sin(q_2)}\right)$$

To implement the PD+ position tracking control algorithm, use the dynamic model defined in Equation (8).

Inertia Matrix $(M(q))$

$$\begin{bmatrix} \left(\begin{array}{c} I_1 + I_2 + I_3 + \frac{l_2^2\,m_3}{2} + \frac{l_{c2}^2\,m_2}{2} + \frac{l_{c3}^2\,m_3}{2} \\ + \frac{l_2^2\,m_2\cos(2q_2)}{2} + \frac{l_2^2\,m_3\cos(2q_2)}{2} \\ + \frac{l_{c3}^2\,m_3\cos(2q_2 + 2q_3)}{2} \\ + l_2\,l_{c3}\,m_3\cos(2q_2 + q_3) \\ + l_2\,l_{c3}\,m_3\cos(q_3) \end{array}\right) & (I_2 + I_3) & I_3 \\ (I_2 + I_3) & \left(\begin{array}{c} I_2 + I_3 + l_{c2}^2\,m_2 \\ + l_{c3}^2 + 2\,l_2\,l_{c3}\,m_3\cos(q_3) \\ + l_2^2\,m_3 \end{array}\right) & \left(\begin{array}{c} I_3 + l_{c3}^2\,m_3 \\ + l_2\,l_{c3}\,m_3\cos(q_3) \end{array}\right) \\ I_3 & \left(\begin{array}{c} I_3 + l_{c3}^2\,m_3 \\ + l_2\,l_{c3}\,m_3\cos(q_3) \end{array}\right) & (I_3 + l_{c3}^2\,m_3) \end{bmatrix}$$

Coriolis Matrix $(C(q,\dot{q}))$

$$\begin{bmatrix} \left(\begin{array}{c} -\dot{q}_2 l_2^2\,m_2\sin(2q_2) \\ -\dot{q}_2 l_2^2 m_3\sin(2q_2) \\ -\dot{q}_2 l_{c3}^2 m_3\sin(2q_2 + 2q_3) \end{array}\right) & (-2\dot{q}_1\,l_2\,l_{c3}\,m_3\sin(2q_2 + q_3)) & \left(\begin{array}{c} -\dot{q}_1\,l_{c3}^2\,m_3\sin(2q_2 + 2q_3) \\ -\dot{q}_1\,l_2\,l_{c3}\,m_3\sin(q_3) \\ -\dot{q}_1\,l_2\,l_{c3}\,m_3\sin(2q_2 + q_3) \end{array}\right) \\ \left(\begin{array}{c} \frac{\dot{q}_1\,l_2^2 m_3\sin(2q_2)}{2} \\ + \frac{-\dot{q}_1 l_{c2}^2\,m_2\sin(2q_2)}{2} \\ + \frac{\dot{q}1\ lc3^2\ m3\sin(2q2+2q3)}{2} \\ + \dot{q}_1\,l_2 l_{c3}\,m_3\sin(2q_2 + q_3) \end{array}\right) & (-2\dot{q}_3\,l_2\,l_{c3}\,m_3\sin(q_3)) & (-\dot{q}_3\,l_2\,l_{c3}\,m_3\sin(q_3)) \\ \left(\begin{array}{c} \frac{\dot{q}_1\,l_{c3}^2\,m_3\sin(2q_2+2q_3)}{2} \\ + \frac{\dot{q}_1\,l_2\,l_{c3}\,m_3\sin(q_3)}{2} \\ + \frac{\dot{q}_1\,l_2\,l_{c3}\,m_3\sin(2q_2+q_3)}{2} \end{array}\right) & (\dot{q}_2\,l_2\,l_{c3}\,m_3\sin(q_3)) & 0 \end{bmatrix}$$

Gravity Vector $(g(q))$

$$\begin{bmatrix} 0 \\ -g\,l_{c3}\,m_3\cos(q_2 + q_3) - g\,l_2\,m_3\cos(q_2) - g\,l_{c2}\,m_2\cos(q_2) \\ -g\,l_{c3}\,m_3\cos(q_2 + q_3) \end{bmatrix}$$

Viscous friction vector (B)

$$B\dot{q} = \begin{bmatrix} B_1\dot{q}_1 \\ B_2\dot{q}_2 \\ B_3\dot{q}_3 \end{bmatrix}$$

Torque Vector

$$\tau = \begin{bmatrix} \tau_1 \\ \tau_2 \\ \tau_3 \end{bmatrix}$$

$$\tau = M(q)\ddot{q} + C(q,\dot{q})\dot{q} + g(q) + B\dot{q} \tag{8}$$

where $M(q)$ is a positive definite symmetric matrix of n x n called the inertia matrix, with I1, I2, I3 being the moments of inertia of the rigid links of the mechanical structure of the robot, $C(q, \dot{q})$ is an n x 1 vector called the vector of centrifugal and Coriolis forces, $B\dot{q}$ is an n x 1 vector that determines the viscous friction, $g(q)$ is an n x 1 vector of gravitational forces and τ is the n x 1 vector that determines the torques and forces applied by the actuators at the joints.

G. *Position Control*

As a result of the cartesian coordinates (px, py, pz) obtained from the classifier by means of a Multilayer Neural Network and assigned to a discrete event by means of a state machine, the desired Cartesian coordinates for the robot are obtained, which are transformed to joint coordinates (q_1, q_2, q_3) from inverse kinematics. These values are the inputs for the PD+ type position control system. [15].

The PD+ control with gravity compensation, defined in Equation (9) by τ_{PD+}, is an algorithm that includes proportional control of the position error $\tilde{q}$ and velocity error proportional control $\dot{\tilde{q}}$, where $K_p, K_v \in \mathbb{R}^{nxn}$ are the proportional and derivative gains, respectively, both are positive definite matrices, and the full dynamics of the robot are added. In the structure of this scheme, the trajectory of position, velocity and desired acceleration is involved, $q_d(t), \dot{q}_d(t), \ddot{q}_d(t) \in \mathbb{R}^n$.

$$\tau_{PD+} = K_p\tilde{q} + K_v\dot{\tilde{q}} + M(q)\ddot{q}_d + C(q, \dot{q})\dot{q}_d + B\dot{q}_d + g(q) \tag{9}$$

The objective of this control is to find a torque value, τ, such that it satisfies the expression indicated in Equation (10).

$$\lim_{t\to\infty} \begin{bmatrix} \tilde{q} \\ \dot{\tilde{q}} \end{bmatrix} = \begin{bmatrix} 0 \\ 0 \end{bmatrix} \in \mathbb{R}^{2n} \tag{10}$$

where $\tilde{q} \in \mathbb{R}^n$ is the following error and is defined as $\tilde{q} = q_d(t) - q(t)$, and $\dot{\tilde{q}} \in \mathbb{R}^n$ is the velocity error, given by $\dot{\tilde{q}} = \dot{q}_d(t) - \dot{q}(t)$. Figure 16 indicates the block diagram of the implemented PD+ control.

Figure 17a shows the behavior of the zero-position error trend in each joint coordinate of the robot whose Cartesian coordinate is assigned by the state machine. The operation of the control when reaching the desired joint position is also presented. Figure 17b shows the virtual simulation of the robot applying the PD+ control for the generation of trajectories through the interaction of the EMG signal.

A graphical user interface is designed as indicated in Figure 18b with visual feedback of the EMG signal, the result of the state machine by means of a green indicator that indicates the position detected of the MNN's classification, the control curves resulting from the implemented PD+ and a simulation of the virtual robot that indicates the position of the end effector. In Figure 18b, the user connection and the operation of the interface to calculate the response time metrics are indicated.

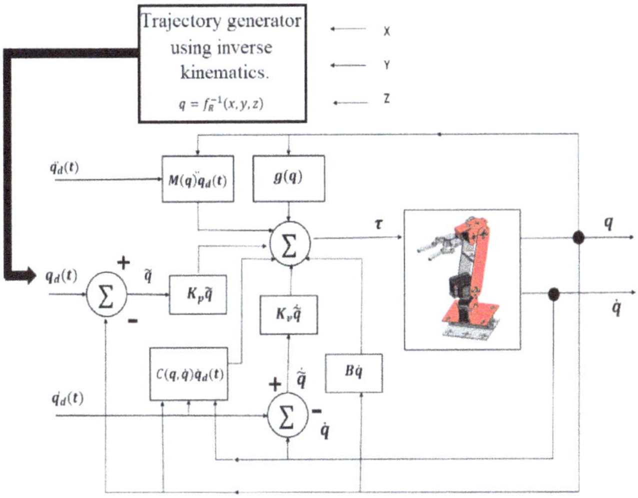

Figure 16. Block diagram of PD + control with gravity compensation.

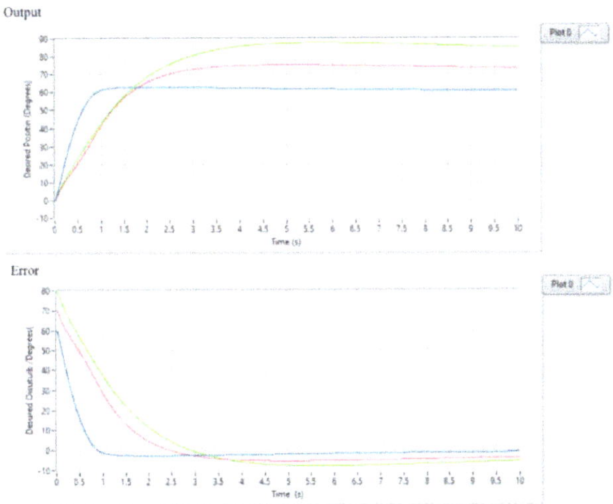

(**a**)

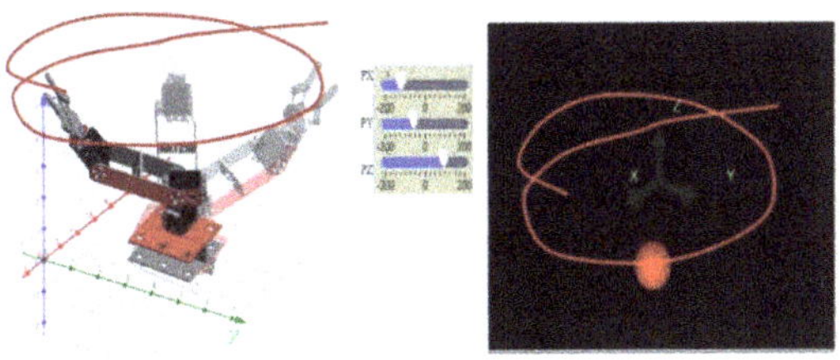

(**b**)

Figure 17. (**a**) Graphs of the position error with a tendency to zero for each of the joint coordinates (q_1, q_2, q_3) and control curves indicating the operation of the PD+ to reach the desired positions, (**b**) Result of the PD+ trajectory control of an anthropomorphic virtual robot.

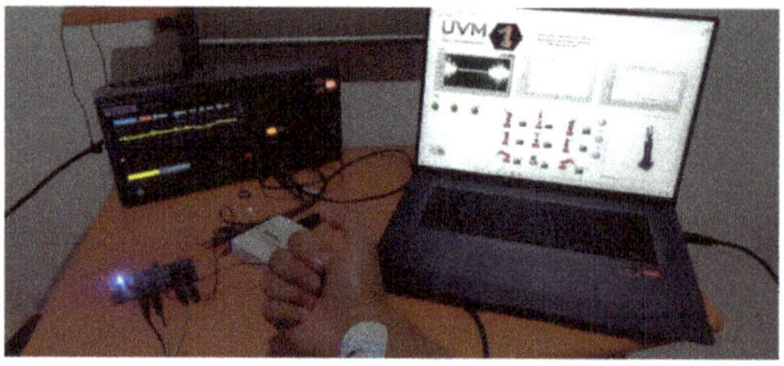

(**a**)

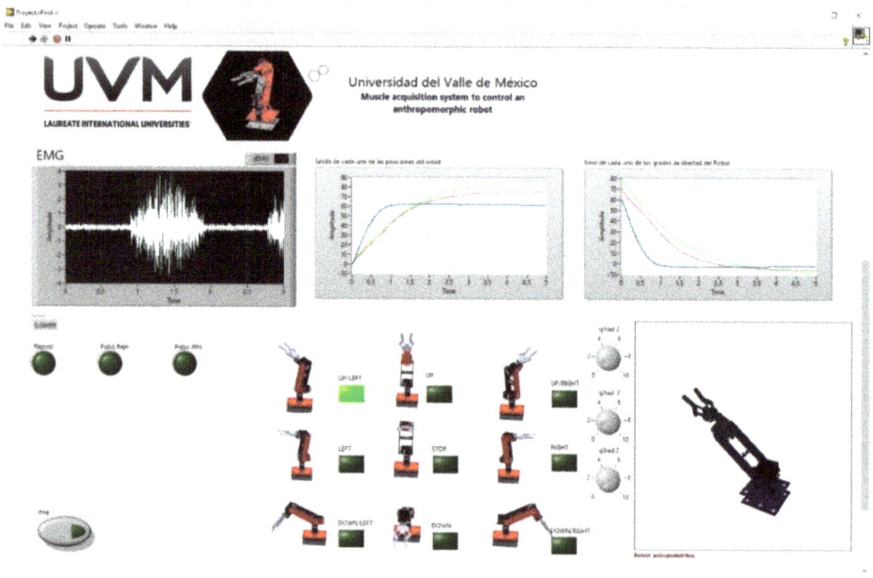

(**b**)

Figure 18. (**a**) Implementation of the real-time acquisition system interacting with the virtual robot simulation, (**b**) Graphical interface designed to record response time metrics.

4. Results and Discussions

The EMG signal classification method that allows for the generation of coordinates for the trajectory control of a manipulator robot has been developed. The user's ability to follow a series of point-to-point coordinates previously determined by colors is measured according to the time of sustained contraction. The yellow dot indicates the starting point of the test, the green dots indicate the path to be followed and the blue dot indicates the end point to which the robot's end effector must reach. Two trajectories are proposed that increase the difficulty indicating a penalty each time the user enters a contraction command other than the one indicated. The time in which the user generates the trajectory is also recorded. The test ends when the user generates the trajectory without penalties. In Figure 19a, the first proposed trajectory is indicated, in Figure 19b, the time and the number of penalties for each test performed by the user are presented and in Figure 19c, a graph of the response time for trajectory 1 is indicated.

- Description of the trajectory in each of the tests. The yellow point is the beginning, passing through each of the green points until reaching the final blue point.

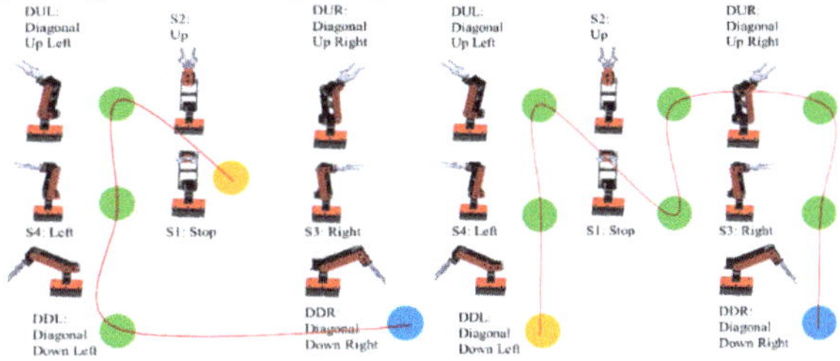

(**a**)

- Description of the time and penalties in each of the tests. When the box is green, it is because the user reached the desired point; when it is red, it is because the user did not reach the indicated point.

Test Number		Correct Trajectory: STOP	DUL	LEFT	DDL	DDR	Total Metrics Result in seconds
1	Penalties	RIGHT	DUL	STOP	DUL	UP	4
	Time (Seconds)	24.3	23.22	26.3	22.4	22.3	118.52
2	Penalties	STOP	UP	STOP	DUL	DDR	2
	Time (Seconds)	21.3	22.12	22.34	22.34	20.19	108.29
3	Penalties	UP	DUL	DUR	RIGHT	STOP	4
	Time (Seconds)	19.2	18.3	19.9	20.22	21.26	98.88
4	Penalties	STOP	LEFT	DDL	DDL	DDR	2
	Time (Seconds)	18.3	18.2	17.3	19.93	20.21	93.94
5	Penalties	STOP	UP	LEFT	STOP	DDR	2
	Time (Seconds)	18.9	19.3	16.4	18.92	19.33	92.85
6	Penalties	STOP	UP	LEFT	DDL	DDR	1
	Time (Seconds)	19.12	17.23	16.23	17.23	18.23	88.04
7	Penalties	STOP	DUL	STOP	DDL	DDR	1
	Time (Seconds)	18.22	18.19	16.23	17.19	18.14	87.97
8	Penalties	STOP	DUL	LEFT	STOP	DDR	1
	Time (Seconds)	17.23	19.23	18.21	18.12	16.12	88.91
9	Penalties	STOP	DUL	LEFT	DDL	STOP	1
	Time (Seconds)	18.12	16.13	17.23	18.14	16.23	85.85
10	Penalties	STOP	DUL	LEFT	DDL	STOP	1
	Time (Seconds)	16.12	16.11	16.01	15.99	16.22	80.45
11	Penalties	STOP	DUL	LEFT	DDL	DDR	0
	Time (Seconds)	15.14	16.12	15.92	15.14	15.22	77.54

Test Number		Correct Trajectory: DDL	LEFT	DUL	STOP	UP	DUR	RIGHT	DDR	Total Metrics Result in seconds
1	Penalties	RIGHT	LEFT	STOP	DUL	UP	DUR	STOP	RIGHT	5
	Time (Seconds)	24.3	23.22	26.3	22.4	22.12	24.5	23.5	22.3	188.64
2	Penalties	DDL	UP	STOP	STOP	UP	DUR	RIGHT	STOP	3
	Time (Seconds)	21.3	22.12	22.34	22.34	21.17	23.12	21.45	20.19	174.03
3	Penalties	DDL	LEFT	DUL	UP	STOP	DUR	RIGHT	DDR	2
	Time (Seconds)	19.2	18.3	19.9	20.22	19.92	20.12	22.34	20.34	160.34
4	Penalties	DDL	LEFT	DUL	LEFT	STOP	DUR	RIGHT	DDR	2
	Time (Seconds)	18.3	18.2	17.3	19.93	18.85	19.29	21.33	19.17	152.37
5	Penalties	STOP	UP	DUL	STOP	UP	DUR	RIGHT	DDR	2
	Time (Seconds)	18.9	19.3	16.4	18.92	18.12	16.34	17.34	19.33	144.65
6	Penalties	DDL	STOP	DUL	STOP	UP	DUR	RIGHT	DDR	1
	Time (Seconds)	19.12	17.23	16.23	17.23	16.23	16.12	18.12	18.22	138.5
7	Penalties	DDL	LEFT	STOP	UP	UP	DUR	RIGHT	DDR	2
	Time (Seconds)	18.22	18.19	16.23	17.19	16.19	15.13	17.12	16.33	134.6
8	Penalties	DL	LEFT	DUL	STOP	UP	STOP	RIGHT	DDR	1
	Time (Seconds)	16.14	18.12	16.33	17.34	15.13	15.22	15.19	15.98	129.45
9	Penalties	DDL	LEFT	DUL	STOP	UP	DUR	RIGHT	STOP	1
	Time (Seconds)	15.13	15.22	15.19	15.02	15.24	14.93	14.96	15.99	121.68
10	Penalties	DDL	LEFT	DUL	STOP	UP	DUR	UP	DUR	1
	Time (Seconds)	15.12	14.99	14.13	14.39	14.29	14.78	14.99	14.87	117.56
11	Penalties	DDL	LEFT	DUL	STOP	UP	DUR	RIGHT	DDR	0
	Time (Seconds)	14.02	14.13	14.12	14.22	13.34	13.99	14.09	14.08	111.99

(**b**)

- The graphic description of the experiment shows that, as the number of tests increases, the time in which the trajectory is performed decreases, as well as the variation between the time of arrival from one point to another. These results are observed after 11 repetitions.

Figure 19. *Cont.*

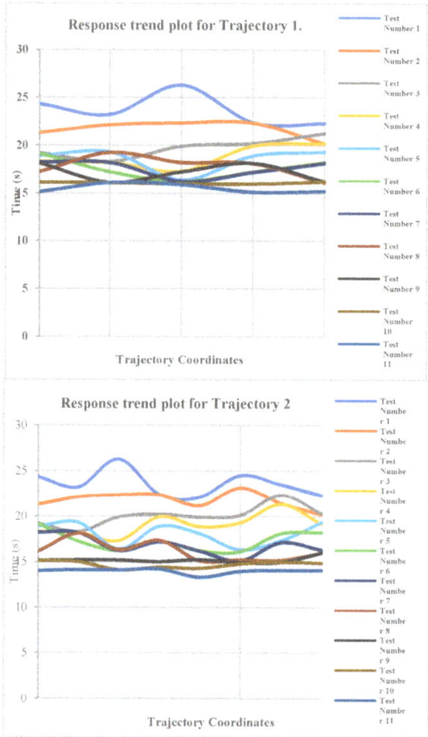

(c)

Figure 19. (**a**) Point-to-point trajectories (Trajectory 1 and Trajectory 2), (**b**) The time and the number of penalties (Trajectory 1 and Trajectory 2) and (**c**) Plot of trend response for each trajectory (Trajectory 1 and Trajectory 2).

A downward trend is observed in this first trajectory in the response time when completing the test with zero penalties. It is shown that 11 repetitions are enough to successfully complete the proposed trajectory. At the beginning, it indicates an initial time of 118.52 s, and, at the end, it indicates an initial time of 77.54 s, which corresponds to a decrease in the response time by 34.58%. In Figure 19a, the second proposed trajectory is indicated. Figure 19b shows the time and the number of penalties for each test performed by the user. Figure 19c indicates a graph of the response time for trajectory 2.

In the second trajectory, a behavior similar to the first trajectory is observed. With 11 repetitions, it is enough to successfully complete the test. At the beginning, an initial time of 188.64 s is indicated, and, at the end, an initial time of 111.99 s is indicated, which corresponds to a decrease in the response time by 40.64%. When performing the test with different points, the same trend is observed in the decrease in response time. By around 11 repetitions, the user has mastery of the HMI. It should be noted that the model is customized for each user according to individual characteristics and muscle contraction time in addition to adding a recognition class for involuntary movements that blocks the operation of the robot and takes it to a "home" state.

5. Conclusions

An HMI that allows for the classification of muscular signals according to the contraction time has been designed. The model implemented through a neural network allows for the personalization and classification in real time for the generation of movement commands of a virtual robot. The HMI can be implemented with inexperienced users who need only 11 repetitions to master the operation of the system, reducing the learning curve. The future work of this project is to implement the classification of multiclass signals in a physical robotic system. In assistive systems or bionic prostheses, although there is the limitation that, being a discrete system, the movement command is determined by a state machine, the improvement consists of implementing neurofuzzy systems that allow for the generation of continuous trajectories in the robot. The development of assistance systems through physiological signals is important for people with disabilities since it allows them to better adapt to their work or personal environment.

Author Contributions: Conceptualization: F.P.-R.; Methodology: F.P.-R. and C.C. Investigation: F.P.-R. Writing—original draft preparation: F.P.-R. Writing—review and editing: F.P.-R. and N.M.-L. Supervision: C.G.-G. and N.F.-V. Project administration: E.L.-N. All authors have read and agreed to the published version of the manuscript.

Funding: This research received no external funding.

Institutional Review Board Statement: The study was conducted in accordance with the Declaration of Helsinki, and approved by Technical Secretary of the internal committee for monitoring and compliance with rationality, discipline and ethics measures of the Universidad del Valle de México, on March 22nd, 2022.

Informed Consent Statement: Ethical review and approval was not required for the study in human participants according to local legislation and institutional requirements. The patients/participants gave their written informed consent to participate in this study. Written informed consent was obtained from the person(s) for the publication of any images or potentially identifiable data included in this article.

Data Availability Statement: All datasets generated for this study are included in the paper.

Conflicts of Interest: The authors declare no conflict of interest.

References

1. Palumbo, A.; Vizza, P.; Calabrese, B.; Ielpo, N. Biopotential Signal Monitoring Systems in Rehabilitation: A Review. *Sensors* **2021**, *21*, 7172. [CrossRef] [PubMed]
2. Laksono, P.; Matsushita, K.; Suhaimi, M.; Kitamura, T.; Njeri, W.; Muguro, J.; Sasaki, M. Mapping Three Electromyography Signals Generated by Human Elbow and Shoulder Movements to Two Degree of Freedom Upper-Limb Robot Control. *Robotics* **2020**, *9*, 83. [CrossRef]
3. Zhu, K.; Guo, W.; Yang, G.; Li, Z.; Wu, H. High-Fidelity Recording of EMG Signals by Multichannel On-Skin Electrode Arrays from Target Muscles for Effective Human–Machine Interfaces. *ACS Appl. Electron. Mater.* **2021**, *3*, 1350–1358. [CrossRef]
4. Aljalal, M.; Ibrahim, S.; Djemal, R.; Ko, W. Comprehensive review on brain-controlled mobile robots and robotic arms based on electroencephalography signals. *Intell. Serv. Robot.* **2020**, *13*, 539–563. [CrossRef]
5. Lu, Z.; Tong, K.-Y.; Shin, H.; Li, S.; Zhou, P. Advanced Myoelectric Control for Robotic Hand-Assisted Training: Outcome from a Stroke Patient. *Front. Neurol.* **2017**, *8*, 8. [CrossRef] [PubMed]
6. Benchabane, S.; Saadia, N.; Ramdane-Cherif, A. Novel algorithm for conventional myocontrol of upper limbs prosthetics. *Biomed. Signal Process. Control* **2020**, *57*, 101791. [CrossRef]
7. Rasool, G.; Iqbal, K.; Bouaynaya, N.; White, G. Real-Time Task Discrimination for Myoelectric Control Employing Task-Specific Muscle Synergies. *IEEE Trans. Neural Syst. Rehabil. Eng.* **2015**, *24*, 98–108. [CrossRef] [PubMed]
8. Karabulut, D.; Ortes, F.; Arslan, Y.Z.; Adli, M.A. Comparative evaluation of EMG signal features for myoelectric controlled human arm prosthetics. *Biocybern. Biomed. Eng.* **2017**, *37*, 326–335. [CrossRef]
9. Hwang, H.-J.; Hahne, J.M.; Müller, K.-R. Real-time robustness evaluation of regression based myoelectric control against arm position change and donning/doffing. *PLoS ONE* **2017**, *12*, e0186318. [CrossRef] [PubMed]
10. Farina, D.; Jiang, N.; Rehbaum, H.; Holobar, A.; Graimann, B.; Dietl, H.; Aszmann, O.C. The Extraction of Neural Information from the Surface EMG for the Control of Upper-Limb Prostheses: Emerging Avenues and Challenges. *IEEE Trans. Neural Syst. Rehabil. Eng.* **2014**, *22*, 797–809. [CrossRef] [PubMed]
11. Scheme, E.; Englehart, K. Electromyogram pattern recognition for control of powered upper-limb prostheses: State of the art and challenges for clinical use. *J. Rehabil. Res. Dev.* **2011**, *48*, 643–659. [CrossRef] [PubMed]
12. Hargrove, L.J.; Scheme, E.J.; Englehart, K.B.; Hudgins, B.S. Multiple Binary Classifications via Linear Discriminant Analysis for Improved Controllability of a Powered Prosthesis. *IEEE Trans. Neural Syst. Rehabil. Eng.* **2010**, *18*, 49–57. [CrossRef] [PubMed]
13. Hahne, J.M.; Graimann, B.; Mueller, K.-R. Spatial Filtering for Robust Myoelectric Control. *IEEE Trans. Biomed. Eng.* **2012**, *59*, 1436–1443. [CrossRef] [PubMed]

14. Vidovic, M.M.-C.; Hwang, H.-J.; Amsuss, S.; Hahne, J.M.; Farina, D.; Muller, K.-R. Improving the Robustness of Myoelectric Pattern Recognition for Upper Limb Prostheses by Covariate Shift Adaptation. *IEEE Trans. Neural Syst. Rehabil. Eng.* **2015**, *24*, 961–970. [CrossRef] [PubMed]
15. Jiang, N.; Tian, L.; Fang, P.; Dai, Y.; Li, G. Motion recognition for simultaneous control of multifunctional transradial prostheses. In Proceedings of the 2013 35th Annual International Conference of the IEEE Engineering in Medicine and Biology Society (EMBC), Osaka, Japan, 3–7 July 2013. [CrossRef]
16. Smith, L.H.; Hargrove, L.J. Comparison of surface and intramuscular EMG pattern recognition for simultaneous wrist/hand motion classification. In Proceedings of the 2013 35th Annual International Conference of the IEEE Engineering in Medicine and Biology Society (EMBC), Osaka, Japan, 3–7 July 2013. [CrossRef]
17. Prahm, C.; Eckstein, K.; Ortiz-Catalan, M.; Dorffner, G.; Kaniusas, E.; Aszmann, O.C. Combining two open source tools for neural computation (BioPatRec and Netlab) improves movement classification for prosthetic control. *BMC Res. Notes* **2016**, *9*, 429. [CrossRef] [PubMed]
18. Abbaspour, S.; Naber, A.; Ortiz-Catalan, M.; GholamHosseini, H.; Lindén, M. Real-Time and Offline Evaluation of Myoelectric Pattern Recognition for the Decoding of Hand Movements. *Sensors* **2021**, *21*, 5677. [CrossRef] [PubMed]
19. Li, X.; Samuel, O.W.; Zhang, X.; Wang, H.; Fang, P.; Li, G. A motion-classification strategy based on sEMG-EEG signal combination for upper-limb amputees. *J. Neuroeng. Rehabil.* **2017**, *14*, 1–13. [CrossRef] [PubMed]
20. Tsai, A.-C.; Hsieh, T.-H.; Luh, J.-J.; Lin, T.-T. A comparison of upper-limb motion pattern recognition using EMG signals during dynamic and isometric muscle contractions. *Biomed. Signal Process. Control* **2014**, *11*, 17–26. [CrossRef]
21. Webster, J.G.; Clark, J.W. *Medical Instrumentation: Application and Design*, 18th ed.; Wiley: New York, NY, USA, 1998.
22. Pérez-Reynoso, F.D.; Rodríguez-Guerrero, L.; Salgado-Ramírez, J.C.; Ortega-Palacios, R. Human–Machine Interface: Multiclass Classification by Machine Learning on 1D EOG Signals for the Control of an Omnidirectional Robot. *Sensors* **2021**, *21*, 5882. [CrossRef] [PubMed]

sensors

MDPI

Article

Evaluating Automatic Body Orientation Detection for Indoor Location from Skeleton Tracking Data to Detect Socially Occupied Spaces Using the Kinect v2, Azure Kinect and Zed 2i †

Violeta Ana Luz Sosa-León * and Angela Schwering

Spatial Intelligence Lab, Institute for Geoinformatics, University of Münster, 48149 Muenster, Germany; schwering@uni-muenster.de

* Correspondence: violetasdev@uni-muenster.de

† This paper is an extended version of our paper "Detecting socially occupied spaces with depth cameras: evaluating location and body orientation as relevant social features". In Proceedings of the 2021 International Conference on Indoor Positioning and Indoor Navigation (IPIN), Lloret del Mar, Spain, 29 November–2 December 2021; pp. 1–8, doi:10.1109/IPIN51156.2021.9662607.

Abstract: Analysing the dynamics in social interactions in indoor spaces entails evaluating spatial–temporal variables from the event, such as location and time. Additionally, social interactions include invisible spaces that we unconsciously acknowledge due to social constraints, e.g., space between people having a conversation with each other. Nevertheless, current sensor arrays focus on detecting the physically occupied spaces from social interactions, i.e., areas inhabited by physically measurable objects. Our goal is to detect the socially occupied spaces, i.e., spaces not physically occupied by subjects and objects but inhabited by the interaction they sustain. We evaluate the social representation of the space structure between two or more active participants, so-called F-Formation for small gatherings. We propose calculating body orientation and location from skeleton joint data sets by integrating depth cameras. The body orientation is derived by integrating the shoulders and spine joint data with head/face rotation data and spatial–temporal information from trajectories. From the physically occupied measurements, we can detect socially occupied spaces. In our user study implementing the system, we compared the capabilities and skeleton tracking datasets from three depth camera sensors, the Kinect v2, Azure Kinect, and Zed 2i. We collected 32 walking patterns for individual and dyad configurations and evaluated the system's accuracy regarding the intended and socially accepted orientations. Experimental results show accuracy above 90% for the Kinect v2, 96% for the Azure Kinect, and 89% for the Zed 2i for assessing socially relevant body orientation. Our algorithm contributes to the anonymous and automated assessment of socially occupied spaces. The depth sensor system is promising in detecting more complex social structures. These findings impact research areas that study group interactions within complex indoor settings.

Keywords: RGB-D sensors; human motion modelling; F-Formation; Kinect v2; Azure Kinect; Zed 2i; socially occupied space

Citation: Sosa-León, V.A.L.; Schwering, A. Evaluating Automatic Body Orientation Detection for Indoor Location from Skeleton Tracking Data to Detect Socially Occupied Spaces Using the Kinect v2, Azure Kinect and Zed 2i. *Sensors* **2022**, *22*, 3798. https://doi.org/10.3390/s22103798

Academic Editors: Tomasz Krzeszowski, Adam Świtoński, Michal Kepski and Carlos Tavares Calafate

Received: 15 March 2022
Accepted: 13 May 2022
Published: 17 May 2022

1. Introduction

While studying how people interact in space, alone or with a companion, the first approximation is to identify variables describing movement and measure them. Specific parameters are straightforward to determine as they can be physically detected in space; for example, the location of people involved in a conversation and their distances can be assessed as properties of physically occupied space. Other aspects describing interactional processes are invisible to the eyes, but still, people inside or outside the group well understand and respect them [1]. While sensors are able to measure the physical properties of people and their location, to date, it is still a challenge to detect their interaction automatically; however, it exists due to social accords in a socially occupied space that is not physically discernible [2]. Sociology studies gatherings of people to identify different roles such as leadership [3], with raised interest to detect them in entertainment to take pictures [4], to help to arrange displays in an interaction encouraging way [5], to improve the communication and design in virtual reality [6], and in computer vision to improve the way robots approach individuals [7,8]. To date, research on identifying interactions among people or between people and their environment often relies on manual observation techniques based on video recording [9]. Other approaches for static scenes analyse videos to detect groups of people automatically by extracting social cues [10]. The distinction between physical and non-physical space is one key unsolved challenge in the automatic interpretation of interactional spaces.

Currently, sensor-based systems focus on spaces physically appropriated by a human body or an object, so-called physically occupied spaces. On the contrary, we aim to detect the socially occupied space, i.e., space occupied not physically by people. Social models such as facing formations, so-called F-Formations, represent this occupancy that occurs due to a social agreement. F-Formations are present when "two or more individuals maintain a spatial and orientational interaction in which the space between them is one with equal, direct and exclusive access" [11]. The model comprises three areas: O-space, the inner transactional space; P-Space, the narrow zone immediate to the O-Space, and R-Space, which protects the system and serves as a transactional space for the participants, as shown in Figure 1. The interactional space is the area in which the interchange occurs, existing between the bodies involved in the exchange [12]. After analysing people's bodies and participation during an interaction, it is then possible to conclude the interactional space. Sociologists have physically distinguished social interaction models by implementing direct observation, interviews, and analysing videos [13,14], concluding that body orientation is crucial in encouraging participation from all members [13]. Nevertheless, the difficulty in detecting these social spaces rises with the number of people, i.e., the size of the gathering; thus, automating the physical features' measurement to describe them is vital. Defining the socially occupied space requires discerning where people stand and their body's direction to detect their interactional space placement.

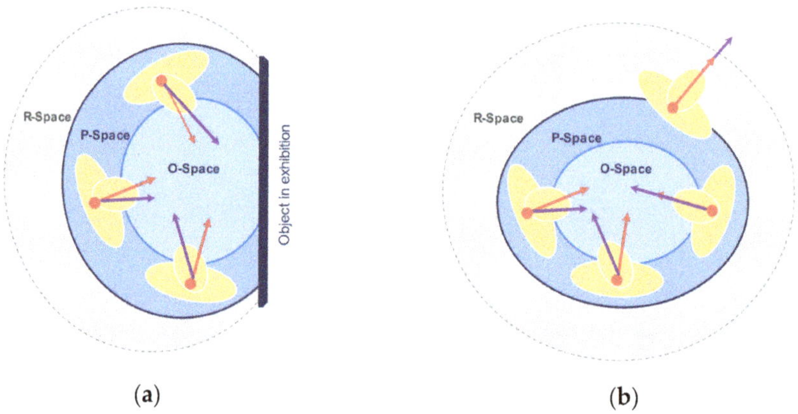

Figure 1. The illustration of the F-Formation model and its three interactional areas are O, P, and R spaces. In (**a**) group–object interaction. In (**b**) group–members interaction.

Moreover, spatial–temporal information is needed to rate the level of engagement in a conversation and describe the encounter's physical dynamics. Tracking technologies such as Bluetooth and Wi-Fi are used to extract position and movement, helping to detect encounter dynamics [14,15]. However, they lack information about bodily signs to identify individuals' interaction with the milieu or everybody else.

Our study concentrates on extracting the data needed to interpret the socially occupied space and defining a methodology to obtain it from different sensor devices. We select a set of depth cameras, with infrared and stereoscopic technology, the Kinect v2, Azure Kinect, and Zed 2i, to tell people's position, body orientation, and viewing direction, which are central in explaining group interaction. Then, we implement the F-Formations model, a social model, by translating the sensors' measurements into an interpretation of the socially occupied space for small size and highly focussed gathering interactions. From the results, we evaluate which device suits better our use case. Our approach does not rely on video storage or trackers' placement, unlike other methods. The emphasis is to assess the different sensor's output data in detecting body orientations. We collect data for eight body orientations in four different walking patterns for each depth camera. The designed algorithm uses the shoulders and spine skeleton joints information collected, together with the trajectories' temporal information, to calculate the body angle. The accuracy evaluation consists of the following methods: evaluating whether the automatic body orientation falls into the correct category with a body orientation category classification, followed by a category deviation analysis, and finally, versus an acceptable social orientation range. Our experiment results show accuracy above 90% for both the Kinect v2 and Zed 2i and 95% for the Azure Kinect for assessing the body angle in the experiment setup, with the different depth sensors' accuracy varying in specific areas for side, back and diagonal body orientations and location to the device. In this paper, our contributions are:

- We compare three different depth sensors to evaluate the use of the skeleton data generated by their depth maps and calculate the body orientation from the skeleton data and assess the sensors' accuracy by analysing the link between location and intended direction. Additionally, we analyse the advantages and disadvantages per device in determining the body orientation.
- We can conclude the spatial extent of the personal interactional space from the body's location and orientation. The focus of attention that intersects allows us to identify people in group interaction and the resulting interactional space.
- We create a system to collect information from physically occupied spaces, analysing the relevant information to interpret socially occupied spaces.

The structure of the paper is as follows: Section 2 introduces related literature to track people and discover groups in indoor spaces. Section 3 depicts the system configuration and our skeleton data processing approach. Sections 4 and 5 describe the experiment setup, the system evaluation, and the discussion of the results for each format and device. Finally, we present in Section 6 our conclusions and future work.

2. Related Work

Social interaction. In analysing human social behaviour, interactions can evolve from a single individual to an increasing number of participants. Social structures such as groups are defined as a social unit with more than one individual and a clear membership that sustain a continuous interaction. Complementary gatherings represent an interaction often in public spaces, defined as a set of two or more co-present individuals sharing a temporal interaction [16]. As the number of individuals decreases, the situation possesses a different level of interaction: large encounters with thirty-one to N participants happen in semi-public and public spaces such as concerts, whereas medium gatherings from seven to thirty participants arise in meetings and classrooms. The larger the number of participants in a gathering, a lower level of common focus exists, whereas lesser members showcase solid social interaction and group belonging [17]. Small gatherings from two to six participants imply a common-focused or jointly focused interaction, where people are involved in a mutual activity [18] encompassing conversational groups, which can be studied within the F-Formation model.

F-Formations exist when "two or more individuals maintain a spatial and orientational interaction in which the space between them is one with equal, direct and exclusive access" [11]. The model comprises three areas: O-space, the inner transactional space in which the focus of attention is present; P-Space, the narrow zone immediate to the O-Space where individuals position themselves; and the R-Space, which protects the system and serves as an entry and exit point for the participants, enclosed by the bodies orientation [19]. The detection of these areas relies on the definition of the *orientational transaction* to assess the intersection of focus of attention in an interactional zone, for which Kendon integrated the concept of social proxemics. Hall defined four physical areas from human observations in social situations: intimate, personal, social and public zones [13]. The interpersonal distance in social zones ranges from 1.20 to 2.10 m. These zones are integrated into the F-Formation model to address its extension and the area in which interactants, and their focus of attention exists as illustrated in Figure 2. The field of view in which the attention spans, is represented by a cone with origin in the frontal body, with a aperture value of around 120°; inside this cone, the inner cone in which humans sustain attention during interaction ranges between 30° and 60° degrees, the so-called gaze area [20]. The different stages of attention can assess the focus during interaction during trajectories [21,22]. During the capture stage, attention is unfocused, and individuals' actions rely on scanning and approaching elements in the environment. Narrow attention arises in the focus stage, where the attention is captured for fewer than three seconds in a single object. Finally, when the attention is deep for more than five seconds, reaching the engagement stage, the bodies are static in a position, and senses are concentrated on reading, discussing, or recalling content, generating a social experience in which the interactional spaces are constructed by the bodies participating in the interchange of information [23]. Our research focuses on small gatherings in indoor spaces, particularly museum exhibitions, by analysing the position and an approximation of the body orientation from static social encounters. The goal is to identify the components of the socially occupied space for highly engaged moments during an interaction.

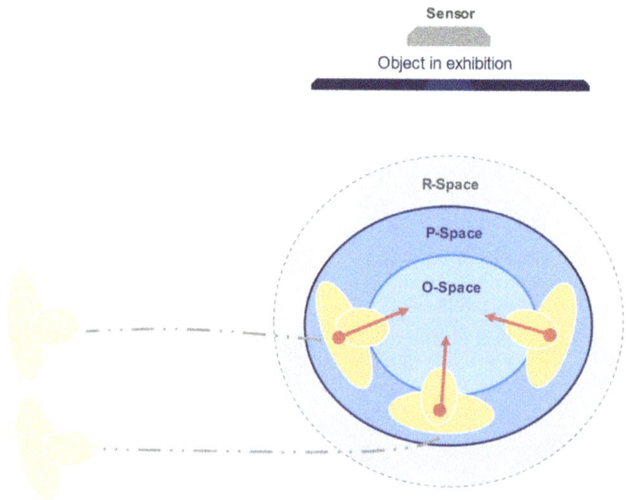

Figure 2. A set of individuals join a third member and construct the interactional space. The position and body orientation establish physically which space is socially occupied. Spatial–temporal variables such as position over time indicate the dynamics of interaction.

Human behaviour tracking approaches. Different studies have been implemented to evaluate individuals' position in interaction in closed spaces [15,24]. For our analysis, these human tracking technologies can be divided into devices with and without physical contact with the user. The first category is unobtrusive because its installation is in the surroundings. For example, Wi-Fi and Bluetooth technologies help identify device interaction and location. However, its utility is limited to positional variables. It does not directly assess body data to analyse social interactions, giving partial information about people's spatial and body arrangement. They need additional data such as video recordings or manual records [25]. LiDAR cameras in museums have similar limitations for measuring social interactions with exhibitions, requiring significant processing tasks to get precise trajectories, deriving information mainly about highly concurred areas [26]. Finally, computer vision techniques to identify human traffic rely on RGB cameras to detect people's bodies and derive their trajectories, incurring privacy concerns and challenges such as occlusion and distortion. The second category is obtrusive as it is installed in peoples' bodies in the form of trackers and markers, interfering with their activities' natural behaviour, especially when users are required to activate beacons to confirm their locations [27]. So far, these technologies focus mainly on spatial data, offering only proximity information to identify groups according to their shared space.

Nevertheless, due to the richness in human interactions, trajectories need to be complemented with relevant data to characterise the interactional space and offer more context for the sociological analysis of models such as F-Formations [9]. The description of social features in group interaction has been studied in museum visits using forms and manual observation, implying expensive and lengthy analysis [23]. Additional techniques involve using cameras to design traditional and interactive displays in closed spaces, reducing the socially occupied space to an area to be physically occupied [28]. Existing computer vision methods use video datasets such as SALSA and Babble and identify attention, proximity, and head orientation to analyse participants in a conversation with the analysis of bodies from video recordings, highlighting the difficulty of the analysis of head rotation as a result of a low-resolution video [29,30]. Other similar studies rely on virtual environments to recreate social dynamics [31]. These approaches focus solely on detecting conversation's physically occupied space, ignoring the surrounding socially occupied space dynamics that led to these groups' construction.

Depth cameras for human interaction. Several depth camera models are available outside the industrial market, such as the Orbbec and the Intel RealSense models used for 3D image extraction, depth map reconstruction and gait analysis [32–34]. Each device provides different software solutions to process scene information, such as semantic segmentation, object detection and skeleton tracking, open to the public or with a fee, as shown in Table 1. Cameras with the skeleton tracking functionality ready to be used without cost for researchers and practitioners in their studies include the Microsoft Kinect series and the Zed 2i.

Table 1. Depth cameras model availability with and without integrated skeleton tracking.

Device	Technique	Range	Skeleton Tracking
Azure Kinect	TOF	0.25–5.46 m	Yes, included
Kinect v2	TOF	0.50–4.50 m	Yes, included
pmd CarmBoard pico monstar	TOF	0.50–6.00 m	No
Intel Realsense D435i	Stereovision	0.30–3.00 m	Yes, to pay for
Intel Realsense D455	Stereovision	0.60–8.00 m	Yes, to pay for
Stereolabs Zed 2i	Stereovision	0.20–20.0 m	Yes, included
Orbbec Astra	Structured Light	0.60–8.00 m	Yes, to pay for
Orbbec Astra Pro Persee	Structured Light	0.40–8.00 m	Yes, to pay for

Approaches using commercial depth cameras include the Kinect v2 camera in an egocentric perspective in robots for conversational participation and events, limiting the analysis to static scenarios evaluating only the physically occupied space by interacting with the artificial participants [35,36]. However, this use demonstrates their great potential in acquiring trajectory and relevant social features due to the processed skeleton data, easiness of installation, and low costs without storing video data from the scene, allowing researchers to exploit these data to extract human behaviour [37–39]. Additional depth cameras available for the public include the Azure Kinect, the successor of the Kinect v2, mainly used for industry and healthcare with promising human activity detection [40]. Studies comparing both devices are limited to joint detection accuracy for medical monitoring, static scenarios, or physical training that does not reflect the natural movement of the body in large trajectories [41,42]. An alternative depth camera to the Time-of-Flight technology from the Microsoft devices is the Zed 2i from Stereolabs, which relies on stereoscopic technology to gather depth information and extract body joints. Nevertheless, only selected studies are available to evaluate the depth map accuracy from the previous model [32,43], and research on the skeleton joints accuracy model is scarce.

This study intends to extend these prior studies by integrating social signals, trajectories, and human behaviour. Social signals describe a set of behavioural attitudes from social intelligence present during interaction [44]. The posture and gesture category highlights the relevance of low-level social features: distance, aperture, and body orientation to assess interaction [6,45]. We use depth cameras as a hybrid technology for tracking individuals and collecting body data during trajectories to automate detecting the invisible space in interaction. To assess which depth camera technology and model is more suitable for detecting socially occupied spaces, we compare the skeleton tracking data generated by two infrared-based and one stereoscopic-based depth camera.

3. System Design

We propose employing depth sensors cameras to extract the body orientation using skeleton tracking data joints, with a series of evaluations assessing the efficacy of detecting F-Formations exploiting spatial–temporal and body cues data. Our methodology comprises five steps: a set of data collection experiments, a coordinates transformation and social cues processing, the estimation of the body angle orientation and evaluation, and finally we test our findings with a group detection algorithm. We process the shoulder left, right, and centre joints, as described in Figure 3, to calculate body angle orientation and the spine joint's coordinates as the position for each skeleton dataset generated collected per device: two infrared-based and one stereoscopic-based technology.

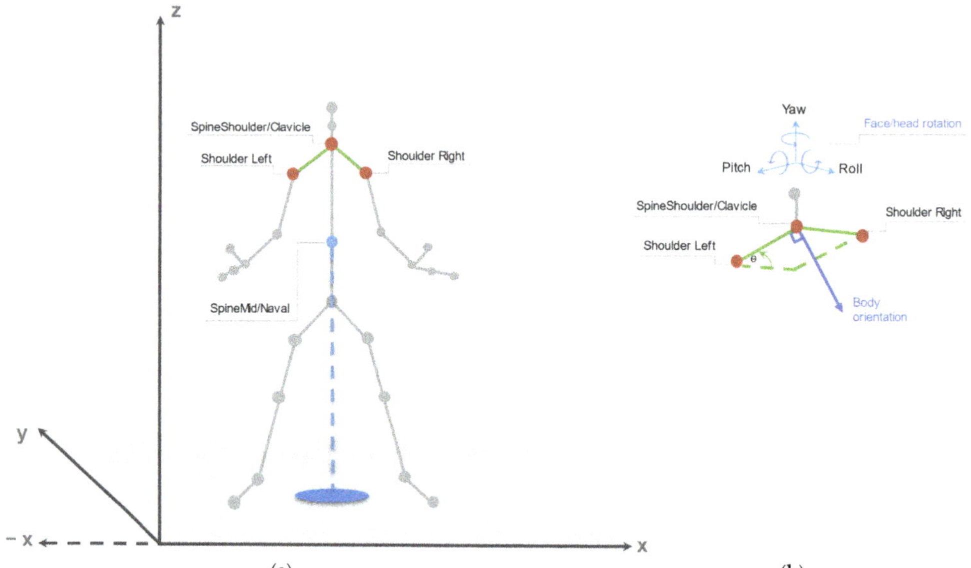

Figure 3. Real−world coordinate system with the skeleton extracted from the depth cameras. On (**a**) the selected skeleton joints are in red, with the positional skeleton joint in blue. On (**b**) the selected upper skeleton joints used to calculate body orientation.

3.1. Depth Sensor Cameras

We selected three depth sensor cameras that are reachable to end-users in terms of the price, market availability to the public, capability to generate skeleton tracking data, easiness of installation and running within different environments. From Microsoft, the Kinect v2 and the Azure Kinect offer end-users a device with capabilities ranging from games to industrial use using Time-of-Flight (TOF) technology. Alternatively, Stereolabs with the Zed 2i offers a stereoscopic camera whose size and configuration make it a good option for developers in robotics and industry.

3.1.1. Kinect v2

Microsoft launched the Kinect v2 in 2016 as an accessory for the XBOX console to track a body's movements for video games. The device extracts the scene depth information by processing the incoming light using an infrared and RGB-D video. The device detects 25 body joints per skeleton for up to six bodies using a set of decision tree-based algorithms with no native information for the head/face elements. The Microsoft Kinect Software Development Kit (SDK) allows to access the device, basic tutorials and depict the camera status, limited to Windows operating systems versions higher than 8. Additionally, the libraries can be implemented in WPF C# projects to access its functionalities [46], adding others, such as a complementary face elements detection, including eyes, mouth and head from the Microsoft.Kinect.Face library.

3.1.2. Azure Kinect

The next generation of Kinect devices came in 2020 with the introduction of the Azure Kinect. The Azure focuses on industrial warehousing, robotics, and health applications compared to the previous generation [37]. The Azure camera uses the highest hardware specification requirements from the three devices. The depth camera implements an amplitude-modulated continuous Wave Time-of-Flight principle, casting illumination in the near-IR spectrum to record the light travelled. The skeleton tracking feature includes 32 body joints including face elements for up to four bodies, employing a neural network algorithm to derive the skeleton bodies from the depth map. Users can access the camera functionalities with the SDK and libraries written in C++ and C# on Windows and Linux operative systems, possibly connecting to Azure Cognitive Services for other processes.

3.1.3. Zed 2i

Stereolabs Zed 2i depth camera has been available in the market since 2021 and is based on stereoscopic technology by using two 4Mpx sensors, calculating the displacement of the pixels between the left and the right images captured. The body tracking is based on a neural network algorithm to detect body joints present on both sides, and it merges the information with the depth and positional tracking model, producing 34 body joints, including face components. The camera requires configuring CUDA and a Zed-specific development environment to access the SDK functionalities, which work on Windows and Linux operative systems [47].

The body skeleton joints structure and the coordinate system for each camera are shown in detail in Figure 4, with a summary and technical comparison in Table 2. The Azure Kinect and the Zed 2i have the highest number of joints, including face and spine ones with slightly different names.

(**a**) (**b**) (**c**)

Figure 4. Skeleton joints map per device. (**a**) Kinect v2, (**b**) Azure Kinect and (**c**) Zed 2i. Greys areas indicate a left-right joint correspondence. Italic joints indicate differences between the devices.

Table 2. Detailed technical comparison of the selected depth cameras.

	Kinect v2	Azure Kinect	Zed 2i
Year	2016	2020	2020
Technology	TOF	TOF	Stereovision
Colour camera resolution	1920 × 1080 px @30 fps	4096 × 3072 @30 fps	2× (2208 × 1242) @15 fps 2× (1920 × 1080) @30 fps 2× (1280 × 720) @60 fps
Depth camera resolution	512 × 424 px @30 fps	Narrow: 654 × 576 @30 fps Wide: 1024 × 1024 @30 fps	
Field of view	70° H–60° V	Narrow: 75° H–65° V Wide: 120° H–129° V	110° H–70° V
Depth extent	0.5 m–4.5 m	0.25 m–5.46 m	0.2 m–20 m
Coding language	C#	C, C#	C, C++, Python
Skeleton joints	25	32	34

3.2. Depth Cameras Coordinates' Transformation

We transform the device's coordinates into the physical space. The devices generate data relative to their positions in their own coordinate system, with x representing the positive or negative horizontal distance, y the height, and z the frontal distance, as shown in Figure 5.

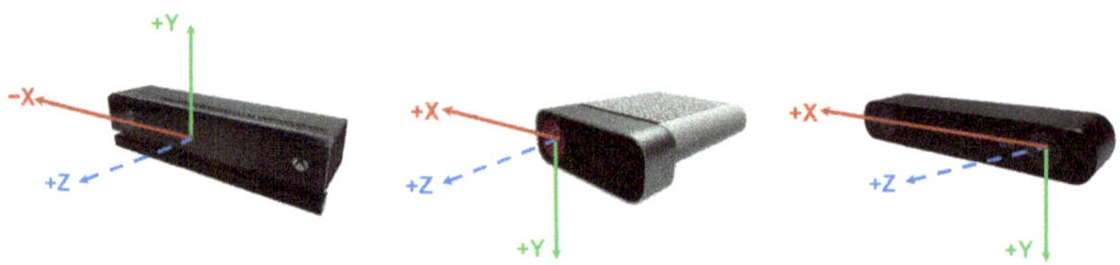

(**a**) (**b**) (**c**)

Figure 5. Depth camera sensors with their coordinate system: (**a**) Kinect v2, (**b**) Azure Kinect, and (**c**) Zed 2i. The Zed 2i has six different coordinate systems.

The coordinates system origin x_o, y_o are set to the device position, calculating the transformed coordinates x_t, y_t, with a translation angle θ, and the camera's original set of coordinates $[x_k, y_k, z_k]$ for each skeleton joint. We apply a matrix product transformation described in Equation (1):

$$[x_t, y_t] = \begin{bmatrix} \cos\theta & -\sin\theta & x_o \\ \sin\theta & \cos\theta & y_o \\ 0 & 0 & 1 \end{bmatrix} \begin{bmatrix} x_k \\ y_k \\ 1 \end{bmatrix} \tag{1}$$

3.3. Processing Social Signals

Humans regularly recognise the direction in which the body is oriented by identifying the head orientation and the upper and lower limbs and reviewing if they are placed left or right. Following this reasoning, the body orientation is calculated from the skeleton joints tracked by each device. The data generated by each sensor in the form of JSON files contain information about every set of joint coordinates, as illustrated in Figure 3. We collect each skeleton data every 200 milliseconds, using the upper limbs as an indicator of the focus of attention to later calculate the body angle. The main methodology is illustrated in Figure 6. The task starts by *processing the input data*. We receive the collected skeleton data in a JSON file, which is analysed and organised by timestamp, creating a dataset with all relevant information for the algorithm, such as body identification, skeleton joints, timestamp, body location, camera, and experiment identifier. Next, the program proceeds with the coordinates' transformation for the body location and each skeleton joint.

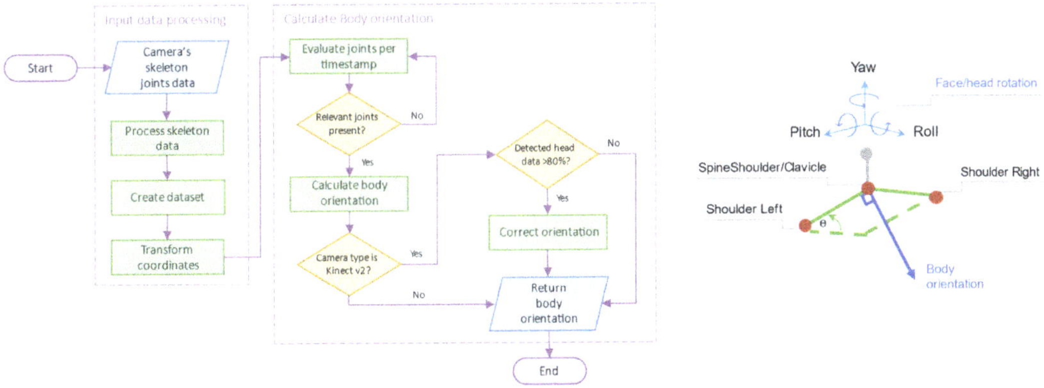

(a) (b)

Figure 6. On (**a**) Methodology to extract body orientation from the skeleton data collected. On (**b**) illustration of upper joints and head rotation data usage to calculate the body orientation.

Once the data is processed, the next step is to *calculate the body orientation*. For each position p_i, we review the availability of the interested upper joints to proceed with the angle calculations. This allows us to warranty the use of complete skeleton data sets as they can be incomplete every other timestamp. The algorithm evaluates if the relevant skeleton joints are present on each timestamp, selecting the upper joints shoulder left and right for all devices, shoulder centre and head orientation detected for the Kinect v2, and clavicle for the Azure Kinect and the Zed 2i. We use Equation (2) to calculate the body angle orientation after applying the corresponding coordinate transformation using Equation (1):

$$\theta_{body\_angle} = \arctan\left(\frac{x_t}{y_t}\right) * \frac{180}{\pi} \tag{2}$$

To assess if the vector perpendicular to the orientation vector should rotate clockwise or counter-clockwise, we evaluate the position of each shoulder joint relative to the sensor's orientation. If the left shoulder joint in the position sl_x is larger than the right shoulder joint position sr_x, the body is looking in the direction of the sensor. This can be observed by plotting the shoulder line and the joint's position as illustrated in Figure 6b.

Lastly, if the camera is the Kinect v2 and the head elements are available for more than 80% of the sample, we apply an additional correction to the orientation using a full 180° rotation as described in Algorithm 1. While Azure Kinect and Zed 2i include information on the face and head joints in detecting skeletal joins related to body orientation, Kinect v2 processes skeletal joints and face and head joints separately. As a result, front and back orientation are often confused, and the device has a strong tendency always to assume a front orientation, even if people are oriented backwards. To make all three approaches comparable, we integrate the face and head joins explicitly into the body orientation processing in the case of Kinect v2, where information from the head is processed. This operation potentially improves assessing the body orientation towards the camera as its absence suggests a non-frontal orientation [2]. On the other hand, the Azure Kinect and Zed 2i offer the face and head joints detected by the algorithm, but as they are already processed to evaluate the joints' left-right correspondence internally, they are not included in the body orientation correction to avoid overfitting. In the end, the calculated angle θ_{pi} is returned. Algorithm 1 describes the mentioned process:

Algorithm 1 Body angle calculation

Input: A dataset N with skeleton joints in the form (x, y) per timestamp
Output: Body orientation angle θ_{srsl}
  for p_i in N:
    if shoulder_joint_pair:
      apply_coordinates_transformation(sl_{xy}, sr_{xy})
      $\theta_{pi} = \tan^{-1}\left(\frac{x_t}{y_t}\right) * \left(\frac{180^\circ}{\pi}\right)$
      if $sr_x < sl_x$:
        $\theta_{pi} = \theta_{pi} - 90^\circ$
      else:
        $\theta_{pi} = \theta_{pi} + 90^\circ$
    else:
      $\theta_{pi} = 1$
    correction_level = analyze_head_data_availability(p_i)
    if camera is Kinect_v2 **and** correction_level > 80%:
      apply_orientation_correction (θ_{pi})
    return θ_{pi}
  end

3.4. An F-Formation Social Model for Group Detection

We integrate the F-Formations model for processing the physically occupied spaces to detect socially occupied spaces. From Kendon's theory of F-Formations [11], the attributes to be extracted from the physical space are related to proximity, spatial–temporal data such as position and time, and focus of attention. We extract Kendo's model attributes and proceed to identify shared stops among subjects in the conducted experiments. Hall defined proxemics with a range of 0.5 to 1.5 m distance between bodies as the personal space for two or more individuals coexisting in a continuous lapse of time [13] and a common focus of attention, to a person or an object, as the intersection of field of views [48].

We develop a group detection algorithm that uses the bodies' trajectories and the calculated body angle on every timestamp as described in Algorithm 2.

Algorithm 2 Group detection

Input: A dataset N with skeleton joints in the form (x, y) per timestamp
an integer *stops_expected* with the number of assigned stops
an integer *groups_expected* representing the number of assigned group locations
Output: A dataset *class_group* with group membership, and stop locations

```
body_identification=spatemp_stop_kmeans(N, stops_expected)
for b_i in body_indentification:
        trajectory_stops = spatemp_stop(b_i, t, r)
shared_stops=intersection_stops(trajectory_stops)
class_group=spatemp_stop_kmeans_time(shared_stops, groups_expected)
return class_group
```

Firstly, we assign a body identifier by applying a spatial K-Means supervised classification algorithm with *stops_expected* parameter, the expected number of members in the scene. Next, we process the trajectory for each skeleton body b_i, and detect individual stopping moments by evaluating the spine joints' temporal and spatial proximity within a radius r and stop time t, generating the *trajectory_stops*. With the individual bodies' long stop detected, we proceed to extract the individual stops intersected, assessing their coexistence in a maximum of 1.5 m personal space. Once the *shared_stops* in the trajectory are extracted, we apply a temporal K-Means supervised classification algorithm to evaluate group temporality with the parameter *groups_expected*, obtaining each *class_group* to assign group membership to each skeleton body. The focus of attention and its intersection is visualised by integrating the body angle calculation results and generating the body's field of view. With this information, is possible to draw the F-Formation model components and thus the socially occupied space during the participants' interaction.

In brief, our system employs three different depth sensor cameras to collect skeleton data during trajectories, from which we can extract the position and orientation of every participant. Once the skeleton joints data is collected and available, we apply a coordinates transformation to have a unified coordinate system as each device possesses its own. Secondly, we calculate the body orientation angle based on three skeleton joints: shoulders (left and right) and shoulder centre. Then, we proceed with an angle correction for the Kinect v2 to adjust the results to the same level as the other devices for a fairer skeleton joint algorithm comparison. To identify the most reliable camera for detecting socially occupied spaces, we assess the results with a set of performance evaluations. Finally, to probe the use of this approach, the attributes extracted from the physically occupied space are exploited to identify when group members are sharing an interactional space and focus of attention, thus the construction of an F-Formation.

4. Body Orientation Angle Evaluation

This section describes the experimental setup for two different configurations and shows the evaluation results, demonstrating the sensors' body orientation accuracy.

4.1. Data and Software Availability

The data collected during these studies are available at https://osf.io/xhwgm/ (accessed on 15 March 2022). The tutorials and code to create a new interface for all devices can be found at https://github.com/violetasdev/bodytrackingdepth_course (accessed on 15 March 2022). For the Azure Kinect we implement a modified version of k4.net, the final version is available at https://github.com/violetasdev/k4a.net (accessed on 15 March 2022). The Kinect v2 body orientation plots can be reproduced at https://osf.io/ghz79/ (accessed on 15 March 2022).

4.2. Experiment Setup

The sensors are positioned in an isolated 4.0 m × 9.0 m area, over a 2.0 m vertical truss with a height of 1.83 m, pointing towards a white wall. In the separated free floor area, coloured feet are placed every 1.2 m to cover each sensor's field of view and guide participants to draw different walking patterns. Regarding the equipment configuration, the Kinect v2 sensor is connected to an Intel Core i7-10 laptop, with 16 GB of DDR4 RAM and a NVIDIA GeForce RTX 3070 Super Max-Q graphics card. The Zed 2i is connected to a laptop with Intel Core i9-11, 32 GB of RAM, and an NVIDIA GeForce RTX 3080 graphics card. The Azure Kinect is connected to an Intel Core i7-10 laptop with 16 GB of RAM and an NVIDIA GeForce GTX 1650 graphics card.

We produce the JSON files containing the skeleton joints data from three different coded solutions. The libraries implemented are in C# for the Kinect v2 and the Azure Kinect devices and in Python for the Zed 2i. A video camera records the computer screens for each trajectory to further review specific timestamps from the scene in search of external factors affecting the experiment. The skeleton data are collected every 200 milliseconds and the exact same scene setup for all devices, with the same starting time and bodies entering the scene simultaneously for the Azure Kinect and the Zed 2i. The resulting Kinect v2 skeleton data is taken from our previous data collection with the same configuration [2]. We use the narrow view configuration for the Azure Kinect for the skeleton tracking algorithm recommended by the fabricant due to the performance results [49]. For the Zed 2i, we select the COORDINATE_SYSTEM_IMAGE as the coordinate system to match all devices [50].

4.3. Participants' Description

Two participants perform the oriented-walking patterns in a single and dyad configuration. First, one female with a height of 1.73 m follows each assigned pattern and body orientation for a total of 32 samples. Secondly, a dyad (a group composed of two participants) with a female and a male with a similar height ranging between 1.73 m and 1.81 m concludes an equal task while keeping a side-to-side configuration.

4.4. Walking Trajectories and Body Orientations Definition

Experiment members are asked to walk in the isolated area in a combined walking pattern and body orientation in front of the camera for one minute per trajectory. For each body orientation representing back, frontal, diagonal, and side orientations, as detailed in Figure 7, four walking patterns should be completed from bottom to top and left to the right direction, as indicated in Figure 8, completing approximately 15 m per trajectory.

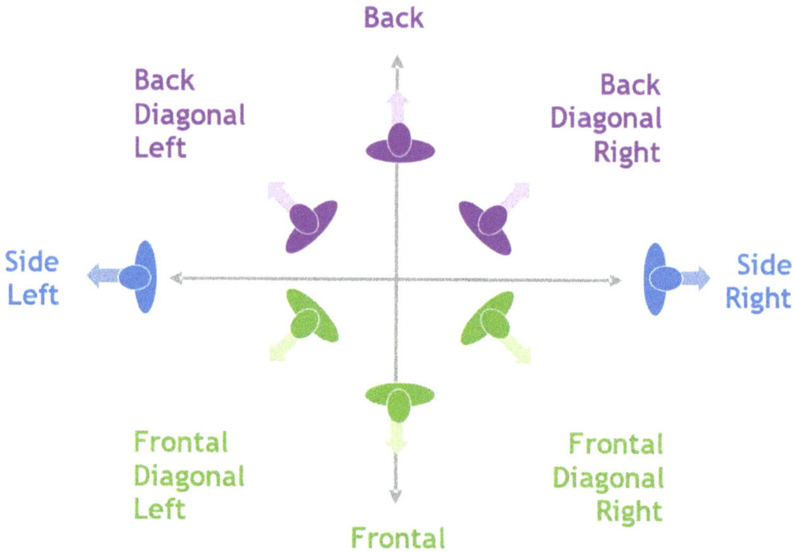

Figure 7. Body orientation categories.

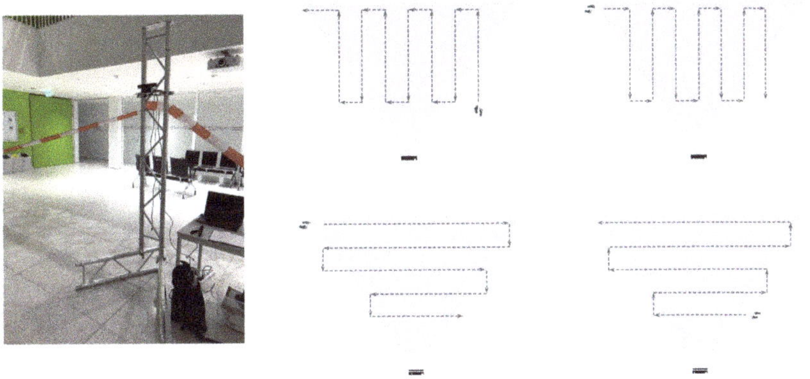

(a) (b)

Figure 8. Experiment set up for all devices: in (**a**) the experiment arrangement; in (**b**) the different walking patterns with the start point and the camera's position and field of view.

We classify the calculated body orientation angle into eight categories in slices of 45° and define an acceptable angle range for each category as displayed in Table 3 to evaluate each calculated body angle. The classification algorithm assigns the category labels with a 100% correspondence to an orientation. The accepted angle range is used to evaluate the margin error in calculating the body angle to identify how much is deviated from the expected result.

Table 3. Description of the body orientations label with the intended and defined acceptable angle range.

Body Orientation Label	Intended Orientation Angle	Accepted Angle Range
Side right	0°	[−22.5°, 0°), [0°, 22.5°)
Back diagonal right	45°	[22.5°, 67.5°)
Back	90°	[67.5°, 112.5°)
Back diagonal left	135°	[112.5°, 157.5°)
Side left	−180°/180°	[157.5°, 180°), (−180°, 157.5°]
Frontal diagonal left	−135°	[−157.5°, −112.5°)
Frontal	−90°	[−112.5°, −67.5°)
Frontal diagonal right	−45°	[−67.5°, −22.5°)
Back diagonal left	0°	[−22.5°, 0°), [0°, 22.5°)

4.5. *Evaluation and Results*

We assess the calculated body angles accuracy in three stages: first, we evaluate whether the automatically detected body orientation falls into the correct category (i.e., the body angle with which the participant walked the experiment). The second evaluation aims to shed light on the accuracy, i.e., how large is the error, particularly for those automatically detected body orientations that did not fall into the correct category. Finally, the third evaluation addresses the context of social interaction in which we assess if the automatically detected body orientation falls into the maximum range of 30° for sustaining social interaction. Due to the findings of outliers in the experiment, we apply an interpolation correction by analysing the socially acceptable angle range and the nature of the outlier, showing the corrected body orientation's angle results. For this evaluation, we compare the Kinect v2 results from our previous work presented at the IPIN 2021 conference [2].

4.5.1. Intended Body Orientation Category Range

The evaluation compares the computed body orientation angle against the acceptable range for the intended orientation defined in Table 3. Figure 8a,b shows each device's corresponding precision and recall for the single configuration. For the Kinect v2, the precision and recall are 0.82, respectively, with back diagonal and side orientations as the least accurate orientations. For the Azure Kinect, the precision and recall are 0.87, respectively, with the back orientation as the weakest. Finally, the Zed 2i possesses precision and recall of 0.83, with back diagonal right, back, and frontal diagonal left orientations with the lowest accuracy.

For the dyad configuration, the precision and recall using the Kinect v2 are 0.79 and 0.80, where back diagonal and side categories have the lowest precision, as shown in Figure 9c,d. The Azure Kinect shows a general precision and recall of 0.81 up to 0.9 for frontal diagonal right, with frontal and back orientations precision below 0.70. The Zed 2i device orientations back, frontal, and back diagonal left show a low precision and recall lower than 0.77 with stronger orientations above 0.80 precision and recall up to 0.94. The most accurate orientations for the Kinect v2 benefit from the availability of the head rotation detection feature. In general, for the Azure Kinect and the Zed 2i, the back orientation is the most challenging orientation to detect, but despite not having the head/face rotation data available, the precision and accuracy results are higher than in the Kinect v2. For the Zed 2i, from the video review, we identified difficulties in detecting both participants, as one person was missing at a time, especially near the borders from the field of view.

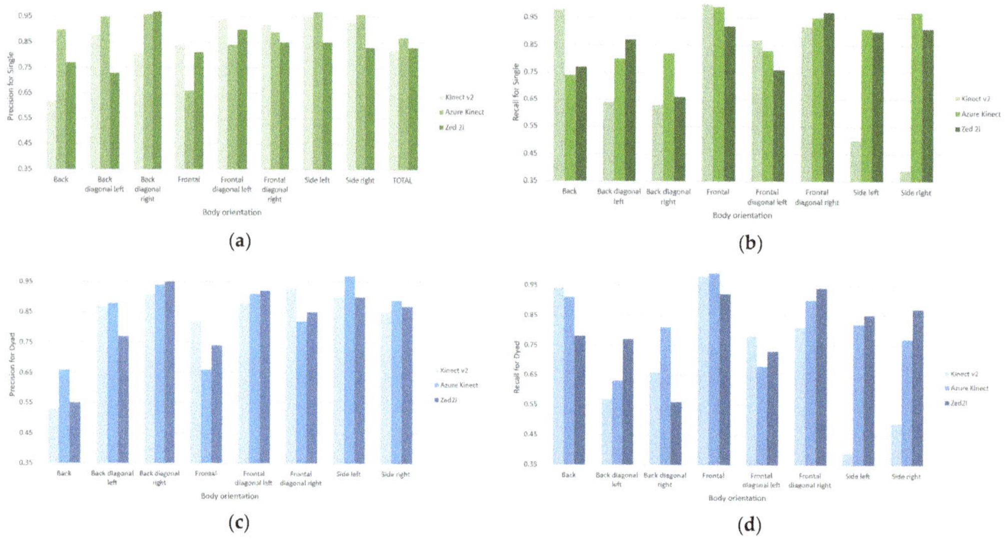

Figure 9. Precision and recall results for body orientation assessment for compiled measurements per device for single configuration in (**a**) and (**b**) and dyad configuration in (**c**) and (**d**), respectively.

4.5.2. Intended Orientation Angle Deviation

For each category label, Tables 4 and 5 show the angle deviation in degrees regarding the intended orientation in which the participants recreated the walking patterns. In general, for all three devices, for body orientations parallel to the depth camera, the average error is low for single and dyad, rising in diagonal body orientation categories. The most significant average error for the Kinect v2 and Zed 2i is the side orientations due to their orthogonality and the back orientation for the Azure Kinect. For the Kinect v2, the side-left and side-right high error results show difficulty collecting accurate skeleton data when the bodies possess a side orientation concerning the sensor's position. The standard deviation in the single and dyad configurations does not surpass all devices' next neighbouring category label. Exceptions are for the single configuration, the side-left and side-right orientations for Kinect v2 and Zed 2i with an error up to 1.5 adjacent classes. In the dyad configuration, the side-right orientation for the Kinect v2 deviates 1.5 adjacent classes. The back orientation deviates three adjacent classes for the Azure Kinect, resulting in the correct frontal orientation and two adjacent classes for the back diagonal right orientation.

Table 4. Single Configuration: Evaluation of body intended orientation angle deviation (IOD) intended orientation angle (IO) in degrees. Bold numbers indicate the highest values.

Body Orientation	Kinect v2		Azure Kinect		Zed 2i	
	IOD AVG	IOD STD	IOD AVG	IOD STD	IOD AVG	IOD STD
Back	6.85	6.51	**30.51**	**58.84**	18.18	**24.12**
Back diagonal left	19.54	11.81	15.72	**29.60**	14.64	22.70
Back diagonal right	21.74	15.82	22.73	**43.11**	21.71	21.45
Frontal	4.81	4.12	5.51	5.01	10.27	7.66
Frontal diagonal left	13.61	8.83	15.20	13.57	18.69	22.32
Frontal diagonal right	11.98	6.95	11.06	16.76	9.45	10.68
Side left	**34.81**	**34.46**	19.78	13.67	**35.29**	10.65
Side right	**35.62**	**31.55**	8.19	6.30	12.38	10.97

Table 5. Dyad Configuration: Evaluation of body intended orientation angle deviation (IOD) against intended orientation angle (IO) in degrees. Bold numbers indicate the highest values.

Body Orientation	Kinect v2		Azure Kinect		Zed 2i	
	IOD AVG	IOD STD	IOD AVG	IOD STD	IOD AVG	IOD STD
Back	8.37	8.37	13.39	**26.62**	15.883	17.499
Back diagonal left	**22.96**	**22.96**	19.53	16.08	15.975	14.067
Back diagonal right	20.35	20.35	14.53	20.39	23.730	18.311
Frontal	6.53	6.53	6.50	5.26	9.690	9.121
Frontal diagonal left	14.8	14.8	17.47	12.84	17.610	19.293
Frontal diagonal right	13.84	13.84	11.58	7.35	9.558	7.610
Side left	**28.1**	**28.1**	**23.08**	18.97	**26.411**	**24.267**
Side right	**36.93**	**36.93**	15.53	10.45	15.005	22.984

The body orientation and the followed patterns are illustrated in Figure 10 for a highly accurate detected body orientation and a low accurate one for all three devices. The Kinect v2 depth camera extracts the head/face rotation and skeleton joints data, and the body disappears once it crosses the camera's centre, recovering the body in a flipped orientation shortly after, as illustrated in Figure 10b. The Azure Kinect shows the designated patterns until the bodies reach the camera's field of view limits in Figure 10c,d. For the Zed 2i, as it has a larger field of view than the other devices, is it possible to continue tracking the participants with difficulties in closer measurements and drawing the distance between each line inconsistently.

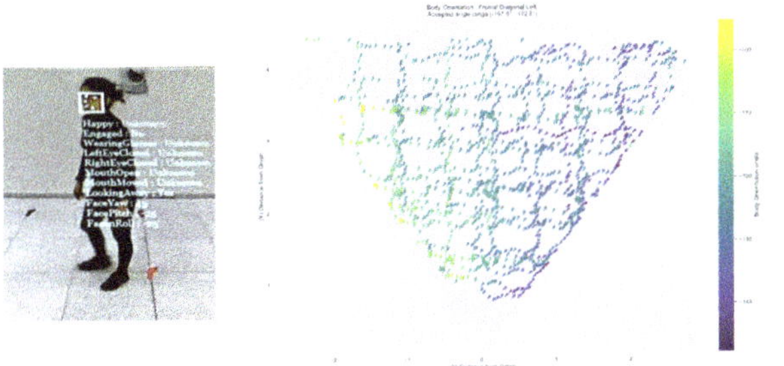

(**a**) Kinect v2: Frontal diagonal left

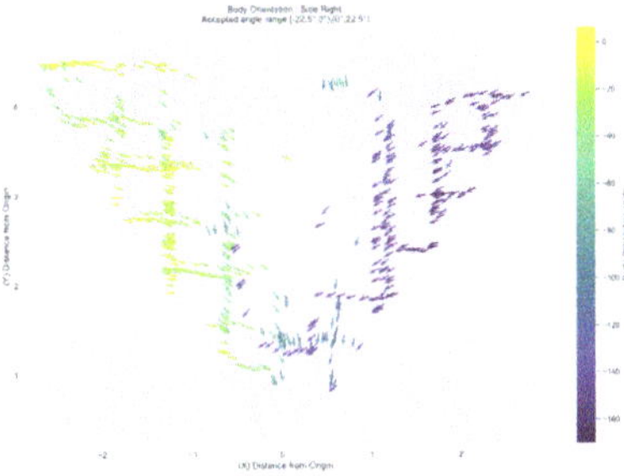

(**b**) Kinect v2: Side right

Figure 10. *Cont.*

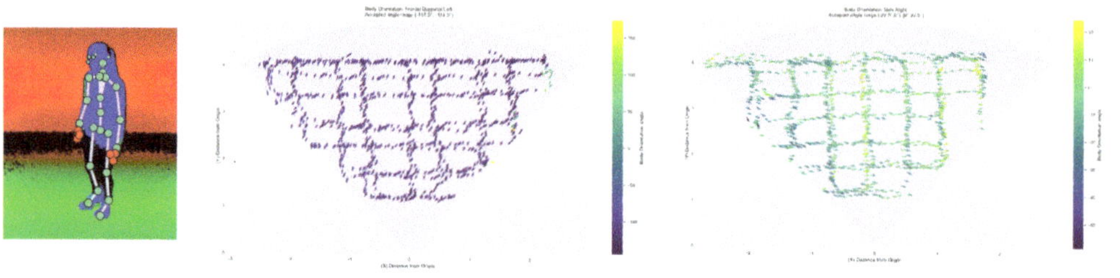

(**c**) Azure Kinect: Frontal diagonal left

(**d**) Azure Kinect: Side right

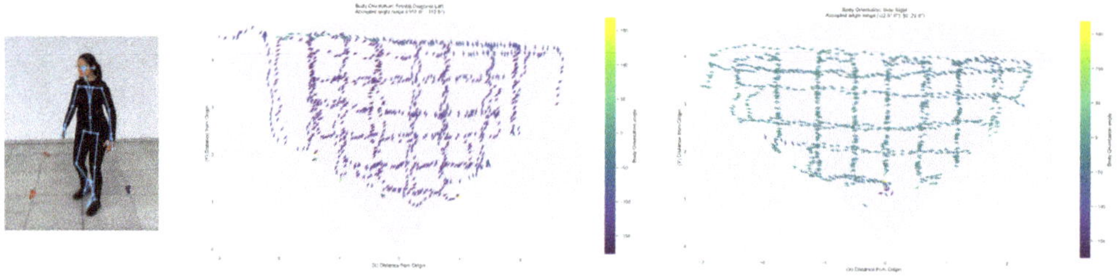

(**e**) Zed 2i: Frontal diagonal left

(**f**) Zed 2i: Side right

Figure 10. Calculated body orientation angles per sensor; (**a**,**c**,**e**) are highly accurate detected Frontal Diagonal orientation with participant view on the left; (**b**,**d**,**f**) show the detected Side Right orientation with the lowest accuracy for the Kinect v2.

4.5.3. Intended Orientation inside the Interactional Space

Because we need the field-of-view extension, we evaluate the body orientation angle to assess the group's focus of attention during an interaction. From the extension of the body orientation, we can define the focus of attention of each participant, and the intersection suggests a shared object of interest. In our case, we extend the participant's field of view to the sides, drawing a 30° cone, and Figure 11 shows the results for the calculated angles classification within the interactional range. The interactional angle is detected around 80% of the time for most categories in all devices for single and dyad configurations. The socially acceptable orientation availability for the Kinect v2 in single configurations is low for side and back diagonal orientations due to the absence of joints and self-occlusion, improving in the dyad configurations by almost 10%. The Azure Kinect has the highest availability, with a socially acceptable range from 94% in the single configuration and 87% in the dyad configuration, up to 100% in both scenarios. The Zed 2i have comparable results to the Azure Kinect, ranging from 85% availability in single configurations and 73% in dyad configurations, with weaknesses in back diagonal right orientations for both configurations.

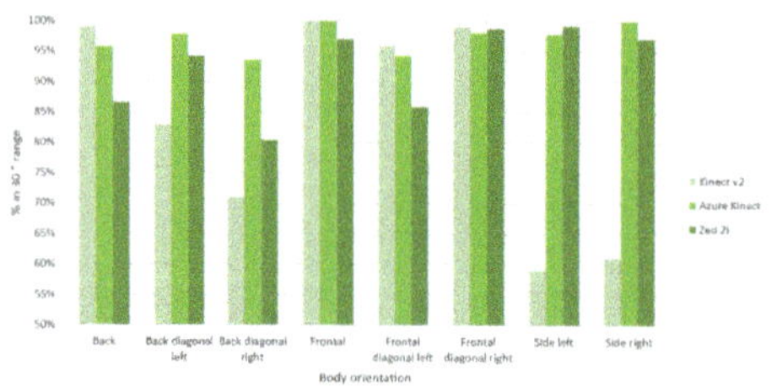

(**a**)

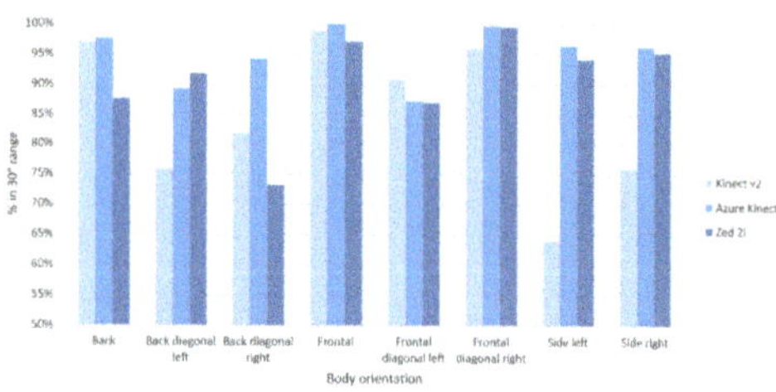

(**b**)

Figure 11. Achieved percentage for acceptable social interaction angle in range per device for single configuration in (**a**) and dyad configuration in (**b**).

4.5.4. Interpolation Correction

We identified outliers in each category during the classification of the body orientation angle. For this reason, to understand better whether inaccurate measurements occur systematically or whether they occur sporadically as isolated outliers, we consider the temporal dimension. In the latter case, we can correct erroneous measurements by considering the prior and subsequent measurements by smoothing out the error. We categorise these outliers into Neighbour outliers and Extreme outliers. Neighbour outliers are continuously wrong predicted angles along the walked trajectory. Extreme outliers are out of the median values with no temporal or spatial reason to appear. We use the recorded videos to examine both situations, review the intended body orientation angle and find an explanation for the wrong calculation. We apply an interpolation median correction to those spatial–temporal continuous values within 400 milliseconds with adjacent properly calculated values, and no external intervention is identified for the outlier to arise. We found that certain outliers followed a spatial–temporal pattern by plotting their location in the corresponding coordinate system. Afterwards, we inspected the video walking trajectories one-by-one to search for factors that might have led to the wrong skeleton-joints data extraction, related to the body's relative position to the camera's field of view, fluctuations in the body orientation while walking, and environment lighting changes.

We identified six distinct causes for an outlier: body entering the scene, body realignment, body proximity to the camera's field of view limits, body with high proximity relative to the camera, depth range limit of camera's field of view, and camera's field of view centre. Body realignment is the natural movement as it moves to the desired location, which creates a forward movement from one shoulder to another as we step on foot at a time. The body entering the scene reveals that the device requires adjusting to the orientation. On the other hand, other outliers expose the weakest areas of the sensors' field of view. We apply a temporal interpolation correction to manage these findings by taking the median value of two temporal and spatially pre and post continuous sample values. We then re-classify the calculated body orientation angles into the corresponding categories, obtaining the results described in Figure 12 with a visual representation of the correction shown in Figure 13. The new corrected body orientation angle values align with the accepted range angle for orientation. For single configurations, precision and recall values increase by 9% for the Kinect v2, 4% and 9% correspondently for the Azure Kinect, and 6% for the Zed 2i.

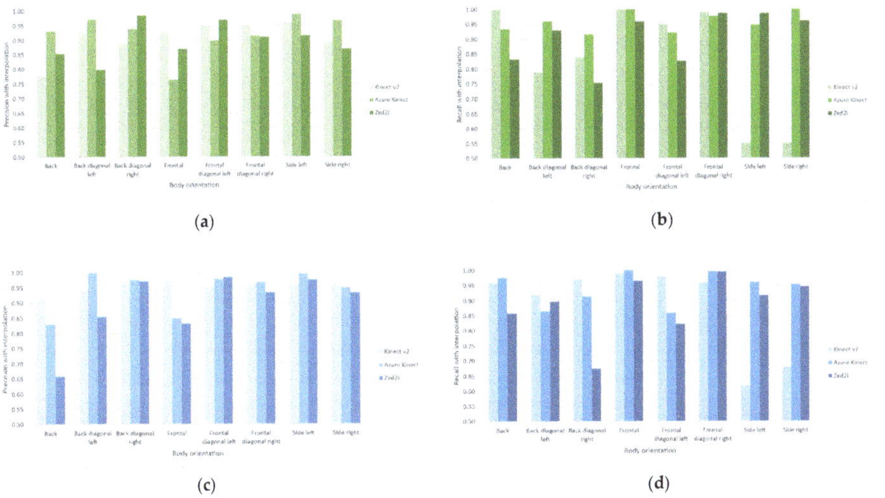

Figure 12. Precision and recall values for single configuration in (**a**,**b**), respectively; precision and recall values for dyad configuration in (**c**,**d**), respectively. Both after temporal interpolation.

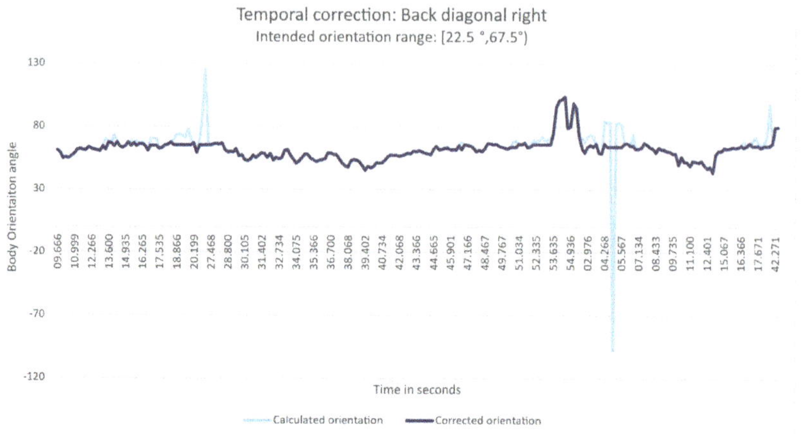

Figure 13. Temporal interpolation correction for the Back Diagonal Right orientation. In light blue, calculated body orientation angle with outliers. In dark blue, the corrected orientation angle.

4.6. Discussion

The extraction of body orientation angles using skeleton data solely from depth cameras shows high accuracy and availability with the integration of spatial–temporal attributes to understand the human body's mobility. Furthermore, there is evidence of the potential of depth sensor cameras to assess diverse body orientations by evaluating the calculated body angles against a set of categories in different walking patterns.

The Kinect v2 is mainly suitable for orientations aligned forward to the camera due to the algorithm training implemented in the device, which was trained primarily for these orientations for playing along with a console. However, it is feasible to have beneficial results for non-frontal orientations with the extracted skeleton joints and the head/face rotation data. The weakness relies on the body's orthogonal orientation to the camera for side orientations, splitting the calculated orientation into two distinct areas. For the side orientation, the head/face rotation and upper skeleton joints are detected differently for each half of the trajectory, reflected in the value of 40% to 50% predicted accuracy.

The Azure Kinect has an accuracy greater than 90% in most orientations for single and dyad configurations, with a weakness in distinguishing between frontal or back orientation, which can be corrected by adding head/face rotation information. During the experiments, it was noticeable that if the body enters the scene from one of the borders, as shown in Figure 10c (top-right) and Figure 10d (bottom-left), the device takes time to adjust the proper orientation, especially in those backwards, recovering rather quickly, in around a second.

The Zed 2i in the single configuration highlights a high accuracy, between 80% to 95% for most orientations, with difficulties in diagonal orientations. As identified with the Azure Kinect, it needs time to adjust the skeleton once the body enters the scene from the borders, but as shown in Figure 10e,f, more spatial information can be recovered with the broad field of view. One primary concern is missing bodies in the scene at times, especially when they come back from leaving the camera's field of view.

Notably, the side orientation was affected by the camera's alignment of the body siding, misclassifying the calculated angle to neighbourhood categories. Concerning the temporal interpolation correction, body orientation angles can be misclassified due to self-occlusion and the delay of the camera algorithm to correct the body orientation, especially in the edges of the field of view, showing the relevance of spatial–temporal data analysis to identify anomalies in the experiment expected pattern.

Lastly, there is an overall precision and recall improvement of 16% for Kinect v2, 13% for Azure Kinect, and 9% for Zed 2i for the dyad configuration, in the lowest category classification: back diagonal and side orientations. The enhancement of more than 10% in the case of the Kinect v2 and the Azure Kinect suggests the possibility of increasing the accuracy of the body orientations by using spatial–temporal information during the movement of several subjects linked to the group's temporality and habitation of a larger area. The positive results, particularly of the Azure Kinect in the interactional range evaluation, with single and multiple body detections, evidence the depth sensor cameras' capability in generating social signals from skeleton joints datasets.

5. Evaluation in the Context of F-Formations

The following section describes the experimental setup for four social arrangements to detect F-Formations and the evaluation results, displaying the system's potential in using measurements from the physically occupied space to interpret the socially occupied space.

5.1. Experiment Setup

This experiment follows the same setup regarding the delimited area, sensor location, data collection software, and configuration from the previous section. However, the task was different. Instead of walking along a path, participants were asked to stand at specific meeting points. Three meeting points are arranged in the free floor area in the camera's field of view.

5.2. Definition of Encounter

The intended positions are marked to form a triangular pattern with vertices separated approximately 1.2 m. Participants stay on each one of the meeting points for 20 s per encounter, with frontal, side, and frontal diagonal orientations for all-frontal, frontal-diagonal and frontal-vis a vis interaction. The pattern is repeated in three distinct positions related to the cameras' field of view and a single static position, as described in Figure 14.

Figure 14. Description of the encounter locations with intended body orientations. From left to right: all-frontal, frontal-diagonal, frontal-vis a vis configuration (colour codes correspond to the Body Orientation angle bar in Figure 7).

5.3. Participants' Description

For studying small gatherings in social encounters, two groups, one with two females and one male, the other one with two males and one female, with heights between 1.73 m and 1.81 m, met at different encounter points with a combination of body orientation and location. Each group concluded the task by producing thirty samples per device.

5.4. Evaluation Metrics and Results

The detection of groups is evaluated by comparing the number of stops identified by the algorithm against the intended meeting point. Additionally, we display the field of view of each participant during interaction to review how they intersect. In general, with the data collected with each depth camera, the algorithm identified groups by analysing the spatial–temporal interactional area's attributes and location with more than 90% for the

Kinect v2, and 100% for the Azure Kinect and the Zed 2i. Details on the number of bodies, stops and groups detected per meeting configuration are described in Table 6. For the Kinect v2, in the case of frontal-diagonal orientations, one-stop could not be detected due to the lack of data in assessing the orientation, which is explained by the orientation's evaluation results. The Azure Kinect did not assign a unique identifier to participants, but the algorithm was able to discriminate them all and their stops thanks to the skeleton data's high spatial–temporal resolution. On the other hand, the Zed 2i deviation for the frontal orientation is more evident in the group interactions and an additional body in the scene in 2 of 8 group interactions. However, similar to the Azure Kinect, the skeleton data resolution is high, allowing the algorithm to identify the authentic participants.

Table 6. Group detection results with the number of bodies and stops per configuration. Bold numbers indicate a lower or higher number of bodies detected.

Stops	Orientation	Kinect v2			Azure Kinect			Zed 2i		
		Bodies	Stops	Groups	Bodies	Stops	Groups	Bodies	Stops	Groups
3	Frontal	3	9	3	3	9	3	**4**	9	3
3	Frontal/Face to face	3	9	3	3	9	3	3	9	3
3	Frontal/Diagonal	**2**	8	2	3	9	3	3	9	3
1	Frontal	3	9	1	3	3	1	3	1	1
1	Frontal/Face to face	3	9	1	3	3	1	**4**	1	1
1	Frontal/Diagonal	3	8	1	3	3	1	3	1	1

5.5. Discussion

With the skeleton data collected per depth camera, the group detection algorithm detected group members' stops and membership in most designated meeting points. The field of view extension is calculated from the location and body orientation. Their intersection suggests the investigated focus of attention displayed in Figure 15 with a first approximation of the O-Space. The accuracy in detecting body orientations increases with differences per device for orientations facing the camera.

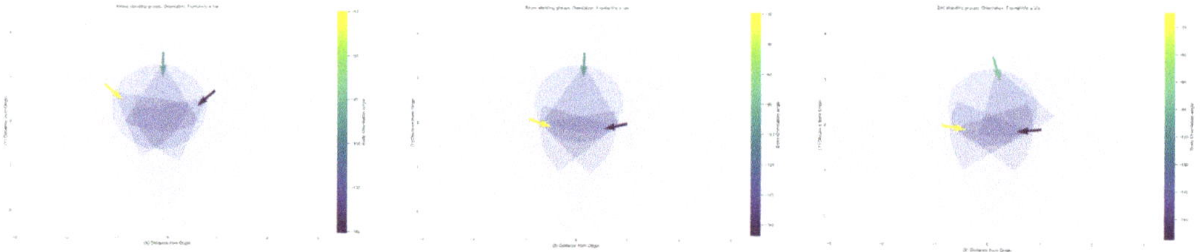

(**a**) Kinect v2 spaces (**b**) Azure Kinect spaces (**c**) Zed 2i spaces

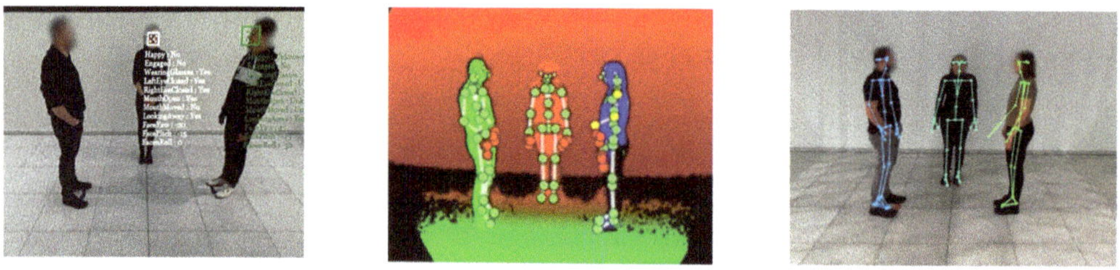

(**d**) Kinect v2 participants view (**e**) Azure Kinect participants view (**f**) Zed 2i participants view

Figure 15. Group meeting points with frontal-vis a vis orientation per device (**a**) Kinect v2, (**b**) Azure Kinect, and (**c**) Zed 2i. Each colour differentiates a person participating in the group with their corresponding orientation angle. From (**d**–**f**), participants view per device.

For the Kinect v2, in the diagonal orientations, a group is incomplete due to one member's unavailability of its skeleton data, as expected for back-diagonal orientation and side orientations result from the classification evaluation. We conclude that the group members are more complicated to detect when the body orientation is sided with the depth camera, restricting the extraction of the skeleton joints. Moreover, for diagonal orientations, we acknowledge body occlusions between participants during trajectories when they move from one meeting point to another while analysing the video recordings, limiting the body angle measurement due to the absence of skeleton joints. On the other hand, despite these limitations and the lack of individual body identification, the Azure Kinect camera produced high spatial–temporal resolution skeleton data facilitating the algorithm detection of all trajectories and stops from the group members as it predicts with a lower level of confidence the occluded areas from other joint data. For each group member, it is possible to see the movement of the upper limbs for resolutions up to 5 cm, as seen in the control video recordings where the participant was moving in a circular motion in a single location. Finally, the Zed 2i possesses a high spatio-temporal resolution and an excellent overall identification of individuals during trajectories, although leaving the scene can leave to missing the body for more extended periods than the Azure Kinect.

With the information derived from the skeleton data, it is possible to identify the different interactional spaces from an F-Formation, displayed in Figure 16. The participant's field of view bounds the O-space, the attention focus. In combination with the bodies' location, the field of view indicates the limits of the P-Space, where the participants sustain the interaction. Finally, the R-Space is constructed with a buffer determined by the social space [13], outside the inner interaction as a transactional space for the arrangement and disarrangement of the group, i.e., people leaving, arriving or standing at the socially occupied space.

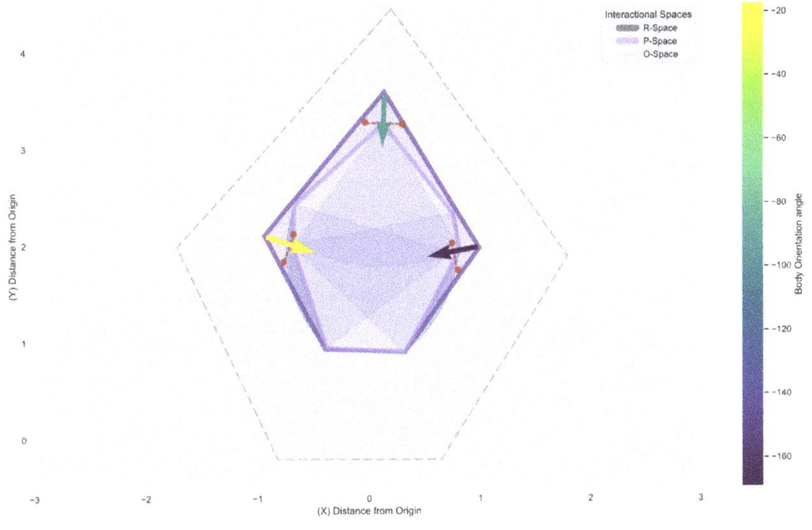

Figure 16. F-Formation's interactional spaces view each body's shoulder line (in red) with the detected orientation (arrows).

By limiting the areas that shape an F-Formation with the physically occupied space, we can observe that the socially occupied space is stiff for tight body angles. With restricted access and more open orientations, the group interaction expands, granting clear access to external participants. This indicates that the interactional space components do not follow a standard shape, and it is related to the body orientation and proximity factors in the interaction.

6. Conclusions

This study performed a series of experiments to measure three different depth sensor cameras' accuracy for assessing body orientation angles and purpose them to detect socially occupied spaces using the F-Formations model. First, we generated three datasets by walking in a combination of four trajectory patterns in eight body orientations in a single and dyad configuration. We observed that the Kinect v2 depth sensor's accuracy is good in frontal, back, and diagonal orientations but weak when the user is aligned orthogonal to the camera in the case of side orientations. For the Azure Kinect, the depth sensor accuracy is higher in most orientations, with difficulties distinguishing frontal from back orientations as it lacks head/face rotation information. The Zed 2i, with its wide range, can collect more information, but it can omit bodies re-entering the scene. For other scenarios, the accuracy for the case of a strict categorisation proves to be 90%, 96%, and 89% for the Kinect v2, Azure Kinect, and Zed 2i, respectively, with a maximum standard deviation of 1.5, 3.0, and 1.5 angle classes. Finally, after the temporal interpolation correction for the socially acceptable interaction, the availability increases to 92.4%, 100.0%, and 99.8% for the single configuration and 94.9%, 100.0%, and 99.8% for the dyad configuration for the Kinect v2, Azure Kinect and Zed 2i, respectively.

Through this system, we can differentiate the components of the socially occupied spaces. For each device skeleton dataset, we reached 90% accuracy for the Kinect v2 and 100% for the Azure Kinect and the Zed 2i. The reached accuracy and socially acceptable angle availability from the Azure Kinect are adequate to detect F-Formations. Additionally, it does not depend on additional software to integrate head/face rotation data to improve the right-left correspondence, as in the case of Kinect v2 or RGB videos in the case of Zed 2i. Regarding our first approximation in detecting F-Formations' interactional spaces, the algorithm identified the group members' positions and assessed each participant's field of view during an interaction. The interactional spaces could be delimited given the participant's position, the study of proxemics, and the body orientation to assess the focus of attention.

Regarding resources and easiness in implementing the system, the hardware for using specific models can limit the performance, especially for the most recent devices, as it requires demanding resources. Nevertheless, the possibility of purchasing the technology is more significant than those specialised body tracking devices. For the particular use case of analysing group behaviour, depth cameras analyse individuals while making their identities more difficult to reveal, whereas the stereoscopic camera requires analysing raw video. Secondly, to collect data, the Zed 2i requires minimum light to discriminate details in the scene. We noticed this necessity in the experiment environment when the lights were dimmed, and the accuracy started to suffer, for which we sustained a proper illumination.

For the upcoming work, we aim to improve the socially occupied spaces detection algorithm and implement it in a desktop application to have live results for experimental group analysis. Improving the algorithm requires integrating parameters to appropriately limit the different interactional areas of the F-Formation model, the O-space, the P-Space, and the R-space from a spatial–temporal perspective and implementing and evaluation against other methods in the literature related to the assessment of social spaces in computer vision. Additionally, we intend to add more information regarding the dynamics of encounters by evaluating factors such as joining, leaving, or avoiding the group to facilitate the automatisation of human behaviour analysis. Lastly, integrating multiple sensors in one synchronised system to improve occlusion is also on the agenda as it may prove helpful in treating occlusions from the participants' bodies and the loss of spatial–temporal data by integrating multiple points of view from the scene.

Author Contributions: Conceptualisation, V.A.L.S.-L. and A.S.; Data curation, V.A.L.S.-L.; Formal analysis, V.A.L.S.-L.; Methodology, V.A.L.S.-L. and A.S.; Project administration, A.S.; Software, V.A.L.S.-L.; Supervision, A.S.; Validation, V.A.L.S.-L.; Visualization, V.A.L.S.-L.; Writing—original draft, V.A.L.S.-L.; Writing—review and editing, V.A.L.S.-L. and A.S. All authors have read and agreed to the published version of the manuscript.

Funding: This research received no external funding.

Institutional Review Board Statement: Not applicable.

Informed Consent Statement: Informed consent was obtained from all subjects involved in the study.

Data Availability Statement: The data collected in these studies are openly available at https://osf.io/xhwgm/ (accessed on 15 March 2022). The tutorials and code to create a new interface for all devices can be found at https://github.com/violetasdev/bodytrackingdepth_course (accessed on 15 March 2022). For the Azure Kinect, we implement a modified version of k4.net, the final version is available at https://github.com/violetasdev/k4a.net (accessed on 15 March 2022). The Kinect v2 graphs can be reproduced at https://osf.io/ghz79/ (accessed on 15 March 2022).

Conflicts of Interest: The authors declare no conflict of interest.

Abbreviations

The following abbreviations are used in this manuscript:

IO	Intended orientation angle
IOD	Intended orientation angle deviation
IR	Infrared
RGB-D	RGB depth
SDK	Software development kit
TOF	Time of Flight

References

1. Garfinkel, H. *Studies in Ethnomethodology*; Wiley: New York, NY, USA, 1991.
2. Leon, V.A.L.S.; Schwering, A. Detecting social spaces with depth cameras: Evaluating location and body orientation as relevant social features. In Proceedings of the 2021 International Conference on Indoor Positioning and Indoor Navigation (IPIN), Lloret de Mar, Spain, 29 November–2 December 2021. [CrossRef]
3. Beyan, C.; Shahid, M.; Murino, V. Investigation of small group social interactions using deep visual activity-based nonverbal features. In Proceedings of the 26th ACM international conference on Multimedia, Seoul, Korea, 22–26 October 2018; pp. 311–319. [CrossRef]
4. Gan, T.; Wong, Y.; Zhang, D.; Kankanhalli, M.S. Temporal encoded F-formation system for social interaction detection. In Proceedings of the 21st ACM international conference on Multimedia, Barcelona, Spain, 21–25 October 2013; pp. 937–946. [CrossRef]
5. Kobayashi, Y.; Yuasa, M.; Katagami, D. Development of an interactive digital signage based on F-formation system. In Proceedings of the First International Conference on Human-Agent Interaction (HAI 2013), Sapporo, Japan, 7–9 August 2013.
6. Kantharaju, R.B.; Pelachaud, C. Social Signals of Cohesion in Multi-party Interactions. In Proceedings of the 21st ACM International Conference on Intelligent Virtual Agents, Virtual Event, 14–17 September 2021; pp. 9–16. [CrossRef]
7. Hedayati, H.; Szafir, D.; Andrist, S. Recognizing F-Formations in the Open World. In Proceedings of the 2019 14th ACM/IEEE International Conference on Human-Robot Interaction (HRI), Daegu, Korea, 11–14 March 2019; pp. 558–559. [CrossRef]
8. Connolly, J.; Tsoi, N.; Vázquez, M. *Perceptions of Conversational Group Membership Based on Robots' Spatial Positioning: Effects of Embodiment*; Association for Computing Machinery: New York, NY, USA, 2021; Volume 1, ISBN 9781450382908.
9. Vom Lehn, D.; Heath, C.; Hindmarsh, J. Exhibiting Interaction: Conduct and Collaboration in Museums and Galleries. *Symb. Interact.* **2001**, *24*, 189–216. [CrossRef]
10. Murino, V.; Cristani, M.; Shah, S.; Savarese, S. *The Group and Crowd Analysis Interdisciplinary Challenge*, 1st ed.; Elsevier Inc.: Amsterdam, The Netherlands, 2017.
11. Kendon, A. Spatial organization in social encounters: The F-formation system. *Man Environ. Syst.* **1976**, *6*, 291–296.
12. Mondada, L. Interactional space and the study of embodied talk-in-interaction. *Space in Language and Linguistics* **2013**, *24*, 247–275. [CrossRef]
13. Hall, E.T. *The Hidden Dimension*; Doubleday: New York, NY, USA, 1966.
14. Yoshimura, Y.; Sobolevsky, S.; Ratti, C.; Girardin, F.; Carrascal, J.P.; Blat, J.; Sinatra, R. An analysis of visitors' behavior in the louvre museum: A study using bluetooth data. *Environ. Plan. B Plan. Des.* **2014**, *41*, 1113–1131. [CrossRef]
15. Kunhoth, J.; Karkar, A.G.; Al-Maadeed, S.; Al-Ali, A. Indoor positioning and wayfinding systems: A survey. *Human-centric Comput. Inf. Sci.* **2020**, *10*, 18. [CrossRef]

16. Goffman, E. *Encounters: Two Studies in the Sociology*; Martino Fine Books: Eastford, CT, USA, 1961.
17. Bassetti, C. *Social Interaction in Temporary Gatherings: A Sociological Taxonomy of Groups and Crowds for Computer Vision Practitioners*, 1st ed.; Elsevier Inc.: Amsterdam, The Netherlands, 2017.
18. Goffman, E. *Behavior in Public Places: Notes on the Social Organization of Gatherings*; The Free Press: New York, NY, USA, 1963.
19. Kendon, A. *Conducting Interaction: Patterns of Behavior in Focused Encounters*; CUP Archive: Cambridge, UK, 1990; ISBN 0-521-38036-7 (Hardcover); 0-521-38938-0 (Paperback).
20. Shi, C.; Shimada, M.; Kanda, T.; Ishiguro, H.; Hagita, N. Spatial formation model for initiating conversation. *Robot. Sci. Syst.* **2012**, *7*, 305–312. [CrossRef]
21. Bitgood, S. An attention-value model of museum visitors. In *Visitor Attention*; Jacksonville State University: Jacksonville, FL, USA, 2010; pp. 1–17.
22. Goffman, E. *Forms of Talk*; University of Pennsylvania Press: Philadelphia, PA, USA, 1981; Incorporated; ISBN 9780812211122, 081221112X.
23. Bitgood, S. An Analysis of Visitor Circulation: Movement Patterns and the General Value Principle. *Curator Mus. J.* **2006**, *49*, 463–475. [CrossRef]
24. Raza, A.; Lolic, L.; Akhter, S.; Liut, M. Comparing and Evaluating Indoor Positioning Techniques. In Proceedings of the 2021 International Conference on Indoor Positioning and Indoor Navigation (IPIN), Lloret de Mar, Spain, 29 November–2 December 2021. [CrossRef]
25. Marquardt, N.; Hinckley, K.; Greenberg, S. Cross-device interaction via micro-mobility and F-formations. In Proceedings of the UIST'12—25th Annual ACM Symposium on User Interface Software and Technology, Cambridge, MA, USA, 7–10 October 2012.
26. Rashed, M.G.; Suzuki, R.; Yonezawa, T.; Lam, A.; Kobayashi, Y.; Kuno, Y. Tracking Visitors in a Real Museum for Behavioral Analysis. In Proceedings of the 2016 Joint 8th International Conference on Soft Computing and Intelligent Systems (SCIS) and 17th International Symposium on Advanced Intelligent Systems (ISIS), Sapporo, Japan, 25–28 August 2016; pp. 80–85. [CrossRef]
27. Dim, E.; Kuflik, T. Automatic detection of social behavior of museum visitor pairs. *ACM Trans. Interact. Intell. Syst.* **2014**, *4*, 17. [CrossRef]
28. Marshall, P.; Rogers, Y.; Pantidi, N. Using F-formations to analyse spatial patterns of interaction in physical environments. In Proceedings of the ACM 2011 conference on Computer supported cooperative work, Hangzhou, China, 19–23 March 2011; pp. 445–454. [CrossRef]
29. Alameda-Pineda, X.; Yan, Y.; Ricci, E.; Lanz, O.; Sebe, N. Analyzing free-standing conversational groups: A multimodal approach. In Proceedings of the 23rd ACM international conference on Multimedia, Brisbane, Australia, 26–30 October 2015; pp. 5–14. [CrossRef]
30. Vascon, S.; Bazzani, L. *Group Detection and Tracking Using Sociological Features*, 1st ed.; Elsevier: Amsterdam, The Netherlands, 2017; ISBN 9780128092804.
31. Zhou, C.; Han, M.; Liang, Q.; Hu, Y.; Kuai, S. A social interaction field model accurately identifies static and dynamic social groupings. *Nat. Hum. Behav.* **2019**, *3*, 847–855. [CrossRef]
32. Marin, G.; Agresti, G.; Minto, L.; Zanuttigh, P. A multi-camera dataset for depth estimation in an indoor scenario. *Data Br.* **2019**, *27*, 104619. [CrossRef]
33. Tsykunov, E.; Ilin, V.; Perminov, S.; Fedoseev, A.; Zainulina, E. Coupling oflocalization and depth data for mapping using Intel RealSense T265 and D435i cameras. *arXiv* **2020**, arXiv:2004.00269.
34. Yeung, L.F.; Yang, Z.; Cheng, K.C.C.; Du, D.; Tong, R.K.Y. Effects of camera viewing angles on tracking kinematic gait patterns using Azure Kinect, Kinect v2 and Orbbec Astra Pro v2. *Gait Posture* **2021**, *87*, 19–26. [CrossRef] [PubMed]
35. Pathi, S.K.; Kristoffersson, A.; Kiselev, A.; Loutfi, A. F-formations for social interaction in simulation using virtual agents and mobile robotic telepresence systems. *Multimodal Technol. Interact.* **2019**, *3*, 69. [CrossRef]
36. Vascon, S.; Mequanint, E.Z.; Cristani, M.; Hung, H.; Pelillo, M.; Murino, V. Detecting conversational groups in images and sequences: A robust game-theoretic approach. *Comput. Vis. Image Underst.* **2016**, *143*, 11–24. [CrossRef]
37. Climent-Pérez, P.; Florez-Revuelta, F. Improved action recognition with separable spatio-temporal attention using alternative skeletal and video pre-processing. *Sensors* **2021**, *21*, 1005. [CrossRef]
38. Ruget, A.; Tyler, M.; Martín, G.M.; Scholes, S.; Zhu, F.; Gyongy, I.; Hearn, B.; McLaughlin, S.; Halimi, A.; Leach, J. Real-time, low-cost multi-person 3D pose estimation. *arXiv* **2021**, arXiv:2110.11414.
39. Nanavati, A.; Doering, M.; Brščic, D.; Kanda, T. Autonomously learning one-to-many social interaction logic from human-human interaction data. In Proceedings of the 2020 ACM/IEEE International Conference on Human-Robot Interaction, Cambridge, UK, 23–26 March 2020; pp. 419–427. [CrossRef]
40. Wilson, A.D. Combating the Spread of Coronavirus by Modeling Fomites with Depth Cameras. *Proc. ACM Hum.-Comput. Interact.* **2020**, *4*, 1–13. [CrossRef]
41. Albert, J.A.; Owolabi, V.; Gebel, A.; Brahms, C.M.; Granacher, U.; Arnrich, B. Evaluation of the pose tracking performance of the azure kinect and kinect v2 for gait analysis in comparison with a gold standard: A pilot study. *Sensors* **2020**, *20*, 5104. [CrossRef]
42. Tölgyessy, M.; Dekan, M.; Chovanec, L'.; Hubinský, P. Evaluation of the azure kinect and its comparison to kinect v1 and kinect v2. *Sensors* **2021**, *21*, 413. [CrossRef]
43. Ortiz, L.E.; Cabrera, E.V.; Gonçalves, L.M. Depth data error modeling of the ZED 3D vision sensor from stereolabs. *Electron. Lett. Comput. Vis. Image Anal.* **2018**, *17*, 1–15. [CrossRef]

44. Vinciarelli, A.; Pantic, M.; Bourlard, H. Social signal processing: Survey of an emerging domain. *Image Vis. Comput.* **2009**, *27*, 1743–1759. [CrossRef]
45. Poggi, I.; Errico, F.D. *Social Signals: A Psychological Perspective*; Springer: London, UK, 2011; ISBN 9780857299949.
46. Microsoft Kinect for Windows. Available online: https://developer.microsoft.com/en-us/windows/kinect/ (accessed on 22 February 2022).
47. Stereolabs Getting Starting with ZED. Available online: https://www.stereolabs.com/docs/ (accessed on 22 February 2022).
48. Bitgood, S. *Attention and Value*; Left Coast Press, Inc.: Walnut Creek, CA, USA, 2016; ISBN 9781611322620.
49. Microsoft Azure Kinect DK Documentation. Available online: https://docs.microsoft.com/en-us/azure/kinect-dk/ (accessed on 22 February 2022).
50. NVIDIA Zed Camera. Available online: https://docs.nvidia.com/isaac/archive/2020.1/packages/sensors/doc/zedcamera.html (accessed on 22 February 2022).

sensors

MDPI

Article

Facial Motion Analysis beyond Emotional Expressions

Manuel Porta-Lorenzo, Manuel Vázquez-Enríquez, Ania Pérez-Pérez, José Luis Alba-Castro * and Laura Docío-Fernández

atlanTTic Research Center, University of Vigo, 36310 Vigo, Spain; mporta@gts.uvigo.es (M.P.-L.); mvazquez@gts.uvigo.es (M.V.-E.); aperez@gts.uvigo.es (A.P.-P.); ldocio@gts.uvigo.es (L.D.-F.)
* Correspondence: jalba@gts.uvigo.es

Abstract: Facial motion analysis is a research field with many practical applications, and has been strongly developed in the last years. However, most effort has been focused on the recognition of basic facial expressions of emotion and neglects the analysis of facial motions related to non-verbal communication signals. This paper focuses on the classification of facial expressions that are of the utmost importance in sign languages (Grammatical Facial Expressions) but also present in expressive spoken language. We have collected a dataset of Spanish Sign Language sentences and extracted the intervals for three types of Grammatical Facial Expressions: negation, closed queries and open queries. A study of several deep learning models using different input features on the collected dataset (LSE_GFE) and an external dataset (BUHMAP) shows that GFEs can be learned reliably with Graph Convolutional Networks simply fed with face landmarks.

Keywords: facial expression recognition; facial landmarks; action units; convolutional neural networks; graph convolutional networks

Citation: Porta-Lorenzo, M.; Vázquez-Enríquez, M.; Pérez-Pérez, A.; Alba-Castro, J.L.; Docío-Fernández, L. Facial Motion Analysis beyond Emotional Expressions. *Sensors* **2022**, *22*, 3839. https://doi.org/10.3390/s22103839

Academic Editors: Tomasz Krzeszowski, Adam Świtoński, Michal Kepski and Carlos Tavares Calafate

Received: 23 April 2022
Accepted: 14 May 2022
Published: 19 May 2022

1. Introduction

Facial expressions are one of the most valuable signals in human interactions. Numerous studies have been conducted on automatic Facial Expression Recognition (FER) due to its practical importance in human–robot interaction, personalized medical treatment, driver fatigue monitoring, customer behavior, etc. In the 1970s, Ekman and Friesen [1] postulated that six basic facial expressions of emotion are perceived by humans in the same way regardless of their culture. Today, these facial expressions are the basis of most computer vision systems for FER. However, these facial expressions of emotion are not sufficient to fully represent the expressiveness of human affection, emotion and communication. Human communication uses nonverbal channels to fully convey a message and its context, and facial expressiveness is one of the most important nonverbal channels. In this context, facial expressions are referred to as Linguistic or Grammatical Facial Expressions (GFEs) [2], as they serve a grammatical function in the sentences. An extreme case of the importance of GFEs is that of sign languages, where they provide adjectivation and can modify the semantics of signs [3]. As shown in [4], many studies have been conducted to recognise basic facial expressions of emotion using a wide variety of input features and classification methods. While static images provide enough information to decently perform such a task [5], more subtle facial expression cues require temporal information [6]. Several large datasets for FER are available both with static and dynamic information. Unfortunately, very few datasets containing video representations of grammatical facial expressions are available, making the development of robust models for GFER extremely difficult. Due to such lack of labelled data, using RGB sequences for automated GFE recognition yields to model overfitting the data. Hence, it is good practice to use previously trained feature extractors for related tasks.

In 1978, Paul Ekman created a taxonomy of facial expressions, the Facial Action Coding System (FACS, [7]). From this system, a set of atomic facial muscles (AUs) is defined, from which any further complex expression or emotional state can be inferred. AUs span a large number of combinations that can be used for the recognition of the 6 basic emotions (happiness, surprise, anger, fear, sadness and disgust), as well as for other complex psychological states, such as depression or pain [8]. AUs have also been studied in combination with vocal prosody, as in the series of AudioVisual Emotion Challenge AVEC2016 [9], highlighting their link with non-verbal communication.

Two of the most used features pre-computed for facial expression recognition are Action Units (AUs) [10] as well as facial landmarks [11] that convey discrete information on head and face muscle movements. Techniques to extract both types of features have been greatly improved in the last years thanks to deep models trained with large datasets [4]. These features have been used to perform FER with standard data classifiers as Support Vector Machines (SVMs) [12] or Multilayer Perceptrons (MLPs) [13], or used as complementary input to derive more complex representations using deep neural networks [14].

In this work we try to push the state of the art on grammatical facial expression recognition systems fed with non-RGB data, namely Action Units or facial landmarks, casting these inputs as graphs. Recently, Graph Neural Networks (GNN) and their convolutional extension to Graph Convolutional Networks (GCN) [15] stand out for their flexibility and their good performance in Human Action Recognition. Furthermore, some works have already combined GNNs and facial landmarks [16] as well as AUs [17], obtaining SOTA results in FER.

The main contributions of this work can be summarized as follows:

- A new dataset specifically acquired for GFE in the context of sign languages providing their face landmarks and AUs.
- A thorough assessment of GCN for Grammatical Facial Expression Recognition (GFER) with two types of input features (facial landmarks and action units) and comparison of this technique over two different GFE datasets and also against classical CNN techniques.
- A comparison of the experimental results with human performance on the new dataset.

The rest of the paper is organized as follows: Section 2 reviews the most recent approaches related to the automated recognition of FER/GFER using facial features extracted from video sequences. Section 3 describes the selected datasets and feature extraction, sampling and pre-processing. Section 4, describes the deep learning models and evaluation metrics. Section 5 presents the experiments performed, and finally, Section 6 draws some conclusions.

2. Related Work

Our research is focused on communicative facial expressions related to sign languages. As pointed out in [2], signers must be able to quickly identify and discriminate between different linguistic and affective facial expressions in order to process and interpret signed sentences. Through fMRI studies, they demonstrated that the parts of the brain that are activated when detecting emotions and language-related facial expressions are different. It is clear then, that head and facial muscle movement in this context could have common features with facial expressions of emotion, but also their own specificities that should be learned from sufficient input samples.

Many more machine learning studies have been conducted on FER than on GFER, so taking advantage of the techniques developed for FER should be a priority when addressing GFER, especially considering that, as presented in the next section, the number of databases prepared for automated GFE analysis is very small. Thus, we analyzed the research articles on FER and GFER available in the SCOPUS database on the basis of the primary keywords indicated in the Table 1. All searches were restricted to journal and conference articles published between 2009 and 2021 (to cover mainly the trend of deep learning approaches), in Computer Science, Engineering and Mathematics subject areas. This search approach retrieved a total of 929 documents. After a review of titles, abstracts and keywords, it was clear that the vast majority was related to FER using RGB video input, CNN-based approaches, detection of Action Units and, in a lesser extent, landmarks detection. As we were primarily interested in non-RGB inputs to avoid the necessity of large datasets, and to include graph-based systems, we run a secondary search, applying an exclusion criteria based on the secondary keywords given in Table 1. This filtering resulted in a total of 120 relevant documents, again most of them related to CNNs. A final filtering including the term "graph convolutional networks" yield nine relevant studies which where analyzed in depth.

Table 1. Bibliographic search keywords.

Primary Keywords	("facial expression recognition" OR "facial emotion recognition" OR ("linguistic" OR "grammatical") AND "facial expressions" AND "recognition") OR ("facial micro-expression recognition") AND ("video" OR "image sequences" OR "dynamic expressions" OR "temporary information" OR "temporal data" OR "spatial-temporal")
Secondary Keywords	("action units" OR "landmarks")
Last filtering	("action units" OR "landmarks" AND "graph convolutional networks"

Existing deep learning approaches for video-based FER from facial landmarks, typically concatenate their coordinates over multiple frames to form a sequence of vectors as the input to Recurrent Neural Networks (RNNs) [18], or rearrange them to form an image-like matrix to feed a Convolutional Neural Networks (CNNs) [14]. As comparative results showed, these methods are not able to fully capture the joint dynamics of spatial and temporal features encoded in a sequence of facial landmarks. In the last years, many of the approaches proposed for human action recognition (HAR) use the estimation of body skeleton keypoints. In order to capture the complex spatial-temporal characteristics of human actions, the best performing approaches plug these keypoints, or a transformation thereof, as features of nodes in a graph convolutional neural network, such as Spatio–Temporal Graph Convolutional Networks (ST-GCNs) [19]. Recently, a GCN-based method has been proposed in [16] which uses only facial landmarks for FER. They showed SOTA performance on three large datasets and better performance when fusing with RGB-based models, highlighting the complementary of the approaches when enough data is available. Also, in [20] the Progressive Spatio–Temporal Bilinear Network (PST-BLN) method was proposed for compact modeling of facial expression recognition. They showed performance only slightly worse than RGB based models over three large FER datasets but with a model one order of magnitude smaller. Node features different to landmarks have been also explored. In [17] nodes are defined in a face graph with features related to AUs and edges related to landmark distances. The solution compares favourably against other FER methods over the same large datasets.

These SOTA methods show that FER can be reliably attained with GCN models that do not use RGB information explicitly, which allows leveraging facial feature extraction techniques from large RGB FER datasets. As we will show in the next section, datasets for GFER are really scarce and small, so GCN based approaches over landmarks and AUs are a promising approach.

3. Materials

In this section the details of the datasets are described, including the extraction and preprocessing of both action units and facial landmarks and the sampling and normalization of videos.

3.1. Datasets

One of the biggest problems in GFER is the lack of extensive and properly collected data. Table 2 summarizes the main datasets collected for dynamic facial expression recognition sorted by number of citations.

Table 2. Main datasets collected for dynamic facial expression recognition.

Name	Cites	Type of Expression	Acquisition Set Up	Classes	Videos	Persons
CK+ [21]	1754	emotions	controlled	7	327	118
OULU-CASIA [22]	413	emotions	controlled (light variation)	6	480	80
MMI [23]	342	induced emotions	natural	6	213	30
AFEW [24]	316	emotions	movies	7	1426	330
MUG [25]	261	induced emotions	semi-controlled	6	1462	52
BU-4DFE [26]	154	emotions	controlled	6	606	101
FABO [27]	85	induced emotions	semi-controlled	9	1900	23
BUHMAP [28]	20	sign-language/emotions	controlled	8	440	11
GFE-LIBRAS [29]	10	sign-language	natural	9	36	2
DFEW [30]	8	emotions	movies	7	16,372	-
LILiR [31]	6	non-verbal communication	natural	4	527	2
SILFA [32]	3	sign-language	semi-controlled	?	230	10

Most of the datasets were acquired for FER. In fact most of the works reviewed in our study have used the four first datasets in the table. From those that had different annotations to the basic six emotions + neutral, FABO, LILiR, SILFA, BUHMAP, and GFE-LIBRAS, only the last three were particularly acquired for Sign Language studies. Unfortunately, GFE-LIBRAS, that has a large number of sign language face motions and expressions (9), comprises only 18 clips from 2 persons and only the landmarks are available for download, SILFA is a quite interesting GFE dataset that can be obtained from the authors but it is annotated with a set of AUs, instead of comprehensible sign-language expressions. On the other hand, BUHMAP contains five sign-language-related expressions + two emotion + neutral and the 440 videos from 11 persons are available in RGB videos, so we chose this dataset to test the approaches developed in this work.

3.1.1. BUHMAP

The Boğaziçi University Head Motion Analysis Project Database (BUHMAP) [28] contains labelled videos of 8 different classes of facial signs. The dataset consists of 440 videos recorded by 11 different subjects (6 women and 5 men) which performed each sign 5 times. The included classes combine basic emotions with head movements acted by the donors when prompted. The classes are defined below:

1. *Neutral*: The neutral state of the face. The subject neither moves his/her face nor makes any facial expressions.
2. *Head L-R*: Shaking the head to right and left sides. The initial side varies among subjects, and the shaking continues about 3–5 times. This sign is frequently used for negation in Turkish Sign Language (TSL).
3. *Head Up*: Raise the head upwards while simultaneously raising the eyebrows. This sign is also frequently used for negation in TSL.
4. *Head F*: Head is moved forward accompanied with raised eyebrows. This sign is frequently used to change the sentence to question form in TSL.
5. *Sadness*: Lips turned down, eyebrows down. It is used to show sadness, e.g., when apologizing. Henceme subjects also move their head downwards.
6. *Head U-D*: Nodding head up and down continuously. Frequently used for agreement.
7. *Happiness*: Lips turned up. Subject smiles.
8. *Happy U-D*: Head U-D + Happiness. The preceding two classes are performed together. It is introduced to be a challenge for the classifier in successfully distinguishing this confusing class with the two preceding ones.

Figure 1 shows an example of frame sequences for the 8 defined classes in BUHMAP dataset.

Figure 1. Example of FE classes in BUHMAP dataset. Rows from top to bottom: 1—Neutral, 2—Head L-R, 3—Head Up, 4—Head F, 5—Sadness, 6—Head U-D, 7—Happiness, 8—Happy U-D.

3.1.2. LSE_GFE

LSE_GFE has been extracted from the LSE_UVIGO [33], a multi-source database designed to foster research on Spanish Sign Language Recognition. The anonymous data needed to reproduce the experiments have been released, jointly with all the code, in the github page https://github.com/mporta-gtm/GrammaticalFacialExpressions (accessed on 1 May 2022). The dataset contains isolated signs, expressive sentences and interviews, all acquired in a controlled lab environment. Also, besides the sign/gloss/sentence labels, 841 videos have also annotations for some grammatical facial expressions, namely:

1. *q.polar*: Yes/no question. Head and body is slightly moved forward accompanied with raising eyebrows. This sign is frequently used to change the sentence to close question form in LSE. 123 samples performed by 19 people.
2. *q.partial*: Open question. Head and body is moved forward accompanied with frown eyebrows. This sign is frequently used to change the sentence to open question form in LSE. 265 samples performed by 13 people.
3. *q.other*: General question form, not assimilable to polar (close) or partial (open). 16 samples from 9 people.
4. *n.L-R*: Typical "no" negation, similar to Head L-R in BUHMAP. 176 samples performed by 22 people.
5. *n.other*: General negation, not assimilable to n.L-R. 53 samples from 19 people.
6. *None*: Samples without any of these questioning or negation components were extracted from the available videos of 24 people, to ensure the capacity of the model to detect the presence of non manual components. This class is quite different to the BUHMAP *Neutral* class, because in the LSE_GFE case, other communicative expressions can be included in the *None* class, i.e, dubitation with complex head and eyebrows movement.

The statistics of this dataset are shown in Table 3. It can be seen that the data distribution is highly unbalanced both in gender and in classes. This problem would be tackled in future research.

Table 3. Gender distribution in LSE_GFE dataset (#samples).

Class	Female	Male
q.polar	68	55
q.partial	174	91
q.other	10	6
n.L-R	105	71
n.other	29	24
None	109	99

This dataset had to be filtered before using it in order to solve some detected issues. First, as classes *q.other* and *n.other* had too few samples and their definition was ambiguous even for expert language interpreters, they were discarded. Second, not all the collaborators which recorded the videos were deaf people so, to maintain the integrity of the dataset, ensuring that all samples were correctly performed, only recordings from deaf people and sign language interpreters were considered. The final number of videos for the dataset in this work was 413, that compares to the 440 videos of BUHMAP.

Figure 2 shows an example of frame sequences for the 4 defined classes of LSE_GFE for this work.

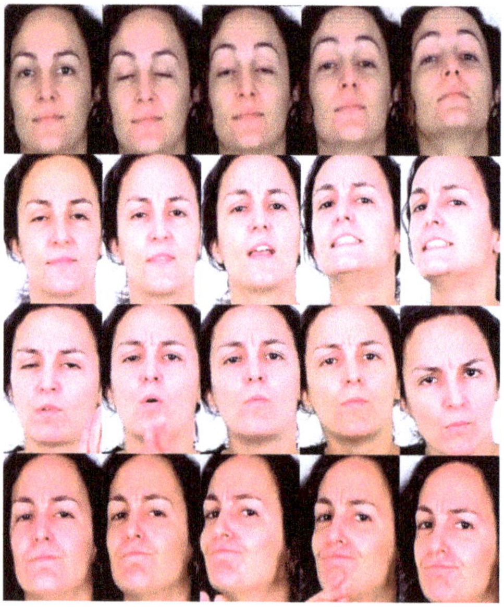

Figure 2. Example of FE classes in LSE_GFE dataset. Rows from top to bottom: None, q.polar, q.partial, n.L-R.

It is important to note that the GFEs of LSE_GFE are extracted from interviews, so facial expressions are expected to be more natural than in the case of BUHMAP, where the 8 classes were forcibly generated. Figure 3 shows a snapshot of the ELAN tool [34] used for annotating the Lex40_UVIGO dataset. In this snapshot an example of annotation of an interval for the class *q.partial* (i.parcial, in spanish) can be observed.

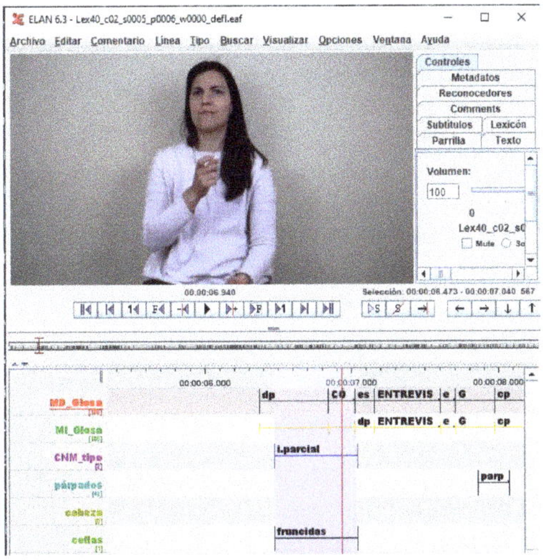

Figure 3. Snapshot of the ELAN annotation tool with an example of interval of an open question class *q.partial*.

3.2. Features

3.2.1. Action Units

Action Units are defined as the fundamental movements of muscles or groups of muscles of the face that correspond to the display of an emotion. These movements are encoded by the Facial Action Coding System (FACS) [7]. Using FACS it is possible to code most anatomically plausible facial expressions by disassembling them into specific Actions Units (AUs). An example of some action units is supplied in Figure 4, where the activation of groups of muscles is labelled with the corresponding AU code.

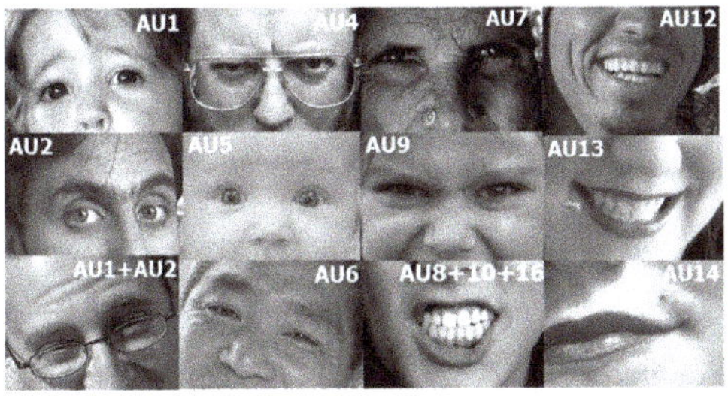

Figure 4. Examples of Action Units. Henceurce: OpenFace.

The extraction of Action Units was carried out using OpenFace [35], a face analysis library that includes a state-of-the-art HOG-based method for detecting AUs. This model provides the presence (binary value) of 18 AUs together with an estimation of the intensity value, from 0 to five, of 17 of them. This data was pre-processed to build a vector of 18 values were the first 17 are the obtained intensities and the last one is the presence value of the remaining AU scaled to the range of the intensities. Then, these vector were concatenated in a matrix where the horizontal axis contains the 18 AUs studied and the vertical axis represents their temporal evolution.

3.2.2. Facial Landmarks

Facial landmarks were extracted also by using OpenFace, which employs a deep learning state-of-the-art method [36]. After the extraction, the x and y coordinates of 68 keypoints are normalized between −1 and 1 with respect to the landmark of the nose tip, where −1 and 1 correspond to the points furthest away from the nose. In addition, in some experiments a data augmentation technique consisting of horizontal flipping of x landmarks coordinates was tested. Finally, the use of 3D coordinates was considered but preliminary tests show that the depth estimation of this method is noisy and does not contribute to enhance the performance on the evaluated models.

3.3. Video Sample Generation and Preprocessing

The assessed classification models require the input samples to have an equal, or at least very similar, size. In order to gain insight of the labelled events of each dataset a small study on their duration was carried out. Figure 5 shows the distribution of such events in both datasets in terms of duration in milliseconds. It can be seen that both datasets follow a non-gaussian distribution with different means and deviations (1.5 s ± 0.9 for LSE_GFE and 1.8 s ± 0.5 for BUHMAP). Furthermore, the longest event in LSE_GFE last 7 s while BUHMAP videos have a maximum length of 4.5 s. These differences talk about the different acquisition setting of both datasets: in BUHMAP volunteers perform a prompted movement/expression, in LSE_UVIGO volunteers respond naturally to questions asked by a deaf interpreter in sign language. In addition, the variation on acquisition frame rate between both sets is also rather significant, as BUHMAP videos were recorded at 30 frames per second while LSE_UVIGO has recordings at 50 fps (38% of the total) and 60 fps (the remaining 62%). Due to all these differences, it is reasonable to conclude that the size of the input samples from both datasets will probably have to be different in order to best fit the kind of events that it pretends to explain. The duration and frame rate were also included in the study.

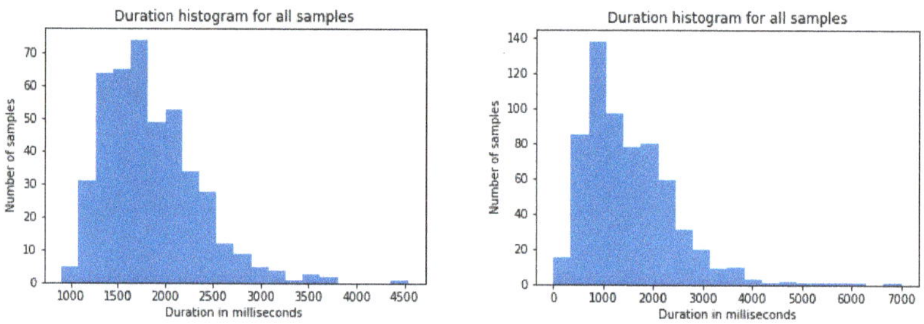

Figure 5. Duration of annotated segments in BUHMAP (**left**) and LSE_GFE (**right**).

In addition to the inter-datasets length differences there also exist inter-class differences for both datasets. Mean and deviation of each class lengths are presented in Figure 6. It is worth noting that the *None* class of LSE_GFE dataset has no deviation as its samples are obtained as fixed size sequences without target events and might contain any other emotional FE or GFE not being studied in this work, so it is not, in general a neutral expression. Furthermore, the classes in LSE_GFE have a higher deviation and particularly class *n.L-R* has a shorter mean duration, which can influence the obtained results.

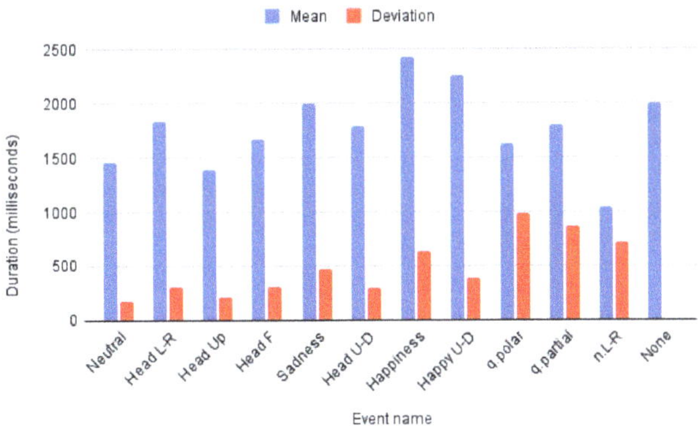

Figure 6. Duration per class in LSE_GFE and BUHMAP.

Video samples were generated by cropping a window with the selected duration centered in each labelled event of each video. An ablation study includes the effect of cropping duration and frame-rate.

4. Methods

This section covers the definition of classification models and the evaluation methods.

4.1. CNN Models

In order to verify the effectiveness of GCNs for GFEs recognition three different CNN architectures are assessed: the *VGG*, model the *MobilenetV2* [37] model and a custom CNN with several convolutional depths.

The rationale behind building a custom CNN instead of just using of-the-shelf models was to accommodate the kernels to the non-image nature of the input. As commented in Section 2, feeding a CNN with sequence of landmarks can be done by arranging them as concatenated rows in time [14], forming a 1-channel image if X-Y(-Z) coordinates are concatenated, or a 2(3)-channels X,Y(,Z). Once this artificial image is built one can use classical CNNs with typical square kernels, like VGG or mobilenet, or rectangular kernels that span one dimension along the length of the feature vector. We have adopted the latter for the custom CNN, both for landmarks and for AU features. Therefore, the custom CNN was composed using a convolutional block with 64 filters of size (5, *size of input features*) and (2, 0) padding, compacting the feature dimension to a single value for each frame while maintaining the temporal resolution. Regardless, more convolutional blocks can be appended to increase the model depth as shown in Figure 7. After an adaptive max pooling, the resulting 64 vectors are flattened and classified using two fully connected layers with a hidden space of 128 neurons. Furthermore, batch normalization is applied after the convolution and a ReLu activation follows the convolution and the first linear layer, while the output of the second linear layer is activated using a softmax function. This model has only $\sim$30 k parameters when only one convolutional block is used.

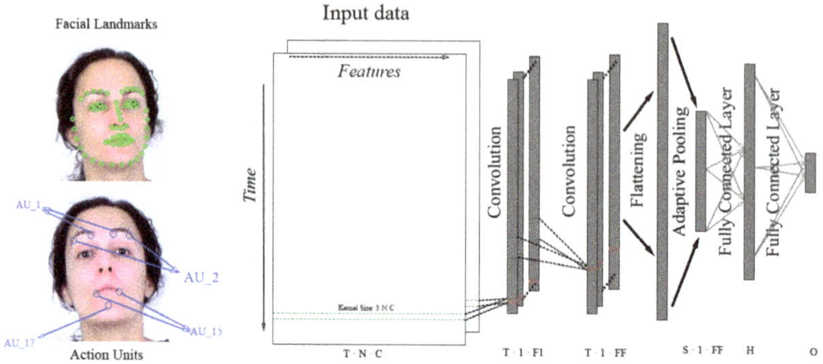

Figure 7. Overview of custom CNN architecture. The input data, whether facial landmarks or action units, is arranged in a matrix of shape $T(temporal\_length) \times N(number\_of\_features) \times C(dimensionality\_of\_features)$. This data is processed using $F1$ convolutional filters of shape $3 \times N \times C$ to obtain $F1$ feature vectors of shape $T \times 1$. Then, extra convolutional blocks with different number of filters can be added resulting in a final set of FF feature vectors with shape $T \times 1$. After that, these feature vectors are flattened and an adaptive pooling reduces their temporal dimension to S, obtaining a set of feature vectors with shape $S \times 1 \times FF$. Finally, those vectors are processed using two fully connected layers with a hidden space of H neurons resulting in O output values.

4.2. GCN Models

GNNs are designed to work with not regularly sorted data and can broadly be classified into two classes: spectral and spatial GNNs. The main difference between them is that spectral GNNs convolve the input graph with a set of learned filters in the graph Fourier domain while spatial GNNs, in general, perform layer-wise updates for each node by, first, selecting neighbors, then merging the features from the selected neighbors with an aggregation function and finally applying a transformation to the merged features. Graph Convolutional Networks (GCNs) are considered to be a spatial GNN variant characterized to perform mean neighborhood aggregation through convolution operations. This networks can be seen as a generalized version of Convolutional Neural Networks (CNNs) in which the data do not need to follow an order and the number and distribution of neighbouring nodes can vary, since the neighbourhoods are not based on spatial constraints but on defined relationships between nodes. To do so, a new element is added in the forward step Equation (1). In such equation the weights W^i of the i-th convolutional filter multiply the input features (nodes) X^i and the adjacency matrix (edges) A, which represents the relationship between the input features. This matrix has shape $N \times N$, where N is the number of input nodes.

$$H^{i+1} = \sigma(W^i X^i A) \quad (1)$$

The GCN selected for this work is based on a recent state-of-the-art model with great success in action recognition [38] (onwards, *msg3d*) and also in sign language recognition [39]. The representational capacity of this spatial-temporal model in HAR and SLR moved us to try testing it for GFER. If the model is able to capture GFE then we can hypothesize that a unified body skeleton and face mesh might be able to find the linguistic relationship between manual and non-manual (including GFE) components in sign languages. This model was adapted in this work to fit both facial landmarks and action units, matching each face landmark or action unit to a node feature of the graph. The structure of the full model is depicted in Figure 8. In short, it stacks r (3 in our case) spatial-temporal graph convolutional (STGC) blocks to process input features and then applies an average pooling and a softmax classifier on a fully connected layer.

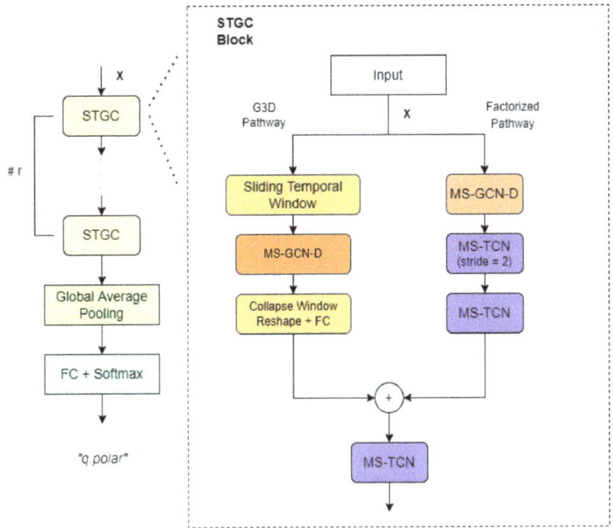

Figure 8. Overview of MSG-3D architecture. "TCN" and "GCN" denote temporal and graph convolutional blocks, and prefix "MS-" and suffix "-D" denote multi-scale and disentangled aggregation.

One of the main advantages of this model is the use of multi-scale disentangled graph convolutions. The goal of this operation is to take into account connections between nodes which are separated by several hops. Previous proposals [40] employed higher-order polynomials of the adjacency matrix to aggregate multi-scale structural information but this method suffers of a bias towards local regions as, even though it takes into account longer paths between nodes, the amount of shorter paths outweigh them. In order to avoid this problem, the authors build the k-adjacency matrix (where k stands for the number of scales used, i.e., the maximum path length between two nodes that is taken into account) as

$$A^{k}_{(i,j)} = \begin{cases} 1 & \text{if } d(v_i, v_j) = k \\ 1 & \text{if } i = j \\ 0 & \text{otherwise} \end{cases} \tag{2}$$

where $d(v_i, v_j)$ gives the shortest distance in number of hops between nodes v_i and v_j.

STGC blocks deploy two paths in order to extract regional as well as long-range spatial and temporal correlations: The first path (called G3D pathway) uses sliding spatial-temporal windows to sample small regions, performs disentangled multi-scale graph convolutions on them and collapses them with a linear layer. On the other hand, the second path (called factorized pathway) chains three different layers to obtain long-range, spatial-only and temporal-only information. The first layer performs disentangled multi-scale graph convolution only in the spatial dimension with the maximum number of graph scales, obtaining a long-range representation of spatial information for each temporal unit. Then, in second and third layers, it performs multi-scale temporal convolutions over the result of the first layer, thus capturing extended temporal information.

The model also includes a trainable mask which is added to the adjacency matrix A before each convolution step, allowing the network to learn directly the best connections between nodes. Unfortunately, the use of multiple scales and temporal windows scale up the size of this mask to tens of thousands of trainable parameters.

In our, pre-computed feature sequences were re-structured to build a graph, using each landmark coordinate or action unit intensity value as a node feature and defining relationships between all nodes. As the proposed method takes into account both spatial and temporal information in a unified model, the input to the model has to be a sequence of feature graphs corresponding to consecutive frames.

In order to test the influence of the input connections, particularly when using facial landmarks as input, two different graphs were defined and used in this work. The first one, named *base graph* and shown in Figure 9, define connections between the key-points with stronger muscle or anatomical relation. The second graph does not define any connection and rely on the ability of the model to learn relevant connections from scratch. In the case of action units, a single fully-connected graph was used in which all nodes were connected to each other.

Finally, an ablation test was carried out to study the impact of the number of trainable parameters of the model in the obtained results. To do so, the same tests were replicated using a single scale in both spatial and temporal dimensions.

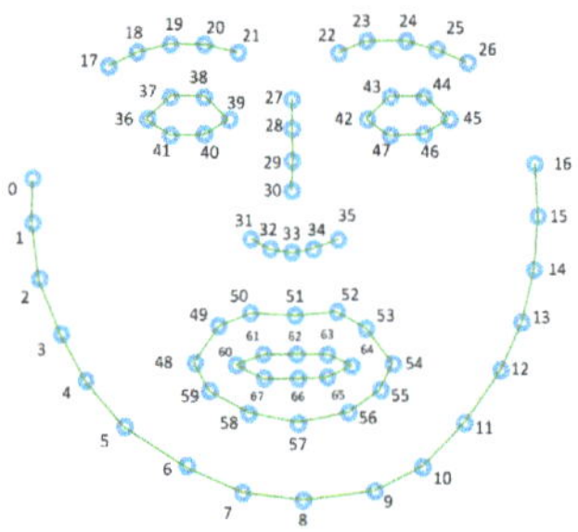

Figure 9. Base graph used with MSG3D.

4.3. Evaluation Strategy and Metrics

To evaluate the performance of each configuration Leave-One-Subject-Out (LOSO) cross-validation was used. This method divides the dataset in as many folds as different persons participate in it. Then, the model is trained from scratch once for each division, using the selected fold as validation data and all the remaining folds as training data.

Regardless, as the LSE_GFE has a large number of collaborators which contributed with different number of samples non-uniformly distributed over the classes, only the 11 persons with more and best distributed samples were used. Remember that BUHMAP is composed by 11 persons. The ids of the selected individuals as well as the number of samples from each class they recorded are depicted in Table 4.

Table 4. Sample distribution (#samples) of LSE_GFE collaborators used in LOSO.

ID	*q.polar*	*q.partial*	*n.L-R*	*None*
p0003	19	47	20	13
p0004	6	4	3	3
p0006	15	49	20	34
p0013	6	15	3	11
p0025	6	14	4	11
p0026	4	12	8	9
p0028	5	18	3	11
p0036	15	30	24	35
p0037	7	18	9	34
p0039	12	19	11	15
p0041	6	13	8	16

In addition, to tackle the randomness of stochastic gradient descent, which resulted in substantial differences between reiterated trainings of the same model, each evaluation needed to be repeated ten times to extract a reliable estimation of the model performance. Mean and $\pm$ standard deviation are provided for every test.

The metrics selected to evaluate the performance of the model are F1 score, for its capacity to assess precision and recall in a single value, and the accuracy. Finally, as each fold could contain a very different number of samples per class and the individual metrics of each split need to be weighted in order to achieve a reliable mean, the output of the model over the validation set of each fold was stored and concatenated in a single matrix. Then, the metrics are computed directly from this matrix, obtaining the equivalent to a weighted mean of all the folds. In addition, as the F1-score is defined for binary classification, it is computed individually for each class versus all others and then the weighted mean among all classes is obtained, taking into account the number of samples of each one in the validation set.

5. Results

In this section, the results obtained for all the relevant experiments carried out are presented and discussed.

Comparative study of classical CNN models.

In this study we compared the performance of VGG-11 and MobilenetV2, as examples of deep networks with different convolutional blocks, with the much smaller custom CNN model presented above. All of them were trained from scratch because there are not pretrained models for these type of inputs. It can be observed from Table 5 that VGG-11 and MobilenetV2 perform worse than the smaller custom CNN model for both datasets and both input features. Deeper custom CNN models (not presented in the Table to avoid overcrowding it) performed worse than the simplest one. It can be argued that VGG-11 and MobileNetV2 were not adapted to the type of input feature and also that the model is too large for the size of the dataset and overfits the data. It is worth noting that allowing a global combination of input features in custom CNN was more beneficial to landmarks than AUs, as the difference is coherent in both datasets. However, the classical local input filters of VGG-11 and MobilenetV2 yields not conclusive results on the advantage of landmarks over AUs, as they depend on the dataset.

Table 5. Comparative study of classical CNN models.

Model	Dataset	Feature	Weighted F1	Accuracy	Parameters
MobilenetV2	BUHMAP	landmarks	75.55% $_{\pm 2.78}$	75.11%±2.75	2233768
MobilenetV2	BUHMAP	AUs	78.30% $_{\pm 2.33}$	78.43%±2.26	2233480
MobilenetV2	LSE_GFE	landmarks	67.16% $_{\pm 2.22}$	66.90%±2.24	2228344
MobilenetV2	LSE_GFE	AUs	62.78% $_{\pm 1.12}$	61.43%±1.17	2228356
VGG-11	BUHMAP	landmarks	77.28% $_{\pm 2.64}$	77.50%±2.53	128804040
VGG-11	BUHMAP	AUs	79.20% $_{\pm 2.13}$	79.02%±2.15	128803464
VGG-11	LSE_GFE	landmarks	66.62% $_{\pm 2.27}$	66.89%±2.14	128787652
VGG-11	LSE_GFE	AUs	63.18% $_{\pm 0.99}$	62.21%±1.01	128787076
custom CNN	BUHMAP	landmarks	88.81% $_{\pm 1.30}$	88.75% $_{\pm 1.31}$	31992
custom CNN	BUHMAP	AUs	85.90% $_{\pm 1.21}$	85.98% $_{\pm 1.18}$	23912
custom CNN	LSE_GFE	landmarks	71.96% $_{\pm 1.45}$	71.89% $_{\pm 1.48}$	31732
custom CNN	LSE_GFE	AUs	70.49% $_{\pm 1.00}$	69.52% $_{\pm 1.06}$	23652

Assessment of MSG3D performance.

The next set of experiments are focused on assessing the performance of Multi-Scale Convolutional Spatial-Temporal Graph Neural Networks when applied to classifying GFEs. The rationale to bring to this work a network that has proven efficiency for action recognition with skeletal-type inputs might be questionable. Facial expressions do not show the same spatial-temporal variability as the full skeleton of a person in action, but given that GCNs can be seen as a generalization of classical CNNs and there are many model meta-parameters that can be adjusted, a rigorous study might throw interesting results.

Table 6 shows the performance of the baseline MSG3D model for both datasets and input features. It is interesting to highlight that MSG3D outperforms the custom CNN in both datasets when using landmark coordinates as the feature of the GCN nodes. This behaviour is even more interesting when looking at the number of free parameters of the baseline model, which is near three times the size of MobilenetV2. This means that MSG3D is a model much more appropriate to explain these kind of spatial–temporal data than MobilenetV2. Comparing with the lighter custom CNN the performance advantage is not so clear. It decreases a bit for BUHMAP but increases largely for LSE_GFE. However the size of the model is more than two orders of magnitude larger. On the other hand, when the features of the nodes are AUs, performance improves for BUHMAP but decrease dramatically for LSE_GFE. The size of the model is three times smaller than using landmarks because, differently to classical CNNs, it scales with the number of graph-nodes. The poor behavior of AUs in LSE_GFE remains unexplained, but our main hypothesis is that LSE_GFE is a more difficult dataset to extract reliable AUs as it only contains GFEs and OpenFace is optimized to extract AUs for facial emotional expressions. BUHMAP contains a mix of GFEs and emotional FEs. As we do not have control on the accuracy of the AUs extractor of OpenFace, and landmarks (trained for more scenarios than FEs) outperforms or equals AUs for this problem, we will do the next ablation study just using landmarks.

Table 6. Performance of baseline MSG3D model.

Dataset	Feature	Weighted F1	Accuracy	Parameters
BUHMAP	landmarks	$87.05\%_{\pm 1.79}$	$86.75\%_{\pm 1.80}$	6527396
BUHMAP	AUs	$87.71\%_{\pm 1.48}$	$87.73\%_{\pm 1.47}$	2913570
LSE_GFE	landmarks	$79.17\%_{\pm 1.45}$	$79.16\%_{\pm 1.50}$	6525856
LSE_GFE	AUs	$59.60\%_{\pm 0.8}$	$58.24\%_{\pm 0.91}$	2912030

5.1. *Ablation Study on MSG3D*

The objective of this ablation study consists of assessing the influence of several meta-parameters of MSG3D and input sample composition in the final performance for both datasets. The first three experiments deal with the bias-variance dilemma and the generalization capacity of the model. Hence, we will test a simple data augmentation technique, change the flexibility of the model and reduce the size of the model. The second set of experiments will test the influence of the temporal duration of the training sample that contains the GFE, and the apparent frame-rate seen by the model. In short:

- Use of data augmentation through horizontal flipping.
- The impact of graph topology.
- Number of spatial and temporal scales.
- The impact of GFE duration and frame-rate

Study on the effect of data augmentation through horizontal flipping.

Table 7 shows the effect of augmenting the input sample with x-flipped node features. Results clearly show that this simple data augmentation improves generalization so, the following tests will include it.

Table 7. Ablation study for data augmentation through horizontal flipping.

Dataset	Flipping	Graph	Weighted F1	Accuracy
BUHMAP	no	base	$87.05\%_{\pm 1.79}$	$86.75_{\pm 1.80}$
BUHMAP	yes	base	$90.64\%_{\pm 1.12}$	$90.45_{\pm 1.25}$
LSE_GFE	no	base	$79.17\%_{\pm 1.45}$	$79.16_{\pm 1.50}$
LSE_GFE	yes	base	$80.38\%_{\pm 0.95}$	$80.37_{\pm 0.93}$

Study on the effect of graph topology.

This study tries to assess if imposing a muscular–anatomical graph from the beginning is better than start from an empty graph. It must be highlighted that the model has the capacity to insert and delete connections in both cases during training. Table 8 shows nothing conclusive as imposing the initial graph is better for LSE_GFE but not for BUHMAP, so we will keep the variable in the next ablation test.

Table 8. Ablation study for graph topology.

Dataset	Graph	Weighted F1	Accuracy
BUHMAP	base-graph	90.64% $_{\pm 1.12}$	90.45 $_{\pm 1.25}$
BUHMAP	empty-graph	91.48% $_{\pm 1.17}$	91.36 $_{\pm 1.23}$
LSE_GFE	base-graph	80.38% $_{\pm 0.95}$	0.8037 $_{\pm 0.93}$
LSE_GFE	empty-graph	79.15% $_{\pm 0.66}$	0.7905 $_{\pm 0.0062}$

Study on the effect of number of spatial and temporal scales.

This study tries to evaluate whether the multiscale approach, which increases complexity and gives extra flexibility to the model to learn spatial-temporal dependencies, is worthwhile for this problem. Results from Table 9 shows that BUHMAP is better explained with a simpler model, but it is not so clear for LSE_GFE. As the complexity of the model is greatly reduced removing spatial and temporal scales (SS, TS) and the difference in LSE_GFE is not statistically significant when using empty-graph, we will keep this model for the last ablation test.

Table 9. Ablation study with one single scale in both spatial and temporal dimensions.

Dataset	Graph	(SS,TS)	Weighted F1	Accuracy	Parameters
BUHMAP	base-graph	(8,8)	90.64% $_{\pm 1.12}$	90.45 $_{\pm 1.25}$	6527396
BUHMAP	base-graph	(1,1)	92.75% $_{\pm 1.60}$	92.66 $_{\pm 1.67}$	2159676
BUHMAP	empty-graph	(8,8)	91.48% $_{\pm 1.17}$	91.36 $_{\pm 1.23}$	6527396
BUHMAP	empty-graph	(1,1)	93.21% $_{\pm 0.82}$	93.09 $_{\pm 0.86}$	2159676
LSE_GFE	base-graph	(8,8)	80.38% $_{\pm 0.95}$	80.37 $_{\pm 0.93}$	6525856
LSE_GFE	base-graph	(1,1)	79.99% $_{\pm 0.92}$	79.86 $_{\pm 0.93}$	2158136
LSE_GFE	empty-graph	(8,8)	79.15% $_{\pm 0.66}$	79.05 $_{\pm 0.62}$	6525856
LSE_GFE	empty-graph	(1,1)	79.93% $_{\pm 0.61}$	79.81 $_{\pm 0.64}$	2158136

Study on the effect of GFE duration and FPS.

The last ablation study is not related to the model itself but to the size and temporal redundancy of the input. We already explained that both datasets have been acquired for different purposes and with different acquisition settings, so there's not a priori information of the optimal duration and frame-rate for feeding the MSG3D in this context. Table 10 shows the result of halving/doubling the duration of the input event inside the limits of the distribution range of both datasets and subsampling the original frame-rate to reduce redundancy (as LSE_GFE is recorded at 50 and 60 fps a slight interpolation effect might be present in some of the tests). Table 10 shows a larger dependency on the duration for LSE_UVIGO than BUHMAP. This can be explained by the larger deviation per class in the former than the latter (see Figure 6). The best results for LSE_GFE are obtained with intervals of 2 s while for BUHMAP there's a slight improvement using 4 s. The influence of frame rate is not very important in general but a slight improvement is observed with less redundancy probably related to a minor overfitting risk.

Table 10. Ablation study on duration and FPS. Empty-graph, horizontal flipping and one single scale in both spatial and temporal dimensions.

Dataset	Duration	FPS	Weighted F1	Accuracy
BUHMAP	4 s	30	$89.87\%_{\pm 0.84}$	$89.0_{\pm 0.93}$
BUHMAP	4 s	15	$93.21\%_{\pm 0.82}$	$93.09_{\pm 0.86}$
BUHMAP	2 s	30	$92.83\%_{\pm 0.96}$	$92.75_{\pm 1.01}$
BUHMAP	2 s	15	$92.68\%_{\pm 0.82}$	$92.57_{\pm 0.89}$
LSE_GFE	4 s	30	$68.67\%_{\pm 1.89}$	$68.23_{\pm 2.00}$
LSE_GFE	4 s	15	$70.42\%_{\pm 1.43}$	$70.20_{\pm 1.43}$
LSE_GFE	2 s	50	$79.93\%_{\pm 0.61}$	$79.81_{\pm 0.64}$
LSE_GFE	2 s	20	$80.70\%_{\pm 0.97}$	$80.45_{\pm 1.01}$

5.2. Accuracy Per Class

In order to understand the difficulty of learning each specific class, the confusion matrices of the best models for each dataset and type of model are presented in Figure 10 only for landmark features. It is worth noting that the two architectures present an apparent complementary behavior, as those classes with worse accuracy in MSG3D have better or similar accuracy using the custom CNN, in both datasets. Complementary of models performance can be exploited for fused decisions, but drawing solid conclusions on fused classifiers must be handle carefully when datasets are as small as in this work. This study is left for future research if LSE_GFE is increased.

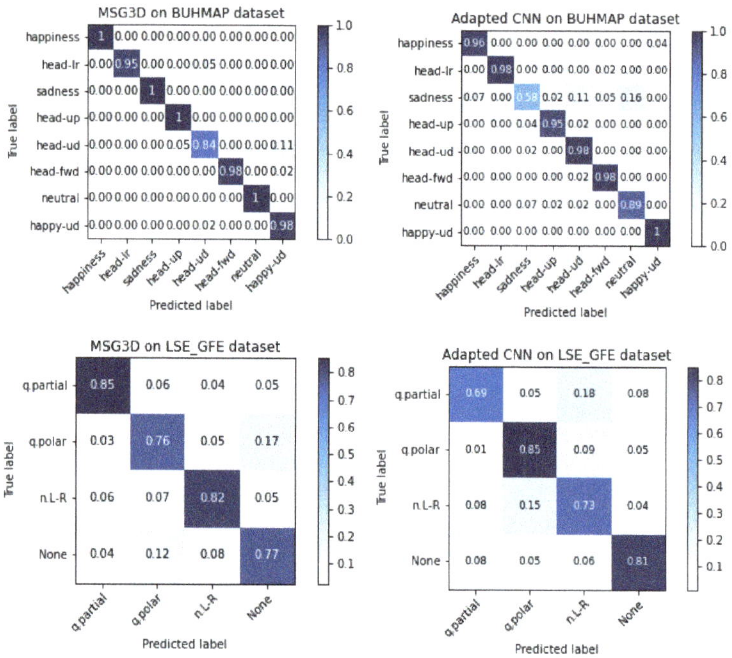

Figure 10. Confusion matrices of MSG3D (**left**) and custom CNN models (**right**) for BUHMAP (**top**) and LSE_GFE (**bottom**).

5.3. Comparison with State-of-the-Art Methods on BUHMAP

The papers published using BUHMAP dataset are a little bit old and most of them use different subsets of it, excluding some of the labelled classes. In addition each study uses different test strategies and report different metrics. In [41] the authors use all the eight labelled classes and report the mean test accuracy over LOSO. In [42] the samples of neutral expression are excluded and the reported results are the mean classification accuracy over LOSO. Ref. [43] uses a subset of the database focused on the three classes related to facial expressions and report the classification accuracy and the F1-score but no explanation about the used evaluation data is given. Finally, ref. [44] performs three different tests: The first one uses samples from seven classes (excluding neutral expression class) from one person to train and the same samples from other subject to test. The second one performs five-fold cross-validation over the samples of seven classes from two different persons. The last one uses three repetitions of four classes related to head movement from nine subjects to train and test on the remaining two subjects. The numerical results reported in the publications of these systems are gathered in Table 11 and compared with the best result obtained for the BUHMAP dataset in this work, which was achieved by training a reduced *MSG3D* model over the facial landmarks augmented through horizontal flipping. The only systems that outperform our proposal are not fully comparable as they remove four classes [44] or just the neutral expression [42]. The latter system is a fusion of several subsystems and an ad hoc selection of feature sets after the merger, which could point to some meta-overfitting. It was not our intention to build the best system for classifying over BUHMAP but presenting a new dataset and propose an alternative deep learning approach that handles spatio-temporal graphs to automate the classification of GFE. Testing over BUHMAP was the only way to make sure that our approaches made sense for this problem, and they could achieve SOTA performance on previous similar datasets.

Table 11. Comparison of performance over BUHMAP against other published works.

System	Classes	Eval. Method	Accuracy	F1-Score
[41]	All	LOSO CV	76.98%	
[42]	2 to 8	LOSO CV	98.2%	
[43]	1, 5 and 7	Unknown	88.50%	88.87%
[44]	2 a 8	1 person test	67.1%	
[44]	2, 3, 4 and 7	1 person test	95.0%	
[44]	2 a 8	5-fold CV	92.5%	
[44]	2, 3, 4 and 7	5-fold CV	96.6%	
[44]	2, 3, 4 and 7	2 persons test	91.6%	
Ours (best)	All	LOSO CV	93.09%	93.21%

5.4. Comparison with Human Performance on LSE_GFE

For the sake of completeness we made a last study on the performance of sign language experts when watching the same video clips extracted for testing the models. The sign language experts were three sign-language interpreters. Two of them had never seen the complete video interview of LSE_UVIGO from where the GFE video clips were extracted. The other interpreter was the same person (also author of this paper) that labeled the LSE_GFE dataset watching the whole sign language sentence. It is important to highlight that at least 12 months passed from the annotation of LSE_GFE to the experiment we are going to explain here, so this interpreter had mostly forgotten the context of the video clips.

A web-based application was prepared for the interpreters to watch a video clip from a random sequence extracted from the LSE_GFE dataset and manually annotate the 4 studied classes. Due to the difficulty of the problem, also alternative responses could be selected from this set:

- q.doubt -> "I am sure that it is question but I do not know whether it is polar or partial",
- doubt -> "I am not sure whether it is a question or a negation but I am sure it is one of them",
- not selected -> "I am not sure of any option. Pass".

With these three options we try to minimize the effect of random responses. Figure 11 shows a screenshot of the web tool accessible to the interpreters (in Spanish). They were free to enter in different moments and annotate as in an Amazon Mechanical Turk task. Every annotation was linked to the annotator and deleted from the random list for that annotator.

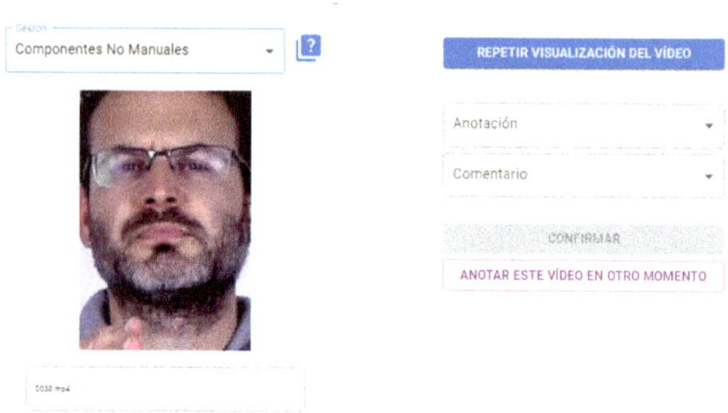

Figure 11. Screenshot of the web tool of LSE_GFE manual annotation.

Figure 12 show the confusion matrices of the different interpreters-annotators. Annotator 0 is the one that 12 months before labeled the dataset watching the whole sign language interviews. The best automated system is again displayed to facilitate comparison with annotators.

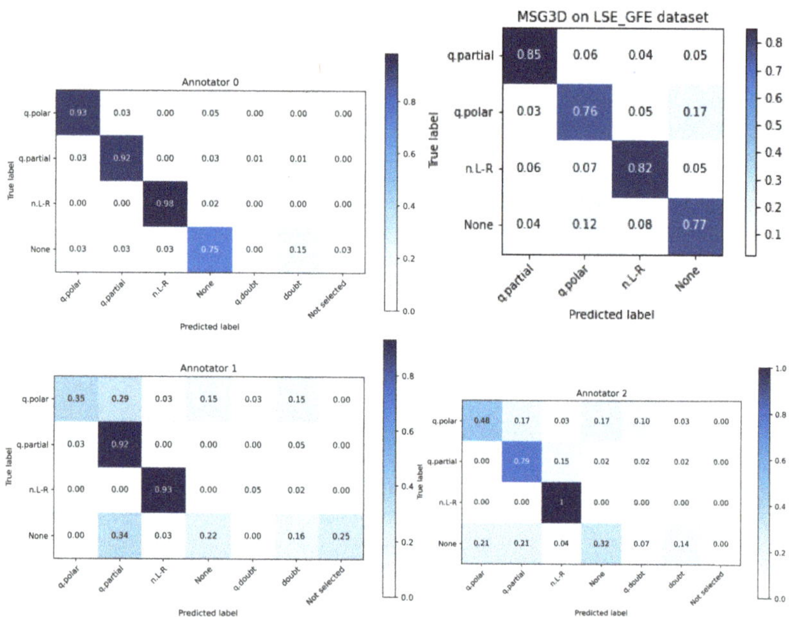

Figure 12. Confusion matrices of the interpreters against the original labels of LSE_GFE. Upper right shows the best automated system.

Several interesting observations can be drawn from this experiment:

- Annotator 0 performs much better than the others, so it is clear that there's an influence on having seen the footage and being involved in the acquisition process and discussions on the experiments.
- The class *None*, that was randomly extracted from interview segments where none of the 3 classes were present, is, by a large margin, the class with worse human

performance. Even the annotator 0 performed worse than the automated system. This is a clear cue that the larger performance of annotator 0 was boosted by the previous knowledge of the dataset.
- The three annotators outperformed the automated system in classes *q.partial* and *n.L-R*, but two of them performed much worse in class *q.polar* where they showed many doubts.

In summary, this experiment demonstrated that LSE_GFE is a challenging dataset worth distributing to the research community to advance the state of the art in GFE recognition and sign language comprehension. Also, the disagreement among annotators and the advantage of having seen the whole sign language sentence tells us that there is much more work to do regarding the duration of needed context in input features for building a successful automated system.

6. Conclusions

This work presented a new dataset of Grammatical Facial Expressions (GFE) acquired in a natural context of Spanish Sign Language (LSE_GFE) and a comparative study of a type of Convolutional Graph Neural Networks for automated classification of GFE classes. This type of GCN has been already successfully applied to FER but, as far as we know this is the first time that this type of complex models, well known in the action recognition arena, are applied to GFE recognition.

In order to assess the model adequacy for the task, all the experiments were carried out with the LSE_GFE and another publicly available dataset, BUHMAP, already gather and studied for FER and GFER in the past. The experiments using landmarks and action units as input features, and a custom CNN and the GCN model called MSG3D, over the two datasets, showed that the best option that surpassed the state of the art was obtained with simplified MSG3D fed with landmarks and augmented data with horizontal coordinate flipping.

Experiments with human expert sign language annotators showed that the simplified MSG3D model is able to compete with them, outperforming class accuracy in two of the four classes. These experiments also showed that the additional temporal context of GFE might be necessary for accurate disambiguation between similar expressive facial movements. Further research on models able to deal with class-dependent input duration could improve accuracy.

Author Contributions: Conceptualization, J.L.A.-C., L.D.-F. and A.P.-P.; methodology, J.L.A.-C., L.D.-F. and M.P.-L.; software, M.P.-L. and M.V.-E.; validation, J.L.A.-C., L.D.-F., M.P.-L. and M.V.-E.; formal analysis, J.L.A.-C. and L.D.-F.; investigation, M.P.-L., J.L.A.-C. and L.D.-F.; resources, M.P.-L. and A.P.-P.; data curation, M.P.-L. and A.P.-P.; writing—original draft preparation, M.P.-L.; writing—review and editing, J.L.A.-C. and Laura Docio; visualization, M.P.-L. and M.V.-E.; supervision, J.L.A.-C. and L.D.-F.; project administration, J.L.A.-C. and L.D.-F.; funding acquisition, J.L.A.-C. and L.D.-F. All authors have read and agreed to the published version of the manuscript.

Funding: This research has been funded by the Spanish Ministry of Science and Innovation, through the project RTI2018-101372-B-I00 "Audiovisual analysis of verbal and nonverbal communication channels (Speech & Signs)" and by the Xunta de Galicia and the European Regional Development Fund through the Consolidated Strategic Group AtlanTTic (2019–2022). M.V.-E. is also funded by the Spanish Ministry of Science and Innovation through the predoc grant PRE2019-088146.

Informed Consent Statement: Informed consent was obtained from all subjects involved in the study.

Data Availability Statement: All the data and code needed to reproduce the experiments of this work can be obtained from https://github.com/mporta-gtm/GrammaticalFacialExpressions (accessed on 1 May 2022).

Acknowledgments: We would like to acknowledge the contribution of the deaf people and interpreters that help to build the dataset and, specially the altruist contribution of the three annotators that made possible the last study of this work: Ania Pérez, Eva Aroca and María del Carmen Cabeza.

Conflicts of Interest: The authors declare no conflict of interest.

References

1. Ekman, P.; Friesen, W.V. Constants across cultures in the face and emotion. *J. Personal. Hencec. Psychol.* **1971**, *17*, 124. [CrossRef] [PubMed]
2. McCullough, S.; Emmorey, K.; Sereno, M. Neural organization for recognition of grammatical and emotional facial expressions in deaf ASL signers and hearing nonsigners. *Cogn. Brain Res.* **2005**, *22*, 193–203. [CrossRef] [PubMed]
3. Da Silva, E.P.; Costa, P.D.P.; Kumada, K.M.O.; De Martino, J.M.; Florentino, G.A. Recognition of Affective and Grammatical Facial Expressions: A Study for Brazilian Sign Language. In Proceedings of the Computer Vision—ECCV 2020 Workshops, Glasgow, UK, 23–28 August 2020; Bartoli, A., Fusiello, A., Eds.; Springer International Publishing: Cham, Switzerland, 2020; pp. 218–236.
4. Li, S.; Deng, W. Deep facial expression recognition: A survey. *IEEE Trans. Affect. Comput.* **2020** [CrossRef]
5. Ouellet, S. Real-time emotion recognition for gaming using deep convolutional network features. *arXiv* **2014**, arXiv:1408.3750.
6. Khor, H.Q.; See, J.; Phan, R.C.W.; Lin, W. Enriched long-term recurrent convolutional network for facial micro-expression recognition. In Proceedings of the 2018 13th IEEE International Conference on Automatic Face & Gesture Recognition (FG 2018), Xi'an, China, 15–19 May 2018; pp. 667–674.
7. Ekman, P.; Friesen, W. *Facial Action Coding System: A Technique for the Measurement of Facial Movement*; Consulting Psychologists Press: Palo Alto, CA, USA, 1978.
8. Zeng, Z.; Pantic, M.; Roisman, G.; Huang, T. A Survey of Affect Recognition Methods: Audio Visual and Spontaneous Expressions. *IEEE Trans. Pattern Anal. Mach. Intell.* **2009**, *31*, 39–58. [CrossRef] [PubMed]
9. Valstar, M.; Gratch, J.; Schuller, B.; Ringeval, F.; Cowie, R.; Pantic, M. AVEC 2016-Depression, Mood, and Emotion Recognition Workshop and Challenge. In Proceedings of the 6th International Workshop on Audio/Visual Emotion Challenge, Amsterdam, The Netherlands, 16 October 2016; pp. 1483–1484. [CrossRef]
10. Lien, J.J.; Kanade, T.; Cohn, J.F.; Li, C.C. Automated facial expression recognition based on FACS action units. In Proceedings of the Third IEEE International Conference on Automatic Face and Gesture Recognition, Nara, Japan, 14–16 April 1998; pp. 390–395.
11. Devries, T.; Biswaranjan, K.; Taylor, G.W. Multi-task Learning of Facial Landmarks and Expression. In Proceedings of the 2014 Canadian Conference on Computer and Robot Vision, Montreal, QC, Canada, 6–9 May 2014; pp. 98–103. [CrossRef]
12. Liu, X.; Cheng, X.; Lee, K. GA-SVM-Based Facial Emotion Recognition Using Facial Geometric Features. *IEEE Sens. J.* **2021**, *21*, 11532–11542. [CrossRef]
13. Qiu, Y.; Wan, Y. Facial Expression Recognition based on Landmarks. In Proceedings of the 2019 IEEE 4th Advanced Information Technology, Electronic and Automation Control Conference (IAEAC), Chengdu, China, 20–22 December 2019; Volume 1, pp. 1356–1360. [CrossRef]
14. Yan, J.; Zheng, W.; Cui, Z.; Tang, C.; Zhang, T.; Zong, Y. Multi-cue fusion for emotion recognition in the wild. *Neurocomputing* **2018**, *309*, 27–35. [CrossRef]
15. Ahmad, T.; Jin, L.; Zhang, X.; Lai, S.; Tang, G.; Lin, L. Graph Convolutional Neural Network for Human Action Recognition: A Comprehensive Survey. *IEEE Trans. Artif. Intell.* **2021**, *2*, 128–145. [CrossRef]
16. Ngoc, Q.T.; Lee, S.; Song, B.C. Facial landmark-based emotion recognition via directed graph neural network. *Electronics* **2020**, *9*, 764. [CrossRef]
17. Liu, Y.; Zhang, X.; Lin, Y.; Wang, H. Facial Expression Recognition via Deep Action Units Graph Network Based on Psychological Mechanism. *IEEE Trans. Cogn. Dev. Syst.* **2020**, *12*, 311–322. [CrossRef]
18. Jung, H.; Lee, S.; Yim, J.; Park, S.; Kim, J. Joint Fine-Tuning in Deep Neural Networks for Facial Expression Recognition. In Proceedings of the 2015 IEEE International Conference on Computer Vision (ICCV), Santiago, Chile, 7–13 December 2015; pp. 2983–2991. [CrossRef]
19. Yan, S.; Xiong, Y.; Lin, D. Spatial Temporal Graph Convolutional Networks for Skeleton-Based Action Recognition. In Proceedings of the Thirty-Second AAAI Conference on Artificial Intelligence and Thirtieth Innovative Applications of Artificial Intelligence Conference and Eighth AAAI Symposium on Educational Advances in Artificial Intelligence, New Orleans, LO, USA, 2–7 February 2018.
20. Heidari, N.; Iosifidis, A. Progressive Spatio–Temporal Bilinear Network with Monte Carlo Dropout for Landmark-based Facial Expression Recognition with Uncertainty Estimation. In Proceedings of the 23rd International Workshop on Multimedia Signal Processing, MMSP 2021, Tampere, Finland, 6–8 October 2021; pp. 1–6. [CrossRef]
21. Lucey, P.; Cohn, J.F.; Kanade, T.; Saragih, J.; Ambadar, Z.; Matthews, I. The Extended Cohn-Kanade Dataset (CK+): A complete dataset for action unit and emotion-specified expression. In Proceedings of the 2010 IEEE Computer Society Conference on Computer Vision and Pattern Recognition workshops, San Francisco, CA, USA, 13–18 June 2010; pp. 94–101. [CrossRef]
22. Zhao, G.; Huang, X.; Taini, M.; Li, S.Z.; Pietikäinen, M. Facial expression recognition from near-infrared videos. *Image Vis. Comput.* **2011**, *29*, 607–619. [CrossRef]
23. Valstar, M.F.; Pantic, M. Induced disgust, happiness and surprise: An addition to the mmi facial expression database. In Proceedings of the 3rd International Workshop on EMOTION (Satellite of LREC): Corpora for Research on Emotion and Affect, Valetta, Malta, 23 May 2010; pp.65–70.

24. Dhall, A.; Goecke, R.; Lucey, S.; Gedeon, T. Collecting Large, Richly Annotated Facial-Expression Databases from Movies. *IEEE Multimed.* **2012**, *19*, 34–41. [CrossRef]
25. Aifanti, N.; Papachristou, C.; Delopoulos, A. The MUG facial expression database. In Proceedings of the 11th International Workshop on Image Analysis for Multimedia Interactive Services WIAMIS 10, Desenzano del Garda, Italy, 12–14 April 2010; pp. 1–4.
26. Yin, L.; Wei, X.; Sun, Y.; Wang, J.; Rosato, M. A 3D facial expression database for facial behavior research. In Proceedings of the 7th International Conference on Automatic Face and Gesture Recognition (FGR06), Southampton, UK, 10–12 April 2006; pp. 211–216. [CrossRef]
27. Gunes, H.; Piccardi, M. A Bimodal Face and Body Gesture Database for Automatic Analysis of Human Nonverbal Affective Behavior. In Proceedings of the 18th International Conference on Pattern Recognition (ICPR'06), Hong Kong, China, 20–24 August 2006; Volume 1, pp. 1148–1153. [CrossRef]
28. Aran, O.; Ari, I.; Guvensan, A.; Haberdar, H.; Kurt, Z.; Turkmen, I.; Uyar, A.; Akarun, L. A Database of Non-Manual Signs in Turkish Sign Language. In Proceedings of the 2007 IEEE 15th Signal Processing and Communications Applications, Eskisehir, Turkey, 11–13 June 2007; pp. 1–4. [CrossRef]
29. Freitas, F.D.A. *Grammatical Facial Expressions;* UCI Machine Learning Repository: Irvine, CA, USA, 2014.
30. Jiang, X.; Zong, Y.; Zheng, W.; Tang, C.; Xia, W.; Lu, C.; Liu, J. DFEW: A Large-Scale Database for Recognizing Dynamic Facial Expressions in the Wild. In Proceedings of the 28th ACM International Conference on Multimedia, Seattle, WA, USA, 12–16 October 2020; pp. 2881–2889.
31. Sheerman-Chase, T.; Ong, E.J.; Bowden, R. Cultural factors in the regression of non-verbal communication perception. In Proceedings of the 2011 IEEE International Conference on Computer Vision Workshops (ICCV Workshops), Barcelona, Spain, 6–13 November 2011; pp. 1242–1249. [CrossRef]
32. Silva, E.P.d.; Costa, P.D.P.; Kumada, K.M.O.; De Martino, J.M. SILFA: Sign Language Facial Action Database for the Development of Assistive Technologies for the Deaf. In Proceedings of the 2020 15th IEEE International Conference on Automatic Face and Gesture Recognition (FG 2020), Buenos Aires, Argentina, 16–20 November 2020; pp. 688–692. [CrossRef]
33. Docío-Fernández, L.; Alba-Castro, J.L.; Torres-Guijarro, S.; Rodríguez-Banga, E.; Rey-Area, M.; Pérez-Pérez, A.; Rico-Alonso, S.; García-Mateo, C. LSE_UVIGO: A Multi-source Database for Spanish Sign Language Recognition. In Proceedings of the LREC2020 9th Workshop on the Representation and Processing of Sign Languages: Sign Language Resources in the Service of the Language Community, Technological Challenges and Application Perspectives, Marseille, France, 11–16 May 2020; pp. 45–52.
34. Max Planck Institute for Psycholinguistics. The Language Archive [Computer Software]. Available online: https://archive.mpi.nl/tla/elan (accessed on 20 November 2020).
35. Baltrušaitis, T.; Mahmoud, M.; Robinson, P. Cross-dataset learning and person-specific normalisation for automatic action unit detection. In Proceedings of the 2015 11th IEEE International Conference and Workshops on Automatic Face and Gesture Recognition (FG), Ljubljana, Slovenia, 4–8 May 2015; Volume 6, pp. 1–6.
36. Zadeh, A.; Lim, Y.C.; Baltrušaitis, T.; Morency, L.P. Convolutional experts constrained local model for 3D facial landmark detection. In Proceedings of the 2017 IEEE International Conference on Computer Vision Workshops, Venice, Italy, 22–29 October 2017; pp. 2519–2528.
37. Sandler, M.; Howard, A.; Zhu, M.; Zhmoginov, A.; Chen, L.C. MobileNetV2: Inverted Residuals and Linear Bottlenecks. *arXiv* **2019**, arXiv:1801.04381.
38. Liu, Z.; Zhang, H.; Chen, Z.; Wang, Z.; Ouyang, W. Disentangling and unifying graph convolutions for skeleton-based action recognition. In Proceedings of the IEEE Computer Society Conference on Computer Vision and Pattern Recognition, Seattle, WA, USA, 13–19 June 2020.
39. Vázquez-Enríquez, M.; Alba-Castro, J.L.; Fernández, L.D.; Banga, E.R. Isolated Sign Language Recognition with Multi-Scale Spatial-Temporal Graph Convolutional Networks. In Proceedings of the 2021 IEEE/CVF Conference on Computer Vision and Pattern Recognition Workshops (CVPRW), Virtual, 19–25 June 2021; pp. 3457–3466.
40. Li, M.; Chen, S.; Chen, X.; Zhang, Y.; Wang, Y.; Tian, Q. Actional-Structural Graph Convolutional Networks for Skeleton-based Action Recognition. *arXiv* **2019**, arXiv:1904.12659.
41. Aran, O.; Akarun, L. A multi-class classification strategy for Fisher scores: Application to signer independent sign language recognition. *Pattern Recognit.* **2010**, *43*, 1776–1788. [CrossRef]
42. Çınar Akakın, H.; Sankur, B. Robust classification of face and head gestures in video. *Image Vis. Comput.* **2011**, *29*, 470–483. [CrossRef]
43. Chouhayebi, H.; Riffi, J.; Mahraz, M.A.; Yahyaouy, A.; Tairi, H.; Alioua, N. Facial expression recognition based on geometric features. In Proceedings of the 2020 International Conference on Intelligent Systems and Computer Vision (ISCV), Fez, Morocco, 9–11 June 2020; pp. 1–6. [CrossRef]
44. Ari, I.; Uyar, A.; Akarun, L. Facial feature tracking and expression recognition for sign language. In Proceedings of the 23rd International Symposium on Computer and Information Sciences, Istanbul, Turkey, 27–29 October 2008. [CrossRef]

sensors

Article

Detection and Classification of Artifact Distortions in Optical Motion Capture Sequences

Przemysław Skurowski [1,*] and Magdalena Pawlyta [2]

1 Department of Graphics, Computer Vision and Digital Systems, Faculty of Automatic Control, Electronics and Computer Science, Silesian University of Technology, Akademicka 16, 44-100 Gliwice, Poland
2 Polish-Japanese Academy of Information Technology, Koszykowa 86, 02-008 Warsaw, Poland
* Correspondence: przemyslaw.skurowski@polsl.pl; Tel.: +48-32-2372151

Abstract: Optical motion capture systems are prone to errors connected to marker recognition (e.g., occlusion, leaving the scene, or mislabeling). These errors are then corrected in the software, but the process is not perfect, resulting in artifact distortions. In this article, we examine four existing types of artifacts and propose a method for detection and classification of the distortions. The algorithm is based on the derivative analysis, low-pass filtering, mathematical morphology, and loose predictor. The tests involved multiple simulations using synthetically-distorted sequences, performance comparisons to human operators (concerning real life data), and an applicability analysis for the distortion removal.

Keywords: motion capture; artifact classification; artifact detection; reconstruction; anomaly detection

Citation: Skurowski, P.; Pawlyta, M. Detection and Classification of Artifact Distortions in Optical Motion Capture Sequences. *Sensors* **2022**, *22*, 4076. https://doi.org/10.3390/s22114076

Academic Editors: Carlo Ricciardi, Tomasz Krzeszowski, Adam Świtoński, Michal Kepski and Carlos Tavares Calafate

Received: 15 March 2022
Accepted: 18 May 2022
Published: 27 May 2022

1. Introduction

Motion capture (mocap) systems [1,2] play important roles in modern computer graphics, where they are applied in gaming and movie FX to generate realistic character animations. Prominent applications of mocap systems could also be found in biomechanics and medical sciences [3]. To date, the most reliable technology is the marker-based optical mocap (OMC)—it is known as the 'gold standard' as it outperforms other mocap technologies. It utilizes visual tracking of active or retro-reflective passive markers. Trajectories of these markers are then used for animation of associated skeletons, which are used as key models in the animation of human-like or animal characters.

The process of acquiring marker locations is error prone. Distortions occurring in the mocap sequences can be simply divided into two classes—random noise and algorithmically-introduced artifact distortions. Random noise is a consequence of the stochastic processes resulting in different kinds of distortions in a mocap sequence. It has been studied in numerous works [4–7]. Among the types of noise, the most prominent [8] is white noise, which can be efficiently filtered out [9] or 'smoothed'. Numerous methods have been proposed [10] utilizing low-pass filtering, interpolating methods, or moving averages.

Artifact distortions are introduced by reconstruction algorithms present in mocap pipelines; they could be regarded as momentary systematic errors. These distortions introduce trajectory modifications of different appearances and of larger amplitudes. At higher levels of mocap processing, when the marker motion is remapped to drive the skeleton animation [11], the false positions of markers result in erroneous poses, which degrade the animation or biomechanical measurements. All distortions, gaps, and artifacts occur commonly in mocap sequences, influencing the praxis of a mocap operator. Since trajectory 'mis'-shapes are poorly filtered out by simple noise removing algorithms, standard industrial quality processing of mocap sequences require visual trajectory screenings and manual trajectory editing by operators. It is a painstaking process that could be assisted with software support for trajectory reconstruction; however, these capabilities are limited, and these methods could also degrade the results if used improperly.

Despite the common knowledge about artifact problems, this topic has not been fully recognized. In related works, researchers have mainly focused on error prevention during gap reconstruction, focusing on the efficiency of error removal (including artifacts) in terms of root mean squared error ($RMSE$). To our knowledge, our proposal is novel as it identifies erroneous intervals and classifies them accordingly.

The key motivation regarding the development of distortion classifiers is that, for each different distortion class, we can select an appropriate method of suppressing (e.g., for the rectangular distortion, which is a result of mistaken marker labeling—it would be enough to find its counterpart marker and swap erroneous parts of the trajectories to achieve perfect reconstruction).

In the article, we propose a marker-wise method for detection and classification of systematic errors—artifacts. The proposed approach is skeleton-free; therefore, it is able to adapt to virtually any vertebrate subject. There are two basic assumptions: the rigid body model and a correlation of marker trajectories. A rigid body model was assumed for the functional body mesh (structured point cloud) [12], which we used to represent the subject's body hierarchy. The next assumption stems from the former—it is the fact that the movements of markers are highly correlated and predictable when they are placed on common body parts (e.g., limbs). We employed a deviation of a trajectory from the prediction as a criterion for classification.

The proposed method is intended to support the mocap operator. It could be used in various ways, either assisting the operator by pointing out potential artifacts, or as a fully automatic method (combined with filtering and capable of detecting and removing artifacts). Our results show that the detection efficiency is on par with operators with intermediate experience, and it outperforms novice ones. Both approaches were verified in the experiments: E2—where we compared recognition abilities to the human operators and E3—where we verified recognition combined with several reconstruction methods.

The article is organized as follows: in Section 2, we disclose the background for the article—the mocap pipeline with distortion sources and former works on the distortions in optical mocap systems; Section 3 describes the proposed method with its rationales and design considerations. In Section 4, we test the method for its performance and discuss the results. Section 5 summarizes the article.

2. Background

2.1. Sources and Types of Distortions

There are two main sources of artifacts occurring in the markers in optical motion capture signals: software-caused (the main scope of the article) and soft tissue-caused artifacts (e.g., [13]). A soft tissue-caused artifact represented the actual marker motion; however, the marker was moved relative to the underlying bone because of local skin deformations.

In optical mocap, marker tracking is obtained by the image registration of the marker position by multiple IR cameras. A multi-view observation allows for the reconstruction of marker trajectories (3D positions over time) through the triangulation of 2D position recordings, which are registered by multiple cameras. The process is error-prone and various sources of the distortions can be identified, as depicted in Figure 1. Besides conventional stochastic noise, two main error sources in marker registration are gaps and erroneous marker matching. Gaps occur when the marker disappears from the camera view, due to locating the body part outside the scene (camera range), covering markers with another body part (occlusion). In such cases, reconstruction algorithms can be sources of errors. Marker matching occurs twice in the mocap pipeline. First, prior to the triangulation, it is necessary to perform marker matching in multiple 2D views of a single frame to identify corresponding 2D locations of markers. Another marker matching procedure is labeling (naming); it is performed among the different frames, where it is necessary to identify corresponding successive positions of 3D markers in the sequences of the frames. These software procedures can result in one trivial and four regular types of distortions, which can be observed in current mocap systems. These are:

1. Simple gap—appears when reconstruction algorithms give up, it is the least type of concern (a trivial case);
2. Single peak—caused by transient erroneous marker matching techniques. It is simple to detect;
3. Heavy noise of a much larger amplitude than ordinary noise introduced by frequent erroneous marker matching techniques;
4. Rectangular distortion—forward (followed by backward) steps caused by mismatching the 3D positions of the markers (part of the 3D trajectory is assigned to another marker) or due to the erroneous marker reconstruction based on a rigid body model;
5. Slow value change—two potential sources—accumulated reconstruction errors in successive frames (e.g., when there is deformation of a body, which is the failure of a commonly assumed rigid body model) or the result of low-pass filtering of peaks.

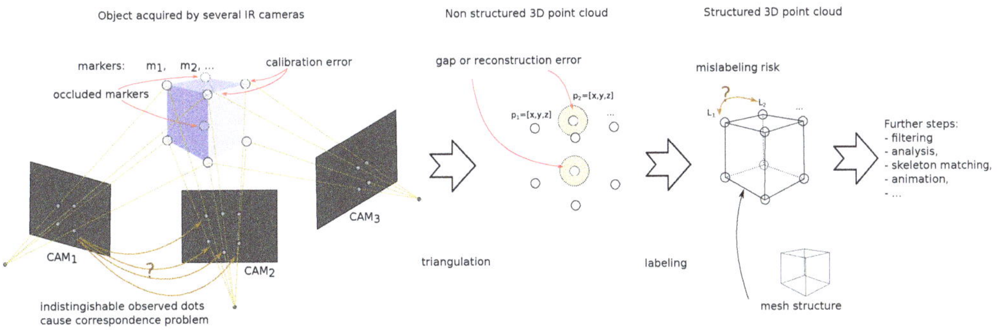

Figure 1. Processing in the early stages of the motion capture pipeline with distortion sources (red) and problems to solve (yellow), question marks (?) indicate ambiguous choices.

All of the above classes are depicted in Figure 2. They can be roughly divided into two basic classes—sudden (2–4) and slow (5) changes to the trajectory. It is worth noting that, aside from software sources, soft tissue over the skeleton is an additional distortion source as it denies the rigid body model. Usually, these artifacts take 'slow change' forms, so they fit into the above classification.

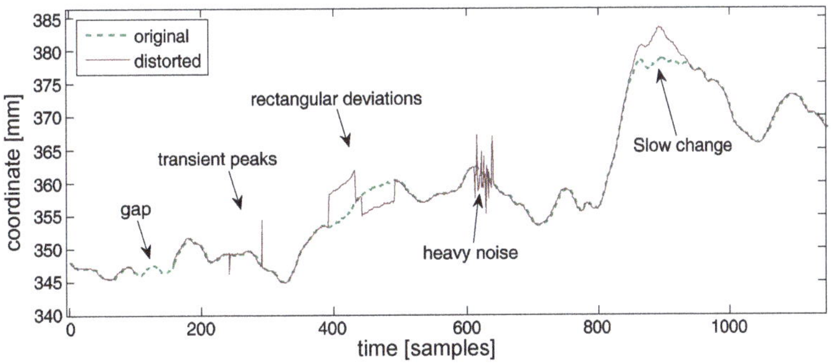

Figure 2. Identified types of distortions inpainted into exemplary data—the first coordinate of the first marker (head) of the IM subject.

2.2. Previous Work

To the authors' knowledge, this work is the first proposal for the identification and classification of artifact distortions in mocap sequences. Of course, the knowledge of reconstruction imperfections and artifacts is present in wider backgrounds of mocap technology and research; artifacts are referenced to in the former works as errors.

Most research studies on mocap signal processing mainly focus on preventing the occurrence of artifacts. Therefore, works on motion capture areas are related mostly to the occlusion gap filling problem and are oriented toward minimizing the gap filling error. They involve various methods for signal reconstruction, when the marker is lost in a recorded sequence. Approaches include interpolation [14], fusion of weak regressors [15], inverse kinematics [16], skeletal model [17], and inter-marker correlations [18,19]. Nowadays, deep neural networks are hot topics) [20–22]. However, such approaches usually require a lot of training data, which might not be available—every new marker configuration, new type of activity, or even individual actor might require retraining the network. It might be difficult, especially for the deep NNs, which may require a lot of training examples; therefore, deep NNs might just be feasible for typical situations.

The methods differ in the assumptions, performances, and complexities; some assume constraints from rigid body constraints and others employ skeletal models (or nothing). Constraint efficiency depends on how adequate and accurate the assumed model is. There could be discrepancies between the rigid body and modeled body segments [23], on the other hand, skeleton-based methods are sensitive to the accurate estimations of model parameters—bone lengths and marker placements—with respect to the underlying bone.

Typically, the errors in reconstruction methods appear in slow changes, which in some cases [17] are elongated so much that they appear as constant biases. The main factor that decides whether a certain method is suitable or not is the length of the gap. Simple signal-based methods (e.g., interpolation) work well for short errors, whereas complex model-based methods are better suited for long gaps.

The other approach, resulting in error/artifact reduction in the mocap pipeline, takes certain stages of the pipeline into consideration. Researchers have focused on partial problems in the motion capture pipeline, and they perfected these individual steps, improving the system configuration [24], e.g., the number and layout of the cameras, calibration [25], and labeling [26–29]. Such approaches undoubtedly reduce errors and improve the overall performance of the mocap pipeline, yet they are not perfect. Errors of reconstruction still occur; therefore, there remains room for improvement.

We identified only one research proposal slightly similar to ours, where erroneous intervals were explicitly identified for further cleaning. In [17], actual markers drove the skeleton first, then virtual markers were placed onto virtual skin. The positions of the actual and virtual markers were compared; if the marker positions did not match, such intervals were assumed to be erroneous and filtered; however, no identifications of distortion types were used.

3. The Proposal

3.1. Premises—Correlation of Trajectory Coordinates

The correlation between the locations and gradients of markers allowed us to propose a method to classify all types of distortions. Since the variables in mocap sequences can be strongly (positively and negatively) correlated within the groups, artifacts introduced by reconstruction algorithms should differ significantly enough to distinguish them on the basis of the proper trajectories of neighboring markers.

In Figure 3a, we present a correlation coefficient (CC) in the form of a distance matrix. It demonstrates structural dependencies in the correlations between the marker positions. One can easily observe the clusters formed by the body parts—hands, torso, legs, and so on. The correlation is high within individual body parts (both positive and negative); on the other hand, the correlates between body parts depend on the registered motions—in case of natural walking, the hand positions would counter-correlate, whereas with a butterfly swimming motion, the hand positions would correlate.

The time aspect of the correlation is depicted in Figure 3b; it presents the correlation and autocorrelation functions for several marker pairs. It clearly illustrates the correlation between successive marker locations and between locations of the markers located in the common body segments and connected body segments (e.g., head and neck).

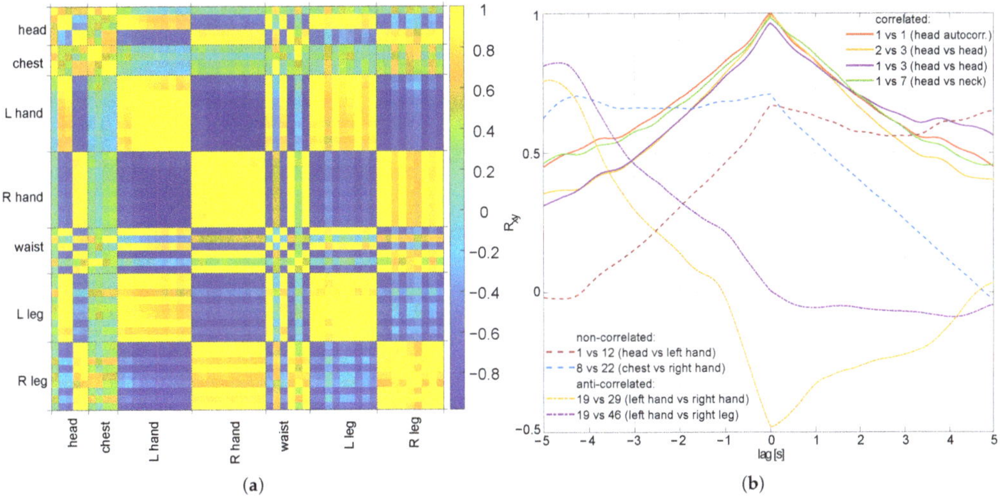

Figure 3. Correlation between X position of markers in exemplary sequence (fast walking HJ subject): (**a**) the whole sequence, all 53 markers; (**b**) inter-marker correlation function for the selected correlated and non-correlated markers.

To conclude this line of reasoning, we suggest that we could reliably identify the outstanding markers on the basis of the markers correlated within groups—from the common body segment or parent body part.

3.2. The Method Overview

The proposal is feasible thanks to the data correlations in mocap sequences, which make the predictions feasible and allow for reliable estimations of the actual positions of markers.

The key idea of the algorithm is to use model (prediction) results as a verification criterion for the data. If the data fall too far from the prediction results, then they are rationally considered distortion, and could be assigned into a distortion class using a pattern recognition method. Each distortion class is identified at a separate stage; it is cleaned from the signal using interpolation. The signal is then passed to the next stage of detection, from the simple distortion (single peaks) to the most difficult (slow changes). The conceptual scheme of the proposed method is depicted in Figure 4; for a detailed view, please refer to the unfolded schematic of the processing pipeline given in Section 3.4.

The method for anomaly identification and classification depends on the type of distortion. Sudden changes to the trajectory are identified on the basis of the differential signals and low-pass filters (as predicting models) with stats-based thresholding and mathematical morphology. This allows distinguishing between the types of sudden changes. Slow change detection is based on the hysteresis thresholding of residuals with backward regrowing of identified segments.

Three predictive models were employed in our pipeline. In case of sudden change detection, which is the simple case, we moved median and Savitzky–Golay filters to identify legitimate changes in the signal. To identify short-term distortions, we employed median filters (as they are estimator-resistant to peak changes). For longer term distortions, we employed Savitzky–Golay, which could follow a low-pass signal waveform in the presence of a high-frequency noise. For the detection of slow changes, we employed neighbor-based predictors—initially we assumed a polynomial predictor based on the least squares method, which we gave up in favor of a feed-forward neural network (FFNN), yet we decided to include it in the description, as it depicts the development process.

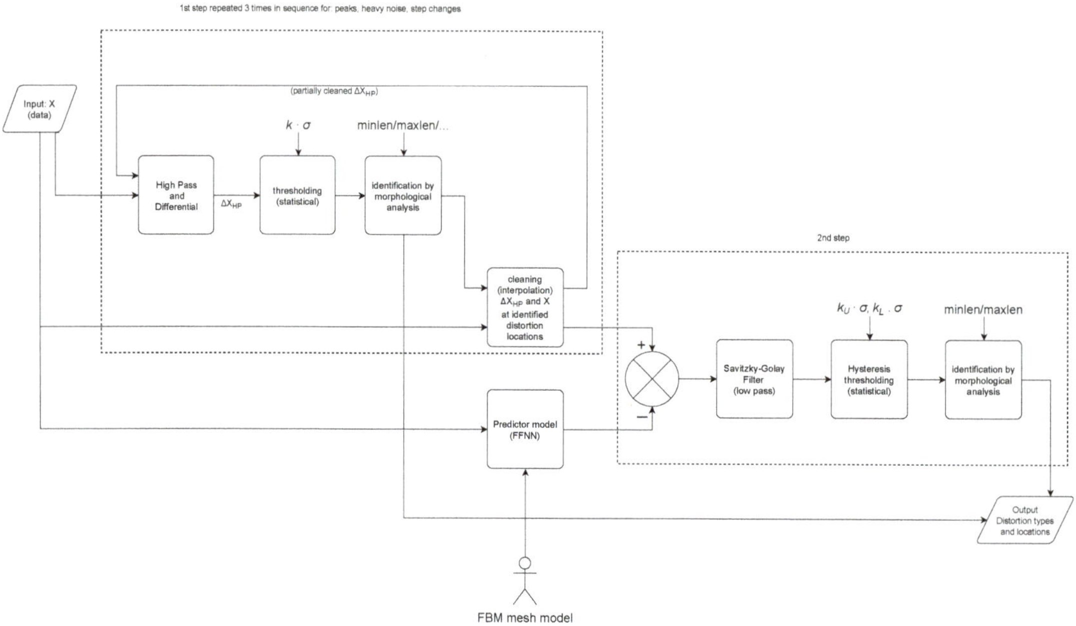

Figure 4. Conceptual schematic of the proposed algorithm.

3.3. Regressive Models

The efficiency of the overall approach depends on the quality of the model predictors and the ability to approximate the real location of a given marker (the location where it should be), on the basis of its own or neighboring markers locations from past, present, or future. In general, the regression model (predictor) [30] has the form of the function:

$$\hat{Y}_i = f(X_i, \beta_i) + R_i, \tag{1}$$

which in the linear [31] case yields:

$$\hat{Y}_i = X_i \beta_i + R_i, \tag{2}$$

where for N observations of M regressor variables, $\hat{Y}_i$ is the N-element long column of the predicted values of the i-th variable, X_i is the N-by-M design matrix for the model, β_i is the N-element column vector of coefficients, R_i is the error (residual). The model coefficients are estimated on the basis of X_i and Y_i, a column vector of goal values with the least squares method (LSM) is denoted as:

$$\beta_i = (X_i' X_i)^{-1} X_i' Y_i. \tag{3}$$

The residual is the remaining value, which is the non-predicted/non-correlated part of the signal, given simply as:

$$R_i = Y_i - \hat{Y}_i. \tag{4}$$

The part that will be further analyzed is the residual. Its probability follows the Laplace (double exponential) distribution:

$$f(x|\mu, b) = \frac{1}{2b} e^{\left(-\frac{|x-\mu|}{b}\right)}, \tag{5}$$

where: μ is the mean value equal to zero in our case, b is a dispersion parameter calculated on the standard deviation as $b = \sigma/\sqrt{2}$. Therefore, the standard deviation of the residual (denoted as σ_R) can be employed to evaluate the quality of the prediction; moreover, it indicates that such a centered distribution allows for an efficient outlier detection using the thresholding based on the standard deviation (e.g., three-sigma rule).

3.3.1. Savitzky–Golay Filter

Savitzky–Golay filter [32] is a smoothing filter that is based conceptually on polynomial fitting in the least squares sense (there are efficient convolution-based implementations). Its output is a value of the polynomial function fit locally to the data. The coefficients (c^l) of the L-th order are fit to the data within the sliding window of size M centered around $x(i)$; the filter output is the polynomial value for the midpoint. In the basic variant, the filter is 'low-pass', but a 'high-pass' can be obtained by a simple difference between the signal and its smoothed variant:

$$p_{LP}(i) = \sum_{l=0}^{L} c_l \cdot x^l(i) \tag{6}$$

$$p_{HP}(i) = c(i) - p_{LP}(i) \tag{7}$$

Its least squares design matrix can be simply noted as:

$$X_i = \begin{pmatrix} 1 & x(i-M) & x^2(i-M) & \cdots & x^L(i-M) \\ \vdots & \vdots & \vdots & \ddots & \vdots \\ 1 & x(i+M) & x^2(i+M) & \cdots & x^L(i+M) \end{pmatrix},$$

3.3.2. Neighbor-Based Linear Least Squares Loose Model

Slow change detection requires the predictor to be able to avoid following slowly-accumulating changes in the signal; hence, we employed loose (weak) prediction, which does not rely on its own momentary positions of the marker, it only uses past and current positions of sibling and parent markers. Initially we employed the polynomial model; it was obtained with ordinary least squares (LS) and was conveniently planned using the Vandermonde matrix X_i, with some caution, as it could be ill-conditioned with growing polynomial orders, as:

$$X_i = \begin{pmatrix} 1 & x_j(1) & x_j^2(1) & \cdots & x_j^L(1) & x_k(1) & x_k^2(1) & \cdots \\ 1 & x_j(2) & x_j^2(2) & \cdots & x_j^L(2) & x_k(2) & x_k^2(2) & \cdots \\ \vdots & \vdots & \vdots & \ddots & \vdots & \vdots & \vdots & \cdots \\ 1 & x_j(N) & x_j^2(N) & \cdots & x_j^L(N) & x_k(N) & x_k^2(N) & \cdots \end{pmatrix},$$

where: $x(n)$ are successive $1 \ldots N$ values of a single regressor variable, L is a polynomial order, $i, j, k, \cdots$ are variable indices, such that $i \neq j, k, \cdots$.

Considering the predictor, the term order appears twice, meaning the context size—number of former values taken into prediction and polynomial order. Hence, to avoid confusion in the paper, we used the following notation for predictors and residuals

$$P_k^L(i,n), \quad R_k^L(i,n) \tag{8}$$

where: L—means polynomial order used in X, k—the number of past values of regressor variables used to construct X, n—the number (time) of frames in the sequence, i is the number of predicted variables.

The selection of proper markers to formulate the predictor for each marker according to Equation (3) is based on the body's hierarchical structure. For that purpose, we used a body structure as depicted in Figure 5a—a functional body mesh (FBM) [12] for an average human subject, inferred for the typical Vicon marker setup. The FBM hierarchy with the corresponding skeleton is shown in Figure 5b. The FBM represents a kinematic structure—group markers located on the structure of the body and a hierarchy as a tree of these groups. The structure-obtaining step must be performed for each class of subjects or different sets of markers separately. The design matrix X was composed of coordinates of parents and siblings in current and k former frames (but it excluded former coordinates of the considered marker). A parent cluster is represented by a single marker—the one that is closest in terms of gradient coherence and distance constancy.

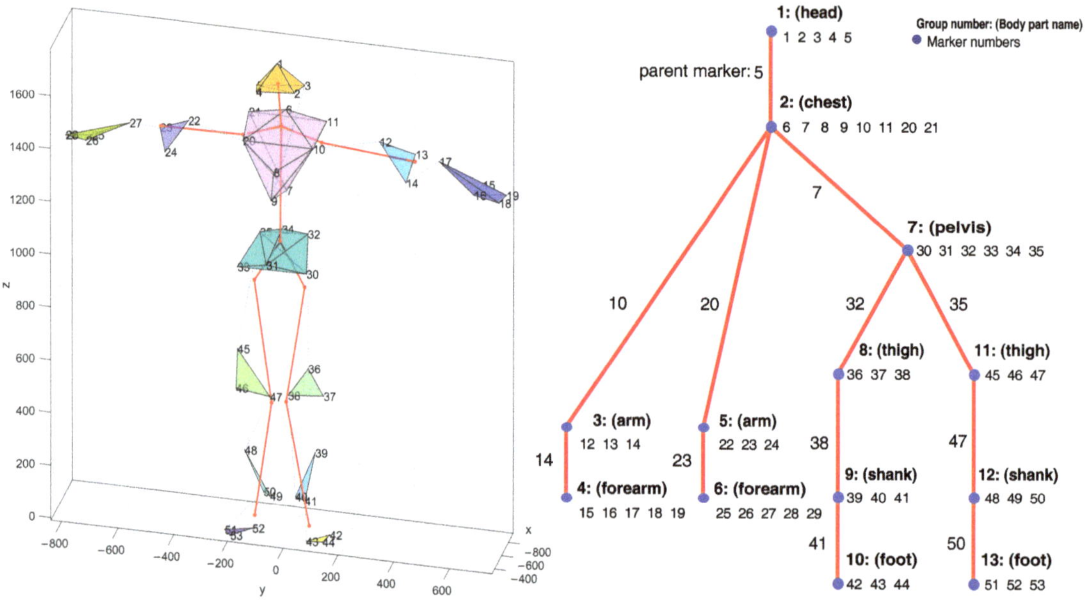

(a) (b)

Figure 5. Outline of the body model, with body parts distinguished with individual random colors and underlying skeleton included (**a**), and corresponding parts hierarchy annotated with parents and siblings (**b**).

Our demands toward the predictor were slightly specific. One obvious requirement was for it to be as precise as possible. Another (but contradictory) requirement was that it would not follow momentary changes induced by artifacts. Such requirements made us choose a special approach to formulate the X matrix; the predictor efficiency of the predictor depended on the data used. We neglected past values of considered markers—this was due to the fact that the largest correlation was between the current and past locations of a marker; therefore, it ensured the accuracy (Figure 6b). Unfortunately, in case of a distortion, it would make the predictor follow the artifact deviation; see Figure 6a. Next, we chose predictor parameters. Usually, the higher the degree of the polynomial and context size, the higher the precision; however, due to ill conditioning of X, it could reach a higher error with the growing polynomial order. Moreover, too large of an increase would not improve the predictor accuracy. The predictor parameters were tuned with numerical testing with preliminary data. We set up parameters to $k = 3$ and $L = 4$, as they appeared during the preliminary model tuning (Figure 7) to be a reasonable trade-off between the predictor accuracy and computation complexity.

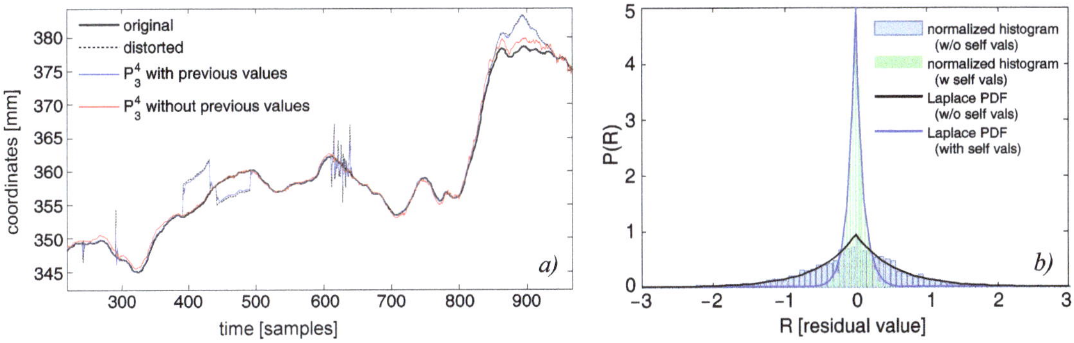

Figure 6. Performance of the two predictor models of P_3^4 (with and without former values of the predicted variables): (**a**) first dimension of the first marker (with artificial distortions); (**b**) residual histograms and corresponding Laplace PDFs for R_3^4 (for explanation of the model construction and parameters, see Equation (8)).

Summarizing each row vector in X is long and is assembled of certain parts as given below:

$$X(n) = \begin{bmatrix} 1, \overbrace{x_p(n), y_p(n), z_p(n), x_p(n-1), \cdots, z_p(n-k)}^{\text{current value and } k \text{ former of parent marker } (p)}, \\ \overbrace{x_{s1}(n), y_{s1}(n), z_{s1}(n), x_{s1}(n-1), \cdots, z_{slast}(n-k)}^{\text{current value and } k \text{ former of first..last siblings}}, \\ \vdots \\ \underbrace{x_{s1}^L(n), y_{s1}^L(n), z_{s1}^L(n), x_{s1}^L(n-1), \cdots, z_{slast}^L(n-k)}_{\text{current value and } k \text{ former of first..last siblings raised to } L\text{th power}} \end{bmatrix}. \quad (9)$$

We studied the polynomial models, thoroughly scanning the parameter space (see Figure 7) to obtain the accuracy, allowing for identification of slow deviations; the residual was noisy and the deviations were visible, but cluttered, making their automatic identification (see Section 3.4.5) work poorly. Therefore, we reviewed a series of various regression techniques—SVM, ANFIS-fuzzy models, lasso, ridge, regression trees, different variants of neural networks.

3.3.3. Regression with Neural Network

We found the solution for the regression problem in neural networks [33], with their ability to solve the regression problems. However, we proposed some additional modifications besides classical feed-forward NN [34] tuning performed during NN engineering, such as the number of layers and neurons, and the selection of the training algorithm.

Random-valued initialization with the scaled conjugate gradient training method made each output replica follow the true values, and the errors (residuals) were not correlated, unless they represented actual distortion. Hence, their averaged residual values exhibited much lower noise levels, so it allowed us to reveal the slow artifacts with thresholding.

The design of the NN structure is a kind of art, as there are no unambiguous rules or guidelines. Usually, it requires simulating with parameter sweeping for a domain of possible (feasible) numbers of layers and neurons, with a critical review of obtained performance (MSE or classification ratio) [35]. We shared that approach and reviewed the performance of NN using the test data. The NN architecture we employed is presented in Figure 8. It is ordinarily a fully connected FFNN, with two hidden layers—first containing 12 sigmoid neurons, second containing $4 \cdot M$ sigmoid neurons. The output is a three-valued x, y, z vector replicated P times—we used a five-fold replication. As input, we used a similar set of neighbor and parent coordinates to Equation (9); additionally, we enhanced it with the moving average of the own value of the marker. The latter could make the NN follow momentary slow changes; therefore, the window of a moving average (MA) should be notably larger than the lengths of the detected distortions (we assumed it to be 200 samples). By extensive testing, we also identified the number of previous values (=1) and the order of power used to raise the input data ($L = 2$). Finally, each input vector X was long and assembled of certain parts, as given below:

$$X(n) = \begin{bmatrix} \overbrace{x_p(n), y_p(n), z_p(n), x_p(n-1), \cdots, z_p(n-k)}^{\text{current value and } k \text{ former of parent marker } (p)}, \\ \overbrace{x_{s1}(n), y_{s1}(n), z_{s1}(n), x_{s1}(n-1), \cdots, z_{slast}(n-k),}^{\text{current value and } k \text{ former of first..last siblings}} \\ \vdots \\ \underbrace{x_{s1}^L(n), y_{s1}^L(n), z_{s1}^L(n), x_{s1}^L(n-1), \cdots, z_{slast}^L(n-k),}_{\text{current value and } k \text{ former of first..last siblings raised to } L\text{th power}} \\ MA_x(n), MA_y(n), MA_z(n) \end{bmatrix}. \quad (10)$$

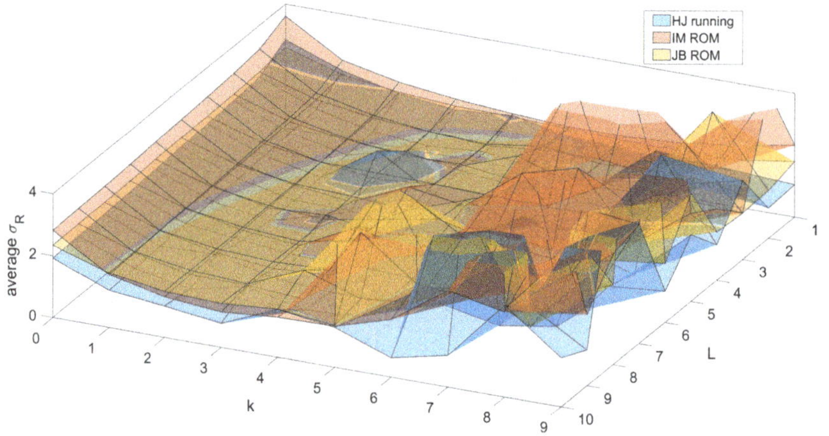

Figure 7. Predictor parameter tuning with preliminary data for three subjects, quality as standard deviation averaged over all markers; the tuned parameters: L—polynomial order, k—context size (lags).

The classical feed-forward NN performed well for the regression of the position of a marker. The residuals should reveal distortions if the NN is not overtrained. However, fluctuations of the residual, which resemble pink noise, could cause false detections when thresholded. Therefore, we decided to mimic multi-start NN training, with P-fold replication of the target output $Y_p = [x, y, z]$ values. Thus, our prediction has the final step:

$$\hat{Y} = \frac{1}{M} \sum_{m=1}^{M} \hat{Y}_m. \quad (11)$$

In Figure 9, we demonstrate the prediction results for the real sequence contaminated with artificially-introduced distortion. We can clearly observe that NN residuals contain the expected changes in signals, whereas residuals from the polynomial model are inconclusive.

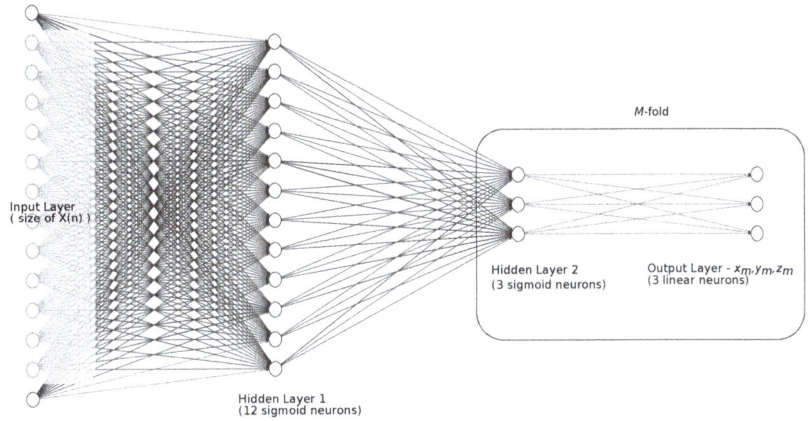

Figure 8. Architecture of the neural network used for regression.

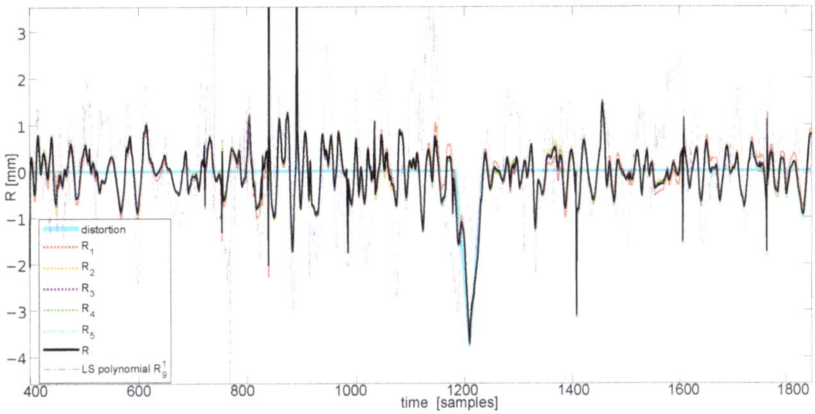

Figure 9. Actual distortion prediction and residuals—component (R_m) and final (R), compared with the polynomial residual R_9^1.

3.4. Recognition and Classification of Distortions

The detection process was organized into a strict pipeline consisting of four stages, where we detected distortions from the simplest to the hardest, along with the removal of the detected distortions (through interpolation after each stage). Such an approach ensures proper classification, otherwise we would have false positive classification due to the fact that subsequent methods could also be sensitive to simpler distortions, i.e., slow change would also be sensitive to rectangular distortion if the amplitude of distortion is sufficient. The detailed schematic of the dataflow in the algorithm is depicted in Figure 10.

The detection worked using the prediction residuals; therefore, we can see the deviations that cannot be explained by expected movements of markers described with the model. The choices of the models depended on the detected anomaly. For the sudden changes, these models were low-pass filters—median (for peaks) and Savitzky–Golay (for longer distortions); therefore, the respective high-pass filters acted as residuals. For the slow change, the model was FFNN and the residual was given explicitly as the difference between the signal and prediction.

To some extent, the residual values as deviations from a model can be considered innovations in the marker positions, resulting in position changes beyond the prediction. Therefore, minor residual values can be interpreted as normal motions, whereas large or sudden changes imply artifacts. Knowledge of the statistical properties of residuals allowed us to evaluate thresholds for detecting outliers of the normal variabilities of residuals.

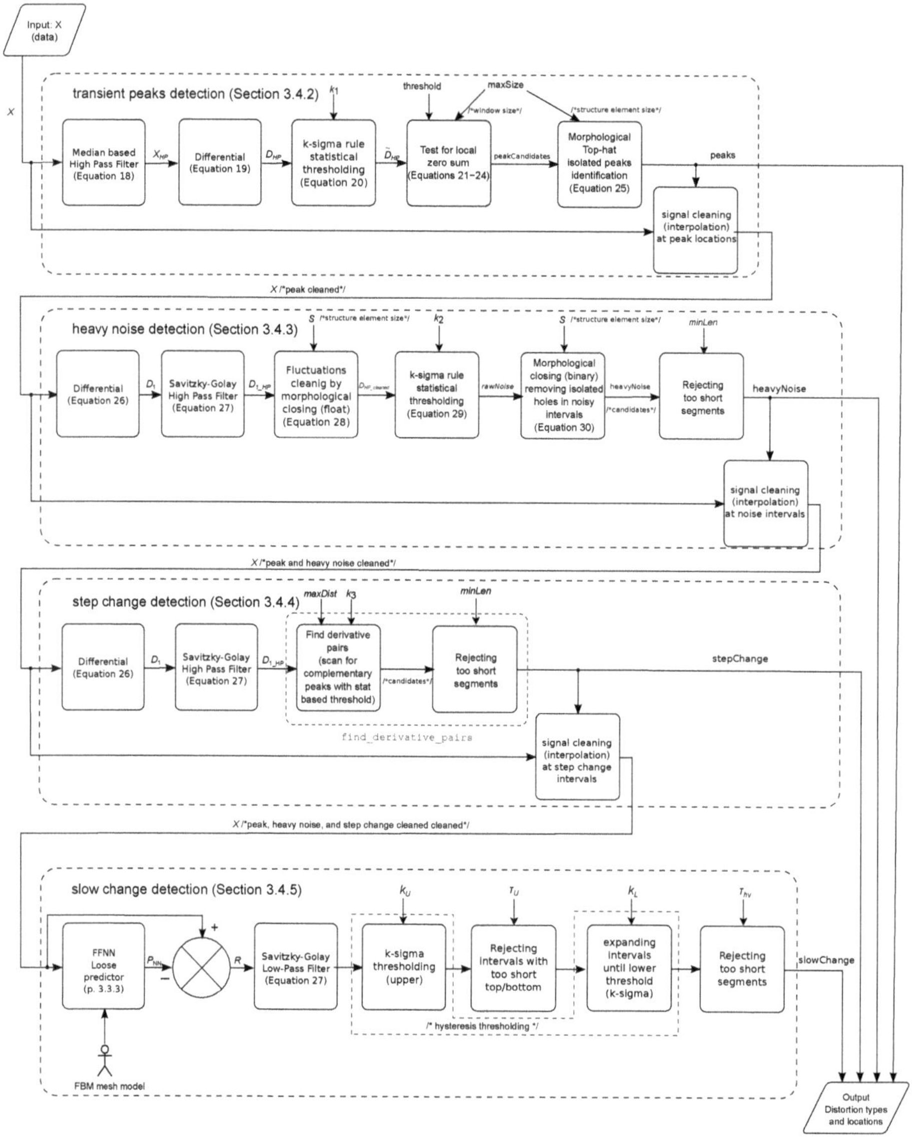

Figure 10. The detailed schematic of the proposed algorithm.

3.4.1. Locating Sudden Changes

Sudden changes are well detectable in the derivative of the basic signal, meanwhile slow changes require use of a base representation of residuals to measure the deviation. For the approximation of derivatives, we used differentials:

$$\Delta X(n) = X(n) - X(n-1). \quad (12)$$

Discrimination of different types of sudden changes cannot solely rely on differentials. This is because a strong peak in ΔX notifies about the existence of a sudden change, but it does not bring information about the structure, i.e., the duration of the change or its neighborhood. Therefore, we employed mathematical morphology methods (MM) to analyze the shapes of those sudden changes. We used the following MM operations:

$$\text{Erosion: } E(n) = x(n) \ominus S = \max(\forall_{j \in S_n} x(n-j)), \tag{13}$$
$$\text{Dilation: } D(n) = x(n) \oplus S = \min(\forall_{j \in S_n} x(n-j)), \tag{14}$$
$$\text{Opening: } O(n) = x(n) \circ S = (x(n) \ominus S) \oplus S, \tag{15}$$
$$\text{Closing: } C(n) = x(n) \bullet S = (x(n) \oplus S) \ominus S, \tag{16}$$
$$\text{Top-hat: } T_w(x(n)) = x(n) - (x(n) \circ S), \tag{17}$$

where: $x()$ is a 1D signal, S is the structuring element defining points to be taken into consideration (j), S_n is the structuring element centered (translated) at n.

An additional morphological method involves seeking sudden changes; we implemented a scanning function (`find_derivate_pairs(dX,T,maxlen)`), which looked for opponent differential pairs `dX`, exceeded the threshold level `T`, and was no more distant than some presumed maximal length `maxlen`. It results=ed in binary decision variables marking located ranges. The function parameters—threshold and distance—depend on the data characteristics and sampling frequency.

3.4.2. Identifying Single Peaks

It is the first stage of processing. Single peaks and heavy noises are two appearances of the same short-term distortion—the key difference is that single peaks are isolated within some neighborhoods, whereas in heavy noise segments, numerous peaks occur next to each other. To identify the isolated peaks, we employed the following sequence of operations.

First, we removed low frequencies (presumed to be legitimate) using a median filter:

$$X_{HP} = X - \text{median}(X, window), \tag{18}$$

where the *window* size should be several times larger than the maximal peak length.

Then, we calculate differential:

$$D_{HP}(n) = X_{HP}(n) - X_{HP}(n-1), \tag{19}$$

which is cleaned of non-interesting low values using thresholding :

$$\tilde{D}_{HP}(n) = \begin{Bmatrix} D_{HP}(n), & \text{if } D_{HP}(n) > threshold_1 \\ 0, & \text{otherwise} \end{Bmatrix} \tag{20}$$

Next, the identification of probable peaks is based on the assumption that the differential sum is local to small values (in theory ≈ 0), whereas the local sum of absolute values of the differential is high. Thus, we calculated these sums within window W_n:

$$movSum(n) = \sum_{j \in W_n} \tilde{D}_{HP}(n-j) \tag{21}$$

and

$$movSumAbs(n) = \sum_{j \in W_n} |\tilde{D}_{HP}(n-j)| \tag{22}$$

which are then tested as:

$$ampRatio(n) = \frac{movSumAbs(n)}{|movSum(n)|} \tag{23}$$

when the $ampRatio$ is larger than the assumed threshold (we assumed 5), it implies there is a peak candidate:

$$peakCandidates(n) = \begin{Bmatrix} 1, & \text{if } ampRatio(n) > threshold_2 \\ 0, & \text{otherwise} \end{Bmatrix} \quad (24)$$

Finally, to identify single peaks only, we employed binary top-hat, which rejects peaks within the neighborhood defined by the structuring element. It allowed us to keep isolated peaks only:

$$peaks = T_w(peakCandidates, S) \quad (25)$$

Tunable parameters of the stage are:

- *window*—for the moving average, we assumed it to be 19 samples long;
- $threshold_1$—it is calculated statistically from the data using $k_1 \cdot \sigma$ of D_{HP}—w employed $3 \cdot \sigma$ as a default value; however, any k_1 can be provided as the parameter;
- $threshold_2$—(anti-sensitivity) for the *ampRatio*;
- *maxSize*—(default 5), which declares the maximal size of the expected peaks; it affects the size of moving sum windows W_n, which is $2 \cdot maxSize + 3$ samples long, it also defines the size of the linear structuring element for morphological operations S.

3.4.3. Heavy Noise

Heavy noise detection is somewhat similar in design to isolated peak detection, but there are differences in the details. Foremost, we assume that the input data are already clear of isolated peaks. First, we calculate the differential:

$$D_1 = X(n) - X(n-1), \quad (26)$$

from which we remove low frequencies using a high pass variant of the Savitzky–Golay filter (Equation (7)):

$$D_{1\_HP} = SavitzkyGolayHiPass(D_1, L, M). \quad (27)$$

Next, we remove small fluctuations within the prospective areas of high values in D_hp with morphological closing (float):

$$D_{1\_HP\_cleaned} = |D_{HP}| \bullet S. \quad (28)$$

These values are now thresholded:

$$rawNoise(n) = \begin{Bmatrix} 1, & \text{if } D_{1\_HP\_cleaned} > threshold \\ 0, & \text{otherwise} \end{Bmatrix}. \quad (29)$$

Raw noise intervals are finally cleaned by removing holes using morphological closing, but the binary variant this time:

$$heavyNoise = rawNoise \bullet S. \quad (30)$$

Finally, heavy noise segments shorter than the *minLen* attribute are rejected.

Tunable parameters of the heavy noise detection stage are:

- *threshold*—this is calculated statistically from the data using $k_2 \cdot \sigma$ of D_{HP}—w employed $2 \cdot \sigma$ as a default value; however, any k can be provided as parameter;
- Minimal length of the segment (*minLen*), which is used to define the linear structuring element S as $2 \cdot minLen - 1$; we assumed $minLen = 20$ samples;
- Default parameters of Savitzky–Golay are $L = 5$, $M = 13$.

3.4.4. Step Change

Step change is another differential-based detection; it resembles the two former detections. It requires removing isolated peaks and heavy noise areas. After that, identification of rectangular-like changes becomes a simple problem.

The first two steps are shared with heavy noise detection. We compute D_1—differential (Equation (26)), which is then high-pass filtered with the Savitzky–Golay filter (Equation (27)), so we have D_{HP}. Next, we employ simple scanning with `find_derivate_pairs`, which seeks the areas between the complementary pairs of differential peaks (above *threshold*); we interpret it as a rectangular distortion, as shown in Figure 11. This scanning requires setting up two parameters—*minLen* and *maxDist*, identifying the minimal length of a step change, and maximal distance of searching.

Tunable parameters for the step change detection stage are:

- *threshold*—this is calculated statistically from D_{HP} using $k_3 \cdot \sigma$ of—we employed $3 \cdot \sigma$ as a default value; however, any k_3 can be provided as parameter,
- Minimal length of the segment (*minLen*); we assumed $minLen = 20$ samples;
- Maximal searching distance *maxDist*; we assumed 200 samples as the default value.
- Default parameters of Savitzky–Golay are the same as for heavy noise $L = 5$, $M = 13$.

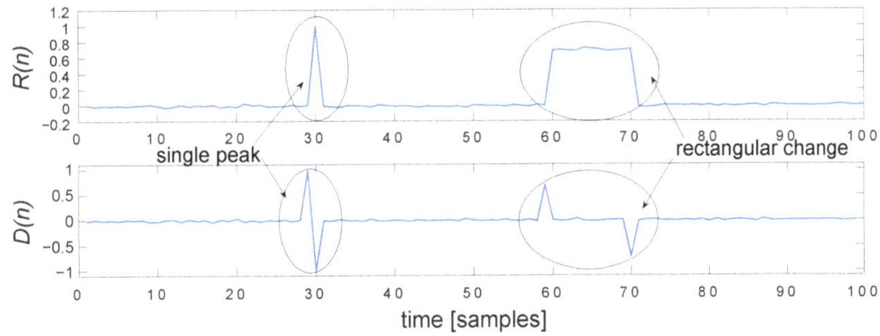

Figure 11. Appearance of sudden distortions in residual/high-pass ($R(n)$)—single peak and step change, and their corresponding peak pairs in differential $D(n)$.

3.4.5. Identifying Slow Changes

The slow changes of the positions (Figure 12a) involve a class of distortions notably different from the other ones, requiring a separate approach for detection due to the fact that its nature makes it hard to detect with differential analysis. We employed a loose NN model predictor (P_{NN}) as described in Section 3.3.3. It predicts the proper marker position on the basis of its neighbor markers (parent and sibling). The deviations from such a model (R—residuals) were analyzed, looking for notably large and long deviations, identified as artifact hills or valleys.

$$R(n) = X(n) - P_{NN}(n) \tag{31}$$

The hills and valleys could be of a different scale; the length and 'differential' can vary significantly. Moreover, predictor fluctuations at the turning points can also seem similar to short-term slow changes (of small amplitudes) appearing when the predictor cannot follow the change of value. Though, based on the statistical properties of the residual of the prediction and on the fact that we know that the distortion should be rather long (as it is an accumulated reconstruction error), one can make certain assumptions allowing for detection of the distortions.

We used 'canning of values' of the Savitzky–Golay smoothed residual of regression with hysteresis thresholding. It can be described in the following steps (see also Figure 12b):

1. Smoothed the R with the Savitzky–Golay low-pass filter ($L = 7$, $M = 11$—parameters heuristically tuned).
2. The upper threshold T_u was set up with a $k_U \cdot \sigma$ rule of a thumb—in our case, three times σ_R was selected ($k_U = 3$) as the default would identify the significant tops and bottoms of the hills and valleys.
3. If the top or bottom lengths were shorter than some minimal τ_U, we skipped it (0.2 s—20 frames in our case), assuming it to be short-term fluctuation.

4. After the identification of a top/bottom value, we looked for the rest of a distortion (below threshold)—the marked range expanded both sides iteratively (in the past and future) until the residual value went below/above the lower threshold T_L, obtained with $k_{hv} \cdot \sigma_R$ with $k_{hv} = 0.5$ as the default value.
5. If the overall located distortion (hill/valley) was shorter than some value τ_{hv} (50 frames —0.5 s), it was omitted, as one can consider it a short-term fluctuation of the predictor.

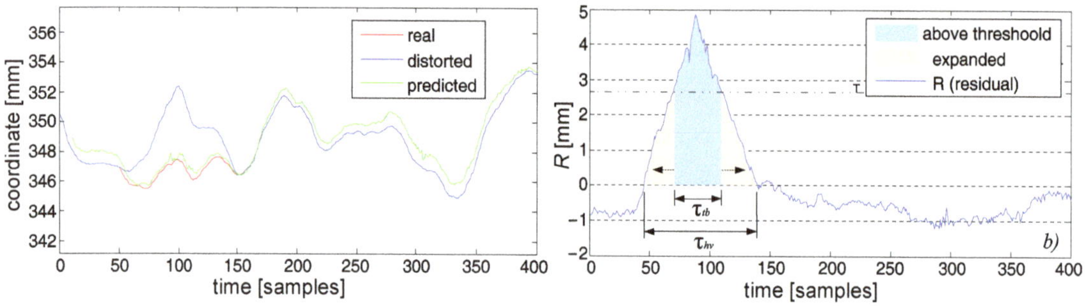

(a)

Figure 12. Slow detection: (**a**) original, predicted, and distorted signal, (**b**) residual with hysteresis thresholding.

4. Verification of the Method

The verification of the efficiency of the proposed approach was three-fold. In the first experiment, we analyzed the efficiency of the distortion classification using synthetically generated distortions in artifact-free sequences, which allowed us to provide some statistics on the classification efficiency. In the second test, we compared the performance of the proposed approach to the human operators of various experiences—from novice to experts. The last test was connected with the applicability of the artifact classification for the data cleaning with a pool of generic reconstruction algorithms.

4.1. Materials and Methods

4.1.1. The Data

For testing purposes, we used data sets acquired for professional applications in the motion capture laboratory. The sequences were obtained at the PJAIT human motion laboratory using the industrial-grade Vicon MX system. The system capture volume was 9 × 5 × 3 m. To minimize the impact of external interference, such as infrared interference from sunlight or vibrations, all windows were permanently darkened and cameras were mounted on scaffolding instead of tripods. The system was equipped with 30 NIR cameras manufactured by Vicon—10 pieces of each kind: MX-T40, Bonita10, Vantage V5.

During the recording, we employed two system configurations—a standard animation pipeline, where data were obtained with Vicon Blade software (using a 53-marker setup) and a typical biomechanical setup using Vicon Nexus software (using a 39-marker setup). The trajectories were acquired at 100 Hz; by default, they were processed in a standard, industrial quality way, which included manual data reviewing, cleaning, and denoising, so they could be considered distortion-free. However, depending on the experiment, different variants of the recordings were used in experiments; these were raw unprocessed data, processed (cleaned), and artificially-modified variants with controlled amounts and locations of distortions. Information, on which variant was used is provided in the detailed description of the experimental protocols.

The two recordings used in the experiment illustrate the ability to adapt to the different marker settings used in different application areas. The first sequence was clean with no errors, but relatively comprehensive—all the limbs were moved and the feet freely swung in random directions; therefore, it could be challenging for the predictor. The second sequence contains quite a lot of reconstruction errors in its raw form, so we had material to compare the results to the human operators.

4.1.2. Experimental Protocols

We planned the first experiment (E1) to test the performance of the method proposed in Section 3, using default parameters for a controlled dataset, with a perfectly clean sequence and controlled artificial distortions. It involved the first recording from the Table 1—'Static', which was manually cleaned by an expert, so it was artifact-free ground truth. Next, we introduced distortions at random locations (randomly drawn markers) and random amplitudes—see noise contamination procedures given further in Section 4.1.3.

Table 1. List of mocap sequence scenarios used for the testing.

No.	Name	Scenario	Duration	Difficulty
1	Static	Actor stands in the T-pose in the middle of the scene, looks around, and shifts from one foot to another	22 s	easy, static
2	Sitting	Actor stands in the middle of the scene and then sits on a chair; actor stands again after a few seconds and repeats this three times	29 s	occlusions

Companion results were additionally acquired in the experiment to compare the results obtained in E1 to the existing methods of anomaly detection. Since artifact classification is a completely new approach, it has been a bit difficult to select suitable methods to compare with. We refer to the two generic methods [36] for anomaly detection in the time series: three-Sigma move (M3S), which employs mean and variance moving; the other one is the Hampel filter (HF), which is based on the moving median and median absolute deviation (MAD), which are more robust measures.

During the experiments, we kept track of the distortions and their types; therefore, we were able to verify whether the error classification was correct. The criteria for evaluation in the artifact recognition task are pretty straightforward—classification rates (true and false recognition) presented as a confusion matrix. The simulations were performed for three distortion shares 5%, 10%, and 20%; each was executed 1000 times and the results are aggregated as average confusion matrices (rounded). For each class, we calculated the following measures:

$$\text{sensitivity (true positive rate)}: TPR = \frac{\text{tp}}{\text{tp}+\text{fn}} \cdot 100\%, \tag{32}$$

$$\text{miss-rate (false negative rate)}: FNR = \frac{\text{fn}}{\text{tp}+\text{fn}} \cdot 100\%, \tag{33}$$

$$\text{fall-out (false positive rate)}: FPR = \frac{\text{fp}}{\text{tp}+\text{fp}} \cdot 100\%, \tag{34}$$

$$\text{precision (positive predictive value)}: PPV = \frac{\text{tp}}{\text{tp}+\text{fp}} \cdot 100\%, \tag{35}$$

Additionally, two more measures were employed to quantify performance. The F-score is a scalar describing the efficiency of overall classification for all classes [37]—its values are between 0 (no proper classification) and 1 (perfect classification). From various equivalent formulas, we chose the following one, because it was simple to adapt to the multiclass problem:

$$F = \frac{\text{tp}}{\text{tp}+\frac{1}{2}(\text{fp}+\text{fn})}. \tag{36}$$

The other measure was Matthews correlation coefficient (MCC) [38], which is a quality measure intended for characterizing the classification efficiency for imbalanced populations of classes (as in our cases). Its values scale between −1 for no classification and 1 for perfect classification. The formula is given as:

$$MCC = \frac{\mathrm{tp} \cdot \mathrm{tn} - \mathrm{fp} \cdot \mathrm{fn}}{\sqrt{(\mathrm{tp}+\mathrm{fp})(\mathrm{tp}+\mathrm{fn})(\mathrm{tn}+\mathrm{fp})(\mathrm{tn}+\mathrm{fn})}}, \quad (37)$$

where cardinality of classifications are denoted as: tp—true positive classifications, fp—false positive, tn—true negative, fn—false negative.

The second experiment (E2) involved comparing the performance of the proposed method using default parameters to four operators of the mocap facility, with different levels of experience—two beginners (2 and 3 months of experience), one intermediate (1.5 years experience), and one expert (10 years of experience). The number of respondents was small and imbalanced because it was hard to find volunteers of that expertise. Additionally, we preferred to control and monitor the reconstruction process—including the software version and its setting—so we needed them in our lab facility. The results we present here represent all the people who worked in the lab and who agreed to do some work for us. Since a mocap operator is a rare profession, every response is informative; therefore, we preferred to present the imbalanced number of operators than to remove one of the respondents.

The test is intended to be a qualitative verification of the proposed approach and to verify the proposal using real life data. We used the raw form (not cleaned) of the 'Sitting' recording, which contained all kinds of distortions. Such data were used against the proposed detection algorithm. Apart from automatic processing, the four operators conducted normal data screening and cleaning. These manual processing steps, using Vicon Nexus software, were a standard approach in the lab, which is in everyday usage in the facility. In the final step, the results obtained by the algorithm and four operators were reviewed by an expert—a human mocap system operator with long experience in data editing and cleaning. The results are reported as raw numbers of distortions located, compared, and verified against human judgment.

The last experiment (E3) involved verification of the applicability of the proposed approach. It was intended to be a proof-of-concept of the targeted distortion cleaning. Therefore, it used different variants of static person sequences with the distortions of variable intensities and durations introduced into the recording—taking 5%, 10%, and 20% of the overall length—similar to E1. In the tests, we employed our algorithm with the default settings as a detector, which then was combined with the following reconstruction methods: Savitzky–Golay (13th order polynomial over 101 samples window), linear interpolation, spline interpolation, and FFNN prediction (as given in Section 3.3). Each method was applied in the locations of the detected artifacts only, the rest of the signal remained intact. All distortions were simulated separately in this case, with a randomized location (marker), time, duration, and an amplitude with 10 mm of average value and 4 mm of std deviation. The simulations of contamination–detection–reconstruction were performed 200 times; for each fold we obtained a quality measure, which was finally averaged. We assumed the root-mean-square error ($RMSE$) as the measure of quality; it was computed over all the coordinates and samples in the considered sequence:

$$RMSE = \sqrt{\frac{1}{K \cdot N} \sum_{k=1}^{K} \sum_{n=1}^{N} (\hat{x}_k(n) - x_k(n))^2}\ , \quad (38)$$

where: K is a number of variables in the time series, N is the number of samples, $x()$ is the original value, $\hat{x}()$ is the reconstructed value.

4.1.3. Artifact Contamination Procedure

The procedure of distortion contamination, which was employed in E1 and E3, introduced artifact distortions into the sequences in a controlled way—we logged the type, duration, location, and amplitude of the distortion. The contamination could include mixtures of all kinds in equal proportions. The key parameter characterizing the experiment was time share (distorted time fraction), for which distortions were generated. For the interpretation clarity, we ensured that only one distortion at a time occurred; therefore, the time share denotes that a distortion occurs at a given fraction of time. The sequence of distorting the signal is as follows: first, we drew locations to contaminate with 'bulk' distortions—slow, step changes, and heavy noise; next, we seeded randomly isolated peaks. Distortion parameters were set up randomly for each instance of distortions:

- The sign was a $+1/-1$ value drawn with equal probabilities;
- The amplitude was a Gaussian random variable with assumed amplitude and standard deviation (in the tested cases: $\mu = 5$ or 10 mm and $\sigma = 0.4 \cdot \mu$); these values were used to scale the peak of the rectangle or triangle distortion and as the standard deviations in the heavy noise area;
- Distortion durations and intervals were part of the Poisson process; an average length of distortion was set up to 50 samples, and the interval length was adjusted according to the duration of the sequence and the target amount of the given distortion.

The distortions introduced were quite demanding for the detection procedure. The amplitudes were on average small (5 and 10 mm) and of short (0.5 s) duration; therefore, we could assume that the synthetic tests were rather rigorous and more difficult to detect than in real life scenarios.

4.2. *Results and Discussion*

The outcomes of all three experiments are provided in the successive sub-paragraphs. They are accompanied by interpretations and discussions.

4.2.1. Synthetic Distortion Classification

The classification for synthetic noise outcomes are demonstrated in Figures 13 and 14 for average amplitudes of 5 and 10 mm, respectively. The raw results in confusion matrices demonstrate the average number of samples (rounded toward the whole sample) assigned to specific classes (correct or not) for 1000 simulations. They present averaged confusion matrices for three distortion shares 5%, 10%, and 20% of the overall time of the sequences with two average amplitudes—5 and 10 mm. According to the length of the recording in a simulation, the contamination procedure should produce approximately 160, 320, and 640 distorted samples, respectively, of each distortion type, and should be in equal proportions.

The comparative results for the generic anomaly detectors, HF and M3S in different configurations of the moving window, are demonstrated in Figure 15 as confusion matrices. The true anomalies are subdivided into classes, whereas the output is binary, whether it is detected as an outlier or not. The figure presents only the best of the results for 10 mm of amplitude and a 20% share of distortions, so it should be compared with the results in Figure 14c. We did not include the results for other distortion configurations, because they were either very similar (about recognition ratios) for 10 mm amplitudes or significantly worse for 5 mm.

Regardless of the number of distortions (shares), the results were pretty consistent; they were also very similar for numerous additional simulation runs, which are not included here. The fractions of true and false classifications hold across the runs. The same almost holds for F-scores and MCC values to a lesser extent. Therefore, the confusion matrix is the most informative presentation of results as it is near 0.999 for both amplitudes and all distortion shares—these large F-score/MCC values are due to the dominance of the properly classified clear signal samples. Nevertheless, they offer some insight into the results, with an increase in the share of distortions in the test sequence, we observe a very slight decrease in the classification performance expressed with the F-score/MCCs. These differences stem mainly from a slightly increased number of clear samples falsely classified as artifact-contaminated, since the classification rates remain on par between the artifact shares.

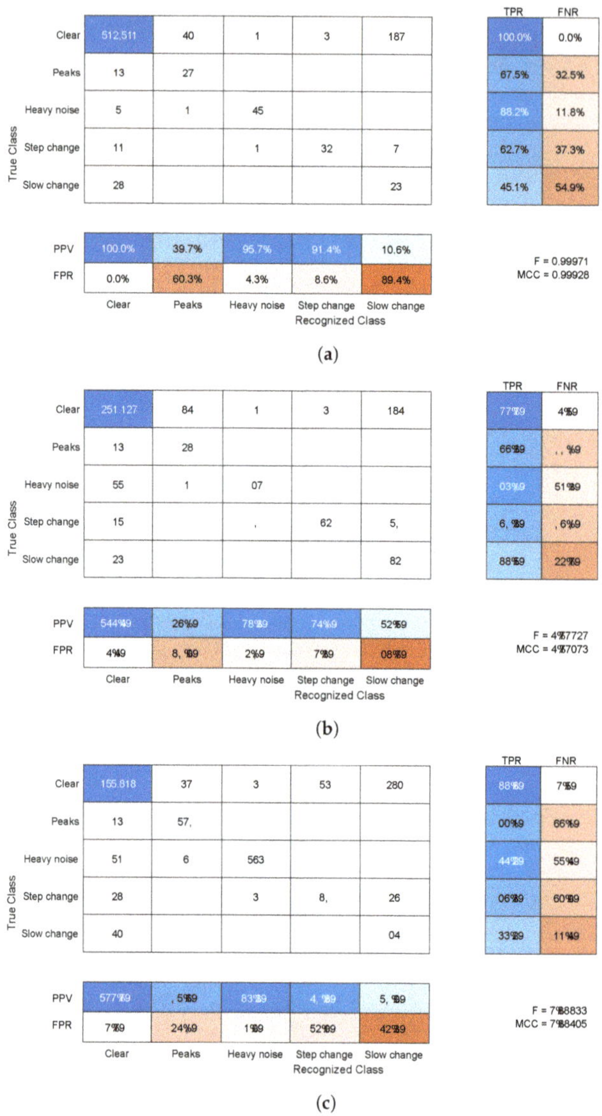

Figure 13. Average confusion matrix for detection of synthetic noises for a 1000-fold simulation with 5-mm average amplitudes of distortions and shares: (**a**) 5%, (**b**) 10%, (**c**) 20% of time (blue—successful results, red—faulty).

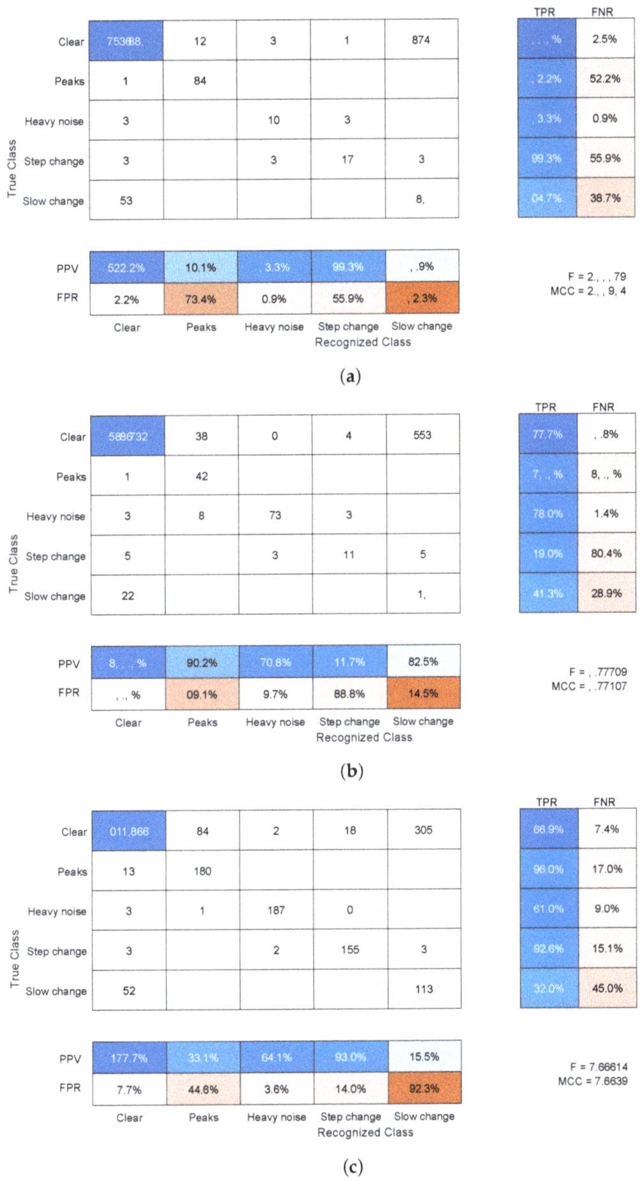

Figure 14. Average confusion matrix for detection of synthetic noises for a 1000-fold simulation with 10-mm average amplitudes of distortions and shares: (a) 5%, (b) 10%, (c) 20% of time (blue—successful results, red—faulty).

Each specific class requires separate insight into the results. These are as follows:

- The clear signal was identified properly for more than 99% of samples; a negligibly small amount of distorted samples was erroneously identified as clean signals (compared to the overall cardinality of the class).

- For the peak change, sensitivity was approximately 66% and 90%, and the main misclassification was in a clear signal; this class was not a cause of confusion for the other classes compared to a clean signal (usually below $FPR = 50\%$).
- Heavy noise sensitivity was above 88%; the main confusions were step change and a clear signal; this class was rarely erroneously recognized in place of the others (FPR = 4–8%); the main confused class was a clear signal.
- For the step change, sensitivity was approximately 70% and the main confusion was slow change; this class was erroneously recognized in place of others at a moderate rate (FPR = 12–27%)—here, a clean signal and heavy noise were wrongly identified.
- For the slow change, TPR was a bit more than 50% and the main confusion was a clear signal; this class was often difficult and erroneously recognized in place of others (FPR = 80–90%)—usually, it was a clear signal, but a step change and heavy noise were also wrongly identified.

Considering the above results, the proposed method has moderate to high sensitivity (depending on the class) and quite high precision for all classes but one (slow change). False negative detection (having relatively small values) was way more undesirable than the others; this was notable from a practical point of view. False positive, or detecting a wrong class of the distortion, would still result in pointing out the operator to the potential error location, or in the case of automatic error filtering, it would lead to the use of a repair procedure (see Section 4.2.3).

The difficulty in identification of slow change was expected; it comes from the fact that this change can be subtle and poorly distinguishable in predictor residuals, which resemble pink noise in our case. The latter is also a cause of high FPR. We analyzed alternative regression methods as a model—neural networks (simple FFNN and NARX-NN models), ridge, lasso, and SVR. However, the results were either poor or impractical (due to long training time), or both. On the other hand, a false positive error (quite frequent) is of much lesser importance than a false negative; the former might result in suggesting additional locations to the operator for reviewing, or using the interpolating method, which should not degrade the signal significantly; whereas the latter might result in preserving the distortion in the signal.

The comparison to the other anomaly detectors shows that the proposed solution outperforms these methods. In the best-tested configurations, they were capable to identify (as outliers) approximately 2/3 of peaks, and a small part (5–10%) of heavy noise-contaminated samples (usually the initial ones before the local variance or MAD value increased in the presence of the noise, so much that the thresholds did not intercept the noise anymore). The two other signal anomalies—step and slow change—remained invisible to these methods (excluding a single accidental case in M3S. Meanwhile, our approach (Figure 14c) was capable of identifying approximately 90% of the first two anomalies and above 75% of the latter two, which were 'invisible' to the generic detectors.

Regarding false detections, in most cases (all but one), both the generic methods returned small numbers of false positive detections. We attributed this to the overall low sensitivity of these methods, and in cases when the detection rates were a bit higher, false positives were also elevated (Figure 15a,f,g).

Both the M3S and Hampel filter are methods that consider each coordinate separately; therefore, they do not use inter-marker dependencies. This allowed us to identify individual anomalous coordinates. Regrettably, because of the latter, we had to reject the other (more sophisticated) methods of anomaly detection based on machine learning, such as clustering, one-class SVM, or autoencoders as insufficient. In their basic variants, they considered the whole frame of a recording as a single observation with coordinates as features. This implies that they could potentially point out anomalous frames in sequence, but they would not identify what marker/dimension is the problem. One might think of adopting such approaches to anomaly detections; however, it would require a separate in-depth study.

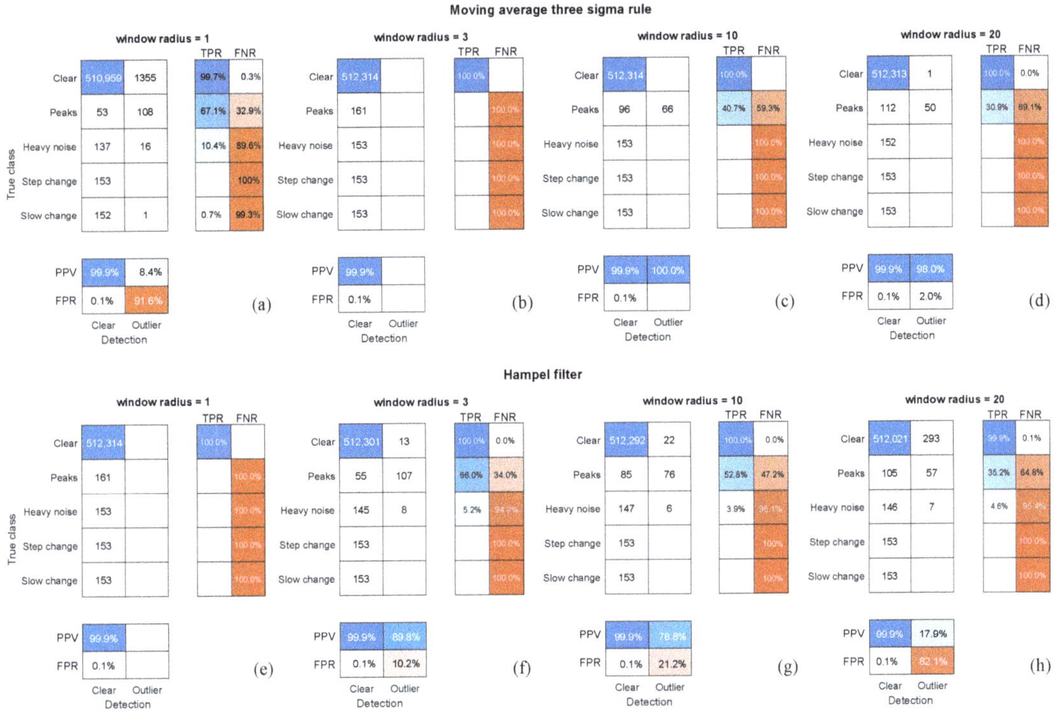

Figure 15. Averaged confusion matrix for detection with M3S (**a**–**d**) and the Hampel filer (**e**–**h**) of synthetic anomalies for a 1000-fold simulation with 10-mm average amplitudes of distortions and a share 20% of the time (blue—successful results, red—faulty).

4.2.2. Comparing to Human Operators

Table 2 comprises the numbers of detected distortions in the recording processed in a standard pipeline, and the recording processed by each of the four operators. The detected distortions were compared and verified by human operators, who either approved the classifications or rejected them.

In the recording processed by the machine, the proposed algorithm found 29 errors—peaks, step, and slow changes; 16 of these errors were correctly classified. The algorithm classified sudden, very dynamic hand movements in a relatively static recording as slow changes (13 incorrectly classified errors). In addition, the algorithm did not detect four errors; after careful analysis, it turned out that these errors did not belong to any of the previously defined classes—they were a combination of a single peak and a slow change. An example of such an error is shown in Figure 16. Such an artifact (having a relatively large amplitude) could be intercepted if the slow change detector had different parameters, but it would require setting up a smaller value for a minimal length of an artifact. It could result in an increased number of false positives for this class, as this parameter prevents false alerts of the short-term fluctuations of the NN predictor. It is worth mentioning that the slow change detector reacts properly for a mixed class of artifacts—step changes followed by slow relaxing to the proper trajectory (or slow error accumulation followed by immediate correction) do not cause detection for any of the detectors based on finding derivative pairs, but they properly trigger detection for the deviations from the predictor if they are long enough. In fact, all the deviations that are large and long enough would be identified as slow changes. It is a matter of human interpretation whether we can name them as slow; however, that name distinguishes them well from all of a sudden changes we identified based on the differential analysis.

Table 2. Comparing the number of distortions located by the proposed method to the human operator (E2).

Operator	Seq. No	Recording	Errors Identified by		Error Verification		
			Human	Algorithm	Approved	Rejected	Missed
None	2	Sitting	—	29	16	13	4
Expert	2	Sitting	20	9	0	0	0
Intermediate	2	Sitting	18	11	2	9	0
Beginner 1	2	Sitting	10	37	20	17	2
Beginner 2	2	Sitting	11	46	26	20	1

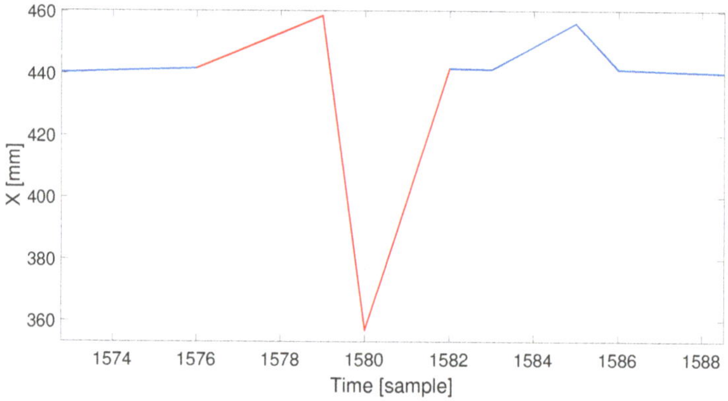

Figure 16. Additional combined distortion types (red line indicates faulty values).

In the recording processed by the expert, the algorithm again incorrectly classified the hand movements as slow changes. The result was similar for the intermediate operator, with the difference that the algorithm found two errors omitted by the human (two small peaks).

In the case of a recording repaired by beginners, both the algorithm and the expert found more errors after the repair than before. This was due to the selection of an inappropriate method of repairing a given artifact. For example, when the distortion occurred only on one axis, the person, in order to correct the error, removed the marker trajectory for those few frames when the error occurred. This resulted in the creation of an additional gap, which the beginner operator filled using simple interpolation. In the case of longer artificial gaps, interpolation caused the data to be incorrectly reconstructed, and the errors no longer appeared on one axis, but on all three.

4.2.3. Applicability Testing

The results are presented in Tables 3–5. Each field presents the averaged *RMSE* of a 200-fold repeated simulation process, distorting the test sequence and its reconstruction using various procedures. Each distortion type was considered separately. It allowed us to quantify how each reconstruction method reduced the distortion. Figure 17 illustrates the reconstruction results, demonstrating the ground truth, distorted signal value, and outcomes of four variants of reconstruction. In the tables, for each distortion type and reconstruction method, we compare two *RMSE* values: hypothetical perfect classification and actual classification with the proposed algorithm.

Table 3. *RMSE* after reconstruction with different methods (with perfect and algorithmic artifact classifications) for the mocap sequence with a 5% distorted time in the sequence.

	Peaks	Heavy Noise	Step Change	Slow Change
Distorted	0.19065	0.17717	0.18069	0.10129
Linear interpolation (perfect)	0.00136	0.18322	0.15939	0.18795
Linear interpolation (classified)	0.03947	0.18514	0.15623	0.29339
Savitzky–Golay filter (perfect)	0.01900	0.04963	0.17440	0.10130
Savitzky–Golay filter (classified)	0.08159	0.06855	0.17553	0.10173
Spline interpolation (perfect)	*0.00025*	0.11041	0.09780	0.10972
Spline interpolation (classified)	*0.03429*	0.68692	0.10895	0.19935
FFNN predictor (perfect)	0.01841	*0.01953*	*0.01933*	*0.01875*
FFNN predictor (classified)	0.03944	*0.04547*	*0.04409*	*0.11000*

Table 4. *RMSE* after reconstruction with different methods (with perfect and algorithmic artifact classifications) for the mocap sequence with a 10% distorted time in the sequence.

	Peaks	Heavy Noise	Step Change	Slow Change
Distorted	0.26906	0.26340	0.26282	0.15037
Linear interpolation (perfect)	0.00186	0.25487	0.28409	0.27662
Linear interpolation (classified)	0.05144	0.26360	0.27618	0.38908
Savitzky–Golay filter (perfect)	0.02678	0.07211	0.25361	0.15046
Savitzky–Golay filter (classified)	0.08959	0.08895	0.25510	0.15083
Spline interpolation (perfect)	*0.00039*	0.14679	0.15207	0.15955
Spline interpolation (classified)	*0.04754*	0.95512	0.16358	0.23918
FFNN predictor (perfect)	0.02785	*0.02468*	*0.02581*	*0.02612*
FFNN predictor (classified)	0.05537	*0.05201*	*0.06349*	*0.14717*

Table 5. *RMSE* after reconstruction with different methods (with perfect and algorithmic artifact classifications) for the mocap sequence with a 20% distorted time in the sequence.

	Peaks	Heavy Noise	Step Change	Slow Change
Distorted	0.38228	0.39623	0.39247	0.21676
Linear interpolation (perfect)	0.00273	0.44107	0.44015	0.39172
Linear interpolation (classified)	0.07007	0.45679	0.45630	0.55692
Savitzky–Golay filter (perfect)	0.03880	0.11228	0.37892	0.21704
Savitzky–Golay filter (classified)	0.10383	0.16007	0.38120	0.21748
Spline interpolation (perfect)	*0.00058*	0.24337	0.28188	0.25274
Spline interpolation (classified)	*0.06753*	1.58979	0.37534	0.35679
FFNN predictor (perfect)	0.04169	*0.03711*	*0.03972*	*0.03735*
FFNN predictor (classified)	0.07903	*0.11433*	*0.10284*	*0.20549*

In the results, we recognize the different efficiencies of the tested reconstruction methods for different distortions. We could also clearly observe that the efficiency of detection of the artifacts directly affects the ability to restore the signal. The key observations are summarized in a few points:

- Peak changes were effectively removed with interpolation methods—simple linear or spline (piecewise cubic polynomial); the other two methods in perfect detection would not offer even comparable efficiency, yet in actual classification, they offered just slightly worse performances.

- FFNN offered the best performance for all 'bulky' distortions (of longer durations), both hypothetical and classified cases.
- Heavy noise, aside from FFNN, was well cleaned with the Savitzky–Golay filter (see Figure 17b).
- Step changes could be effectively removed with FFNN only.
- Slow changes were the most contradictory—the only appropriate reconstruction method was FFNN; in the case of perfect detection, the efficiency was high, but due to the limited actual detection, the results were quite poor. These results correspond well to the detection of slow changes in E1—low sensitivity and high fall out.

The above outcomes of reconstruction are just preliminary results. They should be further analyzed in a separate study for the possible reconstruction methods involving other predictors, interpolations, the rigid body model, and projections on geometric constraints.

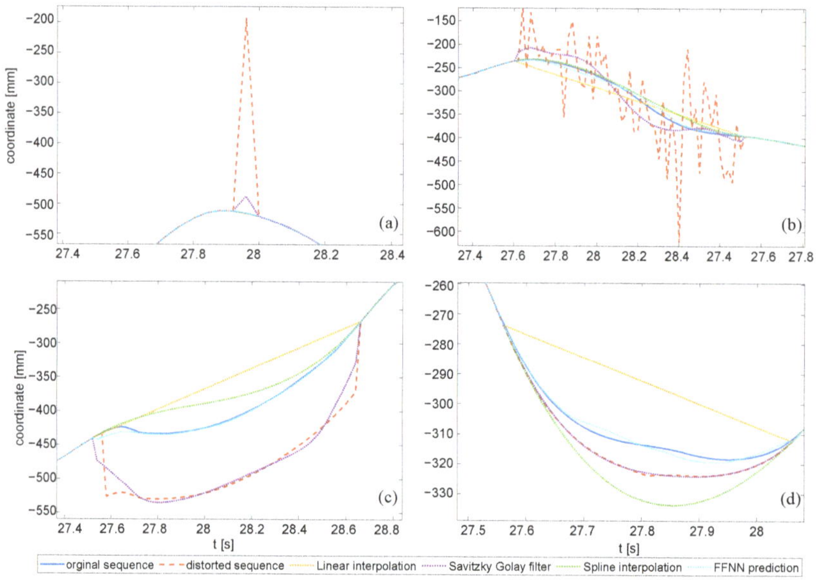

Figure 17. Artifacts and the removal with methods tested in E3: (**a**) single peaks, (**b**) heavy noise, (**c**) step change, (**d**) slow change. Please mind the various scales in the axes.

4.3. Results Recap

All above outcomes indicate that the proposed approach offers reasonable results. Experiment E1 proves that sudden changes are way easier to detect than slow changes, but slowly accumulated errors are also detectable (with less efficiency). It is mainly a matter of predictor efficiency, so the amplitude of distortions is a key factor affecting artifact locating.

Another disputable aspect is the fact that there is no simple possibility to compare the detection and classification efficiencies to the other solutions, as this work is the first proposal in this area. The only method that is somewhat comparable to ours is publicly unavailable. Moreover, it is capable of segmenting artifacts only and cannot classify distortions, so the performance comparison, if possible, would be very limited.

Comparing the reconstruction efficiency (as in E3) to the existing repairing method—the classification efficiency would not only be evaluated, but also (foremost) the quality of the reconstruction algorithm.

However, we were able to compare the efficiency of our solution to the industrial-grade software in an indirect way, as described in E2. Automatic repairing algorithms offered within modern software (Vicon Nexus) could degrade the sequence (increase the number of artifacts)—this is well illustrated by the results of the novice operators, who used automatic repairing, resulting in an increased number of artifacts.

Additionally, referring to our experience with the state-of-the-art software, when using the 'find bad data' function in the Vicon Blade software, it is not the best option to let the software do everything for us. This method requires three parameters: threshold (allowed deviation in millimeters), cut-off (the cut-off frequency), and sensitivity (amplifies the effect of the cut-off frequency). Unfortunately, this method requires one to set different values to find different errors. Moreover, these parameters may be different for each recording, e.g., more dynamic recordings require increasing the threshold. Another drawback is the fact that each marker trajectory is treated separately, with no inter-marker relations, so slow changes are invisible to that method.

Finally, in the proposal, we presumed that (for certain markers) there was one distortion type at a time. This might not be true in some cases; however, this matters in rare cases, since one fault usually happens at a time, or one is dominant and clutters the others. Some combinations of errors can be well distinguished (e.g., peaks/noise during slow change), whereas others cannot (step changes within noise). Therefore, as we demonstrated in E2, such an algorithm cannot replace experienced operators but can be of assistance, making the jobs faster and less burdensome.

5. Summary

In this article, we addressed the issue of artifacts occurring in the mocap signal. We proposed a method for their detection and demonstrated how to employ the detection method to improve signal fidelity. The method proposed in this article seems to be quite effective for sudden changes, and it can detect distortions of relatively small amplitudes. As for the slow changes, their outcomes are moderate, since we observed a relatively large number of false positive detections. However, we expected that this class of distortions might be difficult to detect. This topic is worthy of further study.

Compared to human operators, the proposed solution cannot outperform experienced professionals; however, it offers a notably better performance than a novice operator. On the other hand, even for an expert, it can save time by suggesting locations to review.

The proposal could be adopted in existing software as an optional step of signal refinement and/or for automatic support for the mocap sequence editors. Further improvements are still possible, but require additional research, such as employing better predictive models. Moreover, the engineering approach could be beneficial for detection efficiency. One improvement could be to detect distortions for all three coordinates of a marker, jointly, since distortions usually occur in more than a single coordinate. Moreover, studying the reconstruction methods is a topic that we plan to investigate in the future.

Author Contributions: Conceptualization, P.S.; methodology, P.S. and M.P.; software, P.S. and M.P.; investigation, P.S. and M.P.; resources, M.P.; data curation, M.P.; writing—original draft preparation, P.S. and M.P.; writing—review and editing, P.S. and M.P.; visualization, P.S. and M.P. All authors have read and agreed to the published version of the manuscript.

Funding: The research described in the paper was performed within the statutory project of the Department of Graphics, Computer Vision and Digital Systems at the Silesian University of Technology, Gliwice (RAU-6, 2022). APC were covered from statutory research funds.

Institutional Review Board Statement: Not applicable.

Informed Consent Statement: Not applicable.

Data Availability Statement: Not applicable.

Acknowledgments: The research was supported with motion data by the Human Motion Laboratory of the Polish–Japanese Academy of Information Technology http://bytom.pja.edu.pl/.

Conflicts of Interest: The authors declare no conflict of interest. The funders had no role in the design of the study; in the collection, analyses, or interpretation of data; in the writing of the manuscript, or in the decision to publish the results.

Abbreviations

The following abbreviations are used in this manuscript:

FFNN	feed-forward neural network
HF	Hampel filter
HML	Human Motion Laboratory
LS	least squares
M3S	moving three sigma
mocap	MOtion CAPture
MSE	mean square error
NARX-NN	nonlinear autoregressive exogenous neural network
NN	neural network
PJAIT	Polish–Japanese Academy of Information Technology
RMSE	root mean squared error

References

1. Kitagawa, M.; Windsor, B. *MoCap for Artists: Workflow and Techniques for Motion Capture*; Elsevier/Focal Press: Amsterdam, The Netherlands; Boston, MA, USA, 2008.
2. Menache, A. *Understanding Motion Capture for Computer Animation*, 2nd ed.; Morgan Kaufmann: Burlington, MA, USA, 2011.
3. Mündermann, L.; Corazza, S.; Andriacchi, T.P. The evolution of methods for the capture of human movement leading to markerless motion capture for biomechanical applications. *J. Neuroeng. Rehabil.* **2006**, *3*, 6. [CrossRef] [PubMed]
4. Windolf, M.; Götzen, N.; Morlock, M. Systematic accuracy and precision analysis of video motion capturing systems—Exemplified on the Vicon-460 system. *J. Biomech.* **2008**, *41*, 2776–2780. [CrossRef] [PubMed]
5. Yang, P.F.; Sanno, M.; Brüggemann, G.P.; Rittweger, J. Evaluation of the performance of a motion capture system for small displacement recording and a discussion for its application potential in bone deformation in vivo measurements. *Proc. Inst. Mech. Eng. Part H J. Eng. Med.* **2012**, *226*, 838–847. [CrossRef]
6. Jensenius, A.; Nymoen, K.; Skogstad, S.; Voldsund, A. A Study of the Noise-Level in Two Infrared Marker-Based Motion Capture Systems. In Proceedings of the Proceedings of the 9th Sound and Music Computing Conference, SMC 2012, Copenhagen, Denmark, 1–14 July 2012; pp. 258–263.
7. Eichelberger, P.; Ferraro, M.; Minder, U.; Denton, T.; Blasimann, A.; Krause, F.; Baur, H. Analysis of accuracy in optical motion capture—A protocol for laboratory setup evaluation. *J. Biomech.* **2016**, *49*, 2085–2088. [CrossRef] [PubMed]
8. Skurowski, P.; Pawlyta, M. On the Noise Complexity in an Optical Motion Capture Facility. *Sensors* **2019**, *19*, 4435. [CrossRef] [PubMed]
9. Woltring, H.J. On optimal smoothing and derivative estimation from noisy displacement data in biomechanics. *Hum. Mov. Sci.* **1985**, *4*, 229–245. [CrossRef]
10. Giakas, G.; Baltzopoulos, V. A comparison of automatic filtering techniques applied to biomechanical walking data. *J. Biomech.* **1997**, *30*, 847–850. [CrossRef]
11. Zordan, V.B.; Van Der Horst, N.C. Mapping optical motion capture data to skeletal motion using a physical model. In Proceedings of the Proceedings of the 2003 ACM SIGGRAPH/Eurographics Symposium on Computer Animation, San Diego, CA, USA, 26–27 July 2003; Eurographics Association: Goslar, Germany, 2003; pp. 245–250.
12. Skurowski, P.; Pawlyta, M. Functional Body Mesh Representation, A Simplified Kinematic Model, Its Inference and Applications. *Appl. Math. Inf. Sci.* **2016**, *10*, 71–82. [CrossRef]
13. Barré, A.; Thiran, J.P.; Jolles, B.M.; Theumann, N.; Aminian, K. Soft Tissue Artifact Assessment During Treadmill Walking in Subjects With Total Knee Arthroplasty. *IEEE Trans. Biomed. Eng.* **2013**, *60*, 3131–3140. [CrossRef]
14. Reda, H.E.A.; Benaoumeur, I.; Kamel, B.; Zoubir, A.F. MoCap systems and hand movement reconstruction using cubic spline. In Proceedings of the 2018 5th International Conference on Control, Decision and Information Technologies (CoDIT), Thessaloniki, Greece, 10–13 April 2018; pp. 1–5. [CrossRef]
15. Tits, M.; Tilmanne, J.; Dutoit, T. Robust and automatic motion-capture data recovery using soft skeleton constraints and model averaging. *PLoS ONE* **2018**, *13*, e0199744. [CrossRef]
16. Camargo, J.; Ramanathan, A.; Csomay-Shanklin, N.; Young, A. Automated gap-filling for marker-based biomechanical motion capture data. *Comput. Methods Biomech. Biomed. Eng.* **2020**, *23*, 1180–1189. [CrossRef]
17. Perepichka, M.; Holden, D.; Mudur, S.P.; Popa, T. Robust Marker Trajectory Repair for MOCAP using Kinematic Reference. In Proceedings of the Motion, Interaction and Games, Newcastle upon Tyne, UK, 28–30 October 2019; Association for Computing Machinery: New York, NY, USA, 2019; pp. 1–10. [CrossRef]

18. Gløersen, Ø.; Federolf, P. Predicting Missing Marker Trajectories in Human Motion Data Using Marker Intercorrelations. *PLoS ONE* **2016**, *11*, e0152616. [CrossRef] [PubMed]
19. Liu, G.; McMillan, L. Estimation of missing markers in human motion capture. *Vis. Comput.* **2006**, *22*, 721–728. [CrossRef]
20. Kaufmann, M.; Aksan, E.; Song, J.; Pece, F.; Ziegler, R.; Hilliges, O. Convolutional Autoencoders for Human Motion Infilling. *arXiv* **2020**, arXiv:2010.11531.
21. Zhu, Y. Reconstruction of Missing Markers in Motion Capture Based on Deep Learning. In Proceedings of the 2020 IEEE 3rd International Conference on Information Systems and Computer Aided Education (ICISCAE), Dalian, China, 27–29 September 2020; pp. 346–349. [CrossRef]
22. Skurowski, P.; Pawlyta, M. Gap Reconstruction in Optical Motion Capture Sequences Using Neural Networks. *Sensors* **2021**, *21*, 6115. [CrossRef]
23. Smolka, J.; Lukasik, E. The rigid body gap filling algorithm. In Proceedings of the 2016 9th International Conference on Human System Interactions (HSI), Portsmouth, UK, 6–8 July 2016; pp. 337–343. [CrossRef]
24. Royo Sánchez, A.C.; Aguilar Martín, J.J.; Santolaria Mazo, J. Development of a new calibration procedure and its experimental validation applied to a human motion capture system. *J. Biomech. Eng.* **2014**, *136*, 124502. [CrossRef] [PubMed]
25. Nagymáté, G.; Tuchband, T.; Kiss, R.M. A novel validation and calibration method for motion capture systems based on micro-triangulation. *J. Biomech.* **2018**, *74*, 16–22. [CrossRef] [PubMed]
26. Weber, M.; Amor, H.B.; Alexander, T. Identifying Motion Capture Tracking Markers with Self-Organizing Maps. In Proceedings of the 2008 IEEE Virtual Reality Conference, Reno, NV, USA, 8–12 March 2008; pp. 297–298. [CrossRef]
27. Jiménez Bascones, J.L.; Graña, M.; Lopez-Guede, J.M. Robust labeling of human motion markers in the presence of occlusions. *Neurocomputing* **2019**, *353*, 96–105. [CrossRef]
28. Ghorbani, S.; Etemad, A.; Troje, N.F. Auto-labelling of Markers in Optical Motion Capture by Permutation Learning. In *Computer Graphics International Conference*; Gavrilova, M., Chang, J., Thalmann, N.M., Hitzer, E., Ishikawa, H., Eds.; Lecture Notes in Computer Science; Springer International Publishing: Cham, Switzerland, 2019; pp. 167–178. [CrossRef]
29. Han, S.; Liu, B.; Wang, R.; Ye, Y.; Twigg, C.D.; Kin, K. Online optical marker-based hand tracking with deep labels. *ACM Trans. Graph.* **2018**, *37*, 166:1–166:10. [CrossRef]
30. *Regression Analysis—Encyclopedia of Mathematics*; Springer: Berlin, Germany, 2001.
31. Stapor, K. *Introduction to Probabilistic and Statistical Methods with Examples in R*; Intelligent Systems Reference Library; Springer International Publishing: Cham, Switzerland, 2020. [CrossRef]
32. Savitzky, A.; Golay, M.J.E. Smoothing and Differentiation of Data by Simplified Least Squares Procedures. *Anal. Chem.* **1964**, *36*, 1627–1639. [CrossRef]
33. Hornik, K. Approximation capabilities of multilayer feedforward networks. *Neural Netw.* **1991**, *4*, 251–257. [CrossRef]
34. Insua, D.R.; Müller, P. Feedforward Neural Networks for Nonparametric Regression. In *Practical Nonparametric and Semiparametric Bayesian Statistics*; Dey, D., Müller, P., Sinha, D., Eds.; Lecture Notes in Statistics; Springer: New York, NY, USA, 1998; pp. 181–193. [CrossRef]
35. Czekalski, P.; Łyp, K. Neural network structure optimization in pattern recognition. *Stud. Inform.* **2014**, *35*. [CrossRef] mdpi: please provide pages or doi - PROVIDED
36. Xu, S.; Lu, B.; Baldea, M.; Edgar, T.F.; Wojsznis, W.; Blevins, T.; Nixon, M. Data cleaning in the process industries. *Rev. Chem. Eng.* **2015**, *31*, 453–490. [CrossRef]
37. Pillai, I.; Fumera, G.; Roli, F. Designing multi-label classifiers that maximize F measures: State of the art. *Pattern Recognit.* **2017**, *61*, 394–404. [CrossRef]
38. Gorodkin, J. Comparing two K-category assignments by a K-category correlation coefficient. *Comput. Biol. Chem.* **2004**, *28*, 367–374. [CrossRef] [PubMed]

sensors

MDPI

Article

Top-Down System for Multi-Person 3D Absolute Pose Estimation from Monocular Videos

Amal El Kaid [1,2,3,*], Denis Brazey [3], Vincent Barra [1] and Karim Baïna [2]

1 Université Clermont-Auvergne, CNRS, Mines de Saint-Étienne, Clermont-Auvergne-INP, LIMOS, 63000 Clermont-Ferrand, France; vincent.barra@limos.fr
2 Alqualsadi Research Team, Rabat IT Center, ENSIAS, Mohammed V University in Rabat, Rabat 10112, Morocco; karim.baina@ensias.um5.ac.ma
3 Société Prynel, RD974, 21190 Corpeau, France; dbrazey@pryntec.com
* Correspondence: amal.el_kaid@doctorant.uca.fr

Abstract: Two-dimensional (2D) multi-person pose estimation and three-dimensional (3D) root-relative pose estimation from a monocular RGB camera have made significant progress recently. Yet, real-world applications require depth estimations and the ability to determine the distances between people in a scene. Therefore, it is necessary to recover the 3D absolute poses of several people. However, this is still a challenge when using cameras from single points of view. Furthermore, the previously proposed systems typically required a significant amount of resources and memory. To overcome these restrictions, we herein propose a real-time framework for multi-person 3D absolute pose estimation from a monocular camera, which integrates a human detector, a 2D pose estimator, a 3D root-relative pose reconstructor, and a root depth estimator in a top-down manner. The proposed system, called Root-GAST-Net, is based on modified versions of GAST-Net and RootNet networks. The efficiency of the proposed Root-GAST-Net system is demonstrated through quantitative and qualitative evaluations on two benchmark datasets, Human3.6M and MuPoTS-3D. On all evaluated metrics, our experimental results on the MuPoTS-3D dataset outperform the current state-of-the-art by a significant margin, and can run in real-time at 15 fps on the Nvidia GeForce GTX 1080.

Keywords: 3D multi-person pose estimation; absolute poses; camera-centric coordinates; computer vision; artificial intelligence; deep-learning

Citation: El Kaid, A.; Brazey, D.; Barra, V.; Baïna, K. Top-Down System for Multi-Person 3D Absolute Pose Estimation from Monocular Videos. *Sensors* **2022**, *22*, 4109. https://doi.org/10.3390/s22114109

Academic Editors: Tomasz Krzeszowski, Adam Świtoński, Michal Kepski, Carlos Tavares Calafate and Gregorij Kurillo

Received: 15 March 2022
Accepted: 24 May 2022
Published: 28 May 2022

1. Introduction

Human pose estimation (HPE) is a popular task in computer vision. It aims to predict and track the location of joints (e.g., elbow, wrist) or body parts of one or more human bodies; it associates them with segments in graphical form (from an image or sequence of images) to represent the human's orientation and it describe the actual posture. This is an important process for understanding human behavior and human–computer interactions. An example of a human posture skeleton is illustrated in Figure 1.

With human pose estimation, tracking a person or multiple people in real space can be done at an incredibly granular level. This powerful capability unlocks a wide range of industrial applications [1–8], including gaming, animation, motion transfer, augmented reality, human–robot cooperation and training, biomechanical analysis for medical/healthcare, sports fields, gesture control, autonomous driving, human fall detection, action prediction, security and surveillance, etc.

Pose estimation can be performed in two ways: in a two-dimensional space to predict XY image coordinates or in a three-dimensional space to predict the XYZ camera or world coordinates. However, most real-life applications require depth estimation, which provides informative knowledge since 2D poses are often confusing. They can appear identical when in fact they represent completely distinct poses. This makes activity recognition difficult and leads researchers to employ 3D pose estimation.

0- Head-top	9- Left knee
1- Neck	10- Left ankle
2- Left shoulder	11- Right hip
3- Left elbow	12- Right knee
4- Left wrist	13- Right ankle
5- Right shoulder	14- Hip
6- Right elbow	15- Spine
7- Right wrist	16- Head
8- Left hip	

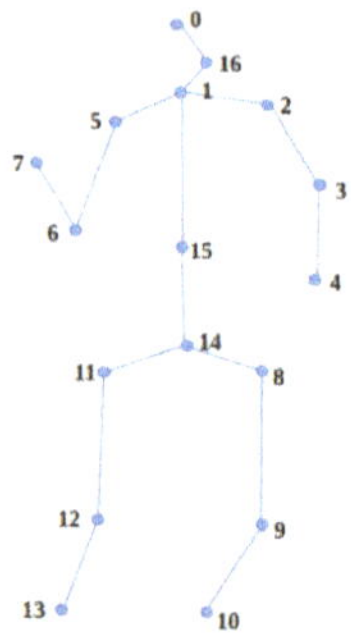

Figure 1. 3D Skeleton model in MuPoTS-3D format and joints names.

Recently, 3D root-relative human pose estimation has shown remarkable progress. Several methods [9–14] propose alleviating the problem by using multi-view images or videos as input. However, multi-view observations are expensive to obtain in daily life scenarios. Thus, the use of 3D human pose estimations from monocular images or videos is in high demand. State-of-the-art approaches that use monocular data [15–22] usually decouple the problem into two main phases: 2D pose estimation for joint detection and localization in the image space, and then lifting of the 2D pixel coordinates to 3D keypoint-position predictions in the camera space. In our research, we followed the same strategy and focused on the second phase, i.e., the 3D pose reconstruction from a sequence of 2D keypoints. Two-dimensional (2D) pose estimation is a popular vision problem that has been studied in many works, e.g., [23–28] and has been greatly improved especially using the deep learning paradigm.

Indeed, 3D pose estimation approaches show promising results on single-person datasets, such as Human3.6M [29] and HumanEva-I [30]. However, they do not perform well in multi-person scenarios, which are the most common cases in real-world applications and surveillance systems. The distances between people can be crucial in the analysis and recognition of their interactions. This introduces the absolute pose [31–33], which aims to locate the root joint (key central point of the person) and estimate its distance from the camera. At present, the 3D multi-person pose estimation still faces a great challenge. When possible, stereo vision calibration is used to determine the exact position of a person from images taken from different points of view. However, these kinds of data are not always available, and they significantly raise the overall costs of the process procedures. Moreover, acquiring such data is impractical in real-time system applications, as we seek to optimize the amount of data that must be captured and processed. This shows the gap between scientific literature and real-world requirements.

The purpose of this study was to present a framework that could accomplish more accurate and robust 3D multi-person pose estimations from a monocular video, from these circumstances and industrial constraints. Thus, we propose an integrated top-down approach that combines GAST-Net for reconstructing 3D root-relative keypoints from 2D keypoints and RootNet for estimating root depth from human bounding boxes. It generates an appropriate 3D multi-skeleton estimation result from a monocular video while maintaining low computational costs and short execution times.

Basically, the system is the result of a series of improvements that boost accuracy by more than 8.8 percentage points on 3D-PCK*abs* on the MuPoTS-3D [34] dataset, when compared to the approaches in the literature [31–33,35,36].

Examples of results from our whole framework are illustrated in Figure 2.

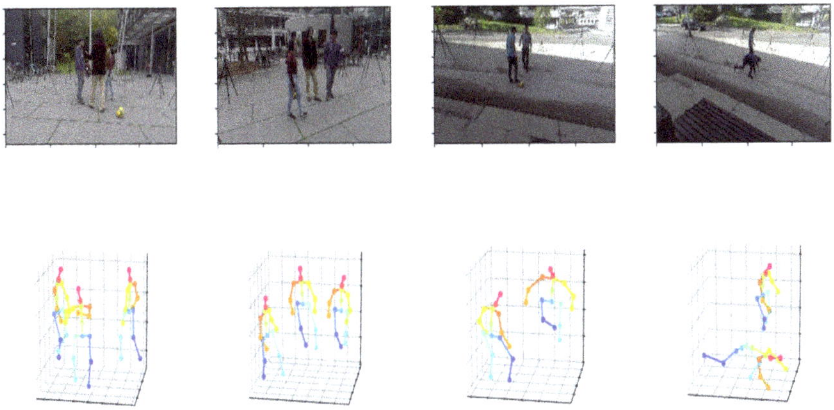

Figure 2. Examples of 3D absolute poses resulting from our whole framework.

The main contributions of this work can be summarized as follows:

- The proposal of an integrated top-down framework based on a modified GAST-Net and RootNet networks for multi-person 3D pose estimation from a monocular RGB video in a short execution time.
- Outperforming existing 3D multi-person absolute pose estimation methods in a MuPoTS-3D dataset by more than 8.8 percentage points on 3D-PCK_{abs} and by more than 12.6 percentage points on AP^{root}_{25} .

The paper is organized as follows. Section 2 illustrates the review of conventional literature on 3D pose estimation based on different levels: the input type (video), the number of instances (multi-person), and the approach following the 3D root-relative pose estimation (two-stage approach). Section 3 demonstrates the proposed framework methodology. Section 4 explains the implementation details, the results and discussion. Section 5 provides a conclusion of the work.

2. Related Works

2.1. Two-Stage Pose Estimation

Several works [22,37–43] apply deep neural networks on 3D pose estimation tasks to learn the direct mapping between RGB images and their corresponding 3D poses in one stage. However, this needs labeled data for supervised training, usually impractical out of MoCap labs. Unsupervised learning algorithms require sophisticated architectures with high computation costs, which are impractical too in realistic applications. To this end, Martinez et al. [44] introduced a two-stage prediction approach. They first predicted the 2D pose from the image and then lifted 2D joint coordinates to the 3D space via a fully connected residual network. Fang et al. [45] introduced a model to encode the mapping function of the human pose from 2D to 3D by explicitly encoding the human body configuration with pose grammar. To improve the generalization of the trained 2D-to-3D pose estimator, Gong et al. [46] proposed a pose augmentation framework (PoseAug) exploiting a differentiable augmentation module based on a neural network. In Ref. [47], the authors created a shape dictionary by collecting all 3D poses in the training set to be aligned by the Procrustes method, to concisely summarize the variability in training data and enable a sparse representation. A convex approach was then proposed to jointly estimate the coefficients of the sparse representation. The same authors [48] predicted the uncertainty heatmaps of the 2D joint locations, then combined these maps with a sparse model of a 3D human pose to retrieve the 3D pose via an EM algorithm. Ref. [49] adopted a large library of 2D keypoints and their 3D representations to match the depths of the 2D poses estimated by the k-nearest neighbor algorithm. Hossain et al. [50] proposed two 2-layered normalized LSTM networks with residual connections to leverage temporal information for lifting 2D joint locations to 3D positions.

2.2. *Video Pose Estimation*

Although 3D coordinates can be determined from a single image, temporal algorithms used in videos have better accuracies than simple frame-by-frame approaches. Most works deploy recurrent neural networks (RNNs) [50,51] to exploit temporal information. Long short-term memory networks (LSTMs) [52] are the most widely used RNN architectures for learning long-term dependencies in pose estimation problems because of their ability to preserve information over time. In [51], propagating LSTM networks (p-LSTMs) were proposed to estimate depth information from 2D keypoints. Ref. [53] presented a two-part spatial–temporal convolutional LSTM model (ST-CLSTM) to capture spatial features and temporal consistency between frames. The authors used ST-CLSTM as the generator and a 3D CNN as the discriminator to output the temporal loss from the estimated and ground truth depth sequences. AnimePose [54] used Scene-LSTM to estimate the person's temporal trajectory and track overlapping postures in obscure frames based on their predictions in prior frames. Temporal convolutional networks (TCNs) [55], on the other hand, give additional benefits, such as convolution sharing and low memory requirements for training; this is very advantageous when dealing with extended input sequences. TCN evaluation and training are hence faster than with RNN. As a result, they are becoming increasingly employed in pose estimation [35,37–39,56], especially in real-time systems [57,58]. Moreover, Ref. [39] proposed employing dilated temporal convolutions in a fully convolutional model; moreover, [59] used it as an automatic framework for semantic motion segmentation. Li et al. [60] captured long-range dependencies using transformer-based architecture.

2.3. *Spatial–Temporal Graph Convolution Network*

Despite the acquired temporal information's ability to anticipate smoother poses, the depths and self-occlusions remain ambiguous. A graph convolutional network (GCN) was used to exploit the spatiotemporal information that allowed to lower these ambiguities. GCNs have greatly improved 3D human pose estimations by representing the human skeleton as an undirected graph. The spatial–temporal graph convolutional network (ST-GCN) [61] was the first approach to use graph CNNs for skeleton-based action recognition. Zhou et al. [22] developed the semantic graph convolutional network (SemGCN) for the 3D human pose regression challenge. The SemGCN aims to learn by capturing semantic information, such as local and global node relationships through end-to-end training. The graph attention spatiotemporal convolutional network (GAST-Net) [57] also combines common convolutional networks to integrate the spatiotemporal information. GAST-Net comprises two types of graph attention blocks: a local spatial attention network (to model the hierarchical and symmetrical structures of the human skeleton) and a global spatial attention network (to extract global semantic information and better encode the human body's spatial characteristics). Cai et al. [62] developed an undirected graph to model the spatial–temporal connections between distinct joints for 3D single-person pose estimation from video data. In Ref. [32], the authors utilized a graphical neural network (GNN) to efficiently aggregate the features corresponding to the different types of articulation, where each type was represented by a graph node. The GCNs based on directed graphs were also adopted by Cheng et al. [35] to model human joint GCNs that refine potentially imperfect poses obtained from 2D pose heatmaps, and human bone GCNs, to model bone connections. The authors also used two TCNs to estimate the 3D root-relative pose and the absolute root depth. Finally, the dynamic graph convolutional module (DGCM) [63] applied GCN for a multi-person 2D pose estimation framework.

2.4. Multi-Person 3D Pose Estimation

Only a few studies were conducted on 3D multi-person pose estimation from a single RGB image. Generally, existing methods can be divided into two categories: top-down and bottom-up approaches.

Top-down 3D human pose estimation methods [64–66] commonly use human detection as an essential part to crop each person in a bounding box and then estimate person-centric 3D full-body joints [31,39,58]. These methods show promising performances, but their main drawbacks still involve the independent detection and process of each person. Hence, they are likely to suffer from inter-person occlusions and close interactions. Rogez et al. [65,67] introduced LCR-Net, which classified bounding boxes generated into a set of K-poses, refined using a regressor. The architecture contains three stages that share the convolutional feature layers and are jointly trained. Likewise, Benzine et al. [68] proposed the pose estimation and detection anchor-based network (PandaNet), an anchor-based single-shot approach. The network predicts the 2D/3D pose regression into a single forward pass for each bounding box detected in a given image.

To predict camera-centric, Moon et al. [31] processed each cropped person's image independently. They produced root-relative 3D joints using PoseNet [21] and estimated the pelvis keypoint localization of each person using the RootNet model. Similarly, hierarchical multi-person ordinal relations (HMOR) [69] is a coarse-to-fine architecture that hierarchically estimates multi-person ordinal relations through instance-level, part-level, and joint-level. The end-to-end HDNet architecture [32] follows the same pipeline, extract pose, and depth data using a pyramidal feature network [70] as the backbone. Features are then propagated and aggregated using GNN for target depth estimation. In [35], after obtaining the 2D poses from the 2D pose estimator, the poses were normalized to be centered on the root point. Then, the authors used three temporal models—joint-TCN, root-TCN, and velocity-TCN—to obtain absolute 3D human poses, but on monocular videos instead of single images.

On the other hand, bottom-up approaches [34,71,72] first produced all body joint locations and depth maps, then associated body parts to each person according to the root depth and part relative depth. Mehta et al. [34] proposed a single forward pass regardless of the number of people in the scene. The authors applied temporal and kinematic constraints in three steps to predict occlusion-robust PoseMaps (ORPM) and part affinity fields [27]. Another bottom-up multi-stage framework was proposed by Zanfir et al. [73], which first estimated the volumetric heatmaps to determine the 3D keypoint locations and limbs using the confidence scores of all possible connections, and then conducted skeleton grouping in order to assign limbs to various people. Likewise, Fabbri et al. [71] proposed estimating the volumetric heatmaps in an encoder–decoder manner. They first produced compressed volumetric heatmaps, which were used as ground truth, and then decompressed at test time to re-obtain the original representation. Zhen et al. [33] proposed estimating 2.5D representations of body parts first and then reconstructed the 3D human pose in a single-shot bottom-up framework. Wang et al. [74] also proposed distribution-aware single-stage models to represent 3D poses with a 2.5D human center, together with 3D center-relative joint offsets in a one pass solution.

TDBU_Net framework [36] combined top-down and bottom-up pipelines to accomplish the multi-person camera-centric 3D human pose estimation.

In this article, we were inspired by all of these proposals in building a top-down framework that could be used in real-world applications. We used monocular video as input, as in [35,36]. Thus, to deal with long-term models, we chose dilated temporal convolutional networks which only required the next images to produce real-time outputs. To respect this constraint, we also needed a system that integrated as few models as possible, unlike [35,36], while maintaining the highest possible accuracy.

3. Framework Overview

The first part of this section presents the basic architectures used in our framework, consisting of four phases: the human detector using Yolo-v3 architecture [75], the 2D human pose estimator employing HrNet network [23], the 3D root-relative pose estimator using the GAST-Net model [76], and the depth root estimator with the RootNet model [31]. The second part describes the overall pipeline of the framework. The last part details the series of enhancements of our framework on the 3D absolute pose estimator and their impacts on the final result.

3.1. *Basic Models Architectures*

Human detector (Yolo-v3): This architecture [75] predicts bounding boxes using dimension clusters as anchor boxes. The network predicts four coordinates for each bounding box (bbox): the 2D image coordinates of the top-left pixel of the bbox, the width and height of the bbox, and the confidence score. Darknet-53 was used for feature extraction.

2D pose estimator (HrNet): The high-resolution network [23] starts from a high-resolution subnetwork and gradually adds high-to-low resolution subnetworks one by one, by decreasing the resolution to half and increasing the width to double in separate branches that connect in parallel. In that way, high-resolution representation is maintained throughout the process. The input image size is 256×192 or 384×288, which produces 17 heatmaps (heatmap per each keypoint) of size 64×48 or 96×72 respectively. The authors proposed a small network (HRNet-W32) with 32 channels and a large one (HRNet-W48) with 48 channels.

3D root-relative pose estimator (GAST-Net): The majority of models that recently analyzed and interpreted input video were based on temporal convolutional networks (TCNs), which were initially introduced to action segmentation by Lea et al. [55]. The GAST-Net (graph attention spatiotemporal network) [76] is inspired by VideoPose3D [39]. The network predicts 3D poses from 2D keypoints. It is designed from dilated temporal convolutional networks (TCNs) to tackle long-term patterns and exhibit extended memory, and from a graph attention block that consists of two spatial attention networks. The local spatial attention network models the hierarchical and symmetrical structures of the human skeleton. The global spatial attention network adaptively extracts global semantic information to better encode the spatial characteristics of the human body.

Depth estimator (RootNet): Moon et al. [31] proposed a top-down system to estimate 3D multi-person poses from a single RGB image, consisting of human detection by the DetectNet model, absolute 3D human root localization by the RootNet model, and root-relative 3D single-person pose estimation by the PoseNet model. Both models adopt ResNet-50 pre-trained on the ImageNet dataset as a backbone to extract the global data. We are particularly interested in the RootNet model, which generates two outputs: the 2D image coordinates of the root's keypoint (x, y) estimated using soft-argmax on the root-heatmap (the central point of the individual), and the root depth absolute determined using a scalar value k, computed using focal lengths divided by the per-pixel distance factors and the human area ratio between the real-world and the image.

3.2. *Taxonomy of the Framework*

Given a sequence of bounding boxes from monocular RGB videos of a person or a group of people in real-time, the goal was to produce a sequence of 3D camera-centric coordinates of everyone in the scene. First, for each person, we assigned a unique ID i to be tracked through the successive frames. Then, we applied a high-resolution network (HRNet) [23] on each frame to produce 17 heatmaps. Each heatmap predicts 2D human joint locations in MS-COCO format P_{2D} for each detected individual.

The 2D-poses P^i_{2D} in 27 frames were collected and given thereafter to a 3D single-pose estimator, GAST-Net, for direct 2D-to-3D mapping and recovering of the 3D root-relative pose P^i_{3Drel}, where all produced joints were represented by their distances from the pelvis keypoints. GAST-Net was applied (as much as the number of people in the frame).

GAST-Net was chosen since it provides the best compromise between the number of frames required to process and the estimation precision. In fact, the methods with the best accuracies on monocular videos from Human3.6M (the largest database of 3D human pose estimation) are: temporal convolution [39] trained in semi-supervision learning, the Attention 3D Human Pose [77], which identifies significant frames and tensor outputs from each layer using the attention mechanism, the RIE paper [43], which improves the accuracy by relative information encoding that yields positional and temporal-enhanced representations, and Anatomy3D [78], which estimates the 3D skeleton by predicting bone orientation and length. These methods reached the MPJPEs (defined in Section 4) of 44.1, 43.3, 45.1, and 46.8 mm, respectively, but required 243 frames as input. This is very costly in terms of memory and processing time; moreover, this increases the delay between the image display and the result, which is not favorable for real-time processing. Furthermore, tracking several individuals on large time scales is more complicated and error-prone. On the other hand, approaches that employ few frames have higher errors. For example, VIBE [79] only used 16 frames but attained an MPJPE error of 65.6 mm, as well as TP-Net [80] which required 20 frames but had an average error of 52.1. Trajectory space factorization [41] scored an error of 46.6 mm from 50 frames; GAST-Net achieved an MPJPE of 46.2 mm using 27 frames. Thus, it presents a good compromise for use in real-world contexts.

For absolute depth estimation of the pelvis keypoint, we employed the RootNet network proposed in [31], due to its adaptability to any 3D root-relative estimator.

The proposed overall pipeline for estimating the absolute camera-centered coordinates of multi-person keypoints from a monocular camera is depicted in Figure 3. The pipeline comprises three boxes. Person detection and 2D keypoint estimation are included in the first box (green). The second box (orange) contains the 2D to 3D lift, and the last box (blue) contains the depth estimation.

3.3. 3D Absolute Pose Estimator

The purpose of this work was to develop a 3D multi-person camera-centric pose estimation system under industrial and real-world settings. Therefore, we started with a hybridization of well-chosen models, GAST-Net for predicting 3D root-relative keypoints and a RootNet network proposed in [31] for predicting absolute root depth (i.e., the depth of pelvis keypoint), obtained by multiplying k (defined above) by the scalar value of the network output. Then, the XY camera coordinates of the root were determined using the camera-intrinsic parameters, the image coordinates of the root, and the predicted absolute root depth. Finally, the absolute coordinates of the rest of the joints were estimated from these two predictions. We call this hybridization the GR method. On the MuPoTS-3D dataset, the system adopting the GR method outperformed previous methods by more than 12.1 percentage points on AP_{25}^{root}, contributing to more than 6.7 percentage points on 3D-PCK_{abs}. However, we observed that the root-relative keypoints were less good by 25.8 percentage points on PCK, which sparked the idea to upgrade the GAST-Net. While the original GAST-Net was trained on single-person databases [29], we chose to retrain our model on both a single-person video database (MPII-3DHP [81]) and a multi-person video database (MuCo-Temp [56]) with the required processing, following [56], to produce direct absolute keypoint coordinates. The TCN-based approaches evaluated on MuPoTS-3D were trained on the MPII-3DHP database, containing videos of a single person recorded in a green-screen studio and/or on the MuCo-3DHP database, composed of MPII-3DHP frames, containing multiple positions copied into a single frame. For this, in order to train the temporal networks, such as GAST-NET_{ABS}, [56] proposed MuCo-Temp, a temporal extension of MuCo-3DHP that was generated, such as MuCo-3DHP, but it is composed of videos instead of frames. As a result, the relative keypoint precision enhanced from 63.8% with the basic GAST-Net to 82.5% on PCK with our modified GAST-Net, which contributes to 1.6 percentage points in absolute points on 3D-PCK_{abs} when compared to the first methodology of hybridization. Note that in the following we name the upgraded

GAST-Net by GAST-NET$_{ABS}$, and this methodology by the GA method. We noticed that although AP_{25}^{root} of GAST-NET$_{ABS}$ (measuring the root depth estimation) has improved compared to the state-of-the-art, it is still not as good as the first hybridization methodology . This pushed us to compute the root-relative keypoints from the absolute keypoints obtained by GAST-NET$_{ABS}$ and employ the RootNet for root depth estimation, generating final absolute joints. We call this methodology the GAR method. In this way, we increased the accuracy (compared to the literature approaches) by more than 8.8 percentage points on 3D-PCK$_{abs}$. Figure 4 presents the structural diagram of the various types of networks used in the framework.

All these experimental results will be presented, detailed, and analyzed in the next section (Section 4).

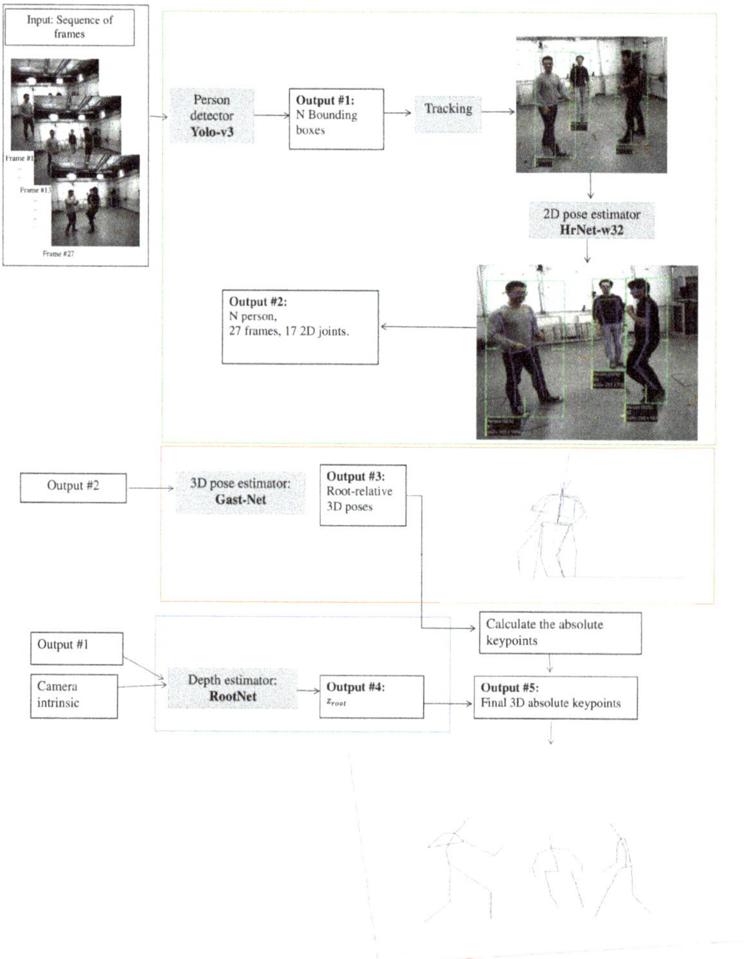

Figure 3. The pipeline of the Root-GAST-Net framework.

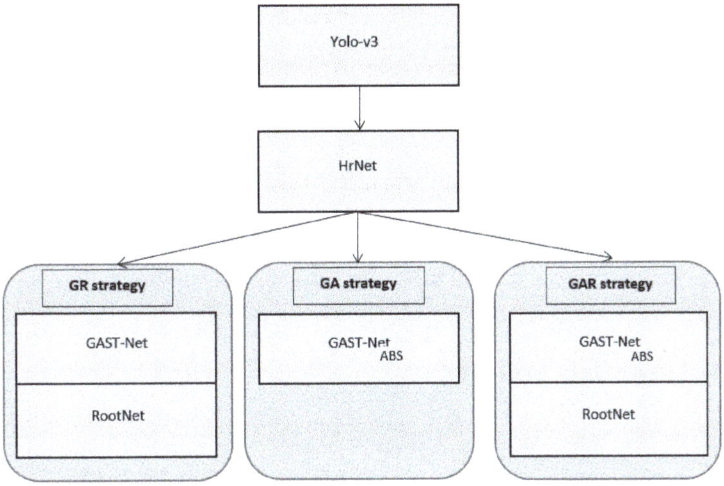

Figure 4. The structural diagram of the various types of networks used in the framework.

4. Experimentation and Results Discussion

This section deals with the experimental details and results of the proposed system. Results are discussed and evaluated using MPJPE, MRPE, 3D-PCK, AP_{25}^{root}, 3D-PCK_{abs} metrics and response times. The proposed Root-GAST-Net system and its three variants (GR, GA, GAR), 3D pose absolute methodologies, were compared to the existing methods grouped in papers_With_Code link of 3D multi-person pose estimations (absolutes) on the MuPoTS-3D page (https://paperswithcode.com/sota/3d-multi-person-pose-estimation-absolute-on, accessed on 1 April 2022). The compared methods are 3D MPPE PoseNet [31], HDNet [32], SMAP [33], HMOR [69], GnTCN [35], and TDBU_Net [36]. The goal of evaluating the three methodologies was to measure the impact of each adjustment.

4.1. Datasets and Evaluation Metrics

Human3.6M is the most popular and biggest dataset/benchmark for 3D human pose estimation [29]. It contains 3.6 million single-person indoor video frames and the corresponding poses of 11 professional actors (6 males, 5 females) captured by the MoCap system from 4 camera viewpoints. Camera extrinsic (rotation and translation with respect to world coordinates) and intrinsic parameters (focal length and principal point) are also available. This could be used to evaluate the single-person-centric pose estimate [39,41,43,57,77–80] as well as the camera-centered coordinate prediction [31–33,35,36,69]. Only subjects 9 and 11 were used for testing, as in prior studies.

For evaluation, we computed the mean per joint position error metric (MPJPE), which is the mean Euclidean error averaged over all joints and all poses, calculated after aligning the human root of the estimated and ground truth 3D poses, calculated on relative poses, as shown in the formula below:

$$MPJPE = \frac{1}{T}\frac{1}{N}\sum_{t=1}^{T}\sum_{i=1}^{N}\left\| J_i^{(t)} - J_i^{*(t)} \right\|_2, \tag{1}$$

where T denotes the total number of test samples and N denotes the number of joints. J and J^* denote the predicted joint and the ground truth joint, respectively.

Another evaluation metric used in this database, proposed in [31], is the mean root position error (MRPE), which is the average error of the absolute root joint (the hip) localization, as follows:

$$MRPE = \frac{1}{T}\sum_{t=1}^{T}\left\|(R^{(t)} - R^{*(t)})\right\|_2, \quad (2)$$

where R and R^* denote the predicted root joint and the ground truth root joint respectively.

MuCo-3DHP and MuPoTS-3D MuCo-3DHP and MuPoTS-3D are two datasets proposed by Mehta et al. [34] for 3D multi-person pose estimation evaluation. MuCo-3DHP is the training dataset that merges randomly sampled 3D poses from a single-person 3D human pose dataset MPI-INF-3DHP [81] to form realistic multi-person scenes. MuPoTS-3D is a dataset used for testing 3D multi-person estimation. It contains 20 videos in both indoor and outdoor scenes. Ground truth is obtained with a multi-view markerless motion capture system.

In order to evaluate person-centric pose estimations, we used the percentage of a correct 3D keypoint (3D-PCK), which treats an estimated joint as correct if it is within a fixed threshold distance from the matched ground truth joint. In the literature, the threshold is set to 15 cm. We also used AUC_{rel}, which is the area under the 3D-PCK curve computed from various thresholds.

We followed [31] to evaluate the absolute camera-centered coordinate estimations. We used average precision AP_{25}^{root} to measure the 3D human root location prediction error, which considers the prediction as correct when the Euclidean distance between the estimated and the ground truth coordinates is smaller than 25 cm. Moreover, we used $3D\text{-}PCK_{abs}$, which is PCK without the root alignment used to evaluate the absolute poses.

MuCo-Temp This dataset was proposed by [56]. It is generated in the same way as MuCo-3DHP. Both use images composited from the MPI-INF-3DHP dataset. The difference is that MuCo-Temp consists of videos instead of frames. So we can use it for temporal network training.

4.2. *Implementation Details*

We adopted Yolo-v3 architecture [75], which is based on the Darknet-53 model as a backbone and is pre-trained on the COCO dataset [82]. The input resolution is 608 × 608.

The cropped image of the bounding box was transformed to 384 × 288 to be used as input for the 2D pose estimator. The transformation applied was an affine transformation that preserves collinearity, parallelism, and the ratio of distances between the points, as in [23]. A unique ID was assigned to each person using the tracking method [83] based on the Hungarian optimization algorithm. Then, we used the small architecture of HrNet (HRNet-w32) pre-trained on the COCO dataset [82], implemented in PyTorch. The output was 17 heatmaps (resolution: 96 × 72). Cropping was resized to 256 × 256 to be processed by RootNet for depth root prediction Z_{abs}^{root}. A unique ID was affected for each person using the tracking method based on the Hungarian optimization algorithm. The 27 consecutive 2D coordinates were collected for each person, to be given to GAST-NET.

All networks, except GAST-NET, were optimized to TensorRT (https://developer.nvidia.com/tensorrt, accessed on 1 April 2022), a Nvidia library allowing to optimize computations on the GPU in order to reach lower computation times. This library also offers lower precision arithmetic but in our experiments, we kept models in the FP32 precision.

For $GAST\text{-}NET_{ABS}$ training, we used the Adam optimizer with a learning rate of 1×10^{-3} and a batch size of 32. We trained the model for 80 epochs on MPII-3DHP [81] and MuCo-Temp [56] datasets. Computations were performed at the supercomputer facilities at Mésocentre Clermont Auvergne University for one week.

Finally, the detected bounding box was resized to 256 × 256 to be processed by RootNet for depth root prediction Z_{abs}^{root}.

4.3. *Results*

4.3.1. Evaluation of Multi-Person Dataset MuPoTS

The results of our system with the three improvements are listed in Table 1, which can be compared to the literature results. We evaluated using the MuPoTS-3D dataset since it has been used to analyze 3D multi-person poses in both person-centric and camera-centric coordinates. Following [31,35], the performance of person-centric 3D pose estimation was evaluated using AUC_{rel} and PCK metrics, while camera-centric 3D pose estimation was evaluated using AP_{25}^{root} and PCK_{abs} metrics. The detailed PCK_{abs} results per sequence are shown in Table 2. We observed an improvement in the estimation accuracy in most of the sequences.

According to both tables, all our strategies outperformed previous 3D multi-person absolute pose estimation approaches by a significant margin, even if the relative poses were weaker.

Table 1. Person-centric and camera-centric evaluations on the MuPoTS-3D dataset. The best is in bold, the second best is underlined.

Method	Year	PCK	AUC_{rel}	3D-PCK_{abs}	AP_{25}^{root}
3D MPPE PoseNet [31]	2019	81.8	39.8	31.5	31.0
HDNet [32]	2020	83.7	-	35.2	39.4
SMAP [33]	2020	80.5	45.5	38.7	45.5
HMOR [69]	2020	82.0	43.5	43.8	-
GnTCN [35]	2021	<u>87.5</u>	<u>48.9</u>	45.7	45.2
TDBU_Net [36]	2021	**89.6**	**50.6**	48.0	46.3
DAS [74]	2022	82.7	-	39.2	-
Root-GAST with GR	-	63.8	30.6	54.7	<u>58.4</u>
Root-GAST with GA	-	82.5	45.3	<u>56.1</u>	56.8
Root-GAST with GAR	-	82.5	45.3	**56.8**	**58.9**

Table 2. Sequence-wise 3D-PCK_{abs} comparison with the state-of-the-art on the MuPoTS-3D dataset. (*) The accuracies of methods are measured on matched ground truths. The best is in bold, the second best is underlined.

Method	S1	S2	S3	S4	S5	S6	S7
3D MPPE PoseNet (*) [31]	59.5	45.3	51.4	46.2	53.0	27.4	23.7
HDNet [32]	21.4	22.7	58.3	27.5	37.3	12.2	49.2
SMAP (*) [33]	42.1	41.4	46.5	16.3	53.0	26.4	47.5
GnTCN (*) [35]	64.7	<u>59.3</u>	<u>59.4</u>	63.1	52.6	<u>42.7</u>	31.9
TDBU_Net [36]	<u>69.2</u>	57.1	49.3	<u>68.9</u>	<u>55.1</u>	36.1	<u>49.4</u>
Root-GAST with GAR (*)	**89.8**	**77.0**	**73.4**	**77.0**	**81.0**	**54.3**	**68.4**
Method	**S8**	**S9**	**S10**	**S11**	**S12**	**S13**	**S14**
3D MPPE PoseNet (*) [31]	26.4	39.1	23.6	8.3	14.9	38.2	29.5
HDNet [32]	<u>40.8</u>	<u>53.1</u>	43.9	43.2	**43.6**	39.7	28.3
SMAP (*) [33]	18.7	36.7	**73.5**	<u>46.0</u>	22.7	24.3	38.9
GnTCN (*) [35]	35.2	53.0	28.3	37.6	26.7	46.3	<u>44.5</u>
TDBU_Net [36]	33.0	43.5	52.8	**48.8**	<u>36.5</u>	<u>51.2</u>	37.1
Root-GAST with GAR (*)	**60.5**	**71.3**	<u>65.4</u>	33.5	26.1	**67.3**	**46.9**

Table 2. *Cont.*

Method	S15	S16	S17	S18	S19	S20	Avg
3D MPPE PoseNet (*) [31]	36.8	23.6	14.4	20.0	18.8	25.4	31.8
HDNet [32]	49.5	23.8	18.0	26.9	25.0	38.8	35.2
SMAP (*) [33]	47.5	34.2	35.0	20.0	38.7	64.8	38.7
GnTCN (*) [35]	<u>50.2</u>	<u>47.9</u>	<u>39.4</u>	23.5	**61.0**	**56.1**	46.3
TDBU_Net [36]	47.3	**52.0**	20.3	**43.7**	<u>57.5</u>	<u>50.4</u>	<u>48.0</u>
Root-GAST with GAR (*)	**66.9**	35.7	**40.1**	<u>38.5</u>	26.0	35.3	**56.8**

The average precisions throughout the entire dataset were then examined using various threshold settings ranging from 25 to 10 cm. AP measured the accuracy of the root key point; we only evaluated the Root-GAST system's performance using the GA approach since GR and GAR methodologies employed RootNet to predict the root joint. They produced the same result as the original paper. Table 3 displays the results. When compared to the state-of-the-art methodology, our method significantly achieves greater AP across all levels of thresholds. We deduce that our method estimates many more correct root keypoints even with a low distance threshold.

Table 3. Average precision of the root keypoint evaluation by different distances on the MuPoTS-3D dataset.

Method	$\mathbf{AP}_{25}^{root}$	$\mathbf{AP}_{20}^{root}$	$\mathbf{AP}_{15}^{root}$	$\mathbf{AP}_{10}^{root}$
3D MPPE PoseNet [31]	31.0	21.5	10.2	2.3
HDNet [32]	39.4	28.0	14.6	4.1
Root-GAST with GA	56.8	47.1	36.8	22.4

To compare with most of the existing methods that evaluate person-centric 3D pose estimations on MuPoTS-3D using MPJPE , we report our results using the same metric in Table 4. Our result was 101.9 mm, the result of [34] was 132 mm, the result of [84] was 120 mm, the result of [56] when adding the pose refinement model was 103 mm. Our method also outperforms the existing methods on this metric.

Table 4. MPJPE of the relative poses on the MuPoTS-3D dataset. The best is in bold, the second best is underlined.

Method	Year	MPJPE (mm)
Temporal smoothing [56]	2020	107
Temporal smoothing + Pose refinement [56]	2020	<u>103</u>
Depth Prediction Network [84]	2019	120
LCR-Net [67]	2017	146
Mehta et al. [34]	2018	132
GAST-Net$_{ABS}$	-	**101.9**

4.3.2. Evaluation on Single-Person Dataset Human3.6M

In order to validate the system, we chose Human3.6M, which contains only single-person videos. Since we compared the results through the mean root position error (MRPE) metric, which measured the accuracy error of the root key point, we only evaluated the Root-GAST system's performance using the GA approach. GR and GAR methodologies employed RootNet to predict the root joint; they produced the same result as the original paper.

The root localization results of our GAST-Net$_{ABS}$ and the RootNet model are shown in Table 5. Even though the evaluation was performed on the Human3.6M dataset, we employed the GA model that was retrained on MPII and the MuCo-Temp dataset, and we compared it to the RootNet model that was trained on the MuCo dataset to make a fair comparison. Our measurement error amounted to 158 mm, while that of [31] was 289.28 mm. However, we could expect greater improvement if we train our model in the Human3.6M dataset.

Table 5. MRPE results comparison with RootNet [31] on the Human3.6M dataset. MRPE$_x$, MRPE$_y$, and MRPE$_z$ are the average MRPE errors in the x, y, and z axes, respectively.

Method	MRPE (mm)	MRPE$_x$ (mm)	MRPE$_y$ (mm)	MRPE$_z$ (mm)
3D MPPE PoseNet [31]	289.28	35.95	58.65	268.49
Root-GAST with GA	178	33	41.9	158

4.3.3. Response Time

The response time is the processing time taken by the algorithm to process its input; it depends on the material configurations. The Root-GAST-Net pipeline was implemented in C++ and executed on a machine equipped with Intel Core i5-9500, with a dedicated memory of 32GB, and the Nvidia GeForce GTX 1080, with a dedicated memory of 8GB.

A comparative analysis of the response times of each network is shown in Table 6. The processing time was measured on batches of monocular images from the Human3.6M dataset, each containing one person. Note that the processing time of the tracking step is negligible.

Table 6. Response time per model.

Model	Min Response Time (ms)	Max Response Time (ms)	Average Response Time (ms)
Yolo-v3	24	30	28
HrNet	9	12	10
GAST-Net	27	33	29
GAST-Net$_{ABS}$	23	29	26
RootNet	4	8	5

Finally, the frame rate of the whole pipeline with each strategy is given in Table 7. The proposed Root-GAST-Net system can run at about 15 frames per second, which is suitable for real-time scenarios. Therefore, improving the metrics does not impact the real-time aspect of the pipeline.

Table 7. Frame rate per strategy.

Strategy	Average Frame Rate (fps)
Root-GAST with GR	13
Root-GAST with GA	16
Root-GAST with GAR	15

4.3.4. Qualitative Results

As the system follows a top-down approach, the final result depends on all previous outputs. If the detection is not correctly done, the 2D keypoints and depths will be wrongly estimated, which will impact the absolute pose. If there are numerous people inside the box or body parts that are partially outside the box's bounds, the full-body joint calculation is likely to be incorrect, as shown in Figure 5. The confusion stems from erroneous 2D point estimations, which have negative impacts on the 3D-lifted process.

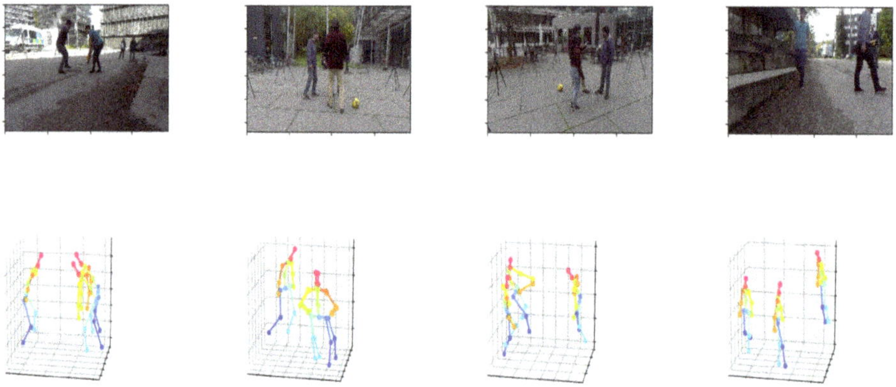

Figure 5. Erroneous 3D multi-person pose estimation. The first two images represent two similar poses of different people because one is completely occluded. In the right two images, one pose is incorrect because the body parts are partially outside of the boxes.

5. Conclusions

In this work, we propose a top-down framework for 3D multi-person absolute pose estimation, reconstructed from 2D poses from a monocular camera. Our framework Root-GAST-Net can combine different models in three strategies. The GR strategy and GAR strategy, which integrate human detection, 2D pose estimation, 3D human root-relative single-person pose estimation, and root depth estimation. Moreover, the GA strategy integrates human detection, 2D pose estimation, and 3D absolute pose estimation.

Experimental results on multiple datasets showed that our framework significantly outperforms the recent approaches in 3D absolute multi-pose estimation. In addition, the system can be used in real-time, as the execution time of each frame containing one person takes around 60 milliseconds using the Nvidia GeForce GTX 1080. This can be reduced using high-performance materials and FP16 precision.

In future works, we plan to retrain the model on the Human3.6M dataset to improve the evaluation accuracy of this database. We also plan to develop a fall detection application based on the absolute and relative 3D postures predicted by the Root-GAST-Net system.

Author Contributions: Conceptualization, all authors; methodology, all authors; software, A.E.K., D.B.; validation, all authors; formal analysis, all authors; investigation, A.E.K.; resources, all authors; data curation, A.E.K.; writing—original draft preparation, A.E.K.; writing—review and editing, D.B., V.B., K.B.; supervision, D.B., V.B., K.B.; project administration, D.B., V.B., K.B.; funding acquisition, V.B., K.B. All authors have read and agreed to the published version of the manuscript.

Funding: This work was funded by a CIFRE France/Morocco grant (2018/1635) with the Prynel Company, within a collaboration with University Mohammed V in Rabat, Morocco, financed by ANRT (Association Nationale de la Recherche et de la Technologie), France and CNRST (Centre National pour la Recherche Scientifique et Technique), Morocco.

Institutional Review Board Statement: Not applicable.

Informed Consent Statement: Not applicable.

Data Availability Statement: We used the MuCo-Temp dataset, generated with the GitHub project at pose_refinement (https://github.com/vegesm/pose_refinement (accessed on 12 February 2022)) for training. For evaluation, we used the Human 3.6M dataset, parsed and available at 3DMPPE_ROOTNET_RELEASE github project https://github.com/mks0601/3DMPPE_ROOTNET_RELEASE (accessed on

12 February 2022), and the MuPoTS-3D dataset is publicly available at website https://vcai.mpi-inf.mpg.de/projects/SingleShotMultiPerson/ (accessed on 12 February 2022).

Acknowledgments: The authors would like to thank Jérôme BROSSAIS for his fruitful contributions.

Conflicts of Interest: The authors declare no conflict of interest.

Abbreviations

The following abbreviations are used in this manuscript:

HPE	human pose estimation
LSTM	long short-term memory
GNN	graph neural network
GCN	graph convolution network
TCN	temporal convolutional network
RNN	recurrent neural network
MPJPE	mean per joint position error
MRPE	mean of the root position error
AUC	area under the curve
3D-PCK	percentage of correct key-points in 3D space
AP^{root}	average precision of the root keypoint
GPU	graphics processing unit
GR	first 3D absolute pose methodology: GAST-Net + RootNet
GA	second 3D absolute pose methodology: GAST-Net_{ABS} trained on MuCo-Temp
GAR	third 3D absolute pose methodology: GAST-Net_{ABS} trained on MuCo-Temp + RootNet
Root-GAST	the whole pipeline: human detector + 2D pose estimator + 3D absolute pose estimator

References

1. Treleaven, P.; Wells, J. 3D body scanning and healthcare applications. *Computer* **2007**, *40*, 28–34. [CrossRef]
2. Grazioso, S.; Selvaggio, M.; Di Gironimo, G. Design and development of a novel body scanning system for healthcare applications. *Int. J. Interact. Des. Manuf.* **2018**, *12*, 611–620. [CrossRef]
3. Chromy, A.; Zalud, L. The RoScan thermal 3D body scanning system: medical applicability and benefits for unobtrusive sensing and objective diagnosis. *Sensors* **2020**, *20*, 6656. [CrossRef] [PubMed]
4. Liberadzki, P.; Adamczyk, M.; Witkowski, M.; Sitnik, R. Structured-light-based system for shape measurement of the human body in motion. *Sensors* **2018**, *18*, 2827. [CrossRef] [PubMed]
5. Nezami, F.N.; Wächter, M.A.; Maleki, N.; Spaniol, P.; Kühne, L.M.; Haas, A.; Pingel, J.M.; Tiemann, L.; Nienhaus, F.; Keller, L.; et al. Westdrive X LoopAR: An Open-Access Virtual Reality Project in Unity for Evaluating User Interaction Methods during Takeover Requests. *Sensors* **2021**, *21*, 1879. [CrossRef]
6. Ku Abd. Rahim, K.N.; Elamvazuthi, I.; Izhar, L.I.; Capi, G. Classification of human daily activities using ensemble methods based on smartphone inertial sensors. *Sensors* **2018**, *18*, 4132. [CrossRef]
7. Michonski, J.; Witkowski, M.; Sitnik, R.; Glinkowski, W.M. Automatic recognition of surface landmarks of anatomical structures of back and posture. *J. Biomed. Opt.* **2012**, *17*, 056015. [CrossRef]
8. Čibiraitė-Lukenskienė, D.; Ikamas, K.; Lisauskas, T.; Krozer, V.; Roskos, H.G.; Lisauskas, A. Passive detection and imaging of human body radiation using an uncooled field-effect transistor-based THz detector. *Sensors* **2020**, *20*, 4087. [CrossRef]
9. Reddy, N.D.; Guigues, L.; Pishchulin, L.; Eledath, J.; Narasimhan, S.G. TesseTrack: End-to-End Learnable Multi-Person Articulated 3D Pose Tracking. In Proceedings of the IEEE/CVF Conference on Computer Vision and Pattern Recognition (CVPR), Nashville, TN, USA, 20–25 June 2021; pp. 15190–15200.
10. He, Y.; Yan, R.; Fragkiadaki, K.; Yu, S.I. Epipolar transformers. In Proceedings of the IEEE/CVF Conference on Computer Vision and Pattern Recognition (CVPR), Seattle, WA, USA, 14–19 June 2020; pp. 7779–7788.
11. Iskakov, K.; Burkov, E.; Lempitsky, V.; Malkov, Y. Learnable triangulation of human pose. In Proceedings of the IEEE/CVF International Conference on Computer Vision, Seoul, Korea, 27–28 October 2019; pp. 7718–7727.
12. Qiu, H.; Wang, C.; Wang, J.; Wang, N.; Zeng, W. Cross view fusion for 3d human pose estimation. In Proceedings of the IEEE/CVF International Conference on Computer Vision, Seoul, Korea, 27–28 October 2019; pp. 4342–4351.
13. Gordon, B.; Raab, S.; Azov, G.; Giryes, R.; Cohen-Or, D. FLEX: Parameter-free Multi-view 3D Human Motion Reconstruction. *arXiv* **2021**, arXiv:2105.01937.
14. Zhang, Y.; Wang, C.; Wang, X.; Liu, W.; Zeng, W. Voxeltrack: Multi-person 3d human pose estimation and tracking in the wild. *IEEE Trans. Pattern Anal. Mach. Intell.* **2022**. [CrossRef]

15. Tekin, B.; Márquez-Neila, P.; Salzmann, M.; Fua, P. Learning to fuse 2d and 3d image cues for monocular body pose estimation. In Proceedings of the 2017 IEEE International Conference on Computer Vision (ICCV), Venice, Italy, 22–29 October 2017; pp. 3941–3950.
16. Moreno-Noguer, F. 3d human pose estimation from a single image via distance matrix regression. In Proceedings of the IEEE Conference on Computer Vision and Pattern Recognition, Honolulu, HI, USA, 21–26 July 2017; pp. 2823–2832.
17. Lee, H.J.; Chen, Z. Determination of 3D human body postures from a single view. *Comput. Vision Graph. Image Process.* **1985**, *30*, 148–168. [CrossRef]
18. Zhou, X.; Zhu, M.; Pavlakos, G.; Leonardos, S.; Derpanis, K.G.; Daniilidis, K. Monocap: Monocular human motion capture using a cnn coupled with a geometric prior. *IEEE Trans. Pattern Anal. Mach. Intell.* **2018**, *41*, 901–914. [CrossRef] [PubMed]
19. Ghezelghieh, M.F.; Kasturi, R.; Sarkar, S. Learning camera viewpoint using CNN to improve 3D body pose estimation. In Proceedings of the 2016 Fourth International Conference on 3D Vision (3DV), Stanford, CA, USA, 25–28 October 2016; pp. 685–693.
20. Wu, J.; Xue, T.; Lim, J.J.; Tian, Y.; Tenenbaum, J.B.; Torralba, A.; Freeman, W.T. Single image 3d interpreter network. In Proceedings of the European Conference on Computer Vision, Amsterdam, The Netherlands, 11–14 October 2016; pp. 365–382.
21. Sun, X.; Xiao, B.; Wei, F.; Liang, S.; Wei, Y. Integral human pose regression. In Proceedings of the European Conference on Computer Vision (ECCV), Munich, Germany, 8–14 September 2018; pp. 529–545.
22. Zhao, L.; Peng, X.; Tian, Y.; Kapadia, M.; Metaxas, D.N. Semantic graph convolutional networks for 3D human pose regression. In Proceedings of the IEEE/CVF Conference on Computer Vision and Pattern Recognition, Long Beach, CA, USA, 15–20 June 2019; pp. 3425–3435.
23. Sun, K.; Xiao, B.; Liu, D.; Wang, J. Deep high-resolution representation learning for human pose estimation. In Proceedings of the IEEE/CVF Conference on Computer Vision and Pattern Recognition, Long Beach, CA, USA, 15–20 June 2019; pp. 5693–5703.
24. Li, J.; Su, W.; Wang, Z. Simple Pose: Rethinking and Improving a Bottom-up Approach for Multi-Person Pose Estimation. *arXiv* **2019**, arXiv:1911.10529.
25. Newell, A.; Yang, K.; Deng, J. Stacked hourglass networks for human pose estimation. In Proceedings of the European Conference on Computer Vision, Amsterdam, The Netherlands, 11–14 October 2016; pp. 483–499.
26. Chen, Y.; Wang, Z.; Peng, Y.; Zhang, Z.; Yu, G.; Sun, J. Cascaded pyramid network for multi-person pose estimation. In Proceedings of the IEEE Conference on Computer Vision and Pattern Recognition, Salt Lake City, UT, USA, 18–23 June 2018; pp. 7103–7112.
27. Cao, Z.; Simon, T.; Wei, S.E.; Sheikh, Y. Realtime multi-person 2d pose estimation using part affinity fields. In Proceedings of the IEEE Conference on Computer Vision and Pattern Recognition, Honolulu, HI, USA, 21–26 July 2017; pp. 7291–7299.
28. Zhang, F.; Zhu, X.; Dai, H.; Ye, M.; Zhu, C. Distribution-aware coordinate representation for human pose estimation. In Proceedings of the IEEE/CVF Conference on Computer Vision and Pattern Recognition (CVPR), Seattle, WA, USA, 14–19 June 2020; pp. 7093–7102.
29. Ionescu, C.; Papava, D.; Olaru, V.; Sminchisescu, C. Human3. 6m: Large scale datasets and predictive methods for 3d human sensing in natural environments. *IEEE Trans. Pattern Anal. Mach. Intell.* **2013**, *36*, 1325–1339. [CrossRef] [PubMed]
30. Sigal, L.; Balan, A.O.; Black, M.J. Humaneva: Synchronized video and motion capture dataset and baseline algorithm for evaluation of articulated human motion. *Int. J. Comput. Vis.* **2010**, *87*, 4. [CrossRef]
31. Moon, G.; Chang, J.Y.; Lee, K.M. Camera distance-aware top-down approach for 3d multi-person pose estimation from a single rgb image. In Proceedings of the IEEE/CVF International Conference on Computer Vision, Seoul, Korea, 27–28 October 2019; pp. 10133–10142.
32. Lin, J.; Lee, G.H. Hdnet: Human depth estimation for multi-person camera-space localization. In Proceedings of the European Conference on Computer Vision, Glasgow, UK, 23–28 August 2020; pp. 633–648.
33. Zhen, J.; Fang, Q.; Sun, J.; Liu, W.; Jiang, W.; Bao, H.; Zhou, X. Smap: Single-shot multi-person absolute 3d pose estimation. In Proceedings of the European Conference on Computer Vision, Glasgow, UK, 23–28 August 2020; pp. 550–566.
34. Mehta, D.; Sotnychenko, O.; Mueller, F.; Xu, W.; Sridhar, S.; Pons-Moll, G.; Theobalt, C. Single-shot multi-person 3d pose estimation from monocular rgb. In Proceedings of the 2018 International Conference on 3D Vision (3DV), Verona, Italy, 5–8 September 2018; pp. 120–130.
35. Cheng, Y.; Wang, B.; Yang, B.; Tan, R.T. Graph and temporal convolutional networks for 3d multi-person pose estimation in monocular videos. *Proc. AAAI Conf. Artif. Intell.* **2021**, *4*, 12.
36. Cheng, Y.; Wang, B.; Yang, B.; Tan, R.T. Monocular 3D multi-person pose estimation by integrating top-down and bottom-up networks. In Proceedings of the IEEE/CVF Conference on Computer Vision and Pattern Recognition (CVPR), Nashville, TN, USA, 20–25 June 2021; pp. 7649–7659.
37. Cheng, Y.; Yang, B.; Wang, B.; Yan, W.; Tan, R.T. Occlusion-aware networks for 3d human pose estimation in video. In Proceedings of the IEEE/CVF International Conference on Computer Vision, Seoul, Korea, 27–28 October 2019; pp. 723–732.
38. Cheng, Y.; Yang, B.; Wang, B.; Tan, R.T. 3d human pose estimation using spatio-temporal networks with explicit occlusion training. *Proc. AAAI Conf. Artif. Intell.* **2020**, *34*, 10631–10638. [CrossRef]
39. Pavllo, D.; Feichtenhofer, C.; Grangier, D.; Auli, M. 3d human pose estimation in video with temporal convolutions and semi-supervised training. In Proceedings of the IEEE/CVF Conference on Computer Vision and Pattern Recognition, Long Beach, CA, USA, 15–20 June 2019; pp. 7753–7762.

40. Chen, H.; Wang, Y.; Zheng, K.; Li, W.; Chang, C.T.; Harrison, A.P.; Xiao, J.; Hager, G.D.; Lu, L.; Liao, C.H.; et al. Anatomy-aware siamese network: Exploiting semantic asymmetry for accurate pelvic fracture detection in x-ray images. In Proceedings of the European Conference on Computer Vision, Glasgow, UK, 23–28 August 2020; pp. 239–255.
41. Lin, J.; Lee, G.H. Trajectory space factorization for deep video-based 3d human pose estimation. *arXiv* **2019**, arXiv:1908.08289.
42. Li, W.; Zhao, Y.; Liu, Y.; Sun, M.; Waterhouse, G.I.; Huang, B.; Zhang, K.; Zhang, T.; Lu, S. Exploiting Ru-induced lattice strain in CoRu nanoalloys for robust bifunctional hydrogen production. *Angew. Chem.* **2021**, *133*, 3327–3335. [CrossRef]
43. Shan, W.; Lu, H.; Wang, S.; Zhang, X.; Gao, W. Improving Robustness and Accuracy via Relative Information Encoding in 3D Human Pose Estimation. In Proceedings of the 29th ACM International Conference on Multimedia, Nice, France, 21–25 October 2021; pp. 3446–3454.
44. Martinez, J.; Hossain, R.; Romero, J.; Little, J.J. A simple yet effective baseline for 3d human pose estimation. In Proceedings of the IEEE International Conference on Computer Vision, Venice, Italy, 22–29 October 2017; pp. 2640–2649.
45. Fang, H.S.; Xu, Y.; Wang, W.; Liu, X.; Zhu, S.C. Learning pose grammar to encode human body configuration for 3d pose estimation. *Proc. AAAI Conf. Artif. Intell.* **2018**, *32*, 1.
46. Gong, K.; Zhang, J.; Feng, J. Poseaug: A differentiable pose augmentation framework for 3d human pose estimation. In Proceedings of the IEEE/CVF Conference on Computer Vision and Pattern Recognition (CVPR), Nashville, TN, USA, 20–25 June 2021; pp. 8575–8584.
47. Zhou, X.; Zhu, M.; Leonardos, S.; Daniilidis, K. Sparse representation for 3D shape estimation: A convex relaxation approach. *IEEE Trans. Pattern Anal. Mach. Intell.* **2016**, *39*, 1648–1661. [CrossRef] [PubMed]
48. Zhou, X.; Zhu, M.; Leonardos, S.; Derpanis, K.G.; Daniilidis, K. Sparseness meets deepness: 3D human pose estimation from monocular video. In Proceedings of the IEEE Conference on Computer Vision and Pattern Recognition, Las Vegas, NV, USA, 27–30 June 2016; pp. 4966–4975.
49. Chen, C.H.; Ramanan, D. 3d human pose estimation= 2d pose estimation+ matching. In Proceedings of the IEEE Conference on Computer Vision and Pattern Recognition, Honolulu, HI, USA, 21–26 July 2017; pp. 7035–7043.
50. Hossain, M.R.I.; Little, J.J. Exploiting temporal information for 3d human pose estimation. In Proceedings of the European Conference on Computer Vision (ECCV), Munich, Germany, 8–14 September 2018; pp. 68–84.
51. Lee, K.; Lee, I.; Lee, S. Propagating lstm: 3d pose estimation based on joint interdependency. In Proceedings of the European Conference on Computer Vision (ECCV), Munich, Germany, 8–14 September 2018; pp. 119–135.
52. Hochreiter, S.; Schmidhuber, J. Long short-term memory. *Neural Comput.* **1997**, *9*, 1735–1780. [CrossRef] [PubMed]
53. Zhang, H.; Shen, C.; Li, Y.; Cao, Y.; Liu, Y.; Yan, Y. Exploiting temporal consistency for real-time video depth estimation. In Proceedings of the IEEE/CVF International Conference on Computer Vision, Seoul, Korea, 27–28 October 2019; pp. 1725–1734.
54. Kumarapu, L.; Mukherjee, P. AnimePose: Multi-person 3D pose estimation and animation. *arXiv* **2020**, arXiv:2002.02792.
55. Lea, C.; Vidal, R.; Reiter, A.; Hager, G.D. Temporal convolutional networks: A unified approach to action segmentation. In Proceedings of the European Conference on Computer Vision, Amsterdam, The Netherlands, 11–14 October 2016; pp. 47–54.
56. Veges, M.; Lőrincz, A. Temporal Smoothing for 3D Human Pose Estimation and Localization for Occluded People. In Proceedings of the International Conference on Neural Information Processing, Bangkok, Thailand, 18–22 November 2020; pp. 557–568.
57. Liu, J.; Guang, Y.; Rojas, J. Gast-net: Graph attention spatio-temporal convolutional networks for 3d human pose estimation in video. *arXiv* **2020**, arXiv:2003.14179.
58. Mehta, D.; Sridhar, S.; Sotnychenko, O.; Rhodin, H.; Shafiei, M.; Seidel, H.P.; Xu, W.; Casas, D.; Theobalt, C. Vnect: Real-time 3d human pose estimation with a single rgb camera. *ACM Trans. Graph.* **2017**, *36*, 1–14. [CrossRef]
59. Cheema, N.; Hosseini, S.; Sprenger, J.; Herrmann, E.; Du, H.; Fischer, K.; Slusallek, P. Dilated temporal fully-convolutional network for semantic segmentation of motion capture data. *arXiv* **2018**, arXiv:1806.09174.
60. Li, W.; Liu, H.; Ding, R.; Liu, M.; Wang, P.; Yang, W. Exploiting temporal contexts with strided transformer for 3d human pose estimation. *IEEE Trans. Multimed.* **2022**. [CrossRef]
61. Yan, S.; Xiong, Y.; Lin, D. Spatial temporal graph convolutional networks for skeleton-based action recognition. In Proceedings of the Thirty-Second AAAI Conference on Artificial Intelligence, New Orleans, LA, USA, 2–7 February 2018.
62. Cai, Y.; Ge, L.; Liu, J.; Cai, J.; Cham, T.J.; Yuan, J.; Thalmann, N.M. Exploiting spatial-temporal relationships for 3d pose estimation via graph convolutional networks. In Proceedings of the IEEE/CVF International Conference on Computer Vision, Seoul, Korea, 27–28 October 2019; pp. 2272–2281.
63. Qiu, Z.; Qiu, K.; Fu, J.; Fu, D. Dgcn: Dynamic graph convolutional network for efficient multi-person pose estimation. *Proc. AAAI Conf. Artif. Intell.* **2020**, *34*, 11924–11931. [CrossRef]
64. Zanfir, A.; Marinoiu, E.; Sminchisescu, C. Monocular 3d pose and shape estimation of multiple people in natural scenes-the importance of multiple scene constraints. In Proceedings of the IEEE Conference on Computer Vision and Pattern Recognition, Salt Lake City, UT, USA, 18–23 June 2018; pp. 2148–2157.
65. Rogez, G.; Weinzaepfel, P.; Schmid, C. Lcr-net++: Multi-person 2d and 3d pose detection in natural images. *IEEE Trans. Pattern Anal. Mach. Intell.* **2019**, *42*, 1146–1161. [CrossRef]
66. Pavlakos, G.; Zhou, X.; Derpanis, K.G.; Daniilidis, K. Coarse-to-fine volumetric prediction for single-image 3D human pose. In Proceedings of the IEEE Conference on Computer Vision and Pattern Recognition, Honolulu, HI, USA, 21–26 July 2017; pp. 7025–7034.

67. Rogez, G.; Weinzaepfel, P.; Schmid, C. Lcr-net: Localization-classification-regression for human pose. In Proceedings of the IEEE Conference on Computer Vision and Pattern Recognition, Honolulu, HI, USA, 21–26 July 2017; pp. 3433–3441.
68. Benzine, A.; Chabot, F.; Luvison, B.; Pham, Q.C.; Achard, C. Pandanet: Anchor-based single-shot multi-person 3d pose estimation. In Proceedings of the IEEE/CVF Conference on Computer Vision and Pattern Recognition (CVPR), Seattle, WA, USA, 14–19 June 2020; pp. 6856–6865.
69. Li, J.; Wang, C.; Liu, W.; Qian, C.; Lu, C. Hmor: Hierarchical multi-person ordinal relations for monocular multi-person 3d pose estimation. *arXiv* **2020**, arXiv:2008.00206.
70. Lin, T.Y.; Dollár, P.; Girshick, R.; He, K.; Hariharan, B.; Belongie, S. Feature pyramid networks for object detection. In Proceedings of the IEEE Conference on Computer Vision and Pattern Recognition, Honolulu, HI, USA, 21–26 July 2017; pp. 2117–2125.
71. Fabbri, M.; Lanzi, F.; Calderara, S.; Alletto, S.; Cucchiara, R. Compressed volumetric heatmaps for multi-person 3d pose estimation. In Proceedings of the IEEE Conference on Computer Vision and Pattern Recognition, Salt Lake City, UT, USA, 18–23 June 2018; pp. 7204–7213.
72. Zhang, C.; Zhan, F.; Chang, Y. Deep monocular 3d human pose estimation via cascaded dimension-lifting. *arXiv* **2021**, arXiv:2104.03520.
73. Zanfir, A.; Marinoiu, E.; Zanfir, M.; Popa, A.I.; Sminchisescu, C. Deep network for the integrated 3d sensing of multiple people in natural images. In Proceedings of the Advances in Neural Information Processing Systems 31 (NeurIPS 2018), Montréal, QC, Canada, 3–8 December 2018.
74. Wang, Z.; Nie, X.; Qu, X.; Chen, Y.; Liu, S. Distribution-Aware Single-Stage Models for Multi-Person 3D Pose Estimation. *arXiv* **2022**, arXiv:2203.07697.
75. Redmon, J.; Farhadi, A. Yolov3: An incremental improvement. *arXiv* **2018**, arXiv:1804.02767.
76. Liu, J.; Rojas, J.; Li, Y.; Liang, Z.; Guan, Y.; Xi, N.; Zhu, H. A graph attention spatio-temporal convolutional network for 3D human pose estimation in video. In Proceedings of the IEEE/CVF Conference on Computer Vision and Pattern Recognition (CVPR), Nashville, TN, USA, 20–25 June 2021; pp. 3374–3380.
77. Liu, R.; Shen, J.; Wang, H.; Chen, C.; Cheung, S.c.; Asari, V. Attention mechanism exploits temporal contexts: Real-time 3d human pose reconstruction. In Proceedings of the IEEE/CVF Conference on Computer Vision and Pattern Recognition (CVPR), Seattle, WA, USA, 14–19 June 2020; pp. 5064–5073.
78. Chen, T.; Fang, C.; Shen, X.; Zhu, Y.; Chen, Z.; Luo, J. Anatomy-aware 3d human pose estimation with bone-based pose decomposition. *IEEE Trans. Circuits Syst. Video Technol.* **2021**, *32*, 198–209. [CrossRef]
79. Kocabas, M.; Athanasiou, N.; Black, M.J. Vibe: Video inference for human body pose and shape estimation. In Proceedings of the IEEE/CVF Conference on Computer Vision and Pattern Recognition (CVPR), Seattle, WA, USA, 14–19 June 2020; pp. 5253–5263.
80. Dabral, R.; Mundhada, A.; Kusupati, U.; Afaque, S.; Sharma, A.; Jain, A. Learning 3d human pose from structure and motion. In Proceedings of the European Conference on Computer Vision (ECCV), Munich, Germany, 8–14 September 2018; pp. 668–683.
81. Mehta, D.; Rhodin, H.; Casas, D.; Fua, P.; Sotnychenko, O.; Xu, W.; Theobalt, C. Monocular 3d human pose estimation in the wild using improved cnn supervision. In Proceedings of the 2017 International Conference on 3D Vision, Qingdao, China, 10–12 October 2017; pp. 506–516.
82. Lin, T.Y.; Maire, M.; Belongie, S.; Hays, J.; Perona, P.; Ramanan, D.; Dollár, P.; Zitnick, C.L. Microsoft coco: Common objects in context. In Proceedings of the European Conference on Computer Vision, Zurich, Switzerland, 6–12 September 2014; pp. 740–755.
83. Galčík, F.; Gargalík, R. Real-time depth map based people counting. In Proceedings of the International Conference on Advanced Concepts for Intelligent Vision Systems, Poznań, Poland, 28–31 October 2013; pp. 330–341.
84. Véges, M.; Lőrincz, A. Absolute human pose estimation with depth prediction network. In Proceedings of the 2019 International Joint Conference on Neural Networks, Budapest, Hungary, 14–19 July 2019; pp. 1–7.

MDPI
St. Alban-Anlage 66
4052 Basel
Switzerland
Tel. +41 61 683 77 34
Fax +41 61 302 89 18
www.mdpi.com

Sensors Editorial Office
E-mail: sensors@mdpi.com
www.mdpi.com/journal/sensors

www.ingramcontent.com/pod-product-compliance
Lightning Source LLC
Chambersburg PA
CBHW061554220926
51850CB00005B/333
* 9 7 8 3 0 3 6 5 5 0 7 3 2 *